COMPLEX ANALYSIS FOR MATHEMATICS AND ENGINEERING

THIRD EDITION

JOHN H. MATHEWS
California State University–Fullerton

RUSSELL W. HOWELL
Westmont College

Wm. C. Brown Publishers

Dubuque, IA Bogota Boston Buenos Aires Caracas Chicago
Guilford, CT London Madrid Mexico City Sydney Toronto

Book Team

Developmental Editor *Daryl Bruflodt*
Designer *Kristyn A. Kalnes*
Art Editor *Joseph P. O'Connell/Miriam J. Hoffman*

Wm. C. Brown Publishers

President and Chief Executive Officer *Beverly Kolz*
Vice President, Publisher *Earl McPeek*
Vice President, Director of Sales and Marketing *Virginia S. Moffat*
Vice President, Director of Production *Colleen A. Yonda*
National Sales Manager *Douglas J. DiNardo*
Marketing Manager *Keri L. Witman*
Advertising Manager *Janelle Keeffer*
Production Editorial Manager *Renée Menne*
Publishing Services Manager *Karen J. Slaght*
Royalty/Permissions Manager *Connie Allendorf*

A Times Mirror Company

Copyedited by Patricia Steele

Library of Congress Catalog Card Number: 95–76589

ISBN 0–697–13548–9

Printed in the United States of America by Times Mirror Higher Education Group, Inc.,
2460 Kerper Boulevard, Dubuque, IA 52001

10 9 8 7 6 5 4 3 2 1

Contents

Preface

Approach This text is intended for undergraduate students in mathematics, physics, and engineering. We have attempted to strike a balance between the pure and applied aspects of complex analysis and to present concepts in a clear writing style that is understandable to students at the junior or senior undergraduate level. A wealth of exercises that vary in both difficulty and substance gives the text flexibility. Sufficient applications are included to illustrate how complex analysis is used in science and engineering. The use of computer graphics gives insight for understanding that complex analysis is a computational tool of practical value. The exercise sets offer a wide variety of choices for computational skills, theoretical understanding, and applications that have been class tested for two editions of the text. Student research projects are suggested throughout the text and citations are made to the bibliography of books and journal articles.

The purpose of the first six chapters is to lay the foundations for the study of complex analysis and develop the topics of analytic and harmonic functions, the elementary functions, and contour integration. If the goal is to study series and the residue calculus and applications, then Chapters 7 and 8 can be covered. If conformal mapping and applications of harmonic functions are desired, then Chapters 9 and 10 can be studied after Chapter 6. A new Chapter 11 on Fourier and Laplace transforms has been added for courses that emphasize more applications.

Proofs are kept at an elementary level and are presented in a self-contained manner that is understandable for students having a sophomore calculus background. For example, Green's theorem is included and it is used to prove the Cauchy-Goursat theorem. The proof by Goursat is included. The development of series is aimed at practical applications.

Features Conformal mapping is presented in a visual and geometric manner so that compositions and images of curves and regions can be understood. Boundary value problems for harmonic functions are first solved in the upper half-plane so that conformal mapping by elementary functions can be used to find solutions in other domains. The Schwarz-Christoffel formula is carefully developed and applications are given. Two-dimensional mathematical models are used for applications in the area of ideal fluid flow, steady state temperatures and electrostatics. Computer drawn figures accurately portray streamlines, isothermals, and equipotential curves.

New for this third edition is a historical introduction of the origin of complex numbers in Chapter 1. An early introduction to sequences and series appears in Chapter 4 and facilitates the definition of the exponential function via series. A new section on the Julia and Mandelbrot sets shows how complex analysis is connected to contemporary topics in mathematics. Many sections have been revised including branches of functions, the elementary functions, and Taylor and Laurent series. New material includes a section on the Joukowski airfoil and an additional chapter on Fourier series and Laplace transforms. Modern computer-generated illustrations have been introduced in the third edition including: Riemann surfaces, contour and surface graphics for harmonic functions, the Dirichlet problem, streamlines involving harmonic and analytic functions, and conformal mapping.

Acknowledgments   We would like to express our gratitude to all the people whose efforts contributed to the development and preparation of the first edition of this book. We want to thank those students who used a preliminary copy of the manuscript. Robert A. Calabretta, California State University–Fullerton, proofread the manuscript. Charles L. Belna, Syracuse University, offered many useful suggestions and changes. Edward G. Thurber, Biola University, used the manuscript in his course and offered encouragement. We also want to thank Richard A. Alo, Lamar University; Arlo Davis, Indiana University of Pennsylvania; Holland Filgo, Northeastern University; Donald Hadwin, University of New Hampshire; E. Robert Heal, Utah State University; Melvin J. Jacobsen, Rensselaer Polytechnic Institute; Charles P. Luehr, University of Florida; John Trienz, United States Naval Academy; William Trench, Drexel University; and Carroll O. Wilde, Naval Postgraduate School, for their helpful reviews and suggestions. And we would like to express our gratitude to all the people whose efforts contributed to the second edition of the book. Our colleagues Vuryl Klassen, Gerald Marley, and Harris Schultz at California State University–Fullerton; Arlo Davis, Indiana University of Pennsylvania; R. E. Williamson, Dartmouth College; Calvin Wilcox, University of Utah; Robert D. Brown, University of Kansas; Geoffrey Price, U.S. Naval Academy; and Elgin H. Johnston, Iowa State University.

For this third edition we thank our colleague C. Ray Rosentrater of Westmont College for class testing the material and for making many valuable suggestions. In addition, we thank T. E. Duncan, University of Kansas; Stuart Goldenberg, California Polytechnic State University–San Luis Obispo; Michael Stob, Calvin College; and Vencil Skarda, Brigham Young University, for reviewing the manuscript.

In production matters we wish to express our appreciation to our copy editor Pat Steele and to the people at Wm. C. Brown Publishers, especially Mr. Daryl Bruflodt, Developmental Editor, Mathematics and Gene Collins, Senior Production Editor, for their assistance.

The use of computer software for symbolic computations and drawing graphs is acknowledged by the authors. Many of the graphs in the text have been drawn with the F(Z)™, MATLAB®, Maple™, and Mathematica™ software. We wish to thank the people at Art Matrix for the color plate pictures in Chapter 4. The authors have developed supplementary materials available for both IBM and Macintosh computers, involving the abovementioned software products. Instructors who use the text may contact the authors directly for information regarding the availability of the F(Z)™, MATLAB®, Maple™, and Mathematica™ supplements. The authors appreciate suggestions and comments for improvements and changes to the text. Correspondence can be made directly to the authors via surface or e-mail.

John Mathews
Dept. of Mathematics
California State University–Fullerton
Fullerton, CA 92634
mathews@fullerton.edu

Russell Howell
Dept. of Mathematics and Computer Science
Westmont College
Santa Barbara, CA 93108
howell@westmont.edu

1

Complex Numbers

1.1   The Origin of Complex Numbers

Complex analysis can roughly be thought of as that subject which applies the ideas of calculus to imaginary numbers. But what exactly are imaginary numbers? Usually, students learn about them in high school with introductory remarks from their teachers along the following lines: ''We can't take the square root of a negative number. But, let's *pretend* we can—and since these numbers are really *imaginary*, it will be convenient notationally to set $i = \sqrt{-1}$.'' Rules are then learned for doing arithmetic with these numbers. The rules make sense. If $i = \sqrt{-1}$, it stands to reason that $i^2 = -1$. On the other hand, it is not uncommon for students to wonder all along whether they are really doing magic rather than mathematics.

If you ever felt that way, congratulate yourself! You're in the company of some of the great mathematicians from the sixteenth through the nineteenth centuries. They too were perplexed with the notion of roots of negative numbers. The purpose of this section is to highlight some of the episodes in what turns out to be a very colorful history of how imaginary numbers were introduced, investigated, avoided, mocked, and, eventually, accepted by the mathematical community. We intend to show you that, contrary to popular belief, there is really nothing *imaginary* about ''imaginary numbers'' at all. In a metaphysical sense, they are just as real as are ''real numbers.''

Our story begins in 1545. In that year the Italian mathematician Girolamo Cardano published *Ars Magna* (*The Great Art*), a 40-chapter masterpiece in which he gave for the first time an algebraic solution to the general cubic equation

$$x^3 + ax^2 + bx + c = 0.$$

His technique involved transforming this equation into what is called a *depressed cubic*. This is a cubic equation without the x^2 term, so that it can be written as

$$x^3 + bx + c = 0.$$

Cardano knew how to handle this type of equation. Its solution had been communicated to him by Niccolo Fontana (who, unfortunately, came to be known as Tartaglia—the stammerer—because of a speaking disorder). The solution was also independently discovered some 30 years earlier by Scipione del Ferro of Bologna. Ferro and Tartaglia showed that one of the solutions to the depressed cubic is

$$(1) \quad x = \sqrt[3]{-\frac{c}{2} + \sqrt{\frac{c^2}{4} + \frac{b^3}{27}}} - \sqrt[3]{\frac{c}{2} + \sqrt{\frac{c^2}{4} + \frac{b^3}{27}}}$$

This value for x could then be used to factor the depressed cubic into a linear term and a quadratic term, the latter of which could be solved with the quadratic formula. So, by using Tartaglia's work, and a clever transformation technique, Cardano was able to crack what had seemed to be the impossible task of solving the general cubic equation.

It turns out that this development eventually gave a great impetus toward the acceptance of imaginary numbers. Roots of negative numbers, of course, had come up earlier in the simplest of quadratic equations such as $x^2 + 1 = 0$. The solutions we know today as $x = \pm\sqrt{-1}$, however, were easy for mathematicians to ignore. In Cardano's time, negative numbers were still being treated with some suspicion, so all the more was the idea of taking square roots of them. Cardano himself, although making some attempts to deal with this notion, at one point said that quantities such as $\sqrt{-1}$ were ''as subtle as they are useless.'' Many other mathematicians also had this view. However, in his 1572 treatise *Algebra*, Rafael Bombeli showed that roots of negative numbers have great utility indeed. Consider the simple depressed cubic equation $x^3 - 15x - 4 = 0$. Letting $b = -15$ and $c = -4$ in the ''Ferro-Tartaglia'' formula (1), we can see that one of the solutions for x is

$$x = \sqrt[3]{2 + \sqrt{-121}} - \sqrt[3]{-2 + \sqrt{-121}}.$$

Bombeli suspected that the two parts of x in the preceding equation could be put in the form $u + v\sqrt{-1}$ and $-u + v\sqrt{-1}$ for some numbers u and v. Indeed, using the well-known identity $(a + b)^3 = a^3 + 3a^2b + 3ab^2 + b^3$, and blindly pretending that roots of negative numbers obey the standard rules of algebra, we can see that

$$(2) \quad \begin{aligned} (2 + \sqrt{-1})^3 &= 2^3 + 3(2^2)\sqrt{-1} + 3(2)(\sqrt{-1})^2 + (\sqrt{-1})^3 \\ &= 8 + 12\sqrt{-1} - 6 - \sqrt{-1} \\ &= 2 + 11\sqrt{-1} \\ &= 2 + \sqrt{-121}. \end{aligned}$$

Bombeli reasoned that if $(2 + \sqrt{-1})^3 = 2 + \sqrt{-121}$, it must be that $2 + \sqrt{-1} = \sqrt[3]{2 + \sqrt{-121}}$. Likewise, he showed $-2 + \sqrt{-1} = \sqrt[3]{-2 + \sqrt{-121}}$. But then we clearly have

$$(3) \quad \sqrt[3]{2 + \sqrt{-121}} - \sqrt[3]{-2 + \sqrt{-121}} = (2 + \sqrt{-1}) - (-2 + \sqrt{-1}) = 4,$$

and this was a bit of a bombshell. Heretofore, mathematicians could easily scoff at imaginary numbers when they arose as solutions to quadratic equations. With cubic equations, they no longer had this luxury. That $x = 4$ was a correct solution to the equation $x^3 - 15x - 4 = 0$ was indisputable, as it could be checked easily. However, to arrive at this very real solution, one was forced to detour through the uncharted territory of "imaginary numbers." Thus, whatever else one might say about these numbers (which, today, we call *complex numbers*), their utility could no longer be ignored.

But even this breakthrough did not authenticate complex numbers. After all, a real number could be represented geometrically on the number line. What possible representation could these new numbers have? In 1673 John Wallis made a stab at a geometric picture of complex numbers that comes close to what we use today. He was interested at the time in representing solutions to general quadratic equations, which we shall write as $x^2 + 2bx + c^2 = 0$ so as to make the following discussion more tractable. Using the quadratic formula, the preceding equation has solutions

$$x = -b - \sqrt{b^2 - c^2} \quad \text{and} \quad x = -b + \sqrt{b^2 - c^2}.$$

Wallis imagined these solutions as displacements to the left and right from the point $-b$. He saw each displacement, whose value was $\sqrt{b^2 - c^2}$, as the length of the sides of the right triangles shown in Figure 1.1.

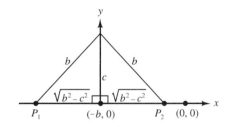

FIGURE 1.1   Wallis' representation of real roots of quadratics.

The points P_1 and P_2 in this figure are the representations of the solutions to our equation. This is clearly correct if $b^2 - c^2 \geq 0$, but how should we picture P_1 and P_2 in the case when negative roots arise—i.e., when $b^2 - c^2 < 0$? Wallis reasoned that if this happened, b would be less than c, so the lines of length b in Figure 1.1 would no longer be able to reach all the way to the x axis. Instead, they would stop somewhere above it, as Figure 1.2 shows. Wallis argued that P_1 and P_2 should represent the geometric locations of the solutions $x = -b - \sqrt{b^2 - c^2}$ and $x = -b + \sqrt{b^2 - c^2}$ in the case when $b^2 - c^2 < 0$. He evidently thought that since b is shorter than c, it could no longer be the hypotenuse of the right triangle as it had been earlier. The side of length c would now have to take that role.

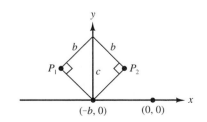

FIGURE 1.2   Wallis' representation of nonreal roots of quadratics.

Wallis' method has the undesirable consequence that $-\sqrt{-1}$ is represented by the same point as is $\sqrt{-1}$. Nevertheless, with this interpretation, the stage was set for thinking of complex numbers as "points in the plane." By 1800, the great Swiss mathematician Leonard Euler (pronounced "oiler") adopted this view concerning the n solutions to the equation $x^n - 1 = 0$. We shall learn shortly that these solutions can be expressed as $\cos \theta + \sqrt{-1} \sin \theta$ for various values of θ; Euler thought of them as being located at the vertices of a regular polygon in the plane. Euler was also the first to use the symbol i for $\sqrt{-1}$. Today, this notation is still the most popular, although some electrical engineers prefer the symbol j instead so that i can be used to represent current.

Perhaps the most influential figure in helping to bring about the acceptance of complex numbers was the brilliant German mathematician Karl Friedrich Gauss, who reinforced the utility of these numbers by using them in his several proofs of the fundamental theorem of algebra (see Chapter 6). In an 1831 paper, he produced a clear geometric representation of $x + iy$ by identifying it with the point (x, y) in the coordinate plane. He also described how these numbers could be added and multiplied.

It should be noted that 1831 was not the year that saw complex numbers transformed into legitimacy. In that same year the prolific logician Augustus De Morgan commented in his book *On the Study and Difficulties of Mathematics*, "We have shown the symbol $\sqrt{-a}$ to be void of meaning, or rather self-contradictory and absurd. Nevertheless, by means of such symbols, a part of algebra is established which is of great utility." To be sure, De Morgan had raised some possible logical problems with the idea of complex numbers. On the other hand, there were sufficient answers to these problems floating around at the time. Even if De Morgan was unaware of Gauss' paper when he wrote his book, others had done work similar to that of Gauss as early as 1806, and the preceding quote illustrates that "raw logic" by itself is often insufficient to sway the entire mathematical community to adopt new ideas. Certainly, logic was a necessary ingredient in the acceptance of complex numbers, but so too was the adoption of this logic by Gauss, Euler, and others of "sufficient clout." As more and more mathematicians came to agree with this new theory, it became socially more and more difficult to raise objections to it. By the end of the nineteenth century, complex numbers were firmly entrenched. Thus, as it is with any new mathematical or scientific theory, the acceptance of complex numbers came through a mixture of sociocultural interactions.

But what is the theory that Gauss and so many others helped produce, and how do we now think of complex numbers? This is the topic of the next few sections.

EXERCISES FOR SECTION 1.1

1. Give an argument to show that $-2 + \sqrt{-1} = \sqrt[3]{-2 + \sqrt{-121}}$.
2. Explain why cubic equations, rather than quadratic equations, played the pivotal role in helping to obtain the acceptance of complex numbers.
3. Find all solutions to $27x^3 - 9x - 2 = 0$. *Hint*: Get an equivalent monic polynomial, then use formula (1).

4. By inspection, one can see that a solution to $x^3 - 6x + 4 = 0$ is $x = 2$. To get an idea of the difficulties Bombeli had in establishing identities (2) and (3) in the text, try and show how the solution $x = 2$ arises when using formula (1).

5. Explain why Wallis' view of complex numbers results in $-\sqrt{-1}$ being represented by the same point as is $\sqrt{-1}$.

6. Is it possible to modify slightly Wallis' picture of complex numbers so that it is consistent with the representation we use today? To assist in your investigation of this question, we recommend the following article: Norton, Alec, and Lotto, Benjamin, ''Complex Roots Made Visible,'' *The College Mathematics Journal*, Vol. 15 (3), June 1984, pp. 248–249.

7. Write a report on the history of complex analysis. Resources include bibliographical items 87, 105, and 179.

1.2   The Algebra of Complex Numbers

We have seen that complex numbers came to be viewed as ordered pairs of real numbers. That is, a complex number z is *defined to be*

$$(1) \quad z = (x, y),$$

where x and y are both real numbers.

The reason we say *ordered* pair is because we are thinking of a point in the plane. The point $(2, 3)$, for example, is not the same as $(3, 2)$. The *order* in which we write x and y in equation (1) makes a difference. Clearly, then, two complex numbers are equal if and only if their x coordinates are equal *and* their y coordinates are equal. In other words,

$$(x, y) = (u, v) \quad \text{iff} \quad x = u \quad and \quad y = v.$$

(Throughout this text, iff means *if and only if*.)

If we are to have a meaningful number system, there needs to be a method for combining these ordered pairs. We need to define algebraic operations in a consistent way so that the sum, difference, product, and quotient of any two ordered pairs will again be an ordered pair. The key to defining how these numbers should be manipulated is to follow Gauss' lead and equate (x, y) with $x + iy$. Then, by letting $z_1 = (x_1, y_1)$ and $z_2 = (x_2, y_2)$ be arbitrary complex numbers, we see that

$$\begin{aligned}
z_1 + z_2 &= (x_1, y_1) + (x_2, y_2) \\
&= (x_1 + iy_1) + (x_2 + iy_2) \\
&= (x_1 + x_2) + i(y_1 + y_2) \\
&= (x_1 + x_2, y_1 + y_2).
\end{aligned}$$

Thus, the following should certainly make sense:

Definition for addition

$$\begin{aligned}
(2) \quad z_1 + z_2 &= (x_1, y_1) + (x_2, y_2) \\
&= (x_1 + x_2, y_1 + y_2).
\end{aligned}$$

Definition for subtraction

$$(3) \quad z_1 - z_2 = (x_1, y_1) - (x_2, y_2)$$
$$= (x_1 - x_2, y_1 - y_2).$$

EXAMPLE 1.1   If $z_1 = (3, 7)$ and $z_2 = (5, -6)$, then

$$z_1 + z_2 = (3, 7) + (5, -6) = (8, 1) \quad \text{and}$$
$$z_1 - z_2 = (3, 7) - (5, -6) = (-2, 13).$$

At this point, it is tempting to define the product z_1z_2 as $z_1z_2 = (x_1x_2, y_1y_2)$. It turns out, however, that this is not a good definition, and you will be asked in the problem set for this section to explain why. How, then, should products be defined? Again, if we equate (x, y) with $x + iy$ and pretend, for the moment, that $i = \sqrt{-1}$ makes sense (so that $i^2 = -1$), we have

$$z_1z_2 = (x_1, y_1)(x_2, y_2)$$
$$= (x_1 + iy_1)(x_2 + iy_2)$$
$$= x_1x_2 + ix_1y_2 + ix_2y_1 + i^2y_1y_2$$
$$= x_1x_2 - y_1y_2 + i(x_1y_2 + x_2y_1)$$
$$= (x_1x_2 - y_1y_2, x_1y_2 + x_2y_1).$$

Thus, it appears we are forced into the following definition.

Definition for multiplication

$$(4) \quad z_1z_2 = (x_1, y_1)(x_2, y_2)$$
$$= (x_1x_2 - y_1y_2, x_1y_2 + x_2y_1).$$

EXAMPLE 1.2   If $z_1 = (3, 7)$ and $z_2 = (5, -6)$, then

$$z_1z_2 = (3, 7)(5, -6) = (15 + 42, -18 + 35) = (57, 17).$$

Note that this is the same answer that would have been obtained if we had used the notation $z_1 = 3 + 7i$ and $z_2 = 5 - 6i$. For then

$$z_1z_2 = (3, 7)(5, -6)$$
$$= (3 + 7i)(5 - 6i)$$
$$= 15 - 18i + 35i - 42i^2$$
$$= 15 - 42(-1) + (-18 + 35)i$$
$$= 57 + 17i$$
$$= (57, 17).$$

Of course, it makes sense that the answer came out as we expected, since we used the notation $x + iy$ as motivation for our definition in the first place.

To motivate our definition for division, we will proceed along the same lines as we did for multiplication, assuming $z_2 \neq 0$.

$$\frac{z_1}{z_2} = \frac{(x_1, y_1)}{(x_2, y_2)}$$
$$= \frac{(x_1 + iy_1)}{(x_2 + iy_2)}.$$

At this point we need to figure out a way to be able to write the preceding quantity in the form $x + iy$. To do this, we use a standard trick and multiply the numerator and denominator by $x_2 - iy_2$. This gives

$$\frac{z_1}{z_2} = \frac{(x_1 + iy_1)}{(x_2 + iy_2)} \frac{(x_2 - iy_2)}{(x_2 - iy_2)}$$
$$= \frac{x_1 x_2 + y_1 y_2 + i(-x_1 y_2 + x_2 y_1)}{x_2^2 + y_2^2}$$
$$= \frac{x_1 x_2 + y_1 y_2}{x_2^2 + y_2^2} + i\frac{-x_1 y_2 + x_2 y_1}{x_2^2 + y_2^2}$$
$$= \left(\frac{x_1 x_2 + y_1 y_2}{x_2^2 + y_2^2}, \frac{-x_1 y_2 + x_2 y_1}{x_2^2 + y_2^2} \right).$$

Thus, we finally arrive at a rather odd definition.

Definition for division

$$(5) \quad \frac{z_1}{z_2} = \frac{(x_1, y_1)}{(x_2, y_2)}$$
$$= \left(\frac{x_1 x_2 + y_1 y_2}{x_2^2 + y_2^2}, \frac{-x_1 y_2 + x_2 y_1}{x_2^2 + y_2^2} \right), \text{ for } z_2 \neq 0.$$

EXAMPLE 1.3   If $z_1 = (3, 7)$ and $z_2 = (5, -6)$, then

$$\frac{z_1}{z_2} = \frac{(3, 7)}{(5, -6)} = \left(\frac{15 - 42}{25 + 36}, \frac{18 + 35}{25 + 36} \right) = \left(\frac{-27}{61}, \frac{53}{61} \right).$$

As we saw with the example for multiplication, we will also get this answer if we use the notation $x + iy$:

$$
\begin{aligned}
\frac{z_1}{z_2} &= \frac{(3, 7)}{(5, -6)} \\
&= \frac{3 + 7i}{5 - 6i} \\
&= \frac{3 + 7i}{5 - 6i}\frac{5 + 6i}{5 + 6i} \\
&= \frac{15 + 18i + 35i + 42i^2}{25 + 30i - 30i - 36i^2} \\
&= \frac{15 - 42 + (18 + 35)i}{25 + 36} \\
&= \frac{-27}{61} + \frac{53}{61}i \\
&= \left(\frac{-27}{61}, \frac{53}{61}\right).
\end{aligned}
$$

The technique most mathematicians would use to perform operations on complex numbers is to appeal to the notation $x + iy$ and perform the algebraic manipulations, as we did here, rather than to apply the complicated looking definitions we gave for those operations on ordered pairs. This is a valid procedure since the $x + iy$ notation was used as a guide to see how we should define the operations in the first place. It is important to remember, however, that the $x + iy$ notation is nothing more than a convenient bookkeeping device for keeping track of how to manipulate ordered pairs. It is the ordered pair algebraic definitions that really form the foundation on which our complex number system is based. In fact, if you were to program a computer to do arithmetic on complex numbers, your program would perform calculations on ordered pairs, using exactly the definitions that we gave.

It turns out that our algebraic definitions give complex numbers all the properties we normally ascribe to the real number system. Taken together, they describe what algebraists call a *field*. In formal terms, a field is a set (in this case, the complex numbers) together with two binary operations (in this case, addition and multiplication) with the following properties:

(P1) Commutative law for addition: $z_1 + z_2 = z_2 + z_1$.

(P2) Associative law for addition: $z_1 + (z_2 + z_3) = (z_1 + z_2) + z_3$.

(P3) Additive identity: There is a complex number ω such that $z + \omega = z$ for all complex numbers z. The number ω is obviously the ordered pair $(0, 0)$.

(P4) Additive inverses: Given any complex number z, there is a complex number η (depending on z) with the property that $z + \eta = (0, 0)$. Obviously, if $z = (x, y) = x + iy$, the number η will be $(-x, -y) = -x - iy$.

(P5) Commutative law for multiplication: $z_1 z_2 = z_2 z_1$.

(P6) Associative law for multiplication: $z_1(z_2 z_3) = (z_1 z_2)z_3$.

(P7) Multiplicative identity: There is a complex number ζ such that $z\zeta = z$ for all complex numbers z. It turns out that $(1, 0)$ is the complex number ζ with this property. You will be asked to verify this in the problem set for this section.

(P8) Multiplicative inverses: Given any complex number z other than the number $(0, 0)$, there is a complex number (depending on z) which we shall denote by z^{-1} with the property that $zz^{-1} = (1, 0)$. Given our definition for division, it seems reasonable that the number z^{-1} would be $z^{-1} = \dfrac{(1, 0)}{z}$.

You will be asked to confirm this in the problem set for this section.

(P9) The distributive law: $z_1(z_2 + z_3) = z_1 z_2 + z_1 z_3$.

None of these properties is difficult to prove. Most of the proofs make use of corresponding facts in the real number system. To illustrate this, we give a proof of property P1.

Proof of the commutative law for addition   Let $z_1 = (x_1, y_1)$ and $z_2 = (x_2, y_2)$ be arbitrary complex numbers. Then,

$$z_1 + z_2 = (x_1, y_1) + (x_2, y_2)$$
$$= (x_1 + x_2, y_1 + y_2) \quad \text{(by definition of addition of complex numbers)}$$
$$= (x_2 + x_1, y_2 + y_1) \quad \text{(by the commutative law for } real \text{ numbers)}$$
$$= (x_2, y_2) + (x_1, y_1) \quad \text{(by definition of addition of complex numbers)}$$
$$= z_2 + z_1.$$

The real number system can actually be thought of as a subset of our complex number system. To see why this is the case, let us agree that since any complex number of the form $(t, 0)$ is on the x axis, we can identify it with the real number t. With this correspondence, it is easy to verify that the definitions we gave for addition, subtraction, multiplication, and division of complex numbers are consistent with the corresponding operations on real numbers. For example, if x_1 and x_2 are real numbers, then

$$x_1 x_2 = (x_1, 0)(x_2, 0) \quad \text{(by our agreed correspondence)}$$
$$= (x_1 x_2 - 0, 0 + 0) \quad \text{(by definition of multiplication of complex numbers)}$$
$$= (x_1 x_2, 0) \quad \text{(confirming the consistency of our correspondence).}$$

It is now time to show specifically how the symbol i relates to the quantity $\sqrt{-1}$. Note that

$$(0, 1)^2 = (0, 1)(0, 1)$$
$$= (0 - 1, 0 + 0) \quad \text{(by definition of multiplication of complex numbers)}$$
$$= (-1, 0)$$
$$= -1 \quad \text{(by our agreed correspondence).}$$

If we use the symbol i for the point $(0, 1)$, the preceding gives $i^2 = (0, 1)^2 = -1$, which means $i = (0, 1) = \sqrt{-1}$. So, the next time you are having a discussion with your friends, and they scoff when you claim that $\sqrt{-1}$ is not imaginary, calmly put your pencil on the point $(0, 1)$ of the coordinate plane and ask them if there is anything imaginary about it. When they agree there isn't, you can tell them that this point, in fact, represents the mysterious $\sqrt{-1}$ in the same way that $(1, 0)$ represents 1.

We can also see more clearly now how the notation $x + iy$ equates to (x, y). Using the preceding conventions, we have

$$
\begin{aligned}
x + iy &= (x, 0) + (0, 1)(y, 0) &&\text{(by our previously discussed conventions,} \\
&&&\text{i.e., } x = (x, 0), \text{ etc.)} \\
&= (x, 0) + (0, y) &&\text{(by definition of multiplication of complex numbers)} \\
&= (x, y) &&\text{(by definition of addition of complex numbers).}
\end{aligned}
$$

Thus, we may move freely between the notations $x + iy$ and (x, y), depending on which is more convenient for the context in which we are working.

We close this section by discussing three standard operations on complex numbers. Suppose $z = (x, y) = x + iy$ is a complex number. Then:

(i)   The *real part* of z, denoted by $\mathrm{Re}(z)$, is the real number x.

(ii)   The *imaginary part* of z, denoted by $\mathrm{Im}(z)$, is the real number y.

(iii) The *conjugate* of z, denoted by $\bar{z}$, is the complex number $(x, -y) = x - iy$.

EXAMPLE 1.4 $\mathrm{Re}(-3 + 7i) = -3$ and $\mathrm{Re}[(9, 4)] = 9$.

EXAMPLE 1.5 $\mathrm{Im}(-3 + 7i) = 7$ and $\mathrm{Im}[(9, 4)] = 4$.

EXAMPLE 1.6 $\overline{-3 + 7i} = -3 - 7i$ and $\overline{(9, 4)} = (9, -4)$.

The following are some important facts relating to these operations that you will be asked to verify in the exercises:

(6) $\mathrm{Re}(z) = \dfrac{z + \bar{z}}{2}$.

(7) $\mathrm{Im}(z) = \dfrac{z - \bar{z}}{2i}$.

(8) $\overline{\left(\dfrac{z_1}{z_2}\right)} = \dfrac{\overline{z_1}}{\overline{z_2}}$ if $z_2 \neq 0$.

(9) $\overline{z_1 + z_2} = \overline{z_1} + \overline{z_2}$.

(10) $\overline{z_1 z_2} = \overline{z_1}\ \overline{z_2}$.

(11) $\overline{\overline{z}} = z$.

(12) $\text{Re}(iz) = -\text{Im}(z)$.

(13) $\text{Im}(iz) = \text{Re}(z)$.

Because of what it erroneously connotes, it is a shame that the term *imaginary* is used in definition (ii). Gauss, who was successful in getting mathematicians to adopt the phrase *complex number* rather than *imaginary number*, also suggested that we use *lateral part* of z in place of *imaginary part* of z. Unfortunately, this suggestion never caught on, and it appears we are stuck with the words history has handed down to us.

EXERCISES FOR SECTION 1.2

1. Perform the required calculation and express the answer in the form $a + ib$.
   (a) $(3 - 2i) - i(4 + 5i)$
   (b) $(7 - 2i)(3i + 5)$
   (c) $\overline{(1 + i)(2 + i)}(3 + i)$
   (d) $(3 + i)/(2 + i)$
   (e) $(i - 1)^3$
   (f) i^5
   (g) $\dfrac{1 + 2i}{3 - 4i} - \dfrac{4 - 3i}{2 - i}$
   (h) $(1 + i)^{-2}$
   (i) $\dfrac{(4 - i)(1 - 3i)}{-1 + 2i}$
   (j) $\overline{(1 + i\sqrt{3})(i + \sqrt{3})}$

2. Find the following quantities.
   (a) $\text{Re}[(1 + i)(2 + i)]$
   (b) $\text{Im}[\overline{(2 + i)(3 + i)}]$
   (c) $\text{Re}\left(\dfrac{4 - 3i}{2 - i}\right)$
   (d) $\text{Im}\left(\dfrac{1 + 2i}{3 - 4i}\right)$
   (e) $\text{Re}[(i - 1)^3]$
   (f) $\text{Im}[(1 + i)^{-2}]$
   (g) $\text{Re}[(x_1 - iy_1)^2]$
   (h) $\text{Im}\left(\dfrac{1}{x_1 - iy_1}\right)$
   (i) $\text{Re}[(x_1 + iy_1)(x_1 - iy_1)]$
   (j) $\text{Im}[(x_1 + iy_1)^3]$

3. Verify identities (6) through (13) given at the end of this section.

4. Let $z_1 = (x_1, y_1)$ and $z_2 = (x_2, y_2)$ be arbitrary complex numbers. Prove or disprove the following.
   (a) $\text{Re}(z_1 + z_2) = \text{Re}(z_1) + \text{Re}(z_2)$
   (b) $\text{Re}(z_1 z_2) = \text{Re}(z_1)\text{Re}(z_2)$
   (c) $\text{Im}(z_1 + z_2) = \text{Im}(z_1) + \text{Im}(z_2)$
   (d) $\text{Im}(z_1 z_2) = \text{Im}(z_1)\text{Im}(z_2)$

5. Prove that the complex number $(1, 0)$ (which, you recall, we identify with the real number 1) is the multiplicative identity for complex numbers. *Hint*: Use the (ordered pair) definition for multiplication to verify that if $z = (x, y)$ is any complex number, then $(x, y)(1, 0) = (x, y)$.

6. Verify that if $z = (x, y)$, with x and y not both 0, then $z^{-1} = \dfrac{(1, 0)}{z}$ $\left(\text{i.e., } z^{-1} = \dfrac{1}{z}\right)$.

   Hint: Use the (ordered pair) definition for division to compute $z^{-1} = \dfrac{(1, 0)}{(x, y)}$. Then, with the result you obtained, use the (ordered pair) definition for multiplication to confirm that $zz^{-1} = (1, 0)$.

7. Show that $z\overline{z}$ is always a real number.

8. From Exercise 6 and basic cancellation laws, it follows that $z^{-1} = \dfrac{1}{z} = \dfrac{\bar{z}}{z\bar{z}}$. The numerator here, $\bar{z}$, is trivial to calculate, and since the denominator $z\bar{z}$ is a real number (Exercise 7), computing the quotient $\bar{z}/(z\bar{z})$ should be rather straightforward. Use this fact to compute z^{-1} if $z = 2 + 3i$ and again if $z = 7 - 5i$.

9. Explain why the complex number $(0, 0)$ (which, you recall, we identify with the real number 0) has no multiplicative inverse.

10. Let's use the symbol $*$ for a new type of multiplication of complex numbers defined by $z_1 * z_2 = (x_1x_2, y_1y_2)$. This exercise shows why this is a bad definition.
    (a) Using the definition given in property P7, state what the new multiplicative identity would be for this new multiplication.
    (b) Show that if we use this new multiplication, nonzero complex numbers of the form $(0, a)$ have no inverse. That is, show that if $z = (0, a)$, there is no complex number z^{-1} with the property that $zz^{-1} = \zeta$, where ζ is the multiplicative identity you found in part (a).

11. Show, by equating the real numbers x_1 and x_2 with $(x_1, 0)$ and $(x_2, 0)$, that the complex definition for division is consistent with the real definition for division. *Hint*: Mimic the argument the text gives for multiplication.

12. Prove property P9, the distributive law for complex numbers.

13. Complex numbers are ordered pairs of real numbers. Is it possible to have a number system for ordered triples, quadruples, etc., of real numbers? To assist in your investigation of this question, we recommend bibliographical items 1, 132, 147, and 173.

14. We have made the statement that complex numbers are, in a metaphysical sense, just as real as are real numbers. But in what sense do numbers exist? It may surprise you that mathematicians hold a variety of views with respect to this question. Write a short paper summarizing the various views on the theme of the existence of number.

1.3   The Geometry of Complex Numbers

Since the complex numbers are ordered pairs of real numbers, there is a one-to-one correspondence between them and points in the plane. In this section we shall see what effect algebraic operations on complex numbers have on their geometric representations.

The number $z = x + iy = (x, y)$ can be represented by a position vector in the xy plane whose tail is at the origin and whose head is at the point (x, y). When the xy plane is used for displaying complex numbers, it is called the *complex plane*, or more simply, the *z plane*. Recall that $\text{Re}(z) = x$ and $\text{Im}(z) = y$. Geometrically, $\text{Re}(z)$ is the projection of $z = (x, y)$ onto the x axis, and $\text{Im}(z)$ is the projection of z onto the y axis. It makes sense, then, that the x axis is also called the *real axis*, and the y axis is called the *imaginary axis*, as Figure 1.3 illustrates.

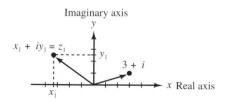

FIGURE 1.3   The complex plane.

Addition of complex numbers is analogous to addition of vectors in the plane. As we saw in Section 1.2, the sum of $z_1 = x_1 + iy_1 = (x_1, y_1)$, and $z_2 = x_2 + iy_2 = (x_2, y_2)$ is $(x_1 + x_2, y_1 + y_2)$. Hence, $z_1 + z_2$ can be obtained vectorially by using the parallelogram law, which Figure 1.4 illustrates.

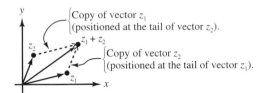

FIGURE 1.4   The sum $z_1 + z_2$.

The difference $z_1 - z_2$ can be represented by the displacement vector from the point $z_2 = (x_2, y_2)$ to the point $z_1 = (x_1, y_1)$, as Figure 1.5 shows.

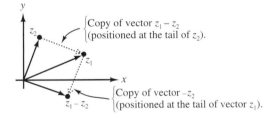

FIGURE 1.5   The difference $z_1 - z_2$.

The *modulus*, or *absolute value*, of the complex number $z = x + iy$ is a nonnegative real number denoted by $|z|$ and is given by the equation

(1) $|z| = \sqrt{x^2 + y^2}$.

The number $|z|$ is the distance between the origin and the point (x, y). The only complex number with modulus zero is the number 0. The number $z = 4 + 3i$ has modulus 5 and is pictured in Figure 1.6. The numbers $|\text{Re}(z)|$, $|\text{Im}(z)|$, and $|z|$ are the lengths of the sides of the right triangle OPQ, which is shown in Figure 1.7. The inequality $|z_1| < |z_2|$ means that the point z_1 is closer to the origin than the point z_2, and it follows that

(2) $|x| = |\text{Re}(z)| \leq |z|$ and $|y| = |\text{Im}(z)| \leq |z|$.

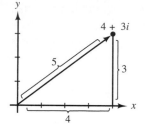

FIGURE 1.6   The real and imaginary parts of a complex number.

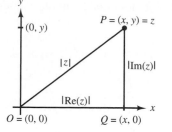

FIGURE 1.7   The moduli of z and its components.

Since the difference $z_1 - z_2$ can represent the displacement vector from z_2 to z_1, it is evident that the *distance between z_1 and z_2* is given by $\left| z_1 - z_2 \right|$. This can be obtained by using identity (3) of Section 1.2 and definition (1) to obtain the familiar formula

$$(3) \quad \text{dist}(z_1, z_2) = \left| z_1 - z_2 \right| = \sqrt{(x_1 - x_2)^2 + (y_1 - y_2)^2}.$$

If $z = (x, y) = x + iy$, then $-z = (-x, -y) = -x - iy$ is the reflection of z through the origin, and $\bar{z} = (x, -y) = x - iy$ is the reflection of z through the x axis, as is illustrated in Figure 1.8.

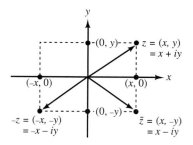

FIGURE 1.8   The geometry of negation and conjugation.

There is a very important algebraic relationship which can be used in establishing properties of the absolute value that have geometric applications. Its proof is rather straightforward, and it is given as Exercise 3.

$$(4) \quad \left| z \right|^2 = z\bar{z}.$$

A beautiful application of equation (4) is its use in establishing the triangle inequality. Figure 1.9 illustrates this inequality, which states that the sum of the lengths of two sides of a triangle is greater than or equal to the length of the third side.

(5)   The triangle inequality: $\left| z_1 + z_2 \right| \le \left| z_1 \right| + \left| z_2 \right|$.

Proof

$$\begin{aligned}
\left| z_1 + z_2 \right|^2 &= (z_1 + z_2)\,\overline{(z_1 + z_2)} && \text{(by equation (4))} \\
&= (z_1 + z_2)\,(\overline{z_1} + \overline{z_2}) && \text{(by identity (9) of Section 1.2)} \\
&= z_1\overline{z_1} + z_1\overline{z_2} + z_2\overline{z_1} + z_2\overline{z_2} \\
&= \left| z_1 \right|^2 + z_1\overline{z_2} + \overline{z_1}z_2 + \left| z_2 \right|^2 && \text{(by equation (4) and the commutative law)} \\
&= \left| z_1 \right|^2 + z_1\overline{z_2} + \overline{(z_1\overline{z_2})} + \left| z_2 \right|^2 && \text{(by identities (10) and (11) of Section 1.2)} \\
&= \left| z_1 \right|^2 + 2\mathrm{Re}(z_1\overline{z_2}) + \left| z_2 \right|^2 && \text{(by identity (6) of Section 1.2)} \\
&\le \left| z_1 \right|^2 + 2\left| \mathrm{Re}(z_1\overline{z_2}) \right| + \left| z_2 \right|^2 \\
&\le \left| z_1 \right|^2 + 2\left| z_1\overline{z_2} \right| + \left| z_2 \right|^2 && \text{(by equations (2))} \\
&= (\left| z_1 \right| + \left| z_2 \right|)^2.
\end{aligned}$$

Taking square roots yields the desired inequality.

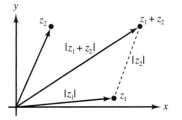

FIGURE 1.9   The triangle inequality.

EXAMPLE 1.7   To produce an example of which Figure 1.9 is a reasonable illustration, let $z_1 = 7 + i$ and $z_2 = 3 + 5i$. Then $\left| z_1 \right| = \sqrt{49 + 1} = \sqrt{50}$ and $\left| z_2 \right| = \sqrt{9 + 25} = \sqrt{34}$. Clearly, $z_1 + z_2 = 10 + 6i$, hence $\left| z_1 + z_2 \right| = \sqrt{100 + 36} = \sqrt{136}$. In this case, we can verify the triangle inequality without recourse to computation of square roots since

$$\left| z_1 + z_2 \right| = \sqrt{136} = 2\sqrt{34} = \sqrt{34} + \sqrt{34} < \sqrt{50} + \sqrt{34} = \left| z_1 \right| + \left| z_2 \right|.$$

Other important identities can also be established by means of the triangle inequality. Note that

$$\begin{aligned}
\left| z_1 \right| &= \left| (z_1 + z_2) + (-z_2) \right| \\
&\le \left| z_1 + z_2 \right| + \left| -z_2 \right| \\
&= \left| z_1 + z_2 \right| + \left| z_2 \right|.
\end{aligned}$$

Subtracting $\left| z_2 \right|$ from the left and right sides of this string of inequalities gives an important relationship that will be used in determining lower bounds of sums of complex numbers.

(6) $|z_1 + z_2| \geq |z_1| - |z_2|$.

From equation (4) and the commutative and associative laws it follows that

$$|z_1 z_2|^2 = (z_1 z_2)\overline{(z_1 z_2)} = (z_1 \overline{z_1})(z_2 \overline{z_2}) = |z_1|^2 |z_2|^2.$$

Taking square roots of the terms on the left and right establishes another important identity.

(7) $|z_1 z_2| = |z_1| |z_2|$.

As an exercise, we ask you to show

(8) $\left| \dfrac{z_1}{z_2} \right| = \dfrac{|z_1|}{|z_2|}$,   provided $z_2 \neq 0$.

EXAMPLE 1.8   If $z_1 = 1 + 2i$ and $z_2 = 3 + 2i$, then $|z_1| = \sqrt{1 + 4} = \sqrt{5}$ and $|z_2| = \sqrt{9 + 4} = \sqrt{13}$. We also see that $z_1 z_2 = -1 + 8i$, hence $|z_1 z_2| = \sqrt{1 + 64} = \sqrt{65} = \sqrt{5}\sqrt{13} = |z_1| |z_2|$.

Figure 1.10 illustrates the multiplication given by Example 1.8. It certainly appears that the length of the $z_1 z_2$ vector equals the product of the lengths of z_1 and z_2, confirming equation (7), but why is it located in the second quadrant, when both z_1 and z_2 are in the first quadrant? The answer to this question will become apparent in Section 1.4.

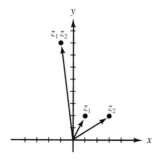

FIGURE 1.10   The geometry of multiplication.

EXERCISES FOR SECTION 1.3

1. Locate the numbers z_1 and z_2 vectorially, and use vectors to find $z_1 + z_2$ and $z_1 - z_2$ when
   (a) $z_1 = 2 + 3i$ and $z_2 = 4 + i$
   (b) $z_1 = -1 + 2i$ and $z_2 = -2 + 3i$
   (c) $z_1 = 1 + i\sqrt{3}$ and $z_2 = -1 + i\sqrt{3}$

2. Find the following quantities.

   (a) $\left|(1 + i)(2 + i)\right|$ (b) $\left|\dfrac{4 - 3i}{2 - i}\right|$ (c) $\left|(1 + i)^{50}\right|$

   (d) $\left|z\bar{z}\right|$, where $z = x + iy$ (e) $\left|z - 1\right|^2$

3. Prove identities (4) and (8).

4. Determine which of the following points lie inside the circle $\left|z - i\right| = 1$.

   (a) $\dfrac{1}{2} + i$ (b) $1 + \dfrac{i}{2}$

   (c) $\dfrac{1}{2} + \dfrac{i\sqrt{2}}{2}$ (d) $\dfrac{-1}{2} + i\sqrt{3}$

5. Show that the point $(z_1 + z_2)/2$ is the midpoint of the line segment joining z_1 to z_2.

6. Sketch the set of points determined by the following relations.

   (a) $\left|z + 1 - 2i\right| = 2$ (b) $\mathrm{Re}(z + 1) = 0$

   (c) $\left|z + 2i\right| \le 1$ (d) $\mathrm{Im}(z - 2i) > 6$

7. Show that the equation of the line through the points z_1 and z_2 can be expressed in the form $z = z_1 + t(z_2 - z_1)$ where t is a real number.

8. Show that the vector z_1 is perpendicular to the vector z_2 if and only if $\mathrm{Re}(z_1\bar{z}_2) = 0$.

9. Show that the vector z_1 is parallel to the vector z_2 if and only if $\mathrm{Im}(z_1\bar{z}_2) = 0$.

10. Show that the four points z, $\bar{z}$, $-z$, and $-\bar{z}$ are the vertices of a rectangle with its center at the origin.

11. Show that the four points z, iz, $-z$, and $-iz$ are the vertices of a square with its center at the origin.

12. Prove that $\sqrt{2}\left|z\right| \ge \left|\mathrm{Re}(z)\right| + \left|\mathrm{Im}(z)\right|$.

13. Show that $\left|z_1 - z_2\right| \le \left|z_1\right| + \left|z_2\right|$.

14. Show that $\left|z_1 z_2 z_3\right| = \left|z_1\right|\left|z_2\right|\left|z_3\right|$.

15. Show that $\left|z^n\right| = \left|z\right|^n$ where n is an integer.

16. Show that $\left|\,\left|z_1\right| - \left|z_2\right|\,\right| \le \left|z_1 - z_2\right|$.

17. Prove that $\left|z\right| = 0$ if and only if $z = 0$.

18. Show that $z_1\bar{z}_2 + \bar{z}_1 z_2$ is a real number.

19. If you study carefully the proof of the triangle inequality, you will note that the reasons for the *inequality* hinge on $\mathrm{Re}(z_1\bar{z}_2) \le \left|z_1\bar{z}_2\right|$. Under what conditions will these two quantities be equal, thus turning the triangle inequality into an equality?

20. Prove that $\left|z_1 - z_2\right|^2 = \left|z_1\right|^2 - 2\,\mathrm{Re}(z_1\bar{z}_2) + \left|z_2\right|^2$.

21. Use mathematical induction to prove that

$$\left|\sum_{k=1}^{n} z_k\right| \le \sum_{k=1}^{n} \left|z_k\right|.$$

22. Let z_1 and z_2 be two distinct points in the complex plane. Let K be a positive real constant that is greater than the distance between z_1 and z_2. Show that the set of points $\{z: \left|z - z_1\right| + \left|z - z_2\right| = K\}$ is an ellipse with foci z_1 and z_2.

23. Use Exercise 22 to find the equation of the ellipse with foci $\pm 2i$ that goes through the point $3 + 2i$.

24. Use Exercise 22 to find the equation of the ellipse with foci $\pm 3i$ that goes through the point $8 - 3i$.

25. Let z_1 and z_2 be two distinct points in the complex plane. Let K be a positive real constant that is less than the distance between z_1 and z_2. Show that the set of points $\{z: \left| \left| z - z_1 \right| - \left| z - z_2 \right| \right| = K\}$ is a hyperbola with foci z_1 and z_2.

26. Use Exercise 25 to find the equation of the hyperbola with foci ± 2 that goes through the point $2 + 3i$.

27. Use Exercise 25 to find the equation of the hyperbola with foci ± 25 that goes through the point $7 + 24i$.

28. Write a report on how complex analysis is used to understand Pythagorean triples. Resources include bibliographical items 94 and 97.

1.4   The Geometry of Complex Numbers, Continued

In Section 1.3 we saw that a complex number $z = x + iy$ could be viewed as a vector in the xy plane whose tail is at the origin and whose head is at the point (x, y). A vector can be uniquely specified by giving its magnitude (i.e., its length) and direction (i.e., the angle it makes with the positive x axis). In this section, we focus on these two geometric aspects of complex numbers.

Let r be the *modulus* of z (i.e., $r = \left| z \right|$), and let θ be the angle that the line from the origin to the complex number z makes with the positive x axis. Then as Figure 1.11 shows,

$$(1) \quad z = (r \cos \theta, r \sin \theta) = r(\cos \theta + i \sin \theta).$$

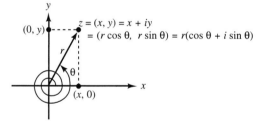

FIGURE 1.11   Polar representation of complex numbers.

Identity (1) is known as a *polar representation* of z, and the values r and θ are called *polar coordinates* of z. The coordinate θ is undefined if $z = 0$, and as Figure 1.11 shows, θ can be any value for which the identities $\cos \theta = x/r$ and $\sin \theta = y/r$ hold true. Thus, θ can take on an infinite number of values for a given complex number and is unique only up to multiples of 2π. We call θ *an argument* of z, and use the notation $\theta = \arg z$. Clearly,

$$(2) \quad \theta = \arg z = \arctan \frac{y}{x} \quad \text{if } x \neq 0,$$

but we must be careful to specify the choice of $\arctan(y/x)$ so that the point z corresponding to r and θ lies in the appropriate quadrant. The reason for this is that $\tan \theta$ has period π, whereas $\cos \theta$ and $\sin \theta$ have period 2π.

EXAMPLE 1.9

$$\sqrt{3} + i = 2 \cos \frac{\pi}{6} + i2 \sin \frac{\pi}{6} = 2 \cos \frac{13\pi}{6} + i2 \sin \frac{13\pi}{6}$$

$$= 2 \cos\left(\frac{\pi}{6} + 2\pi n\right) + i2 \sin\left(\frac{\pi}{6} + 2\pi n\right),$$

where n is any integer.

EXAMPLE 1.10   If $z = -\sqrt{3} - i$, then

$$r = |z| = |-\sqrt{3} - i| = 2 \quad \text{and}$$

$$\theta = \arctan \frac{y}{x} = \arctan \frac{-1}{-\sqrt{3}} = \frac{7\pi}{6}, \text{ so}$$

$$-\sqrt{3} - i = 2 \cos \frac{7\pi}{6} + i2 \sin \frac{7\pi}{6}$$

$$= 2 \cos\left(\frac{7\pi}{6} + 2\pi n\right) + i2 \sin\left(\frac{7\pi}{6} + 2\pi n\right),$$

where n is any integer.

If θ_0 is a value of arg z, then we can display *all* values of arg z as follows:

(3) $\arg z = \theta_0 + 2\pi k$,   where k is an integer.

For a given complex number $z \neq 0$, the value of arg z that lies in the range $-\pi < \theta \leq \pi$ is called the *principal value* of arg z and is denoted by Arg z. Thus,

(4)   Arg $z = \theta$,   where $-\pi < \theta \leq \pi$.

Using equations (3) and (4) we can establish a relation between arg z and Arg z:

(5) $\arg z = \text{Arg } z + 2\pi k$,   where k is an integer.

As we shall see in Chapter 2, Arg z is a discontinuous function of z because it "jumps" by an amount of 2π as z crosses the negative real axis.

In Chapter 5 we will define e^z for any complex number z. You will see that this complex exponential has all the properties of real exponentials that you studied in earlier mathematics courses. That is, $e^{z_1} e^{z_2} = e^{z_1 + z_2}$, and so forth. You will also see that if $z = x + iy$, then

(6) $e^z = e^{x+iy} = e^x(\cos y + i \sin y)$.

We will use these facts freely from now on, and will prove the validity of our actions when we get to Chapter 5.

If we set $x = 0$ and let θ take the role of y in equation (6), we get a famous equation known as *Euler's formula*:

(7) $e^{i\theta} = (\cos \theta + i \sin \theta) = (\cos \theta, \sin \theta)$.

If θ is a real number, $e^{i\theta}$ will be located somewhere on the circle with radius 1 centered at the origin. This is easy to verify since

(8) $\left| e^{i\theta} \right| = \sqrt{\cos^2\theta + \sin^2\theta} = 1.$

Figure 1.12 illustrates the location of the points $e^{i\theta}$ for various values of θ.

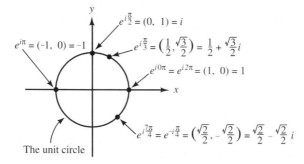

FIGURE 1.12   The location of $e^{i\theta}$ for various values of θ.

Notice that when $\theta = \pi$, we get $e^{i\pi} = (\cos \pi, \sin \pi) = (-1, 0) = -1$, so

(9) $e^{i\pi} + 1 = 0.$

Euler was the first to discover this relationship. It has been labeled by many as the most amazing equation in analysis, and with good reason. Symbols with a rich history are miraculously woven together—the constant π discovered by Hippocrates; e the base of the natural logarithms; the basic concepts of addition ($+$) and equality ($=$); the foundational whole numbers 0 and 1; and i, the number that is the central focus of this book.

Euler's formula (7) is of tremendous use in establishing important algebraic and geometric properties of complex numbers. As a start, it allows us to express a polar form of the complex number z in a more compact way. Recall that if $r = |z|$ and $\theta = \arg z$, then $z = r(\cos \theta + i \sin \theta)$. Using formula (7) we can now write z in its *exponential form*:

(10) $z = re^{i\theta}.$

EXAMPLE 1.11   With reference to Example 1.10, with $z = -\sqrt{3} - i$, we have $z = 2e^{i(7\pi/6)}$.

Together with the rules for exponentiation that we will verify in Chapter 5, equation (10) has interesting applications. If $z_1 = r_1 e^{i\theta_1}$ and $z_2 = r_2 e^{i\theta_2}$, then

(11) $z_1 z_2 = r_1 e^{i\theta_1} r_2 e^{i\theta_2} = r_1 r_2 e^{i(\theta_1 + \theta_2)} = r_1 r_2 [\cos(\theta_1 + \theta_2) + i \sin(\theta_1 + \theta_2)].$

Figure 1.13 illustrates the geometric significance of equation (11). We have already seen that the modulus of the product is the product of the moduli; that is, $|z_1z_2| = |z_1||z_2|$. Identity (11) establishes than an argument of z_1z_2 is an argument of z_1 plus an argument of z_2; that is,

(12) $\arg(z_1z_2) = \arg z_1 + \arg z_2$.

This fact answers the question posed at the end of Section 1.3 regarding why the product z_1z_2 was in a different quadrant than either z_1 or z_2. This also offers an interesting explanation as to why the product of two negative real numbers is a positive real number—the negative numbers, each of which has an angular displacement of π radians, combine to produce a product which is rotated to a point whose argument is $\pi + \pi = 2\pi$ radians, coinciding with the positive real axis.

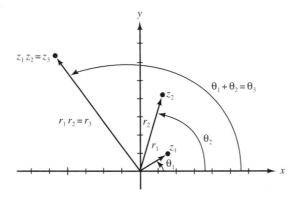

FIGURE 1.13   The product of two complex numbers $z_3 = z_1z_2$.

Using equality (11), we see that $z^{-1} = \dfrac{1}{z} = \dfrac{1}{re^{i\theta}} = \dfrac{1}{r}e^{-i\theta}$. In other words,

(13) $z^{-1} = \dfrac{1}{r}[\cos(-\theta) + i\sin(-\theta)] = \dfrac{1}{r}e^{-i\theta}$.

Notice also that

(14) $\bar{z} = r(\cos\theta - i\sin\theta) = r[\cos(-\theta) + i\sin(-\theta)] = re^{-i\theta}$ and

(15) $\dfrac{z_1}{z_2} = \dfrac{r_1}{r_2}[\cos(\theta_1 - \theta_2) + i\sin(\theta_1 - \theta_2)] = \dfrac{r_1}{r_2}e^{i(\theta_1 - \theta_2)}$.

If z is in the first quadrant, Figure 1.14 shows the numbers z, $\bar{z}$, and z^{-1} in the case where $|z| < 1$. Figure 1.15 depicts the situation when $|z| > 1$.

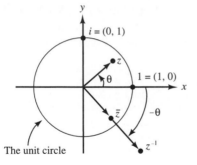

FIGURE 1.14   Relative positions of z, $\bar{z}$, and z^{-1}, when $\left| z \right| < 1$.

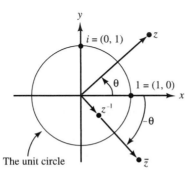

FIGURE 1.15   Relative positions of z, $\bar{z}$, and z^{-1}, when $\left| z \right| > 1$.

EXAMPLE 1.12   If $z = 5 + 12i$, then $r = 13$ and $z^{-1} = \frac{1}{13}[\frac{5}{13} - (12i/13)]$ has modulus $\frac{1}{13}$.

EXAMPLE 1.13   If $z_1 = 8i$ and $z_2 = 1 + i\sqrt{3}$, then the polar forms are $z_1 = 8[\cos(\pi/2) + i \sin(\pi/2)]$ and $z_2 = 2[\cos(\pi/3) + i \sin(\pi/3)]$. So we have

$$\frac{z_1}{z_2} = \frac{8}{2}\left[\cos\left(\frac{\pi}{2} - \frac{\pi}{3} \right) + i \sin\left(\frac{\pi}{2} - \frac{\pi}{3} \right) \right] = 4\left(\cos \frac{\pi}{6} + i \sin \frac{\pi}{6} \right)$$
$$= 2\sqrt{3} + 2i.$$

EXERCISES FOR SECTION 1.4

1. Find Arg z for the following values of z.

   (a) $1 - i$ (b) $-\sqrt{3} + i$ (c) $(-1 - i\sqrt{3})^2$

   (d) $(1 - i)^3$ (e) $\dfrac{2}{1 + i\sqrt{3}}$ (f) $\dfrac{2}{i - 1}$

   (g) $\dfrac{1 + i\sqrt{3}}{(1 + i)^2}$ (h) $(1 + i\sqrt{3})(1 + i)$

2. Represent the following complex numbers in polar form.

   (a) -4 (b) $6 - 6i$ (c) $-7i$

   (d) $-2\sqrt{3} - 2i$ (e) $\dfrac{1}{(1 - i)^2}$ (f) $\dfrac{6}{i + \sqrt{3}}$

   (g) $(5 + 5i)^3$ (h) $3 + 4i$

3. Express the following in $a + ib$ form.

   (a) $e^{i\pi/2}$ (b) $4e^{-i\pi/2}$ (c) $8e^{i7\pi/3}$

   (d) $-2e^{i5\pi/6}$ (e) $2ie^{-i3\pi/4}$ (f) $6e^{i2\pi/3}e^{i\pi}$

   (g) $e^2 e^{i\pi}$ (h) $e^{i\pi/4}e^{-i\pi}$

4. Use the exponential notation to show that

   (a) $(\sqrt{3} - i)(1 + i\sqrt{3}) = 2\sqrt{3} + 2i$ (b) $(1 + i)^3 = -2 + 2i$

   (c) $2i(\sqrt{3} + i)(1 + i\sqrt{3}) = -8$ (d) $8/(1 + i) = 4 - 4i$

5. Show that $\arg(z_1 z_2 z_3) = \arg z_1 + \arg z_2 + \arg z_3$. *Hint*: Use property (12).

6. Let $z = \sqrt{3} + i$. Plot the points z, iz, $-z$, and $-iz$ and describe a relationship among their arguments.

7. Let $z_1 = -1 + i\sqrt{3}$ and $z_2 = -\sqrt{3} + i$. Show that the equation $\text{Arg}(z_1 z_2) = \text{Arg } z_1 + \text{Arg } z_2$ *does not* hold for the specific choice of z_1 and z_2.

8. Show that the equation $\text{Arg}(z_1 z_2) = \text{Arg } z_1 + \text{Arg } z_2$ is true if we require that $-\pi/2 < \text{Arg } z_1 < \pi/2$ and $-\pi/2 < \text{Arg } z_2 < \pi/2$.

9. Show that $\arg z_1 = \arg z_2$ if and only if $z_2 = cz_1$, where c is a positive real constant.

10. Establish the identity $\arg(z_1/z_2) = \arg z_1 - \arg z_2$.

11. Describe the set of complex numbers for which $\text{Arg}(1/z) \neq - \text{Arg}(z)$.

12. Show that $\arg(1/z) = -\arg z$.

13. Show that $\arg(z_1 \overline{z_2}) = \arg z_1 - \arg z_2$.

14. Show that

    (a) $\text{Arg}(z\overline{z}) = 0$ (b) $\text{Arg}(z + \overline{z}) = 0$ when $\text{Re}(z) > 0$.

15. Let $z \neq z_0$. Show that the polar representation $z - z_0 = \rho(\cos \phi + i \sin \phi)$ can be used to denote the displacement vector from z_0 to z as indicated in Figure 1.16.

16. Let z_1, z_2, and z_3 form the vertices of a triangle as indicated in Figure 1.17. Show that

$$\alpha = \arg\left(\frac{z_2 - z_1}{z_3 - z_1}\right) = \arg(z_2 - z_1) - \arg(z_3 - z_1)$$

is an expression for the angle at the vertex z_1.

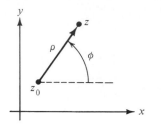

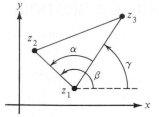

FIGURE 1.16   Accompanies Exercise 15.   **FIGURE 1.17**   Accompanies Exercise 16.

1.5   The Algebra of Complex Numbers, Revisited

The real numbers are deficient in the sense that not all algebraic operations on them produce real numbers. Thus, for $\sqrt{-1}$ to make sense, we must lift our sights to the domain of complex numbers. Do complex numbers have this same deficiency? That is, if we are to make sense out of expressions like $\sqrt{1 + i}$, must we appeal to yet another new number system? The answer to this question is *no*. It turns out that any reasonable algebraic operation we perform on complex numbers gives us complex numbers. In this respect, we say that the complex numbers are *complete*. Later we will learn how to evaluate complicated algebraic expressions such as $(-1)^i$. For now we will be content to study integral powers and roots of complex numbers.

The important players in this regard are the exponential and polar forms of a complex number, $z = re^{i\theta} = r(\cos\theta + i\sin\theta)$. By the laws of exponents (which, you recall, we have promised to prove in Chapter 5!) we clearly have

(1) $z^n = (re^{i\theta})^n = r^n e^{in\theta} = r^n[\cos(n\theta) + i\sin(n\theta)]$, and

(2) $z^{-n} = (re^{i\theta})^{-n} = r^{-n}e^{-in\theta} = r^{-n}[\cos(-n\theta) + i\sin(-n\theta)]$.

EXAMPLE 1.14   Show that $(-\sqrt{3} - i)^3 = 8i$ in two ways.

Solution 1   We appeal to the binomial formula and write

$$(-\sqrt{3} - i)^3 = (-\sqrt{3})^3 + 3(-\sqrt{3})^2(-i) + 3(-\sqrt{3})(-i)^2 + (-i)^3 = 8i.$$

Solution 2   Using identity (1) and Example 1.11, we have

$$(-\sqrt{3} - i)^3 = \left(2e^{i\frac{7\pi}{6}}\right)^3 = \left(2^3 e^{i\frac{21\pi}{6}}\right) = 8\left(\cos\frac{21\pi}{6} + i\sin\frac{21\pi}{6}\right) = 8i.$$

Which of these methods would you use if you were asked to compute $(-\sqrt{3} - i)^{30}$?

EXAMPLE 1.15   Evaluate $(-\sqrt{3} - i)^{-6}$.

Solution $(-\sqrt{3} - i)^{-6} = \left(2e^{i\frac{7\pi}{6}}\right)^{-6} = 2^{-6}e^{-i7\pi} = 2^{-6}(-1) = -\dfrac{1}{64}.$

An interesting application of the laws of exponents comes from putting the equation $(e^{i\theta})^n = e^{in\theta}$ into its polar form. This gives

(3) $(\cos\theta + i\sin\theta)^n = (\cos n\theta + i\sin n\theta),$

which is known as De Moivre's formula, in honor of the French mathematician Abraham De Moivre (1667–1754).

EXAMPLE 1.16   De Moivre's formula (3) can be used to show that

$$\cos 5\theta = \cos^5\theta - 10\cos^3\theta\sin^2\theta + 5\cos\theta\sin^4\theta.$$

If we let $n = 5$, and use the binomial formula to expand the left side of equation (3), then we obtain

$$\cos^5\theta + i5\cos^4\theta\sin\theta - 10\cos^3\theta\sin^2\theta - 10i\cos^2\theta\sin^3\theta$$
$$+ 5\cos\theta\sin^4\theta + i\sin^5\theta.$$

The real part of this expression is

$$\cos^5\theta - 10\cos^3\theta\sin^2\theta + 5\cos\theta\sin^4\theta.$$

Equating this to the real part of $\cos 5\theta + i\sin 5\theta$ on the right side of equation (3) establishes the desired result.

A key ingredient in determining roots of complex numbers turns out to be a corollary to the *fundamental theorem of algebra*. We will prove the theorem in Chapter 6. Our proofs must be independent of conclusions we derive here since we are going to make use of the corollary now:

> **Corollary 1.1**   (*Corollary to the fundamental theorem of algebra*) *If $P(z)$ is a polynomial of degree n ($n > 0$) with complex coefficients, then the equation $P(z) = 0$ has precisely n (not necessarily distinct) solutions.*

EXAMPLE 1.17   Let $P(z) = z^3 + (2 - 2i)z^2 + (-1 - 2i)z - 2$. This polynomial of degree 3 can be written as $P(z) = (z - i)^2(z + 2)$. Hence, the equation $P(z) = 0$ has solutions $z_1 = i$, $z_2 = i$, and $z_3 = -2$. Thus, in accordance with Corollary 1.1, we have three solutions, with z_1 and z_2 being repeated roots.

The corollary to the fundamental theorem of algebra implies that if we can find n *distinct* solutions to the equation $z^n = c$ (or $z^n - c = 0$), we will have found *all* the solutions. We begin our search for these solutions by looking at the simpler equation $z^n = 1$. You will soon see that solving this equation will enable us to handle the more general one quite easily.

To solve $z^n = 1$, let us first note that from identities (5) and (10) of Section 1.4 we can deduce an important condition determining when two complex numbers are equal. Let $z_1 = r_1 e^{i\theta_1}$ and $z_2 = r_2 e^{i\theta_2}$. Then,

(4) $z_1 = z_2$ (i.e., $r_1 e^{i\theta_1} = r_2 e^{i\theta_2}$)   iff $r_1 = r_2$ *and* $\theta_1 = \theta_2 + 2\pi k$,

where k is an integer.

That is, two complex numbers are equal if and only if their moduli agree, and an argument of one equals an argument of the other to within an integral multiple of 2π. Now, suppose $z = re^{i\theta}$ is a solution to $z^n = 1$. Putting the later equation in exponential form gives us $r^n e^{in\theta} = 1 \cdot e^{i0}$, so relation (4) implies

$$r^n = 1 \quad \text{and} \quad n\theta = 0 + 2\pi k,$$

where k is an integer. Clearly, for $z = re^{i\theta}$, if $r = 1$, and $\theta = \dfrac{2\pi k}{n}$, we can generate n *distinct* solutions to $z^n = 1$ (and, therefore, *all* solutions) by setting $k = 0, 1, 2,$ $\ldots, n - 1$. (Note that the solutions for $k = n, n + 1, \ldots$, merely repeat those for $k = 0, 1, \ldots$, since the arguments so generated agree to within an integral multiple of 2π.) As stated in Section 1.1, the n solutions can be expressed as

(5) $z_k = e^{i\frac{2\pi k}{n}} = \cos\dfrac{2\pi k}{n} + i\sin\dfrac{2\pi k}{n}$,   for $k = 0, 1, \ldots, n - 1$.

They are called the *nth roots of unity*. The value ω_n given by

(6) $\omega_n = e^{i\frac{2\pi}{n}} = \cos\dfrac{2\pi}{n} + i\sin\dfrac{2\pi}{n}$

is called a *primitive nth root of unity*. By De Moivre's formula, the nth roots of unity can be expressed as

(7) $1, \omega_n, \omega_n^2, \ldots, \omega_n^{n-1}$.

Geometrically, the nth roots of unity are equally spaced points that lie on the unit circle $\{z: |z| = 1\}$ and form the vertices of a regular polygon with n sides.

EXAMPLE 1.18   The solutions to the equation $z^8 = 1$ are given by the 8 values

$$z_k = e^{i\frac{2\pi k}{8}} = \cos\dfrac{2\pi k}{8} + i\sin\dfrac{2\pi k}{8}, \text{ for } k = 0, 1, \ldots, 7.$$

In Cartesian form these solutions are $\pm 1, \pm i, \pm(\sqrt{2} + i\sqrt{2})/2$, and $\pm(\sqrt{2} - i\sqrt{2})/2$. From expressions (7) it is clear that $\omega_8 = z_1$. Figure 1.18 gives an illustration of this.

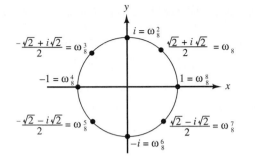

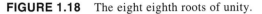

FIGURE 1.18   The eight eighth roots of unity.

The preceding procedure is easy to generalize in solving $z^n = c$ for any nonzero complex number c. If $c = \rho e^{i\phi} = \rho(\cos \phi + i \sin \phi)$, our solutions are given by

$$(8) \quad z_k = \rho^{1/n} e^{i\frac{\phi + 2\pi k}{n}} = \rho^{1/n}\left(\cos \frac{\phi + 2\pi k}{n} + i \sin \frac{\phi + 2\pi k}{n}\right), \quad \text{for}$$

$$k = 0, 1, \ldots, n - 1.$$

Each solution in equation (8) can be considered an nth root of c. Geometrically, the nth roots of c are equally spaced points that lie on the circle $\{z: |z| = \rho^{1/n}\}$ and form the vertices of a regular polygon with n sides. Figure 1.19 pictures the case for $n = 5$.

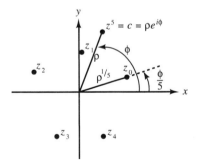

FIGURE 1.19   The five solutions to the equation $z^5 = c$.

It is interesting to note that if ζ is any particular solution to the equation $z^n = c$, then *all* solutions can be generated by multiplying ζ by the various nth roots of unity. That is, the solution set is

$$(9) \quad \zeta, \zeta\omega_n, \zeta\omega_n^2, \ldots, \zeta\omega_n^{n-1}.$$

The reason for this is that for any j, $(\zeta \omega_n^j)^n = \zeta^n (\omega_n^j)^n = \zeta^n (\omega_n^n)^j = \zeta^n = c$, and that multiplying a number by $\omega_n = e^{i \frac{2\pi}{n}}$ increases the argument of that number by $\dfrac{2\pi}{n}$, so that expressions (9) contain n *distinct* values.

EXAMPLE 1.19 Let's find all cube roots of $8i = 8[\cos(\pi/2) + i\,\sin(\pi/2)]$ using formula (8). By direct computation, we see that the roots are

$$z_k = 2\left[\cos\frac{(\pi/2) + 2\pi k}{3} + i\,\sin\frac{(\pi/2) + 2\pi k}{3}\right], \quad \text{for } k = 0, 1, 2.$$

The Cartesian forms of the solutions are $z_0 = \sqrt{3} + i$, $z_1 = -\sqrt{3} + i$, and $z_2 = -2i$, as shown in Figure 1.20.

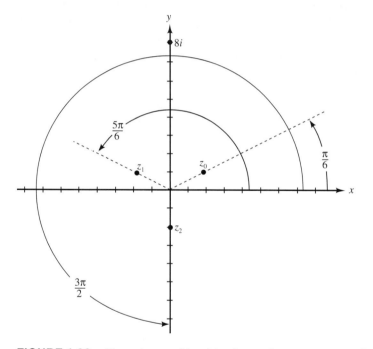

FIGURE 1.20   The point $z = 8i$ and its three cube roots z_0, z_1, and z_2.

EXERCISES FOR SECTION 1.5

1. Show that $(\sqrt{3} + i)^4 = -8 + i8\sqrt{3}$ in two ways:
    (a) by squaring twice
    (b) by using equation (3)

2. Calculate the following.
    (a) $(1 - i\sqrt{3})^3(\sqrt{3} + i)^2$ (b) $\dfrac{(1 + i)^3}{(1 - i)^5}$ (c) $(\sqrt{3} + i)^6$

3. Use De Moivre's formula and establish the following identities.
    (a) $\cos 3\theta = \cos^3\theta - 3\cos\theta\sin^2\theta$ (b) $\sin 3\theta = 3\cos^2\theta\sin\theta - \sin^3\theta$

4. Let z be any nonzero complex number, and let n be an integer. Show that $z^n + (\bar{z})^n$ is a real number.

For Exercises 5–9, find all the roots.

5. $(-2 + 2i)^{1/3}$ **6.** $(-64)^{1/4}$ **7.** $(-1)^{1/5}$

8. $(16i)^{1/4}$ **9.** $(8)^{1/6}$

10. Establish the quadratic formula.

11. Find the solutions to the equation $z^2 + (1 + i)z + 5i = 0$.

12. Solve the equation $(z + 1)^3 = z^3$.

13. Let $P(z) = a_n z^n + a_{n-1} z^{n-1} + \cdots + a_1 z + a_0$ be a polynomial with *real* coefficients $a_n, a_{n-1}, \ldots, a_1, a_0$. If z_1 is a complex root of $P(z)$, show that $\bar{z_1}$ is also a root. *Hint:* Show that $P(\bar{z_1}) = \overline{P(z_1)} = 0$.

14. Find all the roots of the equation $z^4 - 4z^3 + 6z^2 - 4z + 5 = 0$ given that $z_1 = i$ is a root.

15. Let m and n be positive integers that have no common factor. Show that there are n distinct solutions to $w^n = z^m$ and that they are given by

$$w_k = r^{m/n}\left(\cos\frac{m(\theta + 2\pi k)}{n} + i\sin\frac{m(\theta + 2\pi k)}{n}\right) \text{ for } k = 0, 1, \ldots, n - 1.$$

16. Find the three solutions to $z^{3/2} = 4\sqrt{2} + i4\sqrt{2}$.

17. (a) If $z \neq 1$, show that $1 + z + z^2 + \cdots + z^n = \dfrac{1 - z^{n+1}}{1 - z}$.

    (b) Use part (a) and De Moivre's formula to derive *Lagrange's identity*:

$$1 + \cos\theta + \cos 2\theta + \cdots + \cos n\theta = \frac{1}{2} + \frac{\sin\left[\left(n + \dfrac{1}{2}\right)\theta\right]}{2\sin(\theta/2)} \quad \text{where } 0 < \theta < 2\pi.$$

18. Let $z_k \neq 1$ be an nth root of unity. Prove that

$$1 + z_k + z_k^2 + \cdots + z_k^{n-1} = 0.$$

19. If $1 = z_0, z_1, z_2, \ldots, z_{n-1}$ are the nth roots of unity, prove that

$$(z - z_1)(z - z_2)\cdots(z - z_{n-1}) = 1 + z + z^2 + \cdots + z^{n-1}.$$

20. Identity (3), De Moivre's formula, can be established without recourse to properties of the exponential function. Note that this identity is trivially true for $n = 1$, then

    (a) Using basic trigonometric identities, show the identity is valid for $n = 2$.

    (b) Use induction to verify the identity for all positive integers.

    (c) How would you verify this identity for all negative integers?

21. Look up the article on Euler's formula and discuss what you found. Use bibliographical item 169.

22. Look up the article on De Moivre's formula and discuss what you found. Use bibliographical item 103.

23. Look up the article on how complex analysis could be used in the construction of a regular pentagon and discuss what you found. Use bibliographical item 114.

24. Write a report on how complex analysis is used to study roots of polynomials and/or complex functions. Resources include bibliographical items 50, 65, 67, 102, 109, 120, 121, 122, 140, 152, 162, 171, 174, and 178.

1.6   The Topology of Complex Numbers

In this section we investigate some basic ideas concerning sets of points in the plane. The first concept is that of a curve. Intuitively, we think of a curve as a piece of string placed on a flat surface in some type of meandering pattern. More formally, we define a curve to be the range of a continuous complex-valued function $z(t)$ defined on the interval $[a, b]$. That is, a curve C is the range of a function given by $z(t) = (x(t), y(t)) = x(t) + iy(t)$, for $a \leq t \leq b$, where both $x(t)$ and $y(t)$ are continuous real-valued functions. If both $x(t)$ and $y(t)$ are differentiable, we say that the curve is *smooth*. A curve for which $x(t)$ and $y(t)$ are differentiable except for a finite number of points is called *piecewise smooth*. We specify a curve C as

(1) $C: z(t) = x(t) + iy(t)$ for $a \leq t \leq b$,

and say that $z(t)$ is a *parametrization* for the curve C. Notice that with this parameterization, we are specifying a direction to the curve C, and we say that C is a curve that goes from the initial point $z(a) = (x(a), y(a)) = x(a) + iy(a)$ to the terminal point $z(b) = (x(b), y(b)) = x(b) + iy(b)$. If we had another function whose range was the same set of points as $z(t)$ but whose initial and final points were reversed, we would indicate the curve this function defines by $-C$. For example, if $z_0 = x_0 + iy_0$ and $z_1 = x_1 + iy_1$ are two given points, then the straight line segment joining z_0 to z_1 is

(2) $C: z(t) = [x_0 + (x_1 - x_0)t] + i[y_0 + (y_1 - y_0)t]$ for $0 \leq t \leq 1$,

and is pictured in Figure 1.21. One way to derive formula (2) is to use the vector form of a line. A point on the line is $z_0 = x_0 + iy_0$ and the direction of the line is $z_1 - z_0$; hence the line C in formula (2) is given by

$C: z(t) = z_0 + (z_1 - z_0)t$ for $0 \leq t \leq 1$.

Clearly one parametrization for $-C$ is

$-C: \gamma(t) = z_1 + (z_0 - z_1)t$ for $0 \leq t \leq 1$.

It is worth noting that $\gamma(t) = z(1 - t)$. This illustrates a general principle: If C is a curve parameterized by $z(t)$ for $0 \leq t \leq 1$, then one parameterization for $-C$ will be $z(1 - t)$, $0 \leq t \leq 1$.

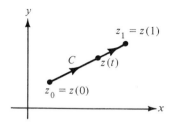

FIGURE 1.21   The straight line segment C joining z_0 to z_1.

A curve C with the property that $z(a) = z(b)$ is said to be a *closed curve*. The line segment (2) is not a closed curve. The curve $x(t) = \sin 2t \cos t$, and $y(t) = \sin 2t \sin t$ for $0 \le t \le 2\pi$ forms the four-leaved rose in Figure 1.22. Observe carefully that as t goes from 0 to $\pi/2$, the point is on leaf 1; from $\pi/2$ to π it is on leaf 2; between π and $3\pi/2$ it is on leaf 3; and finally, for t between $3\pi/2$ and 2π it is on leaf 4. Notice that the curve has crossed over itself at the origin.

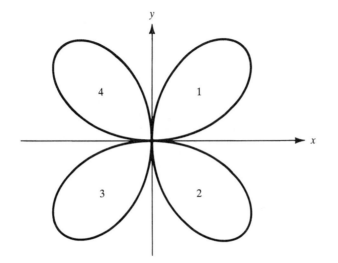

FIGURE 1.22   The curve $x(t) = \sin 2t \cos t$, $y(t) = \sin 2t \sin t$ for $0 \le t \le 2\pi$, which forms a four-leaved rose.

Remark   In calculus the curve in Figure 1.22 was given the polar coordinate parameterization $r = \sin 2\theta$.

We want to be able to distinguish when a curve does not cross over itself. The curve C is called simple if it does not cross over itself, which is expressed by requiring that $z(t_1) \ne z(t_2)$ whenever $t_1 \ne t_2$, except possibly when $t_1 = a$ and $t_2 = b$. For example, the circle C with center $z_0 = x_0 + iy_0$ and radius R can be parameterized to form a simple closed curve:

(3) $C: z(t) = (x_0 + R \cos t) + i(y_0 + R \sin t) = z_0 + Re^{it}$

for $0 \le t \le 2\pi$, as shown in Figure 1.23. As t varies from 0 to 2π, the circle is traversed in a counterclockwise direction. If you were traveling around the circle in this manner, its interior would be on your left. When a simple closed curve is parameterized in this fashion, we say that the curve has a *positive orientation*. We will have more to say about this idea shortly.

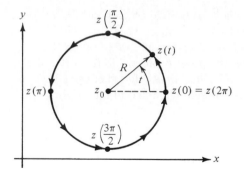

FIGURE 1.23   The simple closed curve $z(t) = z_0 + Re^{it}$ for $0 \leq t \leq 2\pi$.

We need to develop some vocabulary that will help us describe sets of points in the plane. One fundamental idea is that of an ε-*neighborhood* of the point z_0, that is, the set of all points satisfying the inequality

(4) $\left| z - z_0 \right| < \varepsilon$.

This set is the open disk of radius $\varepsilon > 0$ about z_0 shown in Figure 1.24. In particular, the solution sets of the inequalities

$$\left| z \right| < 1, \quad \left| z - i \right| < 2, \quad \left| z + 1 + 2i \right| < 3$$

are neighborhoods of the points 0, i, $-1 - 2i$, respectively, of radius 1, 2, 3, respectively.

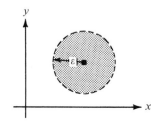

FIGURE 1.24   An ε-neighborhood of the point z_0.

An ε-neighborhood of the point z_0 is denoted by $D_\varepsilon(z_0)$, and is also referred to as the *open disk of radius ε centered at z_0*. Hence,

(5) $D_\varepsilon(z_0) = \{z: \left| z - z_0 \right| < \varepsilon\}$.

We also define the *closed disk of radius ε centered at z_0*,

(6) $\overline{D}_\varepsilon(z_0) = \{z: \left| z - z_0 \right| \leq \varepsilon\}$,

and the *punctured disk of radius ε centered at z_0*,

(7) $D_\varepsilon^*(z_0) = \{z: 0 < \left| z - z_0 \right| < \varepsilon\}$.

The point z_0 is said to be an *interior point* of the set S provided that there exists an ε-neighborhood of z_0 that contains only points of S; z_0 is called an *exterior point* of the set S if there exists an ε-neighborhood of z_0 that contains no points of S. If z_0 is neither an interior point nor an exterior point of S, then it is called a *boundary point* of S and has the property that each ε-neighborhood of z_0 contains both points in S and points not in S. The situation is illustrated in Figure 1.25.

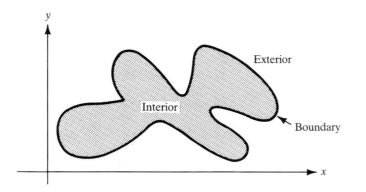

FIGURE 1.25   The interior, exterior, and boundary of a set.

The boundary of $D_R(z_0)$ is the circle depicted in Figure 1.23. We denote this circle by $C_R(z_0)$, and refer to it *as the circle of radius R centered at z_0*. The notation $C_R^+(z_0)$ is used to indicate that the parameterization we chose for this simple closed curve resulted in a positive orientation. $C_R^-(z_0)$ denotes the same circle, but with a negative orientation. (In both cases counterclockwise denotes the positive direction.) Using notation we have already introduced, it is clear that $C_R^-(z_0) = -C_R^+(z_0)$.

EXAMPLE 1.20   Let $S = \{z\colon |z| < 1\}$. Find the interior, exterior, and boundary of S.

Solution   Let z_0 be a point of S. Then $|z_0| < 1$ so that we can choose $\varepsilon = 1 - |z_0| > 0$. If z lies in the disk $|z - z_0| < \varepsilon$, then

$$|z| = |z_0 + z - z_0| \leq |z_0| + |z - z_0| < |z_0| + \varepsilon = 1.$$

Hence the ε-neighborhood of z_0 is contained in S, and z_0 is an interior point of S. It follows that the interior of S is the set $\{z\colon |z| < 1\}$.

Similarly, it can be shown that the exterior of S is the set $\{z: |z| > 1\}$. The boundary of S is the unit circle $\{z: |z| = 1\}$. This is true because if $z_0 = e^{i\theta_0}$ is any point on the circle, then any ε-neighborhood of z_0 will contain the point $(1 - \varepsilon/2)e^{i\theta_0}$, which belongs to S, and $(1 + \varepsilon/2)e^{i\theta_0}$, which does not belong to S.

A set S is called *open* if every point of S is an interior point of S. A set S is called *closed* if it contains all of its boundary points. A set S is said to be *connected* if every pair of points z_1 and z_2 can be joined by a curve that lies entirely in S. Roughly speaking, a connected set consists of a "single piece." The unit disk $D = \{z: |z| < 1\}$ is an open connected set. Indeed, if z_1 and z_2 lie in D, then the straight line segment joining them lies entirely in D. The annulus $A = \{z: 1 < |z| < 2\}$ is an open connected set because any two points in A can be joined by a curve C that lies entirely in A (see Figure 1.26). The set $B = \{z: |z + 2| < 1$ or $|z - 2| < 1\}$ consists of two disjoint disks; hence it is not connected (see Figure 1.27).

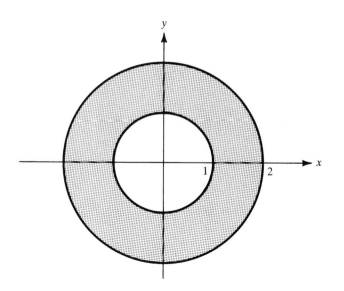

FIGURE 1.26 The annulus $A = \{z: 1 < |z| < 2\}$ is a connected set.

We call an open connected set a *domain*. For example, the right half plane $H = \{z: \mathrm{Re}(z) > 0\}$ is a domain. This is true because if $z_0 = x_0 + iy_0$ is any point in H, then we can choose $\varepsilon = x_0$, and the ε-neighborhood of z_0 lies in H. Also, any two points in H can be connected with the line segment between them. The open unit disk $|z| < 1$ is also a domain. However, the closed unit disk $|z| \leq 1$ is not a domain. It should be noted that the term "domain" is a noun and is a kind of set.

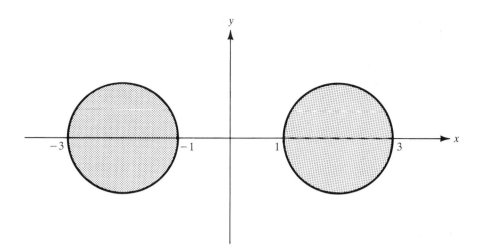

FIGURE 1.27 $B = \{z: |z + 2| < 1 \text{ or } |z - 2| < 1\}$ is not a connected set.

A domain, together with some, none, or all of its boundary points, is called a *region*. For example, the horizontal strip $\{z: 1 < \text{Im}(z) \le 2\}$ is a region. A set that is formed by taking the union of a domain and its boundary is called a *closed region*; that is, the half plane $\{z: x \le y\}$ is a closed region. A set is said to be *bounded* if every point can be enveloped by a circle of some finite fixed radius, that is, there exists an $R > 0$ such that for each z in S we have $|z| \le R$. The rectangle given by $\{z: |x| \le 4 \text{ and } |y| \le 3\}$ is bounded because it is contained inside the circle $|z| = 5$. A set that cannot be enclosed by a circle is called *unbounded*.

We mentioned earlier that a simple closed curve is positively oriented if its interior is on the left when the curve is traversed. How do we know, however, that any given simple closed curve will have an interior and exterior? The following theorem guarantees that this is indeed the case. It is due in part to the work of the French mathematician Camille Jordan.

Theorem 1.1   (*The Jordan Curve Theorem*): *The complement of any simple closed curve C can be partitioned into two mutually exclusive domains I and E in such a way that I is bounded, E is unbounded, and C is the boundary for both I and E. In addition, I ∪ E ∪ C is the entire complex plane. (The domain I is called the interior of C, and the domain E is called the exterior of C.)*

The Jordan curve theorem is a classic example of a result in mathematics that seems obvious but is very hard to demonstrate. Its proof is beyond the scope of this book. Jordan's original argument, in fact, was inadequate, and it was not until 1905 that a correct version was finally given by the American topologist Oswald Veblen. The difficulty lies in describing the interior and exterior of a simple closed curve analytically, and in showing that they are connected sets. For example, in which domain (interior or exterior) do the two points depicted in Figure 1.28 lie? If they are in the same domain, how, specifically, can they be connected with a curve?

Although an introductory treatment of complex analysis can be given without using this theorem, we think it is important for the well-read student at least to be aware of its significance.

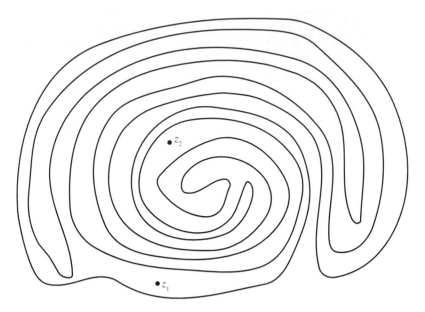

FIGURE 1.28   Are z_1 and z_2 in the interior or exterior of this simple closed curve?

EXERCISES FOR SECTION 1.6

1. Sketch the curve $z(t) = t^2 + 2t + i(t + 1)$
   (a) for $-1 \le t \le 0$.   (b) for $1 \le t \le 2$.
   Hint: Use $x = t^2 + 2t$, $y = t + 1$ and eliminate the parameter t.
2. Find a parameterization of the line that
   (a) joins the origin to the point $1 + i$.   (b) joins the point i to the point $1 + i$.
   (c) joins the point 1 to the point $1 + i$.   (d) joins the point 2 to the point $1 + i$.
3. Find a parameterization of the curve that is a portion of the parabola $y = x^2$ that
   (a) joins the origin to the point $2 + 4i$.   (b) joins the point $-1 + i$ to the origin.
   (c) joins the point $1 + i$ to the origin.
   Hint: For parts (a) and (b), use the parameter $t = x$.
4. Find a parameterization of the curve that is a portion of the circle $|z| = 1$ that joins the point $-i$ to i if
   (a) the curve is the right semicircle.   (b) the curve is the left semicircle.
5. Find a parameterization of the curve that is a portion of the circle $|z| = 1$ that joins the point 1 to i if
   (a) the parameterization is counterclockwise along the quarter circle.
   (b) the parameterization is clockwise.

For Exercises 6–12, refer to the following sets:

(a) $\{z: \text{Re}(z) > 1\}$.
(b) $\{z: -1 < \text{Im}(z) \le 2\}$.
(c) $\{z: |z - 2 - i| \le 2\}$.
(d) $\{z: |z + 3i| > 1\}$.
(e) $\{re^{i\theta}: 0 < r < 1 \text{ and } -\pi/2 < \theta < \pi/2\}$.
(f) $\{re^{i\theta}: r > 1 \text{ and } \pi/4 < \theta < \pi/3\}$.
(g) $\{z: |z| < 1 \text{ or } |z - 4| < 1\}$.

6. Sketch each of the given sets.
7. Which of the sets are open?
8. Which of the sets are connected?
9. Which of the sets are domains?
10. Which of the sets are regions?
11. Which of the sets are closed regions?
12. Which of the sets are bounded?
13. Let $S = \{z_1, z_2, \ldots, z_n\}$ be a finite set of points. Show that S is a bounded set.
14. Let S be the open set consisting of all points z such that $|z + 2| < 1$ or $|z - 2| < 1$. Show that S is not connected.
15. Prove that the neighborhood $|z - z_0| < \varepsilon$ is an open set.
16. Prove that the neighborhood $|z - z_0| < \varepsilon$ is a connected set.
17. Prove that the boundary of the neighborhood $|z - z_0| < \varepsilon$ is the circle $|z - z_0| = \varepsilon$.
18. Prove that the set $\{z: |z| > 1\}$ is the exterior of the set S given in Example 1.20.
19. Prove that the set $\{z: |z| = 1\}$ is the boundary of the set S given in Example 1.20.
20. Look up some articles on teaching complex analysis and discuss what you found. Resources include bibliographical items 7, 11, 24, 27, 33, 43, 74, 84, 90, 101, 102, 103, 105, 114, 123, 134, 137, 160, 171, and 185.

2

Complex Functions

2.1 Functions of a Complex Variable

A function f of the complex variable z is a rule that assigns to each value z in a set D one and only one complex value w. We write

(1) $\quad w = f(z)$

and call w the image of z *under f*. The set D is called the *domain of definition of f*, and the set of all images $R = \{w = f(z): z \in D\}$ is called the *range* of f. Just as z can be expressed by its real and imaginary parts, $z = x + iy$, we write $w = u + iv$, where u and v are the real and imaginary parts of w, respectively. This gives us the representation

(2) $\quad f(x + iy) = u + iv.$

Since u and v depend on x and y, they can be considered to be real functions of the real variables x and y; that is,

(3) $\quad u = u(x, y) \quad$ and $\quad v = v(x, y).$

Combining equations (1), (2), and (3), it is customary to write a complex function f in the form

(4) $\quad f(z) = f(x + iy) = u(x, y) + iv(x, y).$

Conversely, if $u(x, y)$ and $v(x, y)$ are two given real-valued functions of the real variables x and y, then equation (4) can be used to define the complex function f.

EXAMPLE 2.1 Write $f(z) = z^4$ in the form $f(z) = u(x, y) + iv(x, y)$.

Solution Using the binomial formula, we obtain

$$f(z) = (x + iy)^4 = x^4 + 4x^3iy + 6x^2(iy)^2 + 4x(iy)^3 + (iy)^4$$
$$= (x^4 - 6x^2y^2 + y^4) + i(4x^3y - 4xy^3).$$

EXAMPLE 2.2   Express $f(z) = \bar{z} \operatorname{Re}(z) + z^2 + \operatorname{Im}(z)$ in the form of equations (2) and (3) $f(z) = u(x, y) + iv(x, y)$.

   Solution   Using the elementary properties of complex numbers, it follows that

$$f(z) = (x - iy)x + (x^2 - y^2 + i2xy) + y = (2x^2 - y^2 + y) + i(xy).$$

These examples show how to find $u(x, y)$ and $v(x, y)$ when a rule for computing f is given. Conversely, if $u(x, y)$ and $v(x, y)$ are given, then the formulas

$$x = \frac{z + \bar{z}}{2} \quad \text{and} \quad y = \frac{z - \bar{z}}{2i}$$

can be used to find a formula for f involving the variables z and $\bar{z}$.

EXAMPLE 2.3   Express $f(z) = 4x^2 + i4y^2$ by a formula involving the variables z and $\bar{z}$.

   Solution   Calculation reveals that

$$\begin{aligned}
f(z) &= 4\left(\frac{z + \bar{z}}{2}\right)^2 + i4\left(\frac{z - \bar{z}}{2i}\right)^2 \\
&= z^2 + 2z\bar{z} + \bar{z}^2 - i(z^2 - 2z\bar{z} + \bar{z}^2) \\
&= (1 - i)z^2 + (2 + 2i)z\bar{z} + (1 - i)\bar{z}^2.
\end{aligned}$$

It may be convenient to use $z = re^{i\theta}$ in the expression of a complex function f. This gives us the representation

(5) $f(z) = f(re^{i\theta}) = u(r, \theta) + iv(r, \theta),$

where u and v are to be considered as real functions of the real variables r and θ. Note that the functions u and v defined by equations (4) and (5) are *different*, since equation (4) involves Cartesian coordinates and equation (5) involves polar coordinates.

EXAMPLE 2.4   Express $f(z) = z^5 + 4z^2 - 6$ in the polar coordinate form $u(r, \theta) + iv(r, \theta)$.

   Solution   Using equation (1) of Section 1.5, we obtain

$$\begin{aligned}
f(z) &= r^5(\cos 5\theta + i \sin 5\theta) + 4r^2(\cos 2\theta + i \sin 2\theta) - 6 \\
&= (r^5\cos 5\theta + 4r^2\cos 2\theta - 6) + i(r^5\sin 5\theta + 4r^2\sin 2\theta).
\end{aligned}$$

EXERCISES FOR SECTION 2.1

1. Let $f(z) = f(x + iy) = x + y + i(x^3y - y^2)$. Find
   (a) $f(-1 + 3i)$ (b) $f(3i - 2)$
2. Let $f(z) = z^2 + 4z\bar{z} - 5 \operatorname{Re}(z) + \operatorname{Im}(z)$. Find
   (a) $f(-3 + 2i)$ (b) $f(2i - 1)$
3. Find $f(1 + i)$ for the following functions.

   (a) $f(z) = z + z^{-2} + 5$ (b) $f(z) = \dfrac{1}{z^2 + 1}$

4. Find $f(2i - 3)$ for the following functions.

   (a) $f(z) = (z + 3)^3(z - 5i)^2$ (b) $f(z) = \dfrac{z + 2 - 3i}{z + 4 - i}$

5. Let $f(z) = z^{21} - 5z^7 + 9z^4$. Use polar coordinates to find
   (a) $f(-1 + i)$ (b) $f(1 + i\sqrt{3})$
6. Express $f(z) = \bar{z}^2 + (2 - 3i)z$ in the form $u + iv$.
7. Express $f(z) = \dfrac{z + 2 - i}{z - 1 + i}$ in the form $u + iv$.
8. Express $f(z) = z^5 + \bar{z}^5$ in the polar coordinate form $u(r, \theta) + iv(r, \theta)$.
9. Express $f(z) = z^5 + \bar{z}^3$ in the polar coordinate form $u(r, \theta) + iv(r, \theta)$.
10. Let $f(z) = f(x + iy) = e^x\cos y + ie^x\sin y$. Find
    (a) $f(0)$ (b) $f(1)$ (c) $f(i\pi/4)$
    (d) $f(1 + i\pi/4)$ (e) $f(i2\pi/3)$ (f) $f(2 + i\pi)$
11. Let $f(z) = f(x + iy) = (1/2) \ln(x^2 + y^2) + i \arctan(y/x)$. Find
    (a) $f(1)$ (b) $f(1 + i)$ (c) $f(\sqrt{3})$
    (d) $f(\sqrt{3} + i)$ (e) $f(1 + i\sqrt{3})$ (f) $f(3 + 4i)$
12. Let $f(z) = r^2\cos 2\theta + ir^2\sin 2\theta$, where $z = re^{i\theta}$. Find
    (a) $f(1)$ (b) $f(2e^{i\pi/4})$
    (c) $f(\sqrt{2}e^{i\pi/3})$ (d) $f(\sqrt{3}e^{i7\pi/6})$
13. Let $f(z) = \ln r + i\theta$, where $r = |z|$, $\theta = \operatorname{Arg} z$. Find
    (a) $f(1)$ (b) $f(1 + i)$
    (c) $f(-2)$ (d) $f(-\sqrt{3} + i)$
14. A line that carries a charge of $q/2$ coulombs per unit length is perpendicular to the z plane and passes through the point z_0. The electric field intensity $\mathbf{E}(z)$ at the point z varies inversely as the distance from z_0 and is directed along the line from z_0 to z. Show that

    $$\mathbf{E}(z) = \frac{k}{\bar{z} - \bar{z}_0},$$

    where k is some constant. $\left(\text{In Section 10.11 we will see that the answer is in fact } \dfrac{q}{\bar{z} - \bar{z}_0}. \right)$

15. Suppose that three positively charged rods carry a charge of $q/2$ coulombs per unit length and pass through the three points 0, $1 - i$, and $1 + i$. Use the result of Exercise 14 and show that $\mathbf{E}(z) = 0$ at the points $z = (2/3) \pm i(\sqrt{2}/3)$.
16. Suppose that a positively charged rod carrying a charge of $q/2$ coulombs per unit length passes through the point 0 and that positively charged rods carrying a charge of q coulombs per unit length pass through the points $2 + i$ and $-2 + i$. Use the result of Exercise 14 and show that $\mathbf{E}(z) = 0$ at the points $z = \pm\frac{4}{5} + i\frac{3}{5}$.

2.2   Transformations and Linear Mappings

We now take our first look at the geometric interpretation of a complex function. If D is the domain of definition of the real-valued functions $u(x, y)$ and $v(x, y)$, then the system of equations

(1) $u = u(x, y)$ and $v = v(x, y)$

describes a transformation or mapping from D in the xy plane into the uv plane. Therefore, the function

(2) $w = f(z) = u(x, y) + iv(x, y)$

can be considered as a mapping or transformation from the set D in the z plane onto the range R in the w plane. This is illustrated in Figure 2.1.

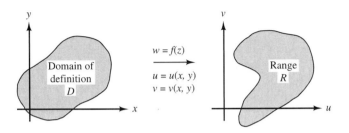

FIGURE 2.1   The mapping $w = f(z)$.

If A is a subset of the domain of definition D, then the set $B = \{f(z): z \in A\}$ is called the *image* of the set A, and f is said to *map A onto B*. The image of a single point is a single point, and the image of the entire domain D is the range R. The mapping $w = f(z)$ is said to be *from A into S* if the image of A is contained in S. The inverse image of a point w is the set of all points z in D such that $w = f(z)$. The inverse image of a point may be one point, several points, or none at all. If the latter case occurs, then the point w is not in the range of f.

The function f is said to be *one-to-one* if it maps distinct points $z_1 \neq z_2$ onto distinct points $f(z_1) \neq f(z_2)$. If $w = f(z)$ maps the set A one-to-one and onto the set B, then for each w in B there exists exactly one point z in A such that $w = f(z)$. Then loosely speaking, we can solve the equation $w = f(z)$ by solving for z as a function of w. That is, the *inverse function* $z = g(w)$ can be found, and the following equations hold:

(3) $g(f(z)) = z$ for all z in A and
       $f(g(w)) = w$ for all w in B.

Conversely, if $w = f(z)$ and $z = g(w)$ are functions that map A into B and B into A, respectively, and equations (3) hold, then $w = f(z)$ maps the set A one-to-one and onto the set B. The one-to-one property is easy to show, for if we have $f(z_1) = f(z_2)$, then $g(f(z_1)) = g(f(z_2))$; and using equation (3), we obtain $z_1 = z_2$.

To show that f is onto, we must show that each point w in B is the image of some point in A. If $w \in B$, then $z = g(w)$ lies in A and $f(g(w)) = w$, and we conclude that f is a one-to-one mapping from A onto B.

We observe that if f is a one-to-one mapping from D onto R and if A is a subset of D, then f is a one-to-one mapping from A onto its image B. One can also show that if $\xi = f(z)$ is a one-to-one mapping from A onto S, and $w = g(\xi)$ is a one-to-one mapping from S onto B, then the composition mapping $w = g(f(z))$ is a one-to-one mapping from A onto B.

It is useful to find the image B of a specified set A under a given mapping $w = f(z)$. The set A is usually described with an equation or inequality involving x and y. A chain of equivalent statements can be constructed that lead to a description of the set B in terms of an equation or an inequality involving u and v.

EXAMPLE 2.5   Show that the function $f(z) = iz$ maps the line $y = x + 1$ onto the line $v = -u - 1$.

Solution   We can write f in the Cartesian form $u + iv = f(z) = i(x + iy) = -y + ix$, and see that the transformation can be given by the equations $u = -y$ and $v = x$. We can substitute these into the equation $y = x + 1$ to obtain $-u = v + 1$, which can be written as $v = -u - 1$.

We now turn our attention to the investigation of some elementary mappings. Let $B = a + ib$ denote a fixed complex number. Then the transformation

(4) $w = T(z) = z + B = x + a + i(y + b)$

is a one-to-one mapping of the z plane onto the w plane and is called a *translation*. This transformation can be visualized as a rigid translation whereby the point z is displaced through the vector $a + ib$ to its new position $w = T(z)$. The inverse mapping is given by

(5) $z = T^{-1}(w) = w - B = u - a + i(v - b)$

and shows that T is a one-to-one mapping from the z plane onto the w plane. The effect of a translation is pictured in Figure 2.2.

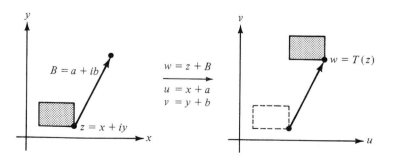

FIGURE 2.2   The translation $w = T(z) = z + B = x + a + i(y + b)$.

Let α be a fixed real number. Then the transformation

(6) $w = R(z) = ze^{i\alpha} = re^{i\theta}e^{i\alpha} = re^{i(\theta+\alpha)}$

is a one-to-one mapping of the z plane onto the w plane and is called a *rotation*. It can be visualized as a rigid rotation whereby the point z is rotated about the origin through an angle α to its new position $w = R(z)$. If we use polar coordinates $w = \rho e^{i\phi}$ in the w plane, then the inverse mapping is given by

(7) $z = R^{-1}(w) = we^{-i\alpha} = \rho e^{i\phi}e^{-i\alpha} = \rho e^{i(\phi-\alpha)}$.

This shows that R is a one-to-one mapping of the z plane onto the w plane. The effect of rotation is pictured in Figure 2.3.

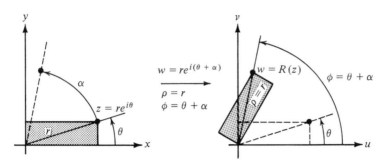

FIGURE 2.3   The rotation $w = R(z) = re^{i(\theta+\alpha)}$.

Let $K > 0$ be a fixed positive real number. Then the transformation

(8) $w = S(z) = Kz = Kx + iKy$

is a one-to-one mapping of the z plane onto the w plane and is called a *magnification*. If $K > 1$, it has the effect of stretching the distance between points by the factor K. If $K < 1$, then it reduces the distance between points by the factor K. The inverse transformation is given by

(9) $z = S^{-1}(w) = \dfrac{1}{K}w = \dfrac{1}{K}u + i\dfrac{1}{K}v$

and shows that S is one-to-one mapping from the z plane onto the w plane. The effect of magnification is shown in Figure 2.4.

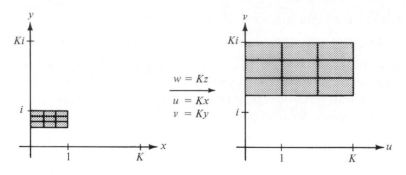

FIGURE 2.4   The magnification $w = S(z) = Kz = Kx + iKy$.

Let $A = Ke^{i\alpha}$ and $B = a + ib$, where $K > 0$ is a positive real number. Then the transformation

(10) $w = W(z) = Az + B$

is a one-to-one mapping of the z plane onto the w plane and is called a *linear transformation*. It can be considered as the composition of a rotation, a magnification, and a translation. It has the effect of rotating the plane through an angle given by $\alpha = \text{Arg } A$, followed by a magnification by the factor $K = |A|$, followed by a translation by the vector $B = a + ib$. The inverse mapping is given by

(11) $z = W^{-1}(w) = \dfrac{1}{A}w - \dfrac{B}{A}$

and shows that W is a one-to-one mapping from the z plane onto the w plane.

EXAMPLE 2.6   Show that the linear transformation $w = iz + i$ maps the right half plane $\text{Re}(z) > 1$ onto the upper half plane $\text{Im}(w) > 2$.

Solution   We can write $w = f(z)$ in Cartesian form $u + iv = i(x + iy) + i = -y + i(x + 1)$ and see that the transformation can be given by the equations $u = -y$ and $v = x + 1$. The substitution $x = v - 1$ can be used in the inequality $\text{Re}(z) = x > 1$ to see that the image values must satisfy $v - 1 > 1$ or $v > 2$, which is the upper half plane $\text{Im}(w) > 2$. The effect of the transformation $w = f(z)$ is a rotation of the plane through the angle $\alpha = \pi/2$ followed by a translation by the vector $B = i$ and is illustrated in Figure 2.5.

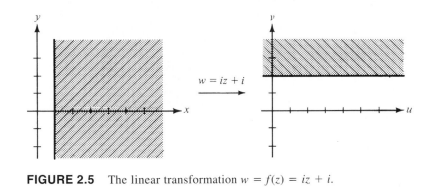

FIGURE 2.5   The linear transformation $w = f(z) = iz + i$.

It is easy to see that translations and rotations preserve angles. Since magnifications rescale distance by a factor K, it follows that triangles are mapped onto similar triangles, and so angles are preserved. Since a linear transformation can be considered as a composition of a rotation, a magnification, and a translation, it follows that linear transformations preserve angles. Consequently, any geometric object is mapped onto an object that is similar to the original object; hence linear transformations can be called *similarity mappings*.

EXAMPLE 2.7   Show that the image of the open disk $|z + 1 + i| < 1$ under the transformation $w = (3 - 4i)z + 6 + 2i$ is the open disk $|w + 1 - 3i| < 5$.

Solution   The inverse transformation is given by

$$z = \frac{w - 6 - 2i}{3 - 4i},$$

and this substitution can be used to show that the image points must satisfy the inequality

$$\left| \frac{w - 6 - 2i}{3 - 4i} + 1 + i \right| < 1.$$

Multiplying both sides by $|3 - 4i| = 5$ results in

$$|w - 6 - 2i + (1 + i)(3 - 4i)| < 5,$$

which can be simplified to obtain the inequality

$$|w + 1 - 3i| < 5.$$

Hence the disk with center $-1 - i$ and radius 1 is mapped one-to-one and onto the disk with center $-1 + 3i$ and radius 5 as pictured in Figure 2.6.

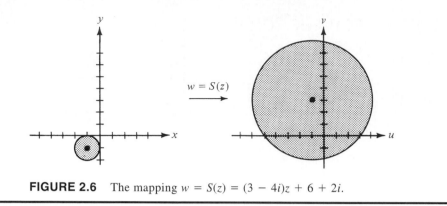

FIGURE 2.6   The mapping $w = S(z) = (3 - 4i)z + 6 + 2i$.

EXAMPLE 2.8   Show that the image of the right half plane $\text{Re}(z) > 1$ under the linear transformation $w = (-1 + i)z - 2 + 3i$ is the half plane $v > u + 7$.

Solution   The inverse transformation is given by

$$z = \frac{w + 2 - 3i}{-1 + i} = \frac{u + 2 + i(v - 3)}{-1 + i},$$

which can be expressed in the component form

$$x + iy = \frac{-u + v - 5}{2} + i\frac{-u - v + 1}{2}.$$

The substitution $x = (-u + v - 5)/2$ can be used in the inequality $\text{Re}(z) = x > 1$ to see that the image points must satisfy $(-u + v - 5)/2 > 1$. This can be simplified to yield the inequality $v > u + 7$. The mapping is illustrated in Figure 2.7.

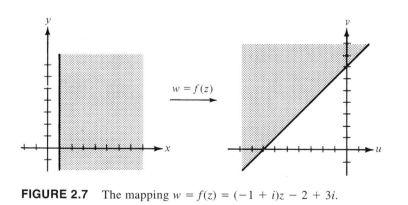

FIGURE 2.7   The mapping $w = f(z) = (-1 + i)z - 2 + 3i$.

EXERCISES FOR SECTION 2.2

1. Let $w = (1 - i)z + 1 - 2i$.
   (a) Find the image of the half-plane $\text{Im}(z) > 1$.
   (b) Sketch the mapping, and indicate the points $z_1 = -1 + i$, $z_2 = i$, and $z_3 = 1 + i$ and their images w_1, w_2, and w_3.

2. Let $w = (2 + i)z - 3 + 4i$. Find the image of the line

   $$x = t, \, y = 1 - 2t \quad \text{for} \, -\infty < t < \infty.$$

3. Let $w = (3 + 4i)z - 2 + i$.
   (a) Find the image of the disk $|z - 1| < 1$.
   (b) Sketch the mapping, and indicate the points $z_1 = 0$, $z_2 = 1 - i$, and $z_3 = 2$ and their images.

4. Let $w = (3 + 4i)z - 2 + i$. Find the image of the circle

   $$x = 1 + \cos t, \quad y = 1 + \sin t \quad \text{for} \, -\pi < t \le \pi.$$

5. Let $w = (2 + i)z - 2i$. Find the triangle onto which the triangle with vertices $z_1 = -2 + i$, $z_2 = -2 + 2i$, and $z_3 = 2 + i$ is mapped.

6. Find the linear transformation $w = f(z)$ that maps the points $z_1 = 2$ and $z_2 = -3i$ onto the points $w_1 = 1 + i$ and $w_2 = 1$, respectively.

7. Find the linear transformation $w = S(z)$ that maps the circle $|z| = 1$ onto the circle $|w - 3 + 2i| = 5$ and satisfies the condition $S(-i) = 3 + 3i$.

8. Find the linear transformation $w = f(z)$ that maps the triangle with vertices $-4 + 2i$, $-4 + 7i$, and $1 + 2i$ onto the triangle with vertices 1, 0, and $1 + i$.

9. Let $S(z) = Kz$, where $K > 0$ is a positive real constant. Show that the equation $|S(z_1) - S(z_2)| = K|z_1 - z_2|$ holds, and interpret this result geometrically.

10. Give a proof that the image of a circle under a linear transformation is a circle. *Hint*: Let the given circle have the parameterization $x = x_0 + R \cos t$, $y = y_0 + R \sin t$.

11. Prove that the composition of two linear transformations is a linear transformation.

12. Show that a linear transformation that maps the circle $|z - z_0| = R_1$ onto the circle $|w - w_0| = R_2$ can be expressed in the form

    $$A(w - w_0)R_1 = (z - z_0) R_2, \quad \text{where} \, |A| = 1.$$

2.3   The Mappings $w = z^n$ and $w = z^{1/n}$

The function $w = f(z) = z^2$ can be expressed in polar coordinates by

(1) $w = f(z) = z^2 = r^2 e^{i 2\theta}$,

where $r > 0$ and $-\pi < \theta \le \pi$. If polar coordinates, $w = \rho e^{i\phi}$ are used in the w plane, then mapping (1) can be given by the system of equations

(2) $\rho = r^2 \quad \text{and} \quad \phi = 2\theta$.

   If we consider the wedge-shaped set $A = \{re^{i\theta} : r > 0 \text{ and } -\pi/4 < \theta < \pi/4\}$, then the image of A under the mapping f is the right half plane described by the inequalities $\rho > 0$, $-\pi/2 < \phi < \pi/2$. Since the argument of the product zz is twice the argument of z, we say that f doubles angles at the origin. Points that lie on the ray $r > 0$, $\theta = \alpha$ are mapped onto points that lie on the ray $\rho > 0$, $\phi = 2\alpha$.

If the domain of definition D for $f(z) = z^2$ is restricted to be the set

$$(3) \quad D = \left\{ re^{i\theta} : r > 0 \quad \text{and} \quad \frac{-\pi}{2} < \theta < \frac{\pi}{2} \right\},$$

then the image of D under the mapping $w = z^2$ consists of all points in the w plane (except the point $w = 0$ and all the points that lie along the negative u axis). The inverse mapping of f is

$$(4) \quad z = f^{-1}(w) = w^{1/2} = \rho^{1/2}e^{i\phi/2}, \quad \text{where } \rho > 0 \text{ and } -\pi < \phi < \pi.$$

The function $f^{-1}(w) = w^{1/2}$ in equation (4) is called the *principal square root function* and shows that f is one-to-one when its domain is restricted by set (3). The mappings $w = z^2$ and $z = w^{1/2}$ are illustrated in Figure 2.8.

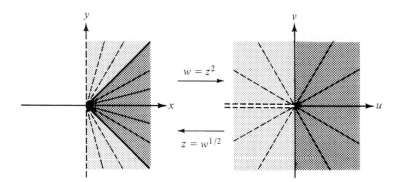

FIGURE 2.8   The mappings $w = z^2$ and $z = w^{1/2}$.

Since $f(-z) = (-z)^2 = z^2$, we see that the image of the left half plane $\text{Re}(z) < 0$ under the mapping $w = z^2$ is the w plane slit along the negative u axis as indicated in Figure 2.9.

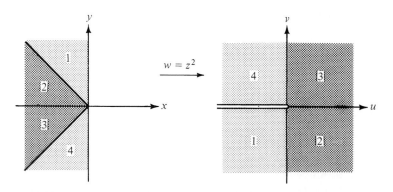

FIGURE 2.9   The mapping $w = z^2$.

Other useful properties of the mapping $w = z^2$ can be investigated if we use the Cartesian form

(5) $w = f(z) = z^2 = x^2 - y^2 + i2xy$

and the resulting system of equations

(6) $u = x^2 - y^2$ and $v = 2xy.$

EXAMPLE 2.9

The transformation $w = f(z) = z^2$ maps vertical and horizontal lines onto parabolas, and this fact is used to find the image of a rectangle. If $a > 0$, then the vertical line $x = a$ is mapped onto the parabola given by the equations $u = a^2 - y^2$ and $v = 2ay$, which can be solved to yield the single equation

(7) $u = a^2 - \dfrac{v^2}{4a^2}.$

If $b > 0$, then the horizontal line $y = b$ is mapped onto the parabola given by the equations $u = x^2 - b^2$ and $v = 2xb$, which can be solved to yield the single equation

(8) $u = -b^2 + \dfrac{v^2}{4b^2}.$

Since quadrant I is mapped onto quadrants I and II by $w = z^2$, we see that the rectangle $0 < x < a, 0 < y < b$ is mapped onto the region bounded by the parabolas (7) and (8) and the u axis. The four vertices $0, a, a + ib$, and ib are mapped onto the four points $0, a^2, a^2 - b^2 + i2ab$, and $-b^2$, respectively, as indicated in Figure 2.10.

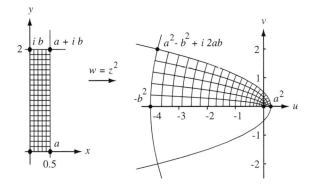

FIGURE 2.10   The transformation $w = z^2$.

The mapping $w = z^{1/2}$ can be expressed in polar form,

(9) $w = f(z) = z^{1/2} = r^{1/2}e^{i\theta/2},$

where the domain of definition D for f is restricted to be $r > 0$, $-\pi < \theta \leq \pi$. If polar coordinates $w = \rho e^{i\phi}$ are used in the w plane, then mapping (9) can be represented by the system

$$(10) \quad \rho = r^{1/2} \quad \text{and} \quad \phi = \frac{\theta}{2}.$$

From equations (10) we see that the argument of the image is half the argument of z and that the modulus of the image is the square root of the modulus of z. Points that lie on the ray $r > 0$, $\theta = \alpha$ are mapped onto the ray $\rho > 0$, $\phi = \alpha/2$. The image of the z plane (with the point $z = 0$ deleted) consists of the right half plane $\text{Re}(w) > 0$ together with the positive v axis, and the mapping is pictured in Figure 2.11.

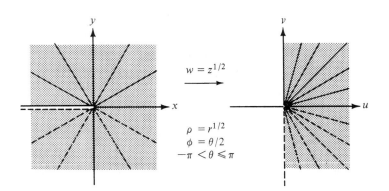

FIGURE 2.11   The mapping $w = z^{1/2}$.

The mapping $w = z^{1/2}$ can be studied through our knowledge about its inverse mapping $z = w^2$. If we use the Cartesian formula

$$(11) \quad z = w^2 = u^2 - v^2 + i2uv,$$

then the mapping $z = w^2$ is given by the system of equations

$$(12) \quad x = u^2 - v^2 \quad \text{and} \quad y = 2uv.$$

EXAMPLE 2.10   The transformation $w = f(z) = z^{1/2}$ maps vertical and horizontal lines onto a portion of a hyperbola, enabling us to find images of half planes. Let $a > 0$. Then system (12) can be used to see that the right half plane given by $\text{Re}(z) = x > a$ is mapped onto the region in the right half plane satisfying $u^2 - v^2 > a$ and lies to the right of the hyperbola $u^2 - v^2 = a$. If $b > 0$, then system (12) can be used to see that the upper half plane $\text{Im}(z) = y > b$ is mapped onto the region in quadrant I satisfying $2uv > b$ and lies above the hyperbola $2uv = b$. The situation is illustrated in Figure 2.12.

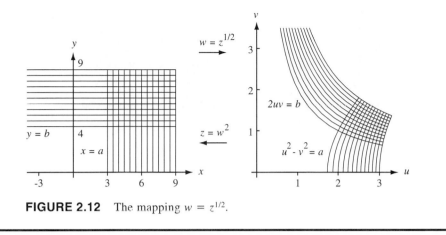

FIGURE 2.12   The mapping $w = z^{1/2}$.

Let n be a positive integer and consider the function $w = f(z) = z^n$, which can be expressed in the polar coordinate form

(13) $w = f(z) = z^n = r^n e^{in\theta}$,   where $r > 0$ and $-\pi < \theta \le \pi$.

If polar coordinates $w = \rho e^{i\phi}$ are used in the w plane, then mapping (13) can be given by the system of equations

(14) $\rho = r^n$ and $\theta = n\theta$.

We see that the image of the ray $r > 0$, $\theta = \alpha$ is the ray $\rho > 0$, $\phi = n\alpha$ and that angles at the origin are increased by the factor n. Since the functions $\cos n\theta$ and $\sin n\theta$ are periodic with period $2\pi/n$, we see that f is in general an n-to-one function; that is, n points in the z plane are mapped onto each point in the w plane (except $w = 0$). If the domain of definition D of f in mapping (13) is restricted to be

(15) $D = \left\{ re^{i\theta}: r > 0, \dfrac{-\pi}{n} < \theta \le \dfrac{\pi}{n} \right\}$,

then the image of D under the mapping $w = f(z) = z^n$ consists of all points in the w plane (except the origin $w = 0$), and the inverse function is given by

(16) $z = f^{-1}(w) = w^{1/n} = \rho^{1/n} e^{i\phi/n}$,   where $\rho > 0$ and $-\pi < \phi \le \pi$.

The function $f^{-1}(w) = w^{1/n}$ is called the *principal nth root function* and shows that f is one-to-one when it is restricted to be the domain set (15). The mappings $w = z^n$ and $z = w^{1/n}$ are shown in Figure 2.13.

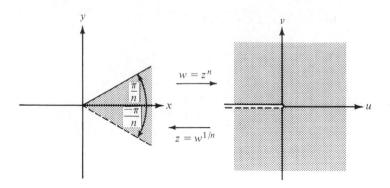

FIGURE 2.13   The mappings $w = z^n$ and $z = w^{1/n}$.

EXERCISES FOR SECTION 2.3

1. Show that the image of the horizontal line $y = 1$ under the mapping $w = z^2$ is the parabola $u = v^2/4 - 1$.

2. Show that the image of the vertical line $x = 2$ under the mapping $w = z^2$ is the parabola $u = 4 - v^2/16$.

3. Find the image of the rectangle $0 < x < 2, 0 < y < 1$ under the mapping $w = z^2$. Sketch the mapping.

4. Find the image of the triangle with vertices 0, 2, and $2 + 2i$ under the mapping $w = z^2$. Sketch the mapping.

5. Show that the infinite strip $1 < x < 2$ is mapped onto the region that lies between the parabolas $u = 1 - v^2/4$ and $u = 4 - v^2/16$ by the mapping $w = z^2$.

6. For what values of z does $(z^2)^{1/2} = z$ hold if the principal value of the square root is to be used?

7. Sketch the set of points satisfying the following relations.
   (a) $\mathrm{Re}(z^2) > 4$ (b) $\mathrm{Im}(z^2) > 6$

8. Show that the region in the right half plane that lies to the right of the hyperbola $x^2 - y^2 = 1$ is mapped onto the right half plane $\mathrm{Re}(w) > 1$ by the mapping $w = z^2$.

9. Show that the image of the line $x = 4$ under the mapping $w = z^{1/2}$ is the right branch of the hyperbola $u^2 - v^2 = 4$.

10. Find the image of the following sets under the mapping $w = z^{1/2}$.
    (a) $\{re^{i\theta} : r > 1 \text{ and } \pi/3 < \theta < \pi/2\}$
    (b) $\{re^{i\theta} : 1 < r < 9 \text{ and } 0 < \theta < 2\pi/3\}$
    (c) $\{re^{i\theta} : r < 4 \text{ and } -\pi < \theta < \pi/2\}$

11. Find the image of the right half plane $\mathrm{Re}(z) > 1$ under the mapping $w = z^2 + 2z + 1$.

12. Show that the infinite strip $2 < y < 6$ is mapped onto the region in the first quadrant that lies between the hyperbolas $uv = 1$ and $uv = 3$ by the mapping $w = z^{1/2}$.

13. Find the image of the region in the first quadrant that lies between the hyperbolas $xy = \frac{1}{2}$ and $xy = 4$ under the mapping $w = z^2$.

14. Show that the region in the z plane that lies to the right of the parabola $x = 4 - y^2/16$ is mapped onto the right half plane $\mathrm{Re}(w) > 2$ by the mapping $w = z^{1/2}$. *Hint*: Use the inverse mapping $z = w^2$.

15. Find the image of the following sets under the mapping $w = z^3$.
    (a) $\{re^{i\theta}: 1 < r < 2 \text{ and } -\pi/4 < \theta < \pi/3\}$
    (b) $\{re^{i\theta}: r > 3 \text{ and } 2\pi/3 < \theta < 3\pi/4\}$
16. Find the image of the sector $r > 2$, $\pi/4 < \theta < \pi/3$ under the following mappings.
    (a) $w = z^3$ (b) $w = z^4$ (c) $w = z^6$
17. Find the image of the sector $r > 0$, $-\pi < \theta < 2\pi/3$ under the following mappings.
    (a) $w = z^{1/2}$ (b) $w = z^{1/3}$ (c) $w = z^{1/4}$
18. Use your knowledge about the complex square root function and explain the fallacy in the following statement: $1 = \sqrt{(-1)(-1)} = \sqrt{-1}\sqrt{-1} = (i)(i) = -1$.

2.4   Limits and Continuity

Let $u = u(x, y)$ be a real-valued function of the two real variables x and y. We say that u has the *limit* u_0 as (x, y) approaches (x_0, y_0) provided that the value of $u(x, y)$ gets close to the value u_0 as (x, y) gets close to (x_0, y_0). We write

(1) $\lim_{(x,y)\to(x_0,y_0)} u(x, y) = u_0.$

That is, u has the limit u_0 as (x, y) approaches (x_0, y_0) if and only if $|u(x, y) - u_0|$ can be made arbitrarily small by making both $|x - x_0|$ and $|y - y_0|$ small. This is like the definition of limit for functions of one variable, except that there are two variables instead of one. Since (x, y) is a point in the xy plane, and the distance between (x, y) and (x_0, y_0) is $\sqrt{(x - x_0)^2 + (y - y_0)^2}$, we can give a precise definition of limit as follows. To each number $\varepsilon > 0$, there corresponds a number $\delta > 0$ such that

(2) $|u(x, y) - u_0| < \varepsilon$,   whenever $0 < \sqrt{(x - x_0)^2 + (y - y_0)^2} < \delta$.

EXAMPLE 2.11   If $u(x, y) = x^3/(x^2 + y^2)$, then

(3) $\lim_{(x,y)\to(0,0)} u(x, y) = 0.$

    Solution   If $x = r \cos\theta$ and $y = r \sin\theta$, then

$$u(x, y) = \frac{r^3\cos^3\theta}{r^2\cos^2\theta + r^2\sin^2\theta} = r\cos^3\theta.$$

Since $\sqrt{(x - 0)^2 + (y - 0)^2} = r$, we see that

$$|u(x, y) - 0| = r|\cos^3\theta| < \varepsilon, \quad \text{whenever } 0 < \sqrt{x^2 + y^2} = r < \varepsilon.$$

Hence for any $\varepsilon > 0$, inequality (2) is satisfied for $\delta = \varepsilon$; that is, $u(x, y)$ has the limit $u_0 = 0$ as (x, y) approaches $(0, 0)$.

    The value u_0 of the limit must *not* depend on how (x, y) approaches (x_0, y_0). So it follows that $u(x, y)$ must approach the value u_0 when (x, y) approaches (x_0, y_0)

along any curve that ends at the point (x_0, y_0). Conversely, if we can find two curves C_1 and C_2 that end at (x_0, y_0) along which $u(x, y)$ approaches the two distinct values u_1 and u_2, respectively, then $u(x, y)$ *does not* have a limit as (x, y) approaches (x_0, y_0).

EXAMPLE 2.12   The function $u(x, y) = xy/(x^2 + y^2)$ *does not* have a limit as (x, y) approaches $(0, 0)$. If we let (x, y) approach $(0, 0)$ along the x axis, then

$$\lim_{(x,0)\to(0,0)} u(x, 0) = \lim_{(x,0)\to(0,0)} \frac{(x)(0)}{x^2 + 0^2} = 0.$$

But if we let (x, y) approach $(0, 0)$ along the line $y = x$, then

$$\lim_{(x,x)\to(0,0)} u(x, x) = \lim_{(x,x)\to(0,0)} \frac{(x)(x)}{x^2 + x^2} = \frac{1}{2}.$$

Since the two values are different, the value of the limit is dependent on how (x, y) approaches $(0, 0)$. We conclude that $u(x, y)$ does not have a limit as (x, y) approaches $(0, 0)$.

Let $f(z)$ be a complex function of the complex variable z that is defined for all values of z in some neighborhood of z_0, except perhaps at the point z_0. We say that f has the *limit* w_0 as z approaches z_0, provided that the value $f(z)$ gets close to the value w_0 as z gets close to z_0; and we write

(4) $\lim_{z\to z_0} f(z) = w_0.$

Since the distance between the points z and z_0 can be expressed by $\left| z - z_0 \right|$, we can give a precise definition of limit (4): For each positive number $\varepsilon > 0$, there exists a $\delta > 0$ such that

(5) $\left| f(z) - w_0 \right| < \varepsilon$,   whenever $0 < \left| z - z_0 \right| < \delta.$

Geometrically, this says that for each ε-neighborhood $\left| w - w_0 \right| < \varepsilon$ of the point w_0 there is a deleted δ-neighborhood $0 < \left| z - z_0 \right| < \delta$ of z_0 such that the image of each point in the δ-neighborhood, except perhaps z_0, lies in the ε-neighborhood of w_0. The image of the δ-neighborhood does not have to fill up the entire ε-neighborhood; but if z approaches z_0 along a curve that ends at z_0, then $w = f(z)$ approaches w_0. The situation is illustrated in Figure 2.14.

If we consider $w = f(z)$ as a mapping from the z plane into the w plane and think about the previous geometric interpretation of a limit, then we are led to conclude that the limit of a function f should be determined by the limits of its real and imaginary parts u and v. This will also give us a tool for computing limits.

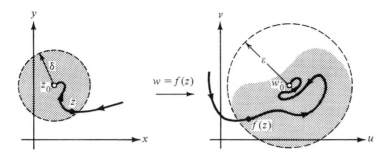

FIGURE 2.14   The limit $f(z) \to w_0$ as $z \to z_0$.

Theorem 2.1   *Let $f(z) = u(x, y) + iv(x, y)$ be a complex function that is defined in some neighborhood of z_0, except perhaps at $z_0 = x_0 + iy_0$. Then*

(6) $\lim\limits_{z \to z_0} f(z) = w_0 = u_0 + iv_0$

   if and only if

(7) $\lim\limits_{(x,y) \to (x_0, y_0)} u(x, y) = u_0 \quad and \quad \lim\limits_{(x,y) \to (x_0, y_0)} v(x, y) = v_0.$

Proof   Let us first assume that statement (6) is true, and show that statement (7) is true. According to the definition of limit, for each $\varepsilon > 0$, there corresponds a $\delta > 0$ such that

$$|f(z) - w_0| < \varepsilon, \quad \text{whenever } 0 < |z - z_0| < \delta.$$

Since $f(z) - w_0 = u(x, y) - u_0 + i(v(x, y) - v_0)$, we can use equations (2) of Section 1.3 to conclude that

$$|u(x, y) - u_0| \le |f(z) - w_0| \quad \text{and} \quad |v(x, y) - v_0| \le |f(z) - w_0|.$$

It now follows that $|u(x, y) - u_0| < \varepsilon$ and $|v(x, y) - v_0| < \varepsilon$ whenever $0 < |z - z_0| < \delta$ so that statement (7) is true.

   Conversely, now let us assume that statement (7) is true. Then for each $\varepsilon > 0$, there exists $\delta_1 > 0$ and $\delta_2 > 0$ so that

$$|u(x, y) - u_0| < \frac{\varepsilon}{2}, \quad \text{whenever } 0 < |z - z_0| < \delta_1 \quad \text{and}$$

$$|v(x, y) - v_0| < \frac{\varepsilon}{2}, \quad \text{whenever } 0 < |z - z_0| < \delta_2.$$

Let δ be chosen to be the minimum of the two values δ_1 and δ_2. Then we can use the triangle inequality

$$|f(z) - w_0| \le |u(x, y) - u_0| + |v(x, y) - v_0|$$

to conclude that

$$|f(z) - w_0| < \frac{\varepsilon}{2} + \frac{\varepsilon}{2} = \varepsilon, \quad \text{whenever } 0 < |z - z_0| < \delta.$$

Hence the truth of statement (7) implies the truth of statement (6), and the proof of the theorem is complete.

For example, $\lim\limits_{z \to 1+i} (z^2 - 2z + 1) = -1$. To show this result, we let

$$f(z) = z^2 - 2z + 1 = x^2 - y^2 - 2x + 1 + i(2xy - 2y).$$

Computing the limits for u and v, we obtain

$$\lim_{(x,y) \to (1,1)} u(x, y) = 1 - 1 - 2 + 1 = -1 \quad \text{and}$$

$$\lim_{(x,y) \to (1,1)} v(x, y) = 2 - 2 = 0.$$

So Theorem 2.1 implies that $\lim\limits_{z \to 1+i} f(z) = -1$.

Limits of complex functions are formally the same as in the case of real functions, and the sum, difference, product, and quotient of functions have limits given by the sum, difference, product, and quotient of the respective limits. We state this result as a theorem and leave the proof as an exercise.

Theorem 2.2   *Let* $\lim\limits_{z \to z_0} f(z) = A$ *and* $\lim\limits_{z \to z_0} g(z) = B$. *Then*

(8) $\lim\limits_{z \to z_0} [f(z) \pm g(z)] = A \pm B.$

(9) $\lim\limits_{z \to z_0} f(z)g(z) = AB.$

(10) $\lim\limits_{z \to z_0} \dfrac{f(z)}{g(z)} = \dfrac{A}{B}, \quad \text{where } B \neq 0.$

Let $u(x, y)$ be a real-valued function of the two real variables x and y. We say that u is *continuous* at the point (x_0, y_0) if the following three conditions are satisfied:

(11) $\lim\limits_{(x,y) \to (x_0,y_0)} u(x, y)$ exists.

(12) $u(x_0, y_0)$ exists.

(13) $\lim\limits_{(x,y) \to (x_0,y_0)} u(x, y) = u(x_0, y_0).$

Condition (13) actually contains conditions (11) and (12), since the existence of the quantity on each side of the equation there is implicitly understood to exist. For example, if $u(x, y) = x^3/(x^2 + y^2)$ when $(x, y) \neq (0, 0)$ and if $u(0, 0) = 0$, then we have already seen that $u(x, y) \to 0$ as $(x, y) \to (0, 0)$ so that conditions (11), (12), and (13) are satisfied. Hence $u(x, y)$ is continuous at $(0, 0)$.

Let $f(z)$ be a complex function of the complex variable z that is defined for all values of z in some neighborhood of z_0. We say that f is *continuous* at z_0 if the following three conditions are satisfied:

(14) $\lim\limits_{z \to z_0} f(z)$ exists.

(15) $f(z_0)$ exists.

(16) $\lim\limits_{z \to z_0} f(z) = f(z_0)$.

A complex function f is continuous if and only if its real and imaginary parts u and v are continuous, and the proof of this fact is an immediate consequence of Theorem 2.1. Continuity of complex functions is formally the same as in the case of real functions, and the sum, difference, and product of continuous functions are continuous; their quotient is continuous at points where the denominator is not zero. These results are summarized by the following theorems, and the proofs are left as exercises.

Theorem 2.3   *Let $f(z) = u(x, y) + iv(x, y)$ be defined in some neighborhood of z_0. Then f is continuous at $z_0 = x_0 + iy_0$ if and only if u and v are continuous at (x_0, y_0).*

Theorem 2.4   *Suppose that f and g are continuous at the point z_0. Then the following functions are continuous at z_0:*

(17)   *Their sum $f(z) + g(z)$.*

(18)   *Their difference $f(z) - g(z)$.*

(19)   *Their product $f(z)g(z)$.*

(20)   *Their quotient $\dfrac{f(z)}{g(z)}$ provided that $g(z_0) \neq 0$.*

(21)   *Their composition $f(g(z))$ provided that $f(z)$ is continuous in a neighborhood of the point $g(z_0)$.*

EXAMPLE 2.13   Show that the polynomial function given by

$$w = P(z) = a_0 + a_1 z + a_2 z^2 + \cdots + a_n z^n$$

is continuous at each point z_0 in the complex plane.

Solution   Observe that if a_0 is the constant function, then $\lim_{z \to z_0} a_0 = a_0$; and if $a_1 \neq 0$, then we can use definition (5) with $f(z) = a_1 z$ and the choice $\delta = \varepsilon / |a_1|$ to prove that $\lim_{z \to z_0} a_1 z = a_1 z_0$. Then using property (9) and mathematical induction, we obtain

(22) $\lim\limits_{z \to z_0} a_k z^k = a_k z_0^k \quad$ for $k = 0, 1, 2, \ldots, n$.

Property (8) can be extended to a finite sum of terms, and we can use the result of equation (22) to obtain

$$(23) \quad \lim_{z \to z_0} P(z) = \lim_{z \to z_0} \left(\sum_{k=0}^{n} a_k z^k \right) = \sum_{k=0}^{n} a_k z_0^k = P(z_0).$$

Since conditions (14), (15), and (16) are satisfied, we can conclude that P is continuous at z_0.

One technique for computing limits is the use of statement (20). Let P and Q be polynomials. If $Q(z_0) \neq 0$, then

$$\lim_{z \to z_0} \frac{P(z)}{Q(z)} = \frac{P(z_0)}{Q(z_0)}.$$

Another technique involves factoring polynomials. If both $P(z_0) = 0$ and $Q(z_0) = 0$, then P and Q can be factored as $P(z) = (z - z_0)P_1(z)$ and $Q(z) = (z - z_0)Q_1(z)$. If $Q_1(z_0) \neq 0$, then the limit is given by

$$\lim_{z \to z_0} \frac{P(z)}{Q(z)} = \lim_{z \to z_0} \frac{(z - z_0)P_1(z)}{(z - z_0)Q_1(z)} = \frac{P_1(z_0)}{Q_1(z_0)}.$$

EXAMPLE 2.14 Show that $\displaystyle \lim_{z \to 1+i} \frac{z^2 - 2i}{z^2 - 2z + 2} = 1 - i.$

Solution   Here P and Q can be factored in the form

$$P(z) = (z - 1 - i)(z + 1 + i) \quad \text{and} \quad Q(z) = (z - 1 - i)(z - 1 + i)$$

so that the limit is obtained by the calculation

$$\lim_{z \to 1+i} \frac{z^2 - 2i}{z^2 - 2z + 2} = \lim_{z \to 1+i} \frac{(z - 1 - i)(z + 1 + i)}{(z - 1 - i)(z - 1 + i)}$$

$$= \lim_{z \to 1+i} \frac{z + 1 + i}{z - 1 + i} = 1 - i.$$

EXERCISES FOR SECTION 2.4

1. Find $\displaystyle \lim_{z \to 2+i} (z^2 - 4z + 2 + 5i)$.

2. Find $\displaystyle \lim_{z \to i} \frac{z^2 + 4z + 2}{z + 1}$.

3. Find $\displaystyle \lim_{z \to i} \frac{z^4 - 1}{z - i}$.

4. Find $\displaystyle \lim_{z \to 1+i} \frac{z^2 + z - 2 + i}{z^2 - 2z + 1}$.

5. Find $\displaystyle \lim_{z \to 1+i} \frac{z^2 + z - 1 - 3i}{z^2 - 2z + 2}$ by factoring.

6. Show that $\displaystyle \lim_{z \to 0} \frac{x^2}{z} = 0$.

7. State why $\lim_{z \to z_0} (e^x \cos y + ix^2 y) = e^{x_0} \cos y_0 + ix_0^2 y_0$.

8. State why $\lim_{z \to z_0} [\ln(x^2 + y^2) + iy] = \ln(x_0^2 + y_0^2) + iy_0$ provided that $|z_0| \neq 0$.

9. Show that $\lim_{z \to 0} \dfrac{|z|^2}{z} = 0$.

10. Let $f(z) = \dfrac{z^2}{|z|^2} = \dfrac{x^2 - y^2 + i2xy}{x^2 + y^2}$.
    (a) Find $\lim f(z)$ as $z \to 0$ along the line $y = x$.
    (b) Find $\lim f(z)$ as $z \to 0$ along the line $y = 2x$.
    (c) Find $\lim f(z)$ as $z \to 0$ along the parabola $y = x^2$.
    (d) What can you conclude about the limit of $f(z)$ as $z \to 0$?

11. Let $f(z) = \bar{z}/z$. Show that $f(z)$ does not have a limit as $z \to 0$.

12. Does $u(x, y) = (x^3 - 3xy^2)/(x^2 + y^2)$ have a limit as $(x, y) \to (0, 0)$?

13. Let $f(z) = z^{1/2} = r^{1/2}[\cos(\theta/2) + i \sin(\theta/2)]$, where $r > 0$ and $-\pi < \theta \leq \pi$. Use the polar form of z and show that
    (a) $f(z) \to i$ as $z \to -1$ along the upper semicircle $r = 1, 0 < \theta < \pi$.
    (b) $f(z) \to -i$ as $z \to -1$ along the lower semicircle $r = 1; -\pi < \theta < 0$.

14. Does $\lim_{z \to -4} \text{Arg } z$ exist? Why? *Hint*: Use polar coordinates and let z approach -4 from the upper and lower half planes.

15. Determine where the following functions are continuous.
    (a) $z^4 - 9z^2 + iz - 2$
    (b) $\dfrac{z + 1}{z^2 + 1}$
    (c) $\dfrac{z^2 + 6z + 5}{z^2 + 3z + 2}$
    (d) $\dfrac{z^4 + 1}{z^2 + 2z + 2}$
    (e) $\dfrac{x + iy}{x - 1}$
    (f) $\dfrac{x + iy}{|z| - 1}$

16. Let $f(z) = [z \, \text{Re}(z)]/|z|$ when $z \neq 0$, and let $f(0) = 0$. Show that $f(z)$ is continuous for all values of z.

17. Let $f(z) = xe^y + iy^2 e^{-x}$. Show that $f(z)$ is continuous for all values of z.

18. Let $f(z) = (x^2 + iy^2)/|z|^2$ when $z \neq 0$, and let $f(0) = 1$. Show that $f(z)$ is not continuous at $z_0 = 0$.

19. Let $f(z) = \text{Re}(z)/|z|$ when $z \neq 0$, and let $f(0) = 1$. Is $f(z)$ continuous at the origin?

20. Let $f(z) = [\text{Re}(z)]^2/|z|$ when $z \neq 0$, and let $f(0) = 1$. Is $f(z)$ continuous at the origin?

21. Let $f(z) = z^{1/2} = r^{1/2}[\cos(\theta/2) + i \sin(\theta/2)]$, where $r > 0$ and $-\pi < \theta \leq \pi$. Show that $f(z)$ is discontinuous at each point along the negative x axis.

22. Let $f(z) = \ln|z| + i \text{ Arg } z$, where $-\pi < \text{Arg } z \leq \pi$. Show that $f(z)$ is discontinuous at $z_0 = 0$ and at each point along the negative x axis.

23. Let A and B be complex constants. Use Theorem 2.1 to prove that
    $$\lim_{z \to z_0} (Az + B) = Az_0 + B.$$

24. Let $\Delta z = z - z_0$. Show that $\lim_{z \to z_0} f(z) = w_0$ if and only if $\lim_{\Delta z \to 0} f(z_0 + \Delta z) = w_0$.

25. Let $|g(z)| \leq M$ and $\lim_{z \to z_0} f(z) = 0$. Show that $\lim_{z \to z_0} f(z)g(z) = 0$.

26. Establish identity (8).   27. Establish identity (9).   28. Establish identity (10).

29. Let $f(z)$ be continuous for all values of z.
    (a) Show that $g(z) = f(\bar{z})$ is continuous for all z.
    (b) Show that $h(z) = \overline{f(z)}$ is continuous for all z.

30. Establish the results of (17) and (18).   31. Establish the result (19).
32. Establish the result (20).   33. Establish the result (21).

2.5   Branches of Functions

In Section 2.3 we defined the principal square root function and investigated some of its properties. We left some unanswered questions concerning the choices of square roots. We now look into this problem because it is similar to situations involving other elementary functions.

In our definition of a function in Section 2.1 we specified that each value of the independent variable in the domain is mapped onto one and only one value of the dependent variable. As a result, one often talks about a *single-valued function*, which emphasizes the *only one* part of the definition and allows us to distinguish such functions from multiple-valued functions, which we now introduce.

Let $w = f(z)$ denote a function whose domain is the set D and whose range is the set R. If w is a value in the range, then there is an associated inverse function $z = g(w)$ that assigns to each value w the value (or values) of z in D for which the equation $f(z) = w$ holds true. But unless f takes on the value w at most once in D, then the inverse function g is necessarily many valued, and we say that g is a *multivalued function*. For example, the inverse of the function $w = f(z) = z^2$ is the square root function $z = g(w) = w^{1/2}$. We see that for each value z other than $z = 0$, the two points z and $-z$ are mapped onto the same point $w = f(z)$; hence g is in general a two-valued function.

The study of limits, continuity, and derivatives loses all meaning if an arbitrary or ambiguous assignment of function values is made. For this reason we did not allow multivalued functions to be considered when we defined these concepts. When working with inverse functions, it is necessary to carefully specify one of the many possible inverse values when constructing an inverse function. The idea is the same as determining implicit functions in calculus. If the values of a function f are determined by an equation that they satisfy rather than by an explicit formula, then we say that the function is defined implicitly or that f is an *implicit function*. In the theory of complex variables we study a similar concept.

Let $w = f(z)$ be a multiple-valued function. A *branch* of f is any single-valued function f_0 that is analytic in *some* domain and, at each point z in the domain, assigns one of the values of $f(z)$.

EXAMPLE 2.15   Let us consider some branches of the two-valued square root function $f(z) = z^{1/2}$. We define the principal square root function as

(1) $f_1(z) = r^{1/2}\cos\dfrac{\theta}{2} + ir^{1/2}\sin\dfrac{\theta}{2} = r^{1/2}e^{i\theta/2},$

where we require that $r > 0$ and $-\pi < \theta \le \pi$. The function f_1 is a branch of f. We can find other branches of the square root function. For example, let

(2) $f_2(z) = r^{1/2}\cos\dfrac{\theta + 2\pi}{2} + ir^{1/2}\sin\dfrac{\theta + 2\pi}{2} = r^{1/2}e^{i(\theta + 2\pi)/2},$

where $r > 0$ and $-\pi < \theta \le \pi$.

If we use the identities $\cos\dfrac{\theta + 2\pi}{2} = -\cos\dfrac{\theta}{2}$ and $\sin\dfrac{\theta + 2\pi}{2} = -\sin\dfrac{\theta}{2}$,
then we see that

$$f_2(z) = -r^{1/2}\cos\frac{\theta}{2} - ir^{1/2}\sin\frac{\theta}{2} = -r^{1/2}e^{i\theta/2} = -f_1(z),$$

so f_1 and f_2 can be thought of as "plus" and "minus" square root functions.

The negative real axis is called a *branch cut* for the functions f_1 and f_2. It is characterized by the fact that each point on the branch cut is a point of discontinuity for both functions f_1 and f_2.

EXAMPLE 2.16   To show that the function f_1 is discontinuous along the negative real axis, let $z_0 = r_0 e^{\pm i\pi}$ denote a negative real number. Now we compute the limit of $f_1(z)$ as z approaches z_0 through the upper half plane $\{z: \text{Im}(z) > 0\}$ and the limit of $f_1(z)$ as z approaches z_0 through the lower half plane $\{z: \text{Im}(z) < 0\}$. In polar coordinates these limits are given by

$$\lim_{(r,\theta)\to(r_0,\pi)} f_1(re^{i\theta}) = \lim_{(r,\theta)\to(r_0,\pi)} r^{1/2}\left(\cos\frac{\theta}{2} + i\sin\frac{\theta}{2}\right) = ir_0^{1/2} \quad \text{and}$$

$$\lim_{(r,\theta)\to(r_0,-\pi)} f_1(re^{i\theta}) = \lim_{(r,\theta)\to(r_0,-\pi)} r^{1/2}\left(\cos\frac{\theta}{2} + i\sin\frac{\theta}{2}\right) = -ir_0^{1/2}.$$

Since the two limits are distinct, the function f_1 is discontinuous at z_0. Likewise, f_2 is discontinuous at z_0. The mappings $w = f_1(z)$ and $w = f_2(z)$ and the branch cut are illustrated in Figure 2.15.

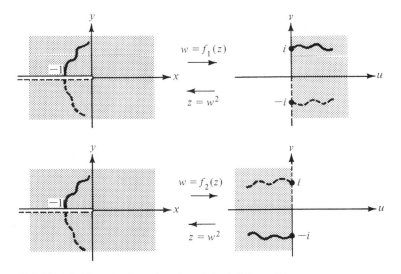

FIGURE 2.15   The branches f_1 and f_2 of $f(z) = z^{1/2}$.

Other branches of the square root function can be constructed by specifying that an argument of z given by $\theta = \arg z$ is to lie in the interval $\alpha < \theta \le \alpha + 2\pi$. Then the branch f_α is given by

$$(3) \quad f_\alpha(z) = r^{1/2}\cos\frac{\theta}{2} + ir^{1/2}\sin\frac{\theta}{2}, \quad \text{where } r > 0 \text{ and } \alpha < \theta \le \alpha + 2\pi.$$

The branch cut for f_α is the ray $r \ge 0$, $\theta = \alpha$, which includes the origin. The point $z = 0$, common to all branch cuts for the multivalued function, is called a *branch point*. The mapping $w = f_\alpha(z)$ and its branch cut are illustrated in Figure 2.16.

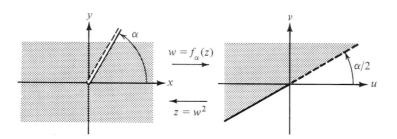

FIGURE 2.16   The branch f_α of $f(z) = z^{1/2}$.

The Riemann Surface for $w = z^{1/2}$

A method for visualizing a multivalued function is provided by using a Riemann surface. These representations were introduced by G. F. B Riemann (1826–1866) in 1851. The idea is ingenious, a geometric construction that permits surfaces to be the domain or range of a multivalued function.

Consider $w = f(z) = z^{1/2}$, which has two values for any given z (except, of course, for $z = 0$). Each function $f_1(z)$ and $f_2(z)$, given in Example 2.15 is single-valued on the domain formed by cutting the z plane along the negative x axis. Let D_1 and D_2 be the domain of $f_1(z)$ and $f_2(z)$, respectively. The range set for $f_2(z)$ is the set H_1 consisting of right half plane H_1 plus the positive v axis, and the range set $f_2(z)$ is the set H_2 consisting of left half plane H_1 plus the negative v axis. The sets H_1 and H_2 are "glued together" along the positive v axis and the negative v axis to form the w plane with the origin deleted.

Stack D_1 to D_2 directly above each other. The edge of D_1 in the upper half plane is joined to the edge of D_2 in the lower half plane, and the edge of D_1 in the lower half plane is joined to the edge of D_2 in the upper half plane. When these domains are "glued" together in this manner they form R, which is a *Riemann surface domain* for the mapping $w = f(z) = z^{1/2}$. The portion of D_1, D_2, and R that satisfy $|z| < 1$ are shown in Figure 2.17.

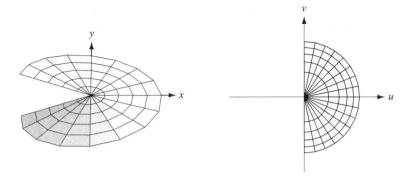

(a) A portion of D_1 and its image under $w = z^{1/2}$.

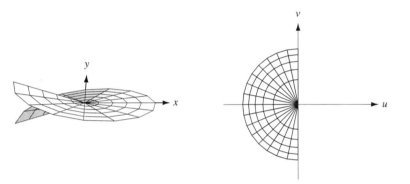

(b) A portion of D_2 and its image under $w = z^{1/2}$.

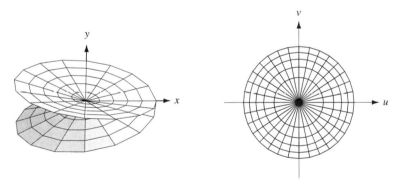

(c) A portion of R and its image under $w = z^{1/2}$.

FIGURE 2.17   Formation of the Riemann surface for $w = z^{1/2}$.

EXERCISES FOR SECTION 2.5

1. Let $f_1(z)$ and $f_2(z)$ be the two branches of the square root function given by equations (1) and (2), respectively. Use the polar coordinate formulas in Section 2.3 to
  (a) Find the image of quadrant II, $x < 0$ and $y > 0$, under the mapping $w = f_1(z)$.
  (b) Find the image of quadrant II, $x < 0$ and $y > 0$, under the mapping $w = f_2(z)$.

    (c) Find the image of the right half plane $\text{Re}(z) > 0$ under the mapping $w = f_1(z)$.

    (d) Find the image of the right half plane $\text{Re}(z) > 0$ under the mapping $w = f_2(z)$.

2. Let $\alpha = 0$ in equation (3), and find the range of the function $w = f(z)$.

3. Let $\alpha = 2\pi$ in equation (3), and find the range of the function $w = f(z)$.

4. Find a branch of the square root function that is continuous along the negative x axis.

5. Let $f_1(z) = r^{1/3}\cos(\theta/3) + ir^{1/3}\sin(\theta/3)$, where $r > 0$ and $-\pi < \theta \le \pi$ denote the principal cube root function.

    (a) Show that f_1 is a branch of the multivalued cube root function $f(z) = z^{1/3}$.

    (b) What is the range of f_1?

    (c) Where is f_1 continuous?

6. Let $f_2(z) = r^{1/3}\cos[(\theta + 2\pi)/3] + ir^{1/3}\sin[(\theta + 2\pi)/3]$, where $r > 0$ and $-\pi < \theta \le \pi$.

    (a) Show that f_2 is a branch of the multivalued cube root function $f(z) = z^{1/3}$.

    (b) What is the range of f_2?

    (c) Where is f_2 continuous?

    (d) What is the branch point associated with f?

7. Find a branch of the multivalued cube root function that is different from those in Exercises 5 and 6. State the domain and range of the branch you find.

8. Let $f(z) = z^{1/n}$ denote the multivalued nth root function, where n is a positive integer.

    (a) Show that f is in general an n-valued function.

    (b) Write down the principal nth root function.

    (c) Write down a branch of the multivalued nth root function that is different from the one in part (b).

9. Describe a Riemann surface for the domain of definition of the multivalued function $w = f(z) = z^{1/3}$.

10. Describe a Riemann surface for the domain of definition of the multivalued function $w = f(z) = z^{1/4}$.

11. Discuss how Riemann surfaces should be used for both the domain of definition and the range to help describe the behavior of the multivalued function $w = f(z) = z^{2/3}$.

12. Show that the principal branch of the argument Arg z is discontinuous at 0 and all points along the negative real axis.

2.6 The Reciprocal Transformation $w = 1/z$ (Prerequisite for Section 9.2)

The mapping $w = 1/z$ is called the *reciprocal transformation* and maps the z plane one-to-one and onto the w plane except for the point $z = 0$, which has no image, and the point $w = 0$, which has no preimage or inverse image. Since $z\bar{z} = |z|^2$, we can express the reciprocal transformation as a composition:

$$(1) \quad w = \bar{Z} \quad \text{and} \quad Z = \frac{z}{|z|^2}.$$

The transformation $Z = z/|z|^2$ is called the *inversion mapping* with respect to the unit circle $|z| = 1$. It has the property that a nonzero point z is mapped onto the point Z such that

$$(2) \quad |Z||z| = 1 \quad \text{and} \quad \arg Z = \arg z.$$

Hence it maps points inside the circle $|z| = 1$ onto points outside the circle $|Z| = 1$, and conversely. Any point of unit modulus is mapped onto itself. The inversion mapping is illustrated in Figure 2.18.

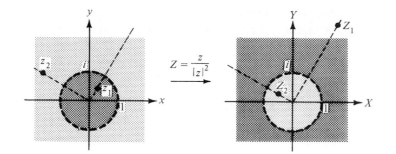

$$Z = \frac{z}{|z|^2}$$

FIGURE 2.18   The inversion mapping.

The geometric description of the reciprocal transformation is now evident from the composition given in expression (1). It is an inversion followed by a reflection through the x axis. If we use the polar coordinate form

(3) $w = \rho e^{i\phi} = \dfrac{1}{r} e^{-i\theta}$,   where $z = re^{i\theta}$,

then we see that the ray $r > 0$, $\theta = \alpha$ is mapped one-to-one and onto the ray $\rho > 0$, $\phi = -\alpha$. Also, points that lie inside the circle $|z| = 1$ are mapped onto points that lie outside the circle $|w| = 1$, and vice versa. The situation is illustrated in Figure 2.19.

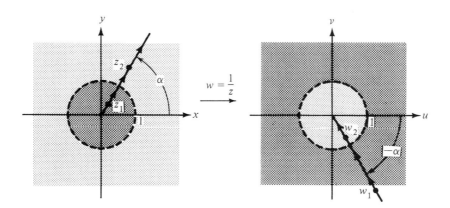

$$w = \frac{1}{z}$$

FIGURE 2.19   The reciprocal transformation $w = 1/z$.

It is convenient to extend the system of complex numbers by joining to it an ''ideal'' point denoted by ∞ and called the *point at infinity*. This new set is called the extended complex plane. The point ∞ has the property that

(4) $\lim\limits_{n \to \infty} z_n = \infty$ if and only if $\lim\limits_{n \to \infty} |z_n| = \infty$.

An ε-neighborhood of the point at infinity is the set $\{z: |z| > 1/\varepsilon\}$. The usual way to visualize the point at infinity is accomplished by using the stereographic projection and is attributed to Riemann. Let Ω be a sphere of diameter 1 that is centered at the $(0, 0, \frac{1}{2})$ in the three-dimensional space where coordinates are denoted by the triple of real numbers (x, y, ξ). Here the complex number $z = x + iy$ will be associated with the point $(x, y, 0)$.

The point $\mathcal{N} = (0, 0, 1)$ on Ω is called the *north pole* of Ω. Let z be a complex number, and consider the line segment L in three-dimensional space that joins z to the north pole $\mathcal{N}$. Then L intersects Ω in exactly one point $\mathcal{L}$. The correspondence $z \leftrightarrow \mathcal{L}$ is called the stereographic projection of the complex z plane onto the *Riemann sphere* Ω. A point $z = x + iy$ of unit modulus will correspond to $\mathcal{L} = \left(\dfrac{x}{2}, \dfrac{y}{2}, \dfrac{1}{2}\right)$. If z has modulus greater than 1, then $\mathcal{L}$ will lie in the upper hemisphere where $\xi > \frac{1}{2}$. If z has modulus less than 1, then $\mathcal{L}$ will lie in the lower hemisphere where $\xi < \frac{1}{2}$. The complex number $z = 0$ corresponds to the *south pole* $\mathcal{S} = (0, 0, 0)$. It is easy to visualize that $z \to \infty$ if and only if $\mathcal{L} \to \mathcal{N}$. Hence $\mathcal{N}$ corresponds to the "ideal" point at infinity. The situation is shown in Figure 2.20.

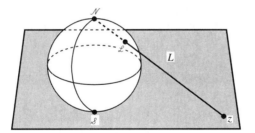

FIGURE 2.20   The Riemann sphere.

Let us reconsider the mapping $w = 1/z$. Let us assign the images $w = \infty$ and $w = 0$ to the points $z = 0$ and $z = \infty$, respectively. The reciprocal transformation can now be written as

$$(5) \quad w = f(z) = \begin{cases} 1/z & \text{when } z \neq 0, z \neq \infty \\ 0 & \text{when } z = \infty \\ \infty & \text{when } z = 0. \end{cases}$$

It is easy to see that the transformation $w = f(z)$ is a one-to-one mapping of the extended complex z plane onto the extended complex w plane. Using property (4) of the point at infinity, it is easy to show that f is a continuous mapping from the extended z plane onto the extended w plane. The details are left for the reader.

EXAMPLE 2.17   Show that the image of the right half plane $\mathrm{Re}(z) > \frac{1}{2}$, under the mapping $w = 1/z$, is the disk $|w - 1| < 1$.

Solution   The inverse mapping $z = 1/w$ can be written as

(6) $x + iy = z = \dfrac{1}{w} = \dfrac{u - iv}{u^2 + v^2}.$

Equating the real and imaginary parts in equation (6), we obtain the equations

(7) $x = \dfrac{u}{u^2 + v^2}$ and $y = \dfrac{-v}{u^2 + v^2}.$

The requirement that $x > \frac{1}{2}$ forces the image values to satisfy the inequality

(8) $\dfrac{u}{u^2 + v^2} > \dfrac{1}{2}.$

It is easy to manipulate inequality (8) to obtain

(9) $u^2 - 2u + 1 + v^2 < 1,$

which is an inequality that determines the set of points in the w plane that lie inside the circle with center $w_0 = 1$ and radius 1. Since the reciprocal transformation is one-to-one, preimages of the points in the disk $|w - 1| < 1$ will lie in the right half plane $\mathrm{Re}(z) > \frac{1}{2}$. The mapping is shown in Figure 2.21.

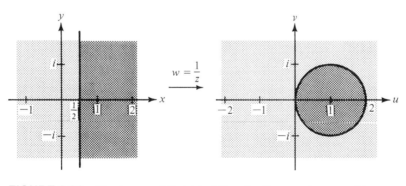

FIGURE 2.21   The image of $\mathrm{Re}(z) > 1/2$ under the mapping $w = 1/z$.

EXAMPLE 2.18   Find the image of the portion of the right half plane $\mathrm{Re}(z) > \frac{1}{2}$ that lies inside the circle $\left|z - \frac{1}{2}\right| < 1$ under the transformation $w = 1/z$.

Solution   Using the result of Example 2.17, we need only find the image of the disk $\left| z - \frac{1}{2} \right| < 1$ and intersect it with the disk $\left| w - 1 \right| < 1$. To start with, we can express the disk $\left| z - \frac{1}{2} \right| < 1$ by the inequality

(10) $x^2 + y^2 - x < \frac{3}{4}$.

We can use the identities (7) to show that the image values of points satisfying inequality (10) must satisfy the inequality

(11) $\dfrac{1}{u^2 + v^2} - \dfrac{u}{u^2 + v^2} < \dfrac{3}{4}$.

Inequality (11) can now be manipulated to yield

$(\frac{4}{3})^2 < (u + \frac{2}{3})^2 + v^2$,

which is an inequality that determines the set of points in the w plane that lie exterior to the circle $\left| w + \frac{2}{3} \right| = \frac{4}{3}$. Therefore, the image is the crescent-shaped region illustrated in Figure 2.22.

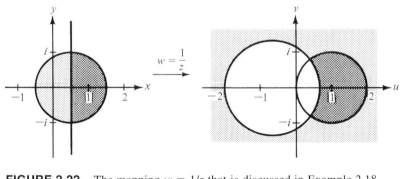

FIGURE 2.22   The mapping $w = 1/z$ that is discussed in Example 2.18.

To study images of ''generalized circles,'' let us consider the equation

(12) $A(x^2 + y^2) + Bx + Cy + D = 0$

where A, B, C, and D are real numbers. Then equation (12) represents either a circle or a line, depending on whether $A \neq 0$ or $A = 0$, respectively. If we use polar coordinates, then equation (12) has the form

(13) $Ar^2 + r(B \cos \theta + C \sin \theta) + D = 0$.

Using the polar coordinate form of the reciprocal transformation given in equation (3), we find that the image of the curve in equation (13) can be expressed by the equation

(14) $A + \rho(B \cos \phi - C \sin \phi) + D\rho^2 = 0$,

which represents either a circle or a line, depending on whether $D \neq 0$ or $D = 0$, respectively. Therefore, we have shown that the reciprocal transformation $w = 1/z$ carries the class of lines and circles onto itself.

EXAMPLE 2.19 Find the images of the vertical lines $x = a$ and the horizontal lines $y = b$ under the mapping $w = 1/z$.

Solution The image of the line $x = 0$ is the line $u = 0$; that is, the y axis is mapped onto the v axis. Similarly, the x axis is mapped onto the u axis.

If $a \neq 0$, then using equations (7), we see that the vertical line $x = a$ is mapped onto the circle

$$(15) \qquad \frac{u}{u^2 + v^2} = a.$$

It is easy to manipulate equation (15) to obtain

$$u^2 - \frac{1}{a}u + \frac{1}{4a^2} + v^2 = \left(u - \frac{1}{2a}\right)^2 + v^2 = \left(\frac{1}{2a}\right)^2,$$

which is the equation of a circle in the w plane with center $w_0 = 1/(2a)$ and radius $1/(2a)$.

Similarly, the horizontal line $y = b$ is mapped onto the circle

$$u^2 + v^2 + \frac{1}{b}v + \frac{1}{4b^2} = u^2 + \left(v + \frac{1}{2b}\right)^2 = \left(\frac{1}{2b}\right)^2,$$

which has center $w_0 = -i/(2b)$ and radius $1/(2b).$ The images of several lines are shown in Figure 2.23.

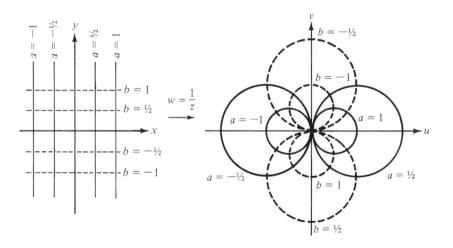

FIGURE 2.23 The images of horizontal and vertical lines under the reciprocal transformation.

EXERCISES FOR SECTION 2.6

For Exercises 1–8, find the image of the given circle or line under the reciprocal transformation $w = 1/z$.

1. The horizontal line $\text{Im}(z) = \frac{1}{5}$.
2. The circle $\left| z + i/2 \right| = \frac{1}{2}$.
3. The vertical line $\text{Re}(z) = -3$.
4. The circle $\left| z + 2 \right| = 1$.
5. The line $2x + 2y = 1$.
6. The circle $\left| z - i/2 \right| = 1$.
7. The circle $\left| z - \frac{3}{2} \right| = 1$.
8. The circle $\left| z + 1 - i \right| = 2$.
9. **(a)** Show that $\lim\limits_{z \to \infty} (1/z) = 0$.   **(b)** Show that $\lim\limits_{z \to 0} (1/z) = \infty$.

10. Show that the reciprocal transformation $w = 1/z$ maps the vertical strip $0 < x < \frac{1}{2}$ onto the region in the right half plane $\text{Re}(w) > 0$ that lies outside the circle $\left| w - 1 \right| = 1$.

11. Find the image of the disk $\left| z + 2i/3 \right| < \frac{4}{3}$ under the reciprocal transformation.

12. Show that the reciprocal transformation maps the disk $\left| z - 1 \right| < 2$ onto the region that lies exterior to the circle $\left| w + \frac{1}{3} \right| = \frac{2}{3}$.

13. Find the image of the half plane $y > \frac{1}{2} - x$ under the mapping $w = 1/z$.

14. Show that the half plane $y < x - \frac{1}{2}$ is mapped onto the disk $\left| w - 1 - i \right| < \sqrt{2}$ by the reciprocal transformation.

15. Find the image of the quadrant $x > 1$, $y > 1$ under the mapping $w = 1/z$.

16. Show that the transformation $w = 2/z$ maps the disk $\left| z - i \right| < 1$ onto the lower half plane $\text{Im}(w) < -1$.

17. Show that the transformation $w = (2 - z)/z = -1 + 2/z$ maps the disk $\left| z - 1 \right| < 1$ onto the right half plane $\text{Re}(w) > 0$.

18. Show that the parabola $2x = 1 - y^2$ is mapped onto the cardioid $\rho = 1 + \cos \phi$ by the reciprocal transformation.

19. Limits involving ∞. The function $f(z)$ is said to have the limit L as z approaches ∞, and we write

$$\lim\limits_{z \to \infty} f(z) = L$$

if for every $\varepsilon > 0$ there exists an $R > 0$ so that

$$\left| f(z) - L \right| < \varepsilon, \quad \text{whenever } \left| z \right| > R.$$

Use this definition to prove that

$$\lim\limits_{z \to \infty} \frac{z + 1}{z - 1} = 1.$$

20. Show that the complex number $z = x + iy$ is mapped onto the point

$$\left(\frac{x}{x^2 + y^2 + 1}, \frac{y}{x^2 + y^2 + 1}, \frac{x^2 + y^2}{x^2 + y^2 + 1} \right)$$

on the Riemann sphere.

21. Explain how are the quantities $+\infty$, $-\infty$, and ∞ different? How are they similar?

22. Write a report on Möbius transformation. Include ideas and examples that are not mentioned in the text. Resources include bibliographical items 12, 23, 24, 30, 36, and 43.

3

Analytic and Harmonic Functions

3.1 Differentiable Functions

Let f be a complex function that is defined at all points in some neighborhood of z_0. The *derivative of f at z_0* is written $f'(z_0)$ and is defined by the equation

$$(1) \quad f'(z_0) = \lim_{z \to z_0} \frac{f(z) - f(z_0)}{z - z_0}$$

provided that the limit exists. When this happens, we say that the function f is *differentiable* at z_0. If we write $\Delta z = z - z_0$, then definition (1) can be expressed in the form

$$(2) \quad f'(z_0) = \lim_{\Delta z \to 0} \frac{f(z_0 + \Delta z) - f(z_0)}{\Delta z}.$$

If we let $w = f(z)$ and $\Delta w = f(z) - f(z_0)$, then the notation dw/dz for the derivative is expressed by

$$(3) \quad f'(z_0) = \frac{dw}{dz} = \lim_{\Delta z \to 0} \frac{\Delta w}{\Delta z}.$$

EXAMPLE 3.1 If $f(z) = z^3$, show we can use definition (1) to get $f'(z) = 3z^2$.

Solution Calculation reveals that

$$f'(z_0) = \lim_{z \to z_0} \frac{z^3 - z_0^3}{z - z_0} = \lim_{z \to z_0} \frac{(z - z_0)(z^2 + z_0 z + z_0^2)}{z - z_0} = 3z_0^2.$$

The subscript on z_0 can be dropped to obtain the general formula $f'(z) = 3z^2$.

We must pay careful attention to the complex value Δz in equation (3), since the value of the limit must be independent of the manner in which $\Delta z \to 0$. If we can find two curves that end at z_0 along which $\Delta w/\Delta z$ approaches distinct values, then $\Delta w/\Delta z$ does *not* have a limit as $\Delta z \to 0$ and f does *not* have a derivative at z_0.

EXAMPLE 3.2   Show that the function $w = f(z) = \bar{z} = x - iy$ is nowhere differentiable.

   Solution   To show this, we choose two approaches to the point $z_0 = x_0 + iy_0$ and compute limits of the difference quotients. First, we approach $z_0 = x_0 + iy_0$ along a line parallel to the x axis by forcing z to be of the form $z = x + iy_0$:

$$\lim_{z \to z_0} \frac{f(z) - f(z_0)}{z - z_0} = \lim_{(x + iy_0) \to (x_0 + iy_0)} \frac{f(x + iy_0) - f(x_0 + iy_0)}{(x + iy_0) - (x_0 + iy_0)}$$

$$= \lim_{(x + iy_0) \to (x_0 + iy_0)} \frac{(x - iy_0) - (x_0 - iy_0)}{(x - x_0) + i(y_0 - y_0)}$$

$$= \lim_{(x + iy_0) \to (x_0 + iy_0)} \frac{x - x_0}{x - x_0}$$

$$= 1.$$

Second, we approach z_0 along a line parallel to the y axis by forcing z to be of the form $z = x_0 + iy$:

$$\lim_{z \to z_0} \frac{f(z) - f(z_0)}{z - z_0} = \lim_{(x_0 + iy) \to (x_0 + iy_0)} \frac{f(x_0 + iy) - f(x_0 + iy_0)}{(x_0 + iy) - (x_0 + iy_0)}$$

$$= \lim_{(x_0 + iy) \to (x_0 + iy_0)} \frac{(x_0 - iy) - (x_0 - iy_0)}{(x_0 - x_0) + i(y - y_0)}$$

$$= \lim_{(x_0 + iy) \to (x_0 + iy_0)} \frac{-i(y - y_0)}{i(y - y_0)}$$

$$= -1.$$

Since the limits along the two approaches are different, there is no computable limit for the right side of equation (1). Therefore $f(z) = \bar{z}$ is not differentiable at the point z_0. Since z_0 was arbitrary, $f(z)$ is nowhere differentiable.

   Our definition of the derivative for complex functions is formally the same as for real functions and is the natural extension from real variables to complex variables. The basic differentiation formulas follow identically as in the case of real functions, and we obtain the same rules for differentiating powers, sums, products, quotients, and compositions of functions. The proof of the differentiation formulas are easily established by using the limit theorems.

   Let C denote a complex constant. From definition (1) and the technique exhibited in the solution to Example 3.1, the following are easily established, just as they were in the real case:

(4) $\dfrac{d}{dz} C = 0$, where C is a constant,   and

(5) $\dfrac{d}{dz} z^n = nz^{n-1}$,   where n is a positive integer.

Furthermore, the rules for finding derivatives of combinations of two differentiable functions f and g are identical to those developed in calculus:

(6) $\dfrac{d}{dz} [Cf(z)] = Cf'(z)$,

(7) $\dfrac{d}{dz} [f(z) + g(z)] = f'(z) + g'(z)$,

(8) $\dfrac{d}{dz} [f(z)g(z)] = f'(z)g(z) + f(z)g'(z)$,

(9) $\dfrac{d}{dz} \dfrac{f(z)}{g(z)} = \dfrac{f'(z)g(z) - f(z)g'(z)}{[g(z)]^2}$,   provided that $g(z) \neq 0$,

(10) $\dfrac{d}{dz} f(g(z)) = f'(g(z))g'(z)$.

Important particular cases of (9) and (10), respectively, are

(11) $\dfrac{d}{dz} \dfrac{1}{z^n} = \dfrac{-n}{z^{n+1}}$,   for $z \neq 0$ and where n is a positive integer,

(12) $\dfrac{d}{dz} [f(z)]^n = n[f(z)]^{n-1}f'(z)$,   where n is a positive integer.

EXAMPLE 3.3   If we use equation (12) with $f(z) = z^2 + i2z + 3$ and $f'(z) = 2z + 2i$, then we see that

$$\frac{d}{dz} (z^2 + i2z + 3)^4 = 8(z^2 + i2z + 3)^3(z + i).$$

Several proofs involving complex functions rely on properties of continuous functions. The following result shows that a differentiable function is a continuous function.

Theorem 3.1   *If f is differentiable at z_0, then f is continuous at z_0.*

Proof   Since f is differentiable at z_0, from definition (1) we obtain

$$\lim_{z \to z_0} \frac{f(z) - f(z_0)}{z - z_0} = f'(z_0).$$

Using the multiplicative property of limits given by formula (9) in Section 2.4, we see that

$$\lim_{z \to z_0} [f(z) - f(z_0)] = \lim_{z \to z_0} \frac{f(z) - f(z_0)}{z - z_0} (z - z_0)$$

$$= \lim_{z \to z_0} \frac{f(z) - f(z_0)}{z - z_0} \lim_{z \to z_0} (z - z_0)$$

$$= f'(z_0) \cdot 0 = 0.$$

Hence $\lim_{z \to z_0} f(z) = f(z_0)$, and f is continuous at z_0.

Using Theorem 3.1, we are able to establish formula (8). Letting $h(z) = f(z)g(z)$ and using definition (1), we write

$$h'(z_0) = \lim_{z \to z_0} \frac{h(z) - h(z_0)}{z - z_0} = \lim_{z \to z_0} \frac{f(z)g(z) - f(z_0)g(z_0)}{z - z_0}.$$

If we add and subtract the term $f(z_0)g(z)$ in the numerator, we can regroup the last term and obtain

$$h'(z_0) = \lim_{z \to z_0} \frac{f(z)g(z) - f(z_0)g(z)}{z - z_0} + \lim_{z \to z_0} \frac{f(z_0)g(z) - f(z_0)g(z_0)}{z - z_0}$$

$$= \lim_{z \to z_0} \frac{f(z) - f(z_0)}{z - z_0} \lim_{z \to z_0} g(z) + f(z_0) \lim_{z \to z_0} \frac{g(z) - g(z_0)}{z - z_0}.$$

Using definition (1) for derivative and the continuity of g, we obtain $h'(z_0) = f'(z_0)g(z_0) + f(z_0)g'(z_0)$. Hence formula (8) is established. The proofs of the other formulas are left as exercises.

The rule for differentiating a polynomial can be extended to complex variables. Let $P(z)$ be a polynomial of degree n:

(13) $P(z) = a_0 + a_1 z + a_2 z^2 + \cdots + a_n z^n$.

Then mathematical induction can be used with formulas (5) and (7) to obtain the derivative of (13):

(14) $P'(z) = a_1 + 2a_2 z + 3a_3 z^2 + \cdots + n a_n z^{n-1}$.

The proof is left as an exercise.

Properties of limits and derivatives can be used to establish L'Hôpital's rule, which has the familiar form that is learned in calculus.

Assume f and g are differentiable at z_0. If $f(z_0) = 0$, $g(z_0) = 0$, and $g'(z_0) \neq 0$, then

$$\lim_{z \to z_0} \frac{f(z)}{g(z)} = \lim_{z \to z_0} \frac{f'(z)}{g'(z)}.$$

Finding limits of the form "0/0" by L'Hôpital's rule is given in Exercise 7.

EXERCISES FOR SECTION 3.1

1. Find the derivatives of the following functions.
   (a) $f(z) = 5z^3 - 4z^2 + 7z - 8$
   (b) $g(z) = (z^2 - iz + 9)^5$
   (c) $h(z) = \dfrac{2z + 1}{z + 2}$ for $z \neq -2$
   (d) $F(z) = (z^2 + (1 - 3i)z + 1)(z^4 + 3z^2 + 5i)$

2. Use definition (1), and show that $\dfrac{d}{dz} \dfrac{1}{z} = \dfrac{-1}{z^2}$.

3. If f is differentiable for all z, then we say that f is an *entire* function. If f and g are entire functions, decide which of the following are entire functions.
   (a) $[f(z)]^3$ (b) $f(z)g(z)$ (c) $f(z)/g(z)$
   (d) $f(1/z)$ (e) $f(z - 1)$ (f) $f(g(z))$

4. Use definition (1) to establish formula (5).

5. Let P be a polynomial of degree n given by $P(z) = a_0 + a_1 z + \cdots + a_n z^n$. Show that $P'(z) = a_1 + 2a_2 z + \cdots + na_n z^{n-1}$.

6. Let P be a polynomial of degree 2, given by

   $$P(z) = (z - z_1)(z - z_2),$$

   where z_1 and z_2 are distinct. Show that

   $$\frac{P'(z)}{P(z)} = \frac{1}{z - z_1} + \frac{1}{z - z_2}.$$

7. Use L'Hôpital's rule to find the following limits.
   (a) $\lim\limits_{z \to i} \dfrac{z^4 - 1}{z - i}$

   (b) $\lim\limits_{z \to 1+i} \dfrac{z^2 - iz - 1 - i}{z^2 - 2z + 2}$

   (c) $\lim\limits_{z \to -i} \dfrac{z^6 + 1}{z^2 + 1}$

   (d) $\lim\limits_{z \to 1+i} \dfrac{z^4 + 4}{z^2 - 2z + 2}$

   (e) $\lim\limits_{z \to 1+i\sqrt{3}} \dfrac{z^6 - 64}{z^3 + 8}$

   (f) $\lim\limits_{z \to -1+i\sqrt{3}} \dfrac{z^9 - 512}{z^3 - 8}$

8. Let f be differentiable at z_0. Show that there exists a function $\eta(z)$, such that

   $$f(z) = f(z_0) + f'(z_0)(z - z_0) + \eta(z)(z - z_0), \quad \text{where } \eta(z) \to 0 \text{ as } z \to z_0.$$

9. Show that $\dfrac{d}{dz} z^{-n} = -nz^{-n-1}$ where n is a positive integer.

10. Establish the identity

    $$\frac{d}{dz} f(z)g(z)h(z) = f'(z)g(z)h(z) + f(z)g'(z)h(z) + f(z)g(z)h'(z).$$

11. Show that the function $f(z) = |z|^2$ is differentiable only at the point $z_0 = 0$. *Hint*: To show that f is *not* differentiable at $z_0 \neq 0$, choose horizontal and vertical lines through the point z_0, and show that $\Delta w / \Delta z$ approaches two distinct values as $\Delta z \to 0$ along those two lines.

12. Establish identity (4).   13. Establish identity (7).
14. Establish identity (9).   15. Establish identity (10).

16. Establish identity (12).

17. Consider the differentiable function $f(z) = z^3$ and the two points $z_1 = 1$ and $z_2 = i$. Show that there does not exist a point c on the line $y = 1 - x$ between 1 and i such that

$$\frac{f(z_2) - f(z_1)}{z_2 - z_1} = f'(c).$$

This shows that the mean value theorem for derivatives does not extend to complex functions.

18. Let $f(z) = z^{1/n}$ denote the multivalued ''nth root function,'' where n is a positive integer. Use the chain rule to show that if $g(z)$ is any branch of the nth root function, then

$$g'(z) = \frac{1}{n} \frac{g(z)}{z}$$

in some suitably chosen domain (which you should specify).

19. Write a report on Rolle's theorem for complex functions. Resources include bibliographical items 64 and 127.

3.2   The Cauchy-Riemann Equations

Let $f(z) = u(x, y) + iv(x, y)$ be a complex function that is differentiable at the point z_0. Then it is natural to seek a formula for computing $f'(z_0)$ in terms of the partial derivatives of $u(x, y)$ and $v(x, y)$. If we investigate this idea, then it is easy to find the required formula; but we will find that there are special conditions that must be satisfied before it can be used. In addition, we will discover two important equations relating the partial derivatives of u and v, which were discovered independently by the French mathematician A. L. Cauchy* and the German mathematician G. F. B. Riemann.

First, let us reconsider the derivative of $f(z) = z^2$. The limit given in formula (1) of Section 3.1 must not depend on how z approaches z_0. We investigate two such approaches, a horizontal and a vertical approach to z_0. Recall from our graphics analysis of $w = z^2$ that the image of a square is a ''curvilinear quadrilateral.'' For convenience, let the square have vertices $z_0 = 2 + i$, $z_1 = 2.01 + i$, $z_2 = 2 + 1.01i$, and $z_3 = 2.01 + 1.01i$. Then the image points are $w_0 = 3 + 4i$, $w_1 = 3.0401 + 4.02i$, $w_2 = 2.9799 + 4.04i$, and $w_3 = 3.02 + 4.0602i$, as shown in Figure 3.1.

---

*A. L. Cauchy (1789–1857) played a prominent role in the development of complex analysis, and you will see his name several times throughout this text. The last name is *not* pronounced as ''kaushee.'' The beginning syllable has a long ''o'' sound, like the word kosher, but with the second syllable having a long ''e'' instead of ''er'' at the end. Thus, we pronounce Cauchy as ''kōshē.''

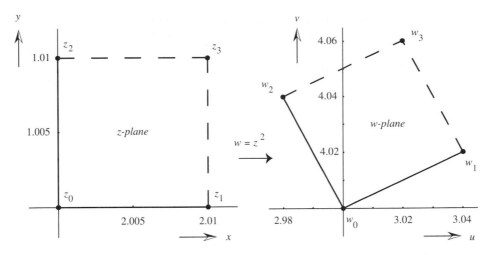

FIGURE 3.1   The image of a small square with vertex $z_0 = 2 + i$ using $w = z^2$.

Approximations for $f'(2 + i)$ are made using horizontal or vertical increments in z:

$$f'(2 + i) \approx \frac{f(2.01 + i) - f(2 + i)}{(2.01 + i) - (2 + i)} = \frac{0.0401 + 0.02i}{0.01} = 4.01 + 2i$$

and

$$f'(2 + i) \approx \frac{f(2 + 1.01i) - f(2 + i)}{(2 + 1.01i) - (2 + i)} = \frac{-0.0201 + 0.04i}{0.01i} = 4 + 2.01i.$$

These computations lead to the idea of taking limits along the horizontal and vertical directions, and the results are, respectively,

$$f'(2 + i) = \lim_{h \to 0} \frac{f(2 + h + i) - f(2 + i)}{h} = \lim_{h \to 0} \frac{4h + h^2 + i2h}{h} = 4 + 2i$$

and

$$f'(2 + i) = \lim_{h \to 0} \frac{f(2 + i + ih) - f(2 + i)}{ih} = \lim_{h \to 0} \frac{-2h - h^2 + i4h}{ih} = 4 + 2i.$$

We now generalize this idea by taking limits of an arbitrary complex function and obtain an important result.

Theorem 3.2 (Cauchy-Riemann Equations)   *Let $f(z) = f(x + iy) = u(x, y) + iv(x, y)$ be differentiable at the point $z_0 = x_0 + iy_0$. Then the partial derivatives of u and v exist at the point (x_0, y_0) and satisfy the equations*

(1) $u_x(x_0, y_0) = v_y(x_0, y_0)$ *and* $u_y(x_0, y_0) = -v_x(x_0, y_0)$.

Proof   We shall choose horizontal and vertical lines that pass through the point (x_0, y_0) and compute the limiting values of $\Delta w / \Delta z$ along these lines. Equating the two resulting limits will result in equations (1). For the horizontal approach to z_0 we set $z = x + iy_0$ and obtain

$$f'(z_0) = \lim_{(x,y_0) \to (x_0,y_0)} \frac{f(x + iy_0) - f(x_0 + iy_0)}{x + iy_0 - (x_0 + iy_0)}$$

$$= \lim_{x \to x_0} \frac{u(x, y_0) - u(x_0, y_0) + i[v(x, y_0) - v(x_0, y_0)]}{x - x_0}$$

$$= \lim_{x \to x_0} \frac{u(x, y_0) - u(x_0, y_0)}{x - x_0} + i \lim_{x \to x_0} \frac{v(x, y_0) - v(x_0, y_0)}{x - x_0}.$$

We see that the last limits are the partial derivatives of u and v with respect to x, and we obtain

(2) $f'(z_0) = u_x(x_0, y_0) + iv_x(x_0, y_0)$.

Along the vertical approach to z_0, we have $z = x_0 + iy$. Calculation reveals that

$$f'(z_0) = \lim_{(x_0,y) \to (x_0,y_0)} \frac{f(x_0 + iy) - f(x_0 + iy_0)}{x_0 + iy - (x_0 + iy_0)}$$

$$= \lim_{y \to y_0} \frac{u(x_0, y) - u(x_0, y_0) + i[v(x_0, y) - v(x_0, y_0)]}{i(y - y_0)}$$

$$= \lim_{y \to y_0} \frac{v(x_0, y) - v(x_0, y_0)}{y - y_0} - i \lim_{y \to y_0} \frac{u(x_0, y) - u(x_0, y_0)}{y - y_0}.$$

We see that the last limits are the partial derivatives of u and v with respect to y, and we obtain

(3) $f'(z_0) = v_y(x_0, y_0) - iu_y(x_0, y_0)$.

Since f is differentiable at z_0, the limits given by equations (2) and (3) must be equal. If we equate the real and imaginary parts in equations (2) and (3), then the result is equations (1), and the proof is complete.

At this stage we may be tempted to use equation (2) or (3) to compute $f'(z_0)$. We now investigate when such a procedure is valid.

EXAMPLE 3.4   The function $f(z) = z^3 = x^3 - 3xy^2 + i(3x^2y - y^3)$ is known to be differentiable. Verify that its derivative satisfies equation (2).

Solution   We can rewrite the function in the form

$$f(z) = u(x, y) + iv(x, y) = x^3 - 3xy^2 + i(3x^2y - y^3),$$

from which it follows that

$$f'(z) = u_x(x, y) + iv_x(x, y) = 3x^2 - 3y^2 + i6xy = 3(x^2 - y^2 + i2xy) = 3z^2.$$

EXAMPLE 3.5   The function defined by

$$f(z) = \frac{(\bar{z})^2}{z} = \frac{x^3 - 3xy^2}{x^2 + y^2} + i\frac{y^3 - 3x^2y}{x^2 + y^2}$$

when $z \neq 0$ and $f(0) = 0$ is *not* differentiable at the point $z_0 = 0$. However, the Cauchy-Riemann equations (1) hold true at $(0, 0)$. To verify this, we must use limits to calculate the partial derivatives at $(0, 0)$. Indeed,

$$u_x(0, 0) = \lim_{x \to 0} \frac{u(x, 0) - u(0, 0)}{x - 0} = \lim_{x \to 0} \frac{\dfrac{x^3 - 0}{x^2 + 0}}{x} = 1.$$

In a similar fashion, one can show that

$$u_y(0, 0) = 0, \qquad v_x(0, 0) = 0, \quad \text{and} \quad v_y(0, 0) = 1.$$

Hence the Cauchy-Riemann equations hold at the point $(0, 0)$.

We now show that f is *not* differentiable at $z_0 = 0$. If we let z approach 0 along the x axis, then

$$\lim_{(x,0) \to (0,0)} \frac{f(x + 0i) - f(0)}{x + 0i - 0} = \lim_{x \to 0} \frac{x - 0}{x - 0} = 1.$$

But if we let z approach 0 along the line $y = x$ given by the parametric equations $x = t$ and $y = t$, then

$$\lim_{(t,t) \to (0,0)} \frac{f(t + it) - f(0)}{t + it - 0} = \lim_{t \to 0} \frac{-t - it}{t + it} = -1.$$

Since the two limits are distinct, we conclude that f is not differentiable at the origin.

Example 3.5 shows that the mere satisfaction of the Cauchy-Riemann equations is not a sufficient criterion to guarantee the differentiability of a function. The next theorem gives us sufficient conditions under which we can use equations (2) and/or (3) to compute the derivative $f'(z_0)$. They are referred to as the *Cauchy-Riemann conditions* for differentiability.

Theorem 3.3 (Sufficient Conditions)   *Let $f(z) = u(x, y) + iv(x, y)$ be a continuous function that is defined in some neighborhood of the point $z_0 = x_0 + iy_0$. If all the partial derivatives u_x, u_y, v_x, and v_y are continuous at the point (x_0, y_0) and if the Cauchy-Riemann equations $u_x(x_0, y_0) = v_y(x_0, y_0)$ and $u_y(x_0, y_0) = -v_x(x_0, y_0)$ hold, then f is differentiable at z_0, and the derivative $f'(z_0)$ can be computed with either formula (2) or (3).*

Proof   Let $\Delta z = \Delta x + i\Delta y$ and $\Delta w = \Delta u + i\Delta v$, and let Δz be chosen small enough that z lies in the ε-neighborhood of z_0 in which the hypotheses hold true. We will show that $\Delta w / \Delta z$ approaches the limit given in equation (2) as Δz approaches zero. The difference Δu can be written as

$$\Delta u = u(x_0 + \Delta x, y_0 + \Delta y) - u(x_0, y_0).$$

If we add and subtract the term $u(x_0, y_0 + \Delta y)$, then the result is

$$(4) \quad \Delta u = [u(x_0 + \Delta x, y_0 + \Delta y) - u(x_0, y_0 + \Delta y)]$$
$$+ [u(x_0, y_0 + \Delta y) - u(x_0, y_0)].$$

Since the partial derivatives u_x and u_y exist, the mean value theorem for real functions of two variables implies that a value x^* exists between x_0 and $x_0 + \Delta x$ such that the first term in brackets on the right side of equation (4) can be written as

$$(5) \quad u(x_0 + \Delta x, y_0 + \Delta y) - u(x_0, y_0 + \Delta y) = u_x(x^*, y_0 + \Delta y)\Delta x.$$

Furthermore, since u_x and u_y are continuous at (x_0, y_0), there exists a quantity ε_1 such that

$$(6) \quad u_x(x^*, y_0 + \Delta y) = u_x(x_0, y_0) + \varepsilon_1,$$

where $\varepsilon_1 \to 0$ as $x^* \to x_0$ and $\Delta y \to 0$. Since $\Delta x \to 0$ forces $x^* \to x_0$, we can use the equation

$$(7) \quad u(x_0 + \Delta x, y_0 + \Delta y) - u(x_0, y_0 + \Delta y) = [u_x(x_0, y_0) + \varepsilon_1]\Delta x,$$

where $\varepsilon_1 \to 0$ as $\Delta x \to 0$ and $\Delta y \to 0$. Similarly, there exists a quantity ε_2 such that the second term in brackets on the right side of equation (4) satisfies the equation

$$(8) \quad u(x_0, y_0 + \Delta y) - u(x_0, y_0) = [u_y(x_0, y_0) + \varepsilon_2]\Delta y,$$

where $\varepsilon_2 \to 0$ as $\Delta x \to 0$ and $\Delta y \to 0$.

Combining equations (7) and (8), we obtain

$$(9) \quad \Delta u = (u_x + \varepsilon_1)\Delta x + (u_y + \varepsilon_2)\Delta y,$$

where the partial derivatives u_x and u_y are evaluated at the point (x_0, y_0) and ε_1 and ε_2 tend to zero as Δx and Δy both tend to zero. Similarly, the change Δv is related to the changes Δx and Δy by the equation

$$(10) \quad \Delta v = (v_x + \varepsilon_3)\Delta x + (v_y + \varepsilon_4)\Delta y$$

where the partial derivatives v_x and v_y are evaluated at the point (x_0, y_0) and ε_3 and ε_4 tend to zero as Δx and Δy both tend to zero. Combining equations (9) and (10), we have

$$(11) \quad \Delta w = u_x\Delta x + u_y\Delta y + i(v_x\Delta x + v_y\Delta y) + \varepsilon_1\Delta x + \varepsilon_2\Delta y + i(\varepsilon_3\Delta x + \varepsilon_4\Delta y).$$

The Cauchy-Riemann equations can be used in equation (11) to obtain

$$\Delta w = u_x\Delta x - v_x\Delta y + i(v_x\Delta x + u_x\Delta y) + \varepsilon_1\Delta x + \varepsilon_2\Delta y + i(\varepsilon_3\Delta x + \varepsilon_4\Delta y).$$

Now the terms can be rearranged to yield

(12) $\Delta w = u_x[\Delta x + i\Delta y] + iv_x[\Delta x + i\Delta y] + \varepsilon_1\Delta x + \varepsilon_2\Delta y + i(\varepsilon_3\Delta x + \varepsilon_4\Delta y).$

Since $\Delta z = \Delta x + i\Delta y$, we can divide both sides of equation (12) by Δz and take the limit as $\Delta z \to 0$:

(13) $\displaystyle \lim_{\Delta z \to 0} \frac{\Delta w}{\Delta z} = u_x + iv_x + \lim_{\Delta z \to 0} \left[\frac{\varepsilon_1\Delta x}{\Delta z} + \frac{\varepsilon_2\Delta y}{\Delta z} + i\frac{\varepsilon_3\Delta x}{\Delta z} + i\frac{\varepsilon_4\Delta y}{\Delta z} \right].$

Using the property of ε_1 mentioned in equation (6), we have

$$\lim_{\Delta z \to 0} \left| \frac{\varepsilon_1\Delta x}{\Delta z} \right| = \lim_{\Delta z \to 0} |\varepsilon_1| \left| \frac{\Delta x}{\Delta z} \right| \le \lim_{\Delta z \to 0} |\varepsilon_1| = 0.$$

Similarly, the limits of the other quantities in equation (13) involving ε_2, ε_3, ε_4 are zero. Therefore the limit in equation (13) becomes

$$\lim_{\Delta z \to 0} \frac{\Delta w}{\Delta z} = f'(z_0) = u_x(x_0, y_0) + iv_x(x_0, y_0),$$

and the proof of the theorem is complete.

EXAMPLE 3.6   The function $f(z) = e^{-y}\cos x + ie^{-y}\sin x$ is differentiable for all z, and its derivative is $f'(z) = -e^{-y}\sin x + ie^{-y}\cos x$. To show this, we first write $u(x, y) = e^{-y}\cos x$ and $v(x, y) = e^{-y}\sin x$ and compute the partial derivatives:

$u_x(x, y) = v_y(x, y) = -e^{-y}\sin x$ and
$v_x(x, y) = -u_y(x, y) = e^{-y}\cos x.$

We see that u, v, u_x, u_y, v_x, and v_y are all continuous functions and that the Cauchy-Riemann equations hold for all values of (x, y). Hence, using equation (2), we write

$$f'(z) = u_x(x, y) + iv_x(x, y) = -e^{-y}\sin x + ie^{-y}\cos x.$$

The Cauchy-Riemann conditions are particularly useful in determining the set of points for which a function f is differentiable.

EXAMPLE 3.7   The function $f(z) = x^3 + 3xy^2 + i(y^3 + 3x^2y)$ is differentiable only at points that lie on the coordinate axes.

Solution   To show this, we write $u(x, y) = x^3 + 3xy^2$ and $v(x, y) = y^3 + 3x^2y$ and compute the partial derivatives:

$u_x(x, y) = 3x^2 + 3y^2,$ $v_y(x, y) = 3x^2 + 3y^2,$
$u_y(x, y) = 6xy,$ $v_x(x, y) = 6xy.$

Here u, v, u_x, u_y, v_x, and v_y are all continuous, and $u_x(x, y) = v_y(x, y)$ holds for all (x, y). But $u_y(x, y) = -v_x(x, y)$ if and only if $6xy = -6xy$, which is equivalent to $12xy = 0$. Therefore the Cauchy-Riemann equations hold only when $x = 0$ or $y = 0$, and according to Theorem 3.3, f is differentiable only at points that lie on the coordinate axes.

When polar coordinates (r, θ) are used to locate points in the plane, it is convenient to use expression (5) of Section 2.1 for a complex function; that is,

$$f(z) = f(re^{i\theta}) = u(r, \theta) + iv(r, \theta).$$

In this case, u and v are real functions of the real variables r and θ. The polar form of the Cauchy-Riemann equations and a formula for finding $f'(z)$ in terms of the partial derivatives of $u(r, \theta)$ and $v(r, \theta)$ are given in the following result which is proved in Exercise 13.

Theorem 3.4 (Polar Form)   *Let $f(z) = u(r, \theta) + iv(r, \theta)$ be a continuous function that is defined in some neighborhood of the point $z_0 = r_0 e^{i\theta_0}$. If all the partial derivatives u_r, u_θ, v_r, and v_θ are continuous at the point (r_0, θ_0) and if the Cauchy-Riemann equations*

(14) $u_r(r_0, \theta_0) = \dfrac{1}{r_0} v_\theta(r_0, \theta_0)$ *and* $v_r(r_0, \theta_0) = \dfrac{-1}{r_0} u_\theta(r_0, \theta_0)$

hold, then f is differentiable at z_0, and the derivative $f'(z_0)$ can be computed by either of the following formulas:

(15) $f'(z_0) = e^{-i\theta_0}[u_r(r_0, \theta_0) + iv_r(r_0, \theta_0)]$ *or*

(16) $f'(z_0) = \dfrac{1}{r_0} e^{-i\theta_0}[v_\theta(r_0, \theta_0) - iu_\theta(r_0, \theta_0)].$

EXAMPLE 3.8   Show that if f is given by

$$f(z) = z^{1/2} = r^{1/2}\cos\frac{\theta}{2} + ir^{1/2}\sin\frac{\theta}{2},$$

where the domain is restricted to be $r > 0$ and $-\pi < \theta < \pi$, then the derivative is given by

$$f'(z) = \frac{1}{2z^{1/2}} = \frac{1}{2}r^{-1/2}\cos\frac{\theta}{2} - i\frac{1}{2}r^{-1/2}\sin\frac{\theta}{2},$$

where $r > 0$ and $-\pi < \theta < \pi$.

Solution   To show this, we write

$$u(r, \theta) = r^{1/2}\cos\frac{\theta}{2} \quad \text{and} \quad v(r, \theta) = r^{1/2}\sin\frac{\theta}{2}.$$

Here,

$$u_r(r, \theta) = \frac{1}{r} v_\theta(r, \theta) = \frac{1}{2} r^{-1/2} \cos \frac{\theta}{2} \quad \text{and}$$

$$v_r(r, \theta) = \frac{-1}{r} u_\theta(r, \theta) = \frac{1}{2} r^{-1/2} \sin \frac{\theta}{2}.$$

Using these results in equation (15), we obtain

$$f'(z_0) = e^{-i\theta} \left(\frac{1}{2} r^{-1/2} \cos \frac{\theta}{2} + i \frac{1}{2} r^{-1/2} \sin \frac{\theta}{2} \right)$$

$$= e^{-i\theta} \left(\frac{1}{2} r^{-1/2} e^{i\theta/2} \right) = \frac{1}{2} r^{-1/2} e^{-i\theta/2} = \frac{1}{2z^{1/2}}.$$

EXERCISES FOR SECTION 3.2

1. Use the Cauchy-Riemann conditions to show that the following functions are differentiable for all z, and find $f'(z)$.
   (a) $f(z) = iz + 4i$ (b) $f(z) = z^3$
   (c) $f(z) = -2(xy + x) + i(x^2 - 2y - y^2)$
2. Let $f(z) = e^x \cos y + i e^x \sin y$. Show that both $f(z)$ and $f'(z)$ are differentiable for all z.
3. Find the constants a and b such that $f(z) = (2x - y) + i(ax + by)$ is differentiable for all z.
4. Show that $f(z) = (y + ix)/(x^2 + y^2)$ is differentiable for all $z \neq 0$.
5. Show that $f(z) = e^{2xy}[\cos(y^2 - x^2) + i \sin(y^2 - x^2)]$ is differentiable for all z.
6. Use the Cauchy-Riemann conditions to show that the following functions are nowhere differentiable.
   (a) $f(z) = \bar{z}$ (b) $g(z) = z + \bar{z}$
   (c) $h(z) = e^x \cos x + i e^y \sin x$
7. Let $f(z) = |z|^2$. Show that f is differentiable at the point $z_0 = 0$ but is not differentiable at any other point.
8. Show that the function $f(z) = x^2 + y^2 + i2xy$ has a derivative only at points that lie on the x axis.
9. Let f be a differentiable function. Establish the identity $|f'(z)|^2 = u_x^2 + v_x^2 = u_y^2 + v_y^2$.
10. Let $f(z) = (\ln r)^2 - \theta^2 + i2\theta \ln r$ where $r > 0$ and $-\pi < \theta \leq \pi$. Show that f is differentiable for $r > 0$, $-\pi < \theta < \pi$, and find $f'(z)$.
11. Let f be differentiable at $z_0 = r_0 e^{i\theta_0}$. Let z approach z_0 along the ray $r > 0$, $\theta = \theta_0$, and use definition (1) of Section 3.1 to show that equation (15) of Section 3.2 holds.
12. A vector field $\mathbf{F}(z) = U(x, y) + iV(x, y)$ is said to be *irrotational* if $U_y(x, y) = V_x(x, y)$. It is said to be *solenoidal* if $U_x(x, y) = -V_y(x, y)$. If $f(z)$ is an analytic function, show that $\mathbf{F}(z) = \overline{f(z)}$ is both irrotational and solenoidal.
13. The polar form of the Cauchy-Riemann equations.
   (a) Use the coordinate transformation

   $$x = r \cos \theta \quad \text{and} \quad y = r \sin \theta$$

   and the chain rules

   $$u_r = u_x \frac{\partial x}{\partial r} + u_y \frac{\partial y}{\partial r} \quad \text{and} \quad u_\theta = u_x \frac{\partial x}{\partial \theta} + u_y \frac{\partial y}{\partial \theta} \quad \text{etc.}$$

to prove that

$$u_r = u_x\cos\theta + u_y\sin\theta \quad \text{and} \quad u_\theta = -u_x r\sin\theta + u_y r\cos\theta \quad \text{and}$$
$$v_r = v_x\cos\theta + v_y\sin\theta \quad \text{and} \quad v_\theta = -v_x r\sin\theta + v_y r\cos\theta.$$

(b) Use the results of part (a) to prove that

$$r u_r = v_\theta \quad \text{and} \quad r v_r = -u_\theta.$$

14. Explain how the limit definition for derivative in complex analysis and the limit definition for the derivative in calculus are different. How are they similar?

15. Write a report on Cauchy-Riemann equations and the other conditions that guarantee that $f(z)$ is analytic. Resources include bibliographical items 21, 39, 62, 72, 86, 155, and 161.

3.3   Analytic Functions and Harmonic Functions

It is seldom of interest to study functions that are differentiable at only a single point. Complex functions that have a derivative at all points in a neighborhood of z_0 deserve further study. In Chapter 7 we will learn that if the complex function f can be represented by a Taylor series at z_0, then it must be differentiable in some neighborhood of z_0. The function f is said to be *analytic at* z_0 if its derivative exists at each point z in some neighborhood of z_0. If f is analytic at each point in the region R, then we say that f is analytic on R. If f is analytic on the whole complex plane, then f is said to be *entire*.

Points of nonanalyticity are called *singular points*. They are important for certain applications in physics and engineering.

EXAMPLE 3.9   The function $f(z) = x^2 + y^2 + i2xy$ is nowhere analytic.

Solution   We identify the functions $u(x, y) = x^2 + y^2$ and $v(x, y) = 2xy$. The equation $u_x = v_y$ becomes $2x = 2x$, which holds everywhere. But the equation $u_y = -v_x$ becomes $2y = -2y$, which holds only when $y = 0$. Thus $f(x)$ is differentiable only at points that lie on the x axis. However, for any point $z_0 = x_0 + 0i$ on the x axis and any δ-neighborhood of z_0, the point $z_1 = x_0 + i\delta/2$ is a point where f is not differentiable. Therefore f is not differentiable in any full neighborhood of z_0, and consequently it is not analytic at z_0.

We have seen that polynomial functions have derivatives at all points in the complex plane; hence polynomials are entire functions. The function $f(z) = e^x\cos y + ie^x\sin y$ has a derivative at all points z, and it is an entire function.

The results in Section 3.2 show that an analytic function must be continuous and must satisfy the Cauchy-Riemann equations. Conversely, if the Cauchy-Riemann conditions hold at all points in a neighborhood of z_0, then f is analytic at z_0. Using properties of derivatives, we see that the sum, difference, and product of two analytic functions are analytic functions. Similarly, the quotient of two analytic functions is analytic, provided that the function in the denominator is not zero. The chain rule can be used to show that the composition $g(f(z))$ of two analytic functions f and g is analytic, provided that g is analytic in a domain that contains the range of f.

The function $f(z) = 1/z$ is analytic for all $z \neq 0$; and if $P(z)$ and $Q(z)$ are polynomials, then their quotient $P(z)/Q(z)$ is analytic at all points where $Q(z) \neq 0$. The square root function is more complicated. If

$$(1) \quad f(z) = z^{1/2} = r^{1/2}\cos\frac{\theta}{2} + ir^{1/2}\sin\frac{\theta}{2}, \quad \text{where } r > 0 \quad \text{and} \quad -\pi < \theta \leq \pi,$$

then f is analytic at all points except $z_0 = 0$ and except at points that lie along the negative x axis. The function $f(z) = z^{1/2}$ defined by equation (1) is not continuous at points that lie along the negative x axis, and for this reason it is not analytic there.

Let $\phi(x, y)$ be a real-valued function of the two real variables x and y. The partial differential equation

$$(2) \quad \phi_{xx}(x, y) + \phi_{yy}(x, y) = 0$$

is known as *Laplace's equation* and is sometimes referred to as the potential equation. If ϕ, ϕ_x, ϕ_y, ϕ_{xx}, ϕ_{xy}, ϕ_{yx}, and ϕ_{yy} are all continuous and if $\phi(x, y)$ satisfies Laplace's equation, then $\phi(x, y)$ is called a *harmonic function*. Harmonic functions are important in the areas of applied mathematics, engineering, and mathematical physics. They are used to solve problems involving steady state temperatures, two-dimensional electrostatics, and ideal fluid flow. An important result for our studies is the fact that if $f(z) = u(x, y) + iv(x, y)$ is an analytic function, then both u and v are harmonic functions. In Chapter 10 we will see how complex variable techniques can be used to solve some problems involving harmonic functions.

Theorem 3.5   *Let $f(z) = u(x, y) + iv(x, y)$ be an analytic function in the domain D. If all second-order partial derivatives of u and v are continuous, then both u and v are harmonic functions in D.*

Proof   Since f is analytic, u and v satisfy the Cauchy-Riemann equations

$$(3) \quad u_x = v_y \quad \text{and} \quad u_y = -v_x.$$

If we differentiate both sides of equations (3) with respect to x, we obtain

$$(4) \quad u_{xx} = v_{yx} \quad \text{and} \quad u_{yx} = -v_{xx}.$$

Similarly, if we differentiate both sides of equations (3) with respect to y, then we obtain

$$(5) \quad u_{xy} = v_{yy} \quad \text{and} \quad u_{yy} = -v_{xy}.$$

Since the partial derivatives u_{xy}, u_{yx}, v_{xy}, and v_{yx} are all continuous, a theorem from the calculus of real functions states that the mixed partial derivatives are equal; that is,

$$(6) \quad u_{xy} = u_{yx} \quad \text{and} \quad v_{xy} = v_{yx}.$$

If we use equations (4), (5), and (6), then it follows that $u_{xx} + u_{yy} = v_{yx} - v_{xy} = 0$, and $v_{xx} + v_{yy} = -u_{yx} + u_{xy} = 0$. Therefore both u and v are harmonic functions.

Remark for Theorem 3.5   Corollary 6.2 in Chapter 6 will show that if $f(z)$ is analytic, then all the partial derivatives of u and v are continuous. Hence Theorem 3.5 holds for all analytic functions.

On the other hand, if we are given a function $u(x, y)$ that is harmonic in the domain D and if we can find another harmonic function $v(x, y)$, where their first partial derivatives satisfy the Cauchy-Riemann equations throughout D, then we say that $v(x, y)$ is the *harmonic conjugate* of $u(x, y)$. It then follows that the function $f(z) = u(x, y) + iv(x, y)$ is analytic in D.

EXAMPLE 3.10   If $u(x, y) = x^2 - y^2$, then $u_{xx}(x, y) + u_{yy}(x, y) = 2 - 2 = 0$; hence u is a harmonic function. We find that $v(x, y) = 2xy$ is also a harmonic function and that

$$u_x = v_y = 2x \quad \text{and} \quad u_y = -v_x = -2y.$$

Therefore v is the harmonic conjugate of u, and the function f given by

$$f(z) = x^2 - y^2 + i2xy = z^2$$

is an analytic function.

Harmonic functions are easily constructed from known analytic functions.

EXAMPLE 3.11   The function $f(z) = z^3 = x^3 - 3xy^2 + i(3x^2y - y^3)$ is analytic for all values of z, hence it follows that

$$u(x, y) = \text{Re}[f(z)] = x^3 - 3xy^2$$

is harmonic, and

$$v(x, y) = \text{Im}[f(z)] = 3x^2y - y^3$$

is the harmonic conjugate of $u(x, y)$. Their graphs are given in Figures 3.2 and 3.3. The partial derivatives are $u_x(x, y) = 3x^2 - 3y^2$, $u_y(x, y) = -6xy$, $v_x(x, y) = 6xy$, and $v_y(x, y) = 3x^2 - 3y^2$, and are easily shown to satisfy the Cauchy-Riemann equations. At the point $(x, y) = (2, -1)$, we have $u_x(2, -1) = v_y(2, -1) = 9$, and these partial derivatives can be seen along the edges of the surfaces for u and v where $x = 2$ and $y = -1$. Similarly, $u_y(2, -1) = 12$ and $v_x(2, -1) = -12$ can also be seen along the edges of the surfaces for u and v where $x = 2$ and $y = -1$.

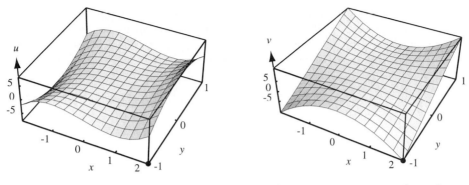

FIGURE 3.2   Graph of $u(x, y) = x^3 - 3xy^2$.   **FIGURE 3.3** $v(x, y) = 3x^2y - y^3$.

Complex variable techniques can be used to show that certain combinations of harmonic functions are harmonic. For example, if v is the harmonic conjugate of u, then their product $\phi(x, y) = u(x, y)v(x, y)$ is a harmonic function. This can be verified directly by computing the partial derivatives and showing that equation (2) holds true, but the details are tedious.

If we use complex variable techniques, we can start with the fact that $f(z) = u(x, y) + iv(x, y)$ is an analytic function. Then we observe that the square of f is also an analytic function and is given by $[f(z)]^2 = [u(x, y)]^2 - [v(x, y)]^2 + i2u(x, y)v(x, y)$. Hence the imaginary part of f^2 is $2u(x, y)v(x, y)$ and is a harmonic function. Since a constant multiple of a harmonic function is harmonic, it follows that ϕ is harmonic. It is left as an exercise to show that if u_1 and u_2 are two harmonic functions that are not related in the preceding fashion, then their product need not be harmonic.

Theorem 3.6 (Construction of a Conjugate)   *Let $u(x, y)$ be harmonic in an ε-neighborhood of the point (x_0, y_0). Then there exists a conjugate harmonic function $v(x, y)$ defined in this neighborhood, and $f(z) = u(x, y) + iv(x, y)$ is an analytic function.*

Proof   The harmonic function u and its conjugate harmonic function v will satisfy the Cauchy-Riemann equations $u_x = v_y$ and $u_y = -v_x$. We can construct $v(x, y)$ in a two-step process. First integrate v_y (which is equal to u_x) with respect to y:

$$(7) \quad v(x, y) = \int u_x(x, y)dy + C(x),$$

where $C(x)$ is a function of x alone (that is, the partial derivative of $C(x)$ with respect to y is zero). Second, we are able to find $C'(x)$ by differentiating equation (7) with respect to x and replacing v_x with $-u_y$ on the left side:

$$(8) \quad -u_y(x, y) = \frac{d}{dx} \int u_x(x, y)dy + C'(x).$$

As a matter of concern, all terms except those involving x in equation (8) will cancel, and a formula for $C'(x)$ involving x alone will be revealed. Now elementary integration of the single-variable function $C'(x)$ can be used to discover $C(x)$.

This technique is a practical method for constructing $v(x, y)$. Notice that both $u_x(x, y)$ and $u_y(x, y)$ are used in the process.

EXAMPLE 3.12

Show that $u(x, y) = xy^3 - x^3y$ is a harmonic function and find the conjugate harmonic function $v(x, y)$.

Solution The first partial derivatives are

(9) $u_x(x, y) = y^3 - 3x^2y$ and $u_y(x, y) = 3xy^2 - x^3$.

To verify that u is harmonic, we use the second partial derivatives and see that $u_{xx}(x, y) + u_{yy}(x, y) = -6xy + 6xy = 0$, which implies that u is harmonic. To construct $v(x, y)$, we start with equation (7) and the first of equations (9) to get

(10) $v(x, y) = \displaystyle\int (y^3 - 3x^2y)dy + C(x) = \frac{1}{4}y^4 - \frac{3}{2}x^2y^2 + C(x)$.

Differentiate the left and right sides of equation (10) with respect to x and use $-u_y(x, y) = v_x(x, y)$ and equations (9) on the left side to get

(11) $-3xy^2 + x^3 = 0 - 3xy^2 + C'(x)$.

Cancel the terms involving both x and y in equation (11) and discover that

(12) $C'(x) = x^3$.

Integrate equation (12) and get $C(x) = \frac{1}{4}x^4 + C$, where C is a constant. Hence the harmonic conjugate of u is

$$v(x, y) = \frac{1}{4}x^4 - \frac{3}{2}x^2y^2 + \frac{1}{4}y^4 + C.$$

EXAMPLE 3.13

Let f be an analytic function in the domain D. If $|f(z)| = K$ where K is a constant, then f is constant in D.

Solution Suppose that $K = 0$. Then $|f(z)|^2 = 0$, and hence $u^2 + v^2 = 0$. It follows that both $u \equiv 0$ and $v \equiv 0$, and therefore $f(z) \equiv 0$ in D.

Now suppose that $K \neq 0$; then we can differentiate the equation $u^2 + v^2 = K^2$ partially with respect to x and then with respect to y to obtain the system of equations

(13) $2uu_x + 2vv_x = 0$ and $2uu_y + 2vv_y = 0$.

The Cauchy-Riemann equations can be used in equations (13) to express the system in the form

(14) $uu_x - vu_y = 0$ and $vu_x + uu_y = 0$.

Treating u and v as coefficients, we easily solve equations (14) for the unknowns u_x and u_y:

$$u_x = \frac{\begin{vmatrix} 0 & -v \\ 0 & u \end{vmatrix}}{\begin{vmatrix} u & -v \\ v & u \end{vmatrix}} = \frac{0}{u^2 + v^2} = 0 \quad \text{and}$$

$$u_y = \frac{\begin{vmatrix} u & 0 \\ v & 0 \end{vmatrix}}{\begin{vmatrix} u & -v \\ v & u \end{vmatrix}} = \frac{0}{u^2 + v^2} = 0.$$

A theorem from the calculus of real functions states that the conditions $u_x \equiv 0$ and $u_y \equiv 0$ together imply that $u(x, y) \equiv c_1$ where c_1 is a constant. Similarly, we find that $v(x, y) \equiv c_2$, and therefore $f(z) \equiv c_1 + ic_2$.

Harmonic functions are solutions to many physical problems. Applications include two-dimensional models of heat flow, electrostatics, and fluid flow. For example, let us see how harmonic functions are used to study fluid flows. We must assume that an incompressible and frictionless fluid flows over the complex plane and that all cross sections in planes parallel to the complex plane are the same. Situations such as this occur when fluid is flowing in a deep channel. The velocity vector at the point (x, y) is

(15) $\mathbf{V}(x, y) = p(x, y) + iq(x, y)$

and is illustrated in Figure 3.4.

FIGURE 3.4   The vector field $\mathbf{V}(x, y) = p(x, y) + iq(x, y)$, which can be considered as a fluid flow.

The assumptions that the flow is irrotational and has no sources or sinks implies that both the curl and divergence vanish, that is, $q_x - p_y = 0$ and $p_x + q_y = 0$. Hence p and q obey the equations

(16) $p_x(x, y) = -q_y(x, y)$ and $p_y(x, y) = q_x(x, y)$.

Equations (16) are similar to the Cauchy-Riemann equations and permit us to define a special complex function:

(17) $f(z) = u(x, y) + iv(x, y) = p(x, y) - iq(x, y)$.

Here we have $u_x = p_x$, $u_y = p_y$, $v_x = -q_x$, and $v_y = -q_y$. Now equations (16) can be used to obtain the Cauchy-Riemann equations for $f(z)$:

(18) $u_x(x, y) = p_x(x, y) = -q_y(x, y) = v_y(x, y)$,
$\quad\quad u_y(x, y) = p_y(x, y) = q_x(x, y) = -v_x(x, y)$.

Therefore the function $f(z)$ defined in equation (17) is analytic, and the fluid flow, equation (15), is the conjugate of an analytic function, that is,

(19) $\mathbf{V}(x, y) = \overline{f(z)}$.

In Chapter 6 we will prove that every analytic function $f(z)$ has an analytic antiderivative $F(z)$; hence we are justified to write

(20) $F(z) = \phi(x, y) + i\psi(x, y)$,   where $F'(z) = f(z)$.

Observe that $\phi(x, y)$ is a harmonic function. If we use the vector interpretation of a complex number, then the gradient of $\phi(x, y)$ can be written as follows:

(21)   grad $\phi(x, y) = \phi_x(x, y) + i\phi_y(x, y)$.

The Cauchy-Riemann equations applied to $F(z)$ give us $\phi_y = -\psi_x$, and equation (21) becomes

(22)   grad $\phi(x, y) = \phi_x(x, y) - i\psi_x(x, y) = \overline{\phi_x(x, y) + i\psi_x(x, y)}$.

Theorem 3.2 says that $\phi_x(x, y) + i\psi_x(x, y) = F'(z)$, which can be substituted in equation (22) to obtain

(23)   grad $\phi(x, y) = \overline{F'(z)}$.

Now use $F'(z) = f(z)$ in equation (23) to conclude that $\phi(x, y)$ is the scalar potential function for the fluid flow in equation (19), that is,

(24) $\mathbf{V}(x, y) = $ grad $\phi(x, y)$.

The curves $\phi(x, y) = $ constant are called *equipotentials*. The curves $\psi(x, y) = $ constant are called *streamlines* and describe paths of fluid flow. In Chapter 10 we will see that the family of equipotentials is orthogonal to the family of streamlines (see Figure 3.5).

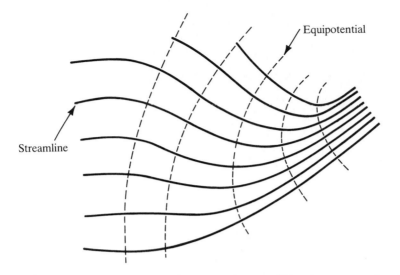

FIGURE 3.5   The families of orthogonal curves $\{\phi(x, y) = \text{constant}\}$ and $\{\psi(x, y) = \text{constant}\}$ for the function $F(z) = \phi(x, y) + i\psi(x, y)$.

EXAMPLE 3.14   Show that the harmonic function $\phi(x, y) = x^2 - y^2$ is the scalar potential function for the fluid flow

$$\mathbf{V}(x, y) = 2x - i2y.$$

Solution   The fluid flow can be written as

$$\mathbf{V}(x, y) = \overline{f(z)} = \overline{2x + i2y} = \overline{2z}.$$

The antiderivative of $f(z) = 2z$ is $F(z) = z^2$, and the real part of $F(z)$ is the desired harmonic function:

$$\phi(x, y) = \text{Re}[F(z)] = \text{Re}[x^2 - y^2 + i2xy] = x^2 - y^2.$$

Observe that the hyperbolas $\phi(x, y) = x^2 - y^2 = C$ are the equipotential curves, and the hyperbolas $\psi(x, y) = 2xy = C$ are the streamline curves; these curves are orthogonal, as is shown in Figure 3.6.

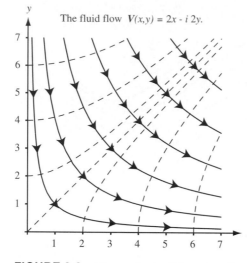

FIGURE 3.6   The equipotential curves $x^2 - y^2 = C$ and streamline curves $2xy = C$ for the function $F(z) = z^2$.

EXERCISES FOR SECTION 3.3

1. Show that the following functions are entire.
   (a) $f(z) = \cosh x \sin y - i \sinh x \cos y$ (b) $g(z) = \cosh x \cos y + i \sinh x \sin y$
2. State why the composition of two entire functions is an entire function.
3. Determine where $f(z) = x^3 + 3xy^2 + i(y^3 + 3x^2y)$ is differentiable. Is f analytic? Why?
4. Determine where $f(z) = 8x - x^3 - xy^2 + i(x^2y + y^3 - 8y)$ is differentiable. Is f analytic? Why?
5. Let $f(z) = x^2 - y^2 + i2|xy|$.
   (a) Where does f have a derivative?   (b) Where is f analytic?
6. Show that $u(x, y) = e^x\cos y$ and $v(x, y) = e^x\sin y$ are harmonic for all values of (x, y).
7. Let $u(x, y) = \ln(x^2 + y^2)$ for $(x, y) \neq (0, 0)$. Compute the partial derivatives of u, and verify that u satisfies Laplace's equation.
8. Let a, b, and c be real constants. Determine a relation among the coefficients that will guarantee that the functions $\phi(x, y) = ax^2 + bxy + cy^2$ is harmonic.
9. Does an analytic function $f(z) = u(x, y) + iv(x, y)$ exist for which $v(x, y) = x^3 + y^3$? Why?
10. Find the analytic function $f(z) = u(x, y) + iv(x, y)$ given the following.
    (a) $u(x, y) = y^3 - 3x^2y$ (b) $u(x, y) = \sin y \sinh x$
    (c) $v(x, y) = e^y\sin x$ (d) $v(x, y) = \sin x \cosh y$
11. Let $v(x, y) = \arctan(y/x)$ for $x \neq 0$. Compute the partial derivatives of v, and verify that v satisfies Laplace's equation.
12. Let $u(x, y)$ be harmonic. Show that $U(x, y) = u(x, -y)$ is harmonic. *Hint:* Use the chain rule for differentiation of real functions.

13. Let $u_1(x, y) = x^2 - y^2$ and $u_2(x, y) = x^3 - 3xy^2$. Show that u_1 and u_2 are harmonic functions and that their product $u_1(x, y)u_2(x, y)$ is not a harmonic function.

14. Let v be the harmonic conjugate of u. Show that $-u$ is the harmonic conjugate of v.

15. Let v be the harmonic conjugate of u. Show that $h = u^2 - v^2$ is a harmonic function.

16. Suppose that v is the harmonic conjugate of u and that u is the harmonic conjugate of v. Show that u and v must be constant functions.

17. Let f be an analytic function in the domain D. If $f'(z) = 0$ for all z in D, then show that f is constant in D.

18. Let f and g be analytic functions in the domain D. If $f'(z) = g'(z)$ for all z in D, then show that $f(z) = g(z) + C$, where C is a complex constant.

19. Let f be a nonconstant analytic function in the domain D. Show that the function $g(z) = \overline{f(z)}$ is *not* analytic in D.

20. Let $f(z) = f(re^{i\theta}) = \ln r + i\theta$ where $r > 0$ and $-\pi < \theta < \pi$. Show that f is analytic in the domain indicated and that $f'(z) = 1/z$.

21. Let $f(z) = f(re^{i\theta}) = u(r, \theta) + iv(r, \theta)$ be analytic in a domain D that does not contain the origin. Use the polar form of the Cauchy-Riemann equations $u_\theta = -rv_r$, and $v_\theta = ru_r$, and differentiate them with respect to θ and then with respect to r. Use the results to establish the *polar form of Laplace's equation:*

$$r^2 u_{rr}(r, \theta) + r u_r(r, \theta) + u_{\theta\theta}(r, \theta) = 0.$$

22. Use the polar form of Laplace's equation given in Exercise 21 to show that $u(r, \theta) = r^n \cos n\theta$ and $v(r, \theta) = r^n \sin n\theta$ are harmonic functions.

23. Use the polar form of Laplace's equation given in Exercise 21 to show that

$$u(r, \theta) = \left(r + \frac{1}{r}\right)\cos\theta \quad \text{and} \quad v(r, \theta) = \left(r - \frac{1}{r}\right)\sin\theta$$

are harmonic functions.

24. Let f be an analytic function in the domain D. Show that if $\text{Re}[f(z)] \equiv 0$ at all points in D, then f is constant in D.

25. Assume that $F(z) = \phi(x, y) + i\psi(x, y)$ is analytic in the domain D and that $F'(z) \neq 0$ in D. Consider the families of level curves $\{\phi(x, y) = \text{constant}\}$ and $\{\psi(x, y) = \text{constant}\}$, which are the equipotentials and streamlines for the fluid flow $\mathbf{V}(x, y) = \overline{F'(z)}$. Prove that the two families of curves are orthogonal. *Hint:* Suppose that (x_0, y_0) is a point common to the two curves $\phi(x, y) = c_1$ and $\psi(x, y) = c_2$. Take the gradient of ϕ and ψ, and show that the normals to the curves are perpendicular.

26. The function $F(z) = 1/z$ is used to determine a field known as a dipole. Express $F(z)$ in the form $F(z) = \phi(x, y) + i\psi(x, y)$ and sketch the equipotentials $\phi = 1, 1/2, 1/4$ and the streamlines $\psi = 1, 1/2, 1/4$.

27. The logarithmic function will be introduced in Chapter 5. Let $F(z) = \log z = \ln|z| + i \arg z$. Here we have $\phi(x, y) = \ln|z|$ and $\psi(x, y) = \arg z$. Sketch the equipotentials $\phi = 0, \ln 2, \ln 3, \ln 4$ and the streamlines $\psi = k\pi/8$ for $k = 0, 1, \ldots, 7$.

28. Discuss and compare the statements "$f(z)$ is analytic" and "$f(z)$ is differentiable."

29. Discuss and compare the statements "$u(x, y)$ is harmonic" and "$u(x, y)$ is the imaginary part of an analytic function."

30. Write a report on analytic functions. Include a discussion of the Cauchy-Riemann equations and the other conditions that guarantee that $f(z)$ is analytic. Resources include bibliographical items 21, 39, 62, 72, 86, 155, and 161.

31. Write a report on harmonic functions. Include ideas and examples that are not mentioned in the text. Resources include bibliographical items 2, 14, 28, 61, 69, 70, 71, 76, 77, 85, 98, 111, 113, 131, 135, 138, 158, and 165.

32. Write a report on how computer graphics are used for graphing harmonic functions and complex functions and conformal mappings. Resources include bibliographical items 33, 34, 109, and 146.

33. Write a report on fluid flow and how it is related to harmonic and analytic functions. Include some ideas not mentioned in the text. Resources include bibliographical items 37, 46, 91, 98, 124, 141, 145, 158, and 166.

34. Write a report on the Polya vector field. Resources include bibliographical items 25, 26, 27, and 83.

4

Sequences, Series, and Julia and Mandelbrot Sets

In this chapter we learn the basics for complex sequences and series. We also explore an application of these ideas in what has popularly come to be known as *chaotic processes*.

4.1 Definitions and Basic Theorems for Sequences and Series

In formal terms, a *complex sequence* is a function whose domain is the positive integers and whose range is a subset of the complex numbers. The following are examples of sequences:

(1) $\quad f(n) = \left(2 - \dfrac{1}{n}\right) + \left(5 + \dfrac{1}{n}\right)i \qquad (n = 1, 2, 3, \ldots),$

(2) $\quad g(n) = e^{i(\pi n/4)} \qquad (n = 1, 2, 3, \ldots),$

(3) $\quad h(n) = 5 + 3i + \left(\dfrac{1}{1+i}\right)^{n} \qquad (n = 1, 2, 3, \ldots),$

(4) $\quad r(n) = \left(\dfrac{1}{4} + \dfrac{i}{2}\right)^{n} \qquad (n = 1, 2, 3, \ldots).$

For convenience, we at times use the term *sequence* rather than *complex sequence*. If we wish a function s to represent an arbitrary sequence, we could specify it by writing $s(1) = z_1$, $s(2) = z_2$, $s(3) = z_3$, and so on. The values $z_1, z_2, z_3, \ldots$, are called the *terms* of a sequence, and mathematicians, being generally lazy when it comes to things like this, often refer to z_1, z_2, z_3, etc., as the sequence itself, even though they are really speaking of the range of the sequence when they do this. You will usually see a sequence written as $\{z_n\}_{n=1}^{\infty}$, $\{z_n\}_{1}^{\infty}$, or, when the indices are understood, as $\{z_n\}$. Mathematicians are also not so fussy about starting a sequence at z_1, so that $\{z_n\}_{n=-1}^{\infty}$, $\{z_n\}_{n=0}^{\infty}$, etc., would also be acceptable notation, provided all terms were defined. For example, the sequence r given by equation (4) could be written in a variety of ways:

$$\left\{\left(\frac{1}{4}+\frac{i}{2}\right)^n\right\}_{n=1}^\infty, \left\{\left(\frac{1}{4}+\frac{i}{2}\right)^n\right\}_1^\infty, \left\{\left(\frac{1}{4}+\frac{i}{2}\right)^n\right\}, \left\{\left(\frac{1}{4}+\frac{i}{2}\right)^{n+3}\right\}_{n=-2}^\infty,$$

$$\left\{\left(\frac{1}{4}+\frac{i}{2}\right)^j\right\}_{j=1}^\infty, \text{ etc.}$$

The sequences f and g given by equations (1) and (2) behave differently as n gets larger and larger. The terms in equation (1) approach $2 + 5i = (2, 5)$, while those in equation (2) do not approach any one particular number, as they simply oscillate around the eight eighth roots of unity on the unit circle. Informally, the sequence $\{z_n\}_1^\infty$ has ζ as its limit as n approaches infinity, provided the terms z_n can be made as close as we please to ζ by making n large enough. When this happens, we write

(5) $\lim_{n\to\infty} z_n = \zeta$,   or $z_n \to \zeta$ as $n \to \infty$.

If $\lim_{n\to\infty} z_n = \zeta$, we say that the sequence $\{z_n\}_1^\infty$ *converges* to ζ.

We need a rigorous definition for statement (5), however, if we are to do honest mathematics. Thus, we have the following.

Definition 4.1 $\lim_{n\to\infty} z_n = \zeta$ *means that for any real number $\varepsilon > 0$ there corresponds a positive integer N_ε (which depends on ε) such that $z_n \in D_\varepsilon(\zeta)$ whenever $n > N_\varepsilon$.*

Note: The reason we use the notation N_ε is to emphasize the fact that this number depends on our choice of ε. Sometimes it will be convenient to drop the subscript. Figure 4.1 illustrates a convergent sequence.

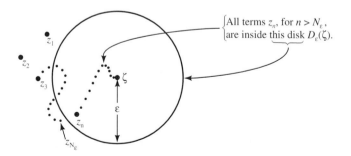

All terms z_n, for $n > N_\varepsilon$, are inside this disk $D_\varepsilon(\zeta)$.

FIGURE 4.1   A sequence that converges to ζ.

In form, Definition 4.1 is exactly the same as the corresponding definition for limits of real sequences. In fact, there is a simple criterion that casts the convergence of complex sequences in terms of the convergence of real sequences.

Theorem 4.1   *Let $z_n = x_n + iy_n$ and $\zeta = u + iv$. Then,*

(6) $\lim\limits_{n \to \infty} z_n = \zeta$ *if and only if*

(7) $\lim\limits_{n \to \infty} x_n = u$ *and*

(8) $\lim\limits_{n \to \infty} y_n = v$.

Proof   First we will assume statement (6) is true, and from this deduce the truth of statements (7) and (8). Let ε be an arbitrary positive real number. To establish statement (7), we must show that there is a positive integer N_ε such that the inequality $\left| x_n - u \right| < \varepsilon$ holds whenever $n > N_\varepsilon$. Since we are assuming statement (6) to be true, we know according to Definition 4.1 that there is a positive integer N_ε such that $z_n \in D_\varepsilon(\zeta)$ if $n > N_\varepsilon$. Recall that $z_n \in D_\varepsilon(\zeta)$ is equivalent to the inequality $\left| z_n - \zeta \right| < \varepsilon$. Thus, whenever $n > N_\varepsilon$, we have

$$
\begin{aligned}
\left| x_n - u \right| &= \left| \mathrm{Re}(z_n - \zeta) \right| \\
&\le \left| z_n - \zeta \right| \quad \text{(by inequality (2) of Section 1.3)} \\
&< \varepsilon,
\end{aligned}
$$

and this proves statement (7). In a similar way, it can be shown that statement (6) implies statement (8), and we leave this verification as an exercise.

To complete the proof of this theorem, we must show that statements (7) and (8) jointly imply statement (6). Let $\varepsilon > 0$ be an arbitrary real number. By statements (7) and (8) there exists positive integers N_ε and M_ε such that

(9) $\left| x_n - u \right| < \dfrac{\varepsilon}{2}$,   whenever $n > N_\varepsilon$,   and

(10) $\left| y_n - v \right| < \dfrac{\varepsilon}{2}$,   whenever $n > M_\varepsilon$.

Let $L_\varepsilon = \max\{N_\varepsilon, M_\varepsilon\}$. Then if $n > L_\varepsilon$, we see that

$$
\begin{aligned}
\left| z_n - \zeta \right| &= \left| (x_n + iy_n) - (u + iv) \right| \\
&= \left| (x_n - u) + i(y_n - v) \right| \\
&\le \left| (x_n - u) \right| + \left| i(y_n - v) \right| \quad \text{(What is the reason for this step?)} \\
&= \left| (x_n - u) \right| + \left| i \right| \left| (y_n - v) \right| \quad \text{(by properties of absolute value)} \\
&= \left| (x_n - u) \right| + \left| (y_n - v) \right| \quad \text{(since } \left| i \right| = 1) \\
&< \frac{\varepsilon}{2} + \frac{\varepsilon}{2} \quad \text{(by statements (9) and (10))} \\
&= \varepsilon.
\end{aligned}
$$

We needed to show the strict inequality $\left| z_n - \zeta \right| < \varepsilon$, and the next to the last line in the preceding proof gives us precisely that. Note also that we have been speaking of *the* limit of a sequence. Strictly speaking, we are not entitled to use this terminology, since we have not proved that a given complex sequence can have only one limit. The proof of this, however, is almost identical to the corresponding result for real sequences, and we have left it as an exercise.

EXAMPLE 4.1 Consider $z_n = [\sqrt{n} + i(n + 1)]/n$. Then we write

$$z_n = x_n + iy_n = \frac{1}{\sqrt{n}} + i\frac{n + 1}{n}.$$

Using results about sequences of real numbers, which are studied in calculus, we find that

$$\lim_{n\to\infty} x_n = \lim_{n\to\infty} \frac{1}{\sqrt{n}} = 0 \quad \text{and} \quad \lim_{n\to\infty} y_n = \lim_{n\to\infty} \frac{n + 1}{n} = 1.$$

Therefore

$$\lim_{n\to\infty} z_n = \lim_{n\to\infty} \frac{\sqrt{n} + i(n + 1)}{n} = i.$$

EXAMPLE 4.2 Let us show that $\{(1 + i)^n\}$ diverges. In this case, we have

$$z_n = (1 + i)^n = (\sqrt{2})^n \cos\frac{n\pi}{4} + i(\sqrt{2})^n \sin\frac{n\pi}{4}.$$

Since the real sequences $\{(\sqrt{2})^n \cos(n\pi/4)\}$ and $\{(\sqrt{2})^n \sin(n\pi/4)\}$ both diverge, we conclude that $\{(1 + i)^n\}$ diverges.

As is the case with the real numbers, we also have

Definition 4.2 *The sequence $\{z_n\}$ is said to be a Cauchy sequence if for every $\varepsilon > 0$ there exists a positive integer N_ε such that if $n, m > N_\varepsilon$, then $|z_n - z_m| < \varepsilon$, or, equivalently, $z_n - z_m$ belongs to the disk $D_\varepsilon(0)$.*

The following should now come as no surprise.

Theorem 4.2 *If $\{z_n\}$ is a Cauchy sequence, $\{z_n\}$ converges.*

Proof Let $z_n = x_n + iy_n$. Using the techniques of Theorem 4.1, it is easy to show that both $\{x_n\}$ and $\{y_n\}$ are Cauchy sequences of real numbers. Since Cauchy sequences of real numbers are convergent, we know that

$$\lim_{n\to\infty} x_n = x_0 \quad \text{and} \quad \lim_{n\to\infty} y_n = y_0$$

for some real numbers x_0 and y_0. By Theorem 4.1, this means

$$\lim_{n\to\infty} z_n = z_0,$$

where $z_0 = x_0 + iy_0$. In other words, the sequence $\{z_n\}$ converges to z_0.

Let $\{z_n\}$ be a complex sequence. We can form a new sequence $\{S_n\}$, called the sequence of *partial sums*, in the following way:

(11) $S_1 = z_1,$
$\qquad S_2 = z_1 + z_2,$

$\qquad\qquad$.
$\qquad\qquad$.
$\qquad\qquad$.

$$S_n = z_1 + z_2 + \cdots + z_n = \sum_{k=1}^{n} z_k,$$

$\qquad\qquad$.
$\qquad\qquad$.
$\qquad\qquad$.

The formal expression $\sum_{k=1}^{\infty} z_k = z_1 + z_2 + \cdots + z_n + \cdots$ is called an *infinite series*, and z_1, z_2, etc., are called the *terms* of the series. If there is a complex number S for which

$$(12) \quad S = \lim_{n \to \infty} S_n = \lim_{n \to \infty} \sum_{k=1}^{n} z_k,$$

we will say that the infinite series $\sum_{k=1}^{\infty} z_k$ *converges* to S, and that S is the *sum* of the infinite series. When this happens, we write

$$(13) \quad S = \sum_{k=1}^{\infty} z_k.$$

The series $\sum_{k=1}^{\infty} z_k$ is said to be *absolutely convergent* provided that the (real) series of magnitudes $\sum_{k=1}^{\infty} |z_k|$ converges. If a series does not converge, we say that it *diverges*.

It is important to note that the first finitely many terms of a series do not affect its convergence or divergence and that in this respect the beginning index of a series is irrelevant. Thus, we will without comment conclude that if a series $\sum_{k=N+1}^{\infty} z_k$ converges, then so does $\sum_{k=0}^{\infty} z_k$, where z_0, z_1, . . . , z_N is *any* finite collection of terms. A similar remark holds for determining divergence of a series.

As one might expect, many of the results concerning real series carry over to the complex case. We give several of the more standard theorems along with examples of how they are used.

Theorem 4.3   *Let $z_n = x_n + iy_n$ and $S = U + iV$. Then*

$$S = \sum_{n=1}^{\infty} z_n = \sum_{n=1}^{\infty} (x_n + iy_n)$$

if and only if both

$$U = \sum_{n=1}^{\infty} x_n \quad and \quad V = \sum_{n=1}^{\infty} y_n.$$

Proof   Let $U_n = \sum_{k=1}^{n} x_k$ and $V_n = \sum_{k=1}^{n} y_k$ and $S_n = U_n + iV_n$. We can use Theorem 4.1 to conclude that

$$\lim_{n\to\infty} S_n = \lim_{n\to\infty} (U_n + iV_n) = U + iV = S$$

if and only if both $\lim_{n\to\infty} U_n = U$ and $\lim_{n\to\infty} V_n = V$, and the completion of the proof follows easily from definitions (12) and (13).

Theorem 4.4   *If $\sum_{n=1}^{\infty} z_n$ is a convergent complex series, then $\lim_{n\to\infty} z_n = 0$.*

The proof of Theorem 4.4 is left as an exercise.

EXAMPLE 4.3   Show that the series

$$\sum_{n=1}^{\infty} \frac{1 + in(-1)^n}{n^2} = \sum_{n=1}^{\infty} \left[\frac{1}{n^2} + i\frac{(-1)^n}{n} \right]$$

is convergent.

Solution   From the calculus it is known that the series

$$\sum_{n=1}^{\infty} \frac{1}{n^2} \quad and \quad \sum_{n=1}^{\infty} \frac{(-1)^n}{n}$$

are convergent. Hence Theorem 4.3 implies that the given complex series is convergent.

EXAMPLE 4.4   The series

$$\sum_{n=1}^{\infty} \frac{(-1)^n + i}{n} = \sum_{n=1}^{\infty} \left[\frac{(-1)^n}{n} + i\frac{1}{n} \right]$$

is divergent.

Solution   From the study of calculus it is known that the series $\sum_{n=1}^{\infty} (1/n)$ is divergent. Hence Theorem 4.3 implies that the given complex series is divergent.

EXAMPLE 4.5   The series $\sum_{n=1}^{\infty} (1 + i)^n$ is divergent.

Solution Here we set $z_n = (1 + i)^n$, and we observe that $\lim_{n \to \infty} |z_n| = \lim_{n \to \infty} (\sqrt{2})^n = \infty$. Hence $\lim_{n \to \infty} z_n \neq 0$, and Theorem 4.4 implies that the given series is not convergent; hence it is divergent.

Theorem 4.5 *Let $\sum_{n=1}^{\infty} z_n$ and $\sum_{n=1}^{\infty} w_n$ be convergent series, and let c be a complex number. Then*

$$\sum_{n=1}^{\infty} cz_n = c\sum_{n=1}^{\infty} z_n \quad and$$

$$\sum_{n=1}^{\infty} (z_n + w_n) = \sum_{n=1}^{\infty} z_n + \sum_{n=1}^{\infty} w_n.$$

Proof The proof of this theorem is left as an exercise.

Definition 4.2 *Let $\sum_{n=0}^{\infty} a_n$ and $\sum_{n=0}^{\infty} b_n$ be convergent series, where a_n and b_n are complex numbers. The Cauchy product of the two series is defined to be the series $\sum_{n=0}^{\infty} c_n$, where $c_n = \sum_{k=0}^{n} a_k b_{n-k}$.*

Theorem 4.6 *If the Cauchy product converges, then*

$$\sum_{n=0}^{\infty} c_n = \left(\sum_{n=0}^{\infty} a_n\right)\left(\sum_{n=0}^{\infty} b_n\right).$$

Proof The proof can be found in a number of texts, for example, *Infinite Sequences and Series*, by Konrad Knopp (translated by Frederick Bagemihl; New York: Dover, 1956).

Theorem 4.7 (Comparison Test) *Let $\sum_{n=1}^{\infty} M_n$ be a convergent series of real nonnegative terms. If $\{z_n\}$ is a sequence of complex numbers and $|z_n| \leq M_n$ holds for all n, then*

$$\sum_{n=1}^{\infty} z_n = \sum_{n=1}^{\infty} (x_n + iy_n)$$

converges.

Proof Using equations (2) of Section 1.3, we see that $|x_n| \leq |z_n| \leq M_n$ and $|y_n| \leq |z_n| \leq M_n$ holds for all n. The comparison test for real sequences can be used to conclude that

$$\sum_{n=1}^{\infty} |x_n| \quad and \quad \sum_{n=1}^{\infty} |y_n|$$

are convergent. A result from calculus states that an absolutely convergent series is convergent. Hence

$$\sum_{n=1}^{\infty} x_n \quad \text{and} \quad \sum_{n=1}^{\infty} y_n$$

are convergent. We can use these results together with Theorem 4.3 to conclude that $\sum_{n=1}^{\infty} z_n = \sum_{n=1}^{\infty} x_n + i \sum_{n=1}^{\infty} y_n$ is convergent.

Corollary 4.1   *If $\sum_{n=0}^{\infty} z_n$ converges absolutely, then $\sum_{n=0}^{\infty} z_n$ converges.*

We leave the proof of this corollary as an exercise.

EXAMPLE 4.6   Show that $\sum_{n=1}^{\infty} (3 + 4i)^n/(5^n n^2)$ converges.

Solution   Calculating the modulus of the terms, we find that $\left| z_n \right| = \left| (3 + 4i)^n/(5^n n^2) \right| = 1/n^2 = M_n$. We can use the comparison test and the fact that $\sum_{n=1}^{\infty} (1/n^2)$ converges to conclude that $\sum_{n=1}^{\infty} (3 + 4i)^n/(5^n n^2)$ converges.

Suppose that we have a series $\sum_{n=0}^{\infty} z_n$, where $z_n = c_n(z - \alpha)^n$. If α and the collection of c_n are fixed complex numbers, we will get different series by selecting different values for z. For example, if $\alpha = 0$, and $c_n = \dfrac{1}{n!}$ for all n, we get the series

$$\sum_{n=0}^{\infty} \frac{1}{n!} \left(\frac{i}{2} \right)^n \text{ if } z = \frac{i}{2} \text{ and } \sum_{n=0}^{\infty} \frac{1}{n!} (4 + i)^n \text{ if } z = 4 + i.$$ The collection of points for which the series $\sum_{n=0}^{\infty} c_n(z - \alpha)^n$ converges will thus be the domain of a function

$f(z) = \sum_{n=0}^{\infty} c_n(z - \alpha)^n$, which is called a *power series function*. Technically, this series is undefined if $z = \alpha$ and $n = 0$, since 0^0 is undefined. We get around this difficulty by stipulating that the series $\sum_{n=0}^{\infty} c_n(z - \alpha)^n$ is really compact notation

for $c_0 + \sum_{n=1}^{\infty} c_n(z - \alpha)^n$.

If $\alpha = 0$ and $c_n = 1$ for all n in the preceding, our series becomes $\sum_{n=0}^{\infty} z^n$. We call this a *geometric series*, one of the most important series in mathematics.

Theorem 4.8 (Geometric Series)   *If $|z| < 1$, the series $\sum\limits_{n=0}^{\infty} z^n$ converges to $f(z) = \dfrac{1}{1-z}$. That is, if $|z| < 1$, then*

$$(14) \quad \sum_{n=0}^{\infty} z^n = 1 + z + z^2 + \cdots + z^k + \cdots = \frac{1}{1-z}.$$

If $|z| \geq 1$, the series diverges.

Proof   Suppose $|z| < 1$. According to equation (12), we must show $\lim\limits_{n\to\infty} S_n = \dfrac{1}{1-z}$, where

$$(15) \quad S_n = 1 + z + z^2 + \cdots + z^{n-1}.$$

Multiplying both sides of equation (15) by z gives

$$(16) \quad zS_n = z + z^2 + z^3 + \cdots + z^{n-1} + z^n.$$

Subtracting equation (16) from equation (15) yields

$$(17) \quad (1 - z)S_n = 1 - z^n,$$

so that

$$(18) \quad S_n = \frac{1}{1-z} - \frac{z^n}{1-z}.$$

Since $|z| < 1$, $\lim\limits_{n\to\infty} z^n = 0$. (Can you *prove* this? You will be asked to do so in the exercises!) Hence $\lim\limits_{n\to\infty} S_n = \dfrac{1}{1-z}$.

If $|z| \geq 1$, then clearly $\lim\limits_{n\to\infty} |z^n| \neq 0$. Hence $\lim\limits_{n\to\infty} z^n \neq 0$ (see problem 24), so by the contrapositive of Theorem 4.4, $\sum\limits_{n=0}^{\infty} z^n$ must diverge.

Corollary 4.2   *If $|z| > 1$, the series $\sum\limits_{n=1}^{\infty} z^{-n}$ converges to $f(z) = \dfrac{1}{z-1}$. That is, if $|z| > 1$, then*

$$(19) \quad \sum_{n=1}^{\infty} z^{-n} = z^{-1} + z^{-2} + \cdots + z^{-n} + \cdots = \frac{1}{z-1}, \quad \textit{or equivalently,}$$

$$(20) \quad -\sum_{n=1}^{\infty} z^{-n} = -z^{-1} - z^{-2} - \cdots - z^{-n} - \cdots = \frac{1}{1-z}.$$

If $|z| \leq 1$, the series diverges.

Proof   If we let $\dfrac{1}{z}$ take the role of z in equation (14), we get

(21) $\displaystyle\sum_{n=0}^{\infty}\left(\dfrac{1}{z}\right)^{n}=\dfrac{1}{1-\dfrac{1}{z}}$, if $\left|\dfrac{1}{z}\right|<1$.

Multiplying both sides of equation (21) by $\dfrac{1}{z}$ gives

(22) $\dfrac{1}{z}\displaystyle\sum_{n=0}^{\infty}\left(\dfrac{1}{z}\right)^{n}=\dfrac{1}{z-1}$,   if $\left|\dfrac{1}{z}\right|<1$,

which, by Theorem 4.5, is the same as

(23) $\displaystyle\sum_{n=0}^{\infty}\left(\dfrac{1}{z}\right)^{n+1}=\dfrac{1}{z-1}$,   if $\left|\dfrac{1}{z}\right|<1$.

But this is equivalent to saying that $\displaystyle\sum_{n=1}^{\infty}\left(\dfrac{1}{z}\right)^{n}=\dfrac{1}{z-1}$, if $1<|z|$, which is what the corollary claims.

It is left as an exercise to show that the series diverges if $|z|\le 1$.

Corollary 4.3   *If $z\ne 1$, then for all n*

$$\dfrac{1}{1-z}=1+z+z^{2}+\cdots+z^{n-1}+\dfrac{z^{n}}{1-z}.$$

Proof   This follows immediately from equation (18).

EXAMPLE 4.7   Show that $\sum_{n=0}^{\infty}[(1-i)^{n}/2^{n}]=1-i$.

Solution   If we set $z=(1-i)/2$, then we see that $|z|=\sqrt{2}/2<1$, so we can use representation (14) for a geometric series. The sum is given by

$$\dfrac{1}{1-\dfrac{1-i}{2}}=\dfrac{2}{2-1+i}=\dfrac{2}{1+i}=1-i.$$

EXAMPLE 4.8   Evaluate $\displaystyle\sum_{n=3}^{\infty}\left(\dfrac{i}{2}\right)^{n}$.

Solution   We can put this expression in the form of a geometric series:

(24) $$\sum_{n=3}^{\infty} \left(\frac{i}{2}\right)^n = \sum_{n=3}^{\infty} \left(\frac{i}{2}\right)^3 \left(\frac{i}{2}\right)^{n-3}$$

$$= \left(\frac{i}{2}\right)^3 \sum_{n=3}^{\infty} \left(\frac{i}{2}\right)^{n-3} \qquad \text{(by Theorem 4.5)}$$

$$= \left(\frac{i}{2}\right)^3 \sum_{n=0}^{\infty} \left(\frac{i}{2}\right)^n \qquad \text{(by reindexing)}$$

$$= \left(\frac{i}{2}\right)^3 \left(\frac{1}{1 - \dfrac{i}{2}}\right) \qquad \left(\text{by Theorem 4.8, since } \left|\frac{i}{2}\right| = \frac{1}{2} < 1\right)$$

$$= \frac{1}{20} - \frac{i}{10} \qquad \text{(by standard simplification procedures)}.$$

The equality given by equation (24) illustrates an important point when evaluating a geometric series whose beginning index is other than zero. The value of $\sum_{n=r}^{\infty} z^n$ will equal $\dfrac{z^r}{1 - z}$. If we think of z as the "ratio" by which a given term of the series is multiplied to generate successive terms, we see that the sum of a geometric series equals $\dfrac{\text{first term}}{1 - \text{ratio}}$, provided $|\text{ratio}| < 1$.

The geometric series is used in the proof of the following theorem, known as the *ratio test*. It is one of the most commonly used tests for determining the convergence or divergence of series. The proof is similar to the one used for real series, and is left for the reader to establish.

Theorem 4.9   (d'Alembert's Ratio Test)   *If $\sum_{n=0}^{\infty} \zeta_n$ is a complex series with the property that*

$$\lim_{n \to \infty} \frac{|\zeta_{n+1}|}{|\zeta_n|} = L,$$

then the series is absolutely convergent if $L < 1$ and divergent if $L > 1$.

EXAMPLE 4.9   Show that $\sum_{n=0}^{\infty} [(1 - i)^n / n!]$ converges.

Solution   Using the ratio test, we find that

$$\lim_{n \to \infty} \frac{\left|\dfrac{(1 - i)^{n+1}}{(n + 1)!}\right|}{\left|\dfrac{(1 - i)^n}{n!}\right|} = \lim_{n \to \infty} \frac{n! \, |1 - i|}{(n + 1)!} = \lim_{n \to \infty} \frac{|1 - i|}{n + 1} = \lim_{n \to \infty} \frac{\sqrt{2}}{n + 1} = 0 = L.$$

Since $L < 1$, the series converges.

EXAMPLE 4.10 Show that the series $\sum_{n=0}^{\infty} [(z - i)^n/2^n]$ converges for all values of z in the disk $|z - i| < 2$ and diverges if $|z - i| > 2$.

Solution   Using the ratio test, we find that

$$\lim_{n\to\infty} \frac{\left|\dfrac{(z - i)^{n+1}}{2^{n+1}}\right|}{\left|\dfrac{(z - i)^n}{2^n}\right|} = \lim_{n\to\infty} \frac{|z - i|}{2} = \frac{|z - i|}{2} = L.$$

If $|z - i| < 2$, then $L < 1$ and the series converges. If $|z - i| > 2$, then $L > 1$, and the series diverges.

Our next result, known as the *root test*, is slightly more powerful than the ratio test. Before we state this test, we need to discuss a rather sophisticated idea that it uses—the *limit supremum*.

Definition 4.3   *Let $\{t_n\}$ be a sequence of positive real numbers. The limit supremum of the sequence (denoted by $\lim\sup_{n\to\infty} t_n$) is the smallest real number L with the property that for any $\varepsilon > 0$ there are at most finitely many terms in the sequence that are larger than $L + \varepsilon$. If there is no such number L, then we set $\lim\sup_{n\to\infty} t_n = \infty$.*

EXAMPLE 4.11   The limit supremum of the sequence

$$\{t_n\} = \{4.1, 5.1, 4.01, 5.01, 4.001, 5.001, \ldots\} \text{ is } \lim\sup_{n\to\infty} t_n = 5,$$

because if we set $L = 5$, then for any $\varepsilon > 0$, there are only finitely many terms in the sequence larger than $L + \varepsilon = 5 + \varepsilon$. Additionally, if L is smaller than 5, then by setting $\varepsilon = 5 - L$, we can find infinitely many terms in the sequence larger than $L + \varepsilon$, since $L + \varepsilon = 5$.

EXAMPLE 4.12   The limit supremum of the sequence

$$\{t_n\} = \{1, 2, 3, 1, 2, 3, 1, 2, 3, 1, 2, 3, \ldots\} \text{ is } \lim\sup_{n\to\infty} t_n = 3,$$

because if we set $L = 3$, then for any $\varepsilon > 0$, there are only finitely many terms (actually, there are none) in the sequence larger than $L + \varepsilon = 3 + \varepsilon$. Additionally, if L is smaller than 3, then by setting $\varepsilon = \dfrac{3 - L}{2}$ we can find infinitely many terms in the sequence larger than $L + \varepsilon$, since $L + \varepsilon < 3$.

$$\left(L + \varepsilon = L + \frac{3 - L}{2} = \frac{3 + L}{2} = \frac{3}{2} + \frac{L}{2} < \frac{3}{2} + \frac{3}{2} = 3. \right)$$

EXAMPLE 4.13 The limit supremum of the Fibonacci sequence

$$\{t_n\} = \{1, 1, 2, 3, 5, 8, 13, 21, 34, \ldots\} \text{ is } \limsup_{n \to \infty} t_n = \infty.$$

(The Fibonacci sequence has the property that for every $n > 2$, $t_n = t_{n-1} + t_{n-2}$.)

The limit supremum is a powerful idea because the limit supremum of a sequence always exists, which is not true for the limit. However, Example 4.14 illustrates the fact that if the limit of a sequence does exist, it will be the same as the limit supremum.

EXAMPLE 4.14 The sequence

$$\{t_n\} = \left\{1 + \frac{1}{n}\right\}$$
$$= \{2, 1.5, 1.333, \ldots, 1.25, 1.2, \ldots\} \text{ has } \limsup_{n \to \infty} t_n = 1.$$

We leave the verification of this as an exercise.

Theorem 4.10 (The Root Test) *Given the series* $\sum_{n=0}^{\infty} \zeta_n$, *suppose*

$$\limsup_{n \to \infty} |\zeta_n|^{1/n} = L.$$

Then the series is absolutely convergent if $L < 1$ and divergent if $L > 1$.

Proof We give a proof assuming $\lim_{n \to \infty} |\zeta_n|^{1/n}$ exists. (A proof of the more general case using the limit supremum can be found in a number of advanced texts.) Since the limit supremum is the same as the limit when the latter exists, we have

(25) $\quad \lim_{n \to \infty} |\zeta_n|^{1/n} = L.$

Suppose first that $L < 1$. We can select a number r such that $L < r < 1$. By equation (25) there exists a positive integer N such that for all $n > N$ we have $|\zeta_n|^{1/n} < r$, and so $|\zeta_n| < r^n$. Since $r < 1$, Theorem 4.8 implies $\sum_{n=N+1}^{\infty} r^n$ converges. But then by Theorem 4.7 and Corollary 4.1 $\sum_{n=N+1}^{\infty} |\zeta_n|$ converges, hence so does $\sum_{n=0}^{\infty} |\zeta_n|$.

Now suppose $L > 1$. We can select a number r such that $1 < r < L$. Again, using equation (25) we conclude that there exists a positive integer N such that for all $n > N$ we have $|\zeta_n|^{1/n} > r$, and so $|\zeta_n| > r^n$. But since $r > 1$, this implies that ζ_n does not converge to 0, and so by Theorem 4.4, $\sum_{n=0}^{\infty} \zeta_n$ does not converge.

Note that in applying either Theorems 4.9 and 4.10, if $L = 1$, the convergence or divergence of the series is unknown, and further analysis is required to determine the true state of affairs.

EXERCISES FOR SECTION 4.1

1. Find the following limits.

   (a) $\lim\limits_{n\to\infty} \left(\dfrac{1}{2} + \dfrac{i}{4}\right)^n$

   (b) $\lim\limits_{n\to\infty} \dfrac{n + (i)^n}{n}$

   (c) $\lim\limits_{n\to\infty} \dfrac{n^2 + i2^n}{2^n}$

   (d) $\lim\limits_{n\to\infty} \dfrac{(n + i)(1 + ni)}{n^2}$

2. Show that $\lim_{n\to\infty} (i)^{1/n} = 1$, where $(i)^{1/n}$ is the principal value of the nth root of i.

3. Let $\lim_{n\to\infty} z_n = z_0$. Show that $\lim_{n\to\infty} \bar{z}_n = \bar{z}_0$.

4. Let $\sum_{n=1}^{\infty} z_n = S$. Show that $\sum_{n=1}^{\infty} \bar{z}_n = \bar{S}$.

5. Show that $\displaystyle\sum_{n=0}^{\infty} \left(\dfrac{1}{2 + i}\right)^n = \dfrac{3 - i}{2}$.

6. Show that $\displaystyle\sum_{n=0}^{\infty} \left(\dfrac{1}{n + 1 + i} - \dfrac{1}{n + i}\right) = i$.

7. Show that $\displaystyle\sum_{n=1}^{\infty} \left(\dfrac{1}{n} + \dfrac{i}{2^n}\right)$ diverges.

8. Does $\lim\limits_{n\to\infty} \left(\dfrac{1 + i}{\sqrt{2}}\right)^n$ exist? Why?

9. Let $\{r_n\}$ and $\{\theta_n\}$ be two convergent sequences of real numbers such that

   $$\lim_{n\to\infty} r_n = r_0 \quad \text{and} \quad \lim_{n\to\infty} \theta_n = \theta_0.$$

   Show that $\lim_{n\to\infty} r_n e^{i\theta_n} = r_0 e^{i\theta_0}$.

10. Show that $\displaystyle\sum_{n=0}^{\infty} \dfrac{(1 + i)^n}{2^n} = 1 + i$.

11. Show that $\sum_{n=0}^{\infty} [(z + i)^n/2^n]$ converges for all values of z in the disk $|z + i| < 2$ and diverges if $|z + i| > 2$.

12. Is the series $\displaystyle\sum_{n=1}^{\infty} \dfrac{(4i)^n}{n!}$ convergent? Why?

13. Use the ratio test and show that the following series converge.

   (a) $\displaystyle\sum_{n=0}^{\infty} \left(\dfrac{1 + i}{2}\right)^n$ (b) $\displaystyle\sum_{n=1}^{\infty} \dfrac{(1 + i)^n}{n2^n}$ (c) $\displaystyle\sum_{n=1}^{\infty} \dfrac{(1 + i)^n}{n!}$ (d) $\displaystyle\sum_{n=0}^{\infty} \dfrac{(1 + i)^{2n}}{(2n + 1)!}$

14. Use the ratio test to find a disk in which the following series converge.

   (a) $\displaystyle\sum_{n=0}^{\infty} (1 + i)^n z^n$ (b) $\displaystyle\sum_{n=0}^{\infty} \dfrac{z^n}{(3 + 4i)^n}$ (c) $\displaystyle\sum_{n=0}^{\infty} \dfrac{(z - i)^n}{(3 + 4i)^n}$ (d) $\displaystyle\sum_{n=0}^{\infty} \dfrac{(z - 3 - 4i)^n}{2^n}$

15. Show that if $\sum_{n=1}^{\infty} z_n$ converges, then $\lim_{n\to\infty} z_n = 0$. Hint: $z_n = S_n - S_{n-1}$.

16. Is the series $\displaystyle\sum_{n=1}^{\infty} \dfrac{(i)^n}{n}$ convergent? Why?

17. Let $\sum_{n=1}^{\infty} (x_n + iy_n) = U + iV$. If $c = a + ib$ is a complex constant, show that

   $$\sum_{n=1}^{\infty} (a + ib)(x_n + iy_n) = (a + ib)(U + iV).$$

18. Let $f(z) = z + z^2 + z^4 + \cdots + z^{2^n} + \cdots$. Show that $f(z) = z + f(z^2)$.

19. If $\sum_{n=0}^{\infty} z_n$ converges, show that $\left| \sum_{n=0}^{\infty} z_n \right| \le \sum_{n=0}^{\infty} |z_n|$.

20. Prove that statement (6) implies statement (8) in Theorem 4.1.

21. You were asked to justify one of the inequalities in the proof of Theorem 4.1. Give the justification.

22. Prove that a sequence can have only one limit. *Hint*: Suppose there is a sequence $\{z_n\}$ such that $z_n \to \zeta_1$ and $z_n \to \zeta_2$. Show this implies $\zeta_1 = \zeta_2$ by proving that for all $\varepsilon > 0$, $|\zeta_1 - \zeta_2| < \varepsilon$.

23. Prove Corollary 4.1.

24. Prove $\lim_{n \to \infty} z_n = 0$ iff $\lim_{n \to \infty} |z_n| = 0$.

25. Establish the claim in the proof of Theorem 4.8 that if $|z| < 1$, then $\lim_{n \to \infty} z^n = 0$.

26. In the geometric series, show that if $|z| > 1$, then $\lim_{n \to \infty} |S_n| = \infty$. *Hint*:

$$|S_n| = \left| \frac{1}{1-z} - \frac{z^n}{1-z} \right| \ge \left| \frac{z^n}{1-z} \right| - \left| \frac{1}{1-z} \right| = |z^n| \left| \frac{1}{1-z} \right| - \left| \frac{1}{1-z} \right|.$$

27. Prove the series in Corollary 4.2 diverges if $|z| \le 1$.

28. Prove Theorem 4.9.

29. Give a rigorous argument to show that $\limsup_{n \to \infty} t_n = 1$ in Example 4.14.

30. (a) Use the formula for geometric series with $z = re^{i\theta}$ where $r < 1$ to show that

$$\sum_{n=0}^{\infty} r^n e^{in\theta} = \frac{1 - r\cos\theta + ir\sin\theta}{1 + r^2 - 2r\cos\theta}.$$

(b) Use part (a) to obtain

$$\sum_{n=0}^{\infty} r^n \cos n\theta = \frac{1 - r\cos\theta}{1 + r^2 - 2r\cos\theta} \quad \text{and}$$

$$\sum_{n=0}^{\infty} r^n \sin n\theta = \frac{r\sin\theta}{1 + r^2 - 2r\cos\theta}.$$

31. Show that $\sum_{n=0}^{\infty} e^{inz}$ converges for $\operatorname{Im} z > 0$.

4.2   Power Series Functions

In this section we list some results that will be useful in helping us establish properties of functions defined by power series.

> **Theorem 4.11**   *Suppose* $f(z) = \sum_{n=0}^{\infty} c_n(z - \alpha)^n$. *The set of points z for which the series converges is one of the following*:
>
> (i) *The single point* $z = \alpha$.
> (ii) *The disk* $D_\rho(\alpha)$ $\{z: |z - \alpha| < \rho\}$, *along with part (either none, some, or all) of the circle* $C_\rho(\alpha) = \{z: |z - \alpha| = \rho\}$.
> (iii) *The entire complex plane.*

Proof   We give a proof assuming $\lim_{n\to\infty}|c_n|^{1/n}$ exists. (A completely general proof making use of the limit supremum can be found in a number of advanced texts.) By Theorem 4.10, the series converges absolutely at those values of z for which $\lim_{n\to\infty}|c_n(z-\alpha)^n|^{1/n} < 1$. This is the same as requiring

(1) $|z - \alpha|\,\lim_{n\to\infty}|c_n|^{1/n} < 1.$

There are three possibilities to consider for the value of $\lim_{n\to\infty}|c_n|^{1/n}$: If the limit equals ∞, inequality (1) holds true iff $z = \alpha$, which puts us in case (i). We will be in case (ii) if $0 < \lim_{n\to\infty}|c_n|^{1/n} < \infty$, since inequality (1) then holds iff $|z - \alpha| < \dfrac{1}{\lim_{n\to\infty}|c_n|^{1/n}}$, i.e., iff $z \in D_\rho(\alpha)$, where $\rho = \dfrac{1}{\lim_{n\to\infty}|c_n|^{1/n}}$. Finally, if the limit equals 0, we will be in case (iii), as the left-hand side of inequality (1) will be 0 for any value of z. Notice we are unable to say for sure what happens with respect to convergence on $C_\rho(\alpha)$. You will see in the exercises that there are various possibilities.

Another way to phrase case (ii) of Theorem 4.11 is to say that the power series $f(z) = \sum_{n=0}^{\infty} c_n(z - \alpha)^n$ converges if $|z - \alpha| < \rho$, and diverges if $|z - \alpha| > \rho$. We call the number ρ the *radius of convergence* of the power series (see Figure 4.2). If we are in case (i) of Theorem 4.11, we say that the radius of convergence is zero, and that the radius of convergence is infinity if we are in case (iii).

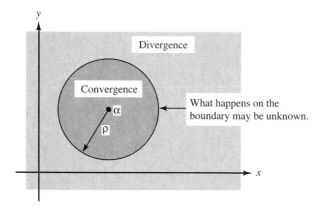

FIGURE 4.2   The radius of convergence of a power series.

Theorem 4.12   *The radius of convergence, ρ, of the power series function*
$f(z) = \displaystyle\sum_{n=0}^{\infty} c_n(z - \alpha)^n$ *can be found by any of the following methods:*

(i) *Cauchy's root test:* $\rho = \dfrac{1}{\lim\limits_{n\to\infty} |c_n|^{1/n}}$ *(Provided the limit exists).*

(ii) *Cauchy-Hadamard formula:* $\rho = \dfrac{1}{\limsup\limits_{n\to\infty} |c_n|^{1/n}}$. *(This limit always*
exists.)

(iii) *d'Alembert's ratio test:* $\rho = \lim\limits_{n\to\infty} \left| \dfrac{c_n}{c_{n+1}} \right|$ *(Provided the limit exists).*

In cases (i) *and* (ii) *we set* $\rho = \infty$ *if the limit equals* 0, *and* $\rho = 0$ *if the limit*
equals ∞.

Proof   If you examine carefully the proof of Theorem 4.11, you will see
that we have already proved case (i). It follows directly from inequality (1). Case
(ii) is left for more advanced courses, and case (iii) can be established by appealing
to the ratio test.

EXAMPLE 4.15   Find the radius of convergence of
$$f(z) = \sum_{n=0}^{\infty} \left(\frac{n+2}{3n+1} \right)^n (z-4)^n.$$

Solution   By Cauchy's root test, $\lim\limits_{n\to\infty} |c_n|^{1/n} = \lim\limits_{n\to\infty} \dfrac{n+2}{3n+1} = \dfrac{1}{3}$, so the
radius of convergence is 3.

EXAMPLE 4.16   The series $\displaystyle\sum_{n=0}^{\infty} c_n z^n = 1 + 4z + 5^2 z^2 + 4^3 z^3 + 5^4 z^4 + 4^5 z^5$

$+ \cdots$ has radius of convergence $\dfrac{1}{5}$ by the Cauchy-Hadarmard formula because

$\limsup\limits_{n\to\infty} |c_n|^{1/n} = 5.$

EXAMPLE 4.17   Find the radius of convergence of $f(z) = \displaystyle\sum_{n=0}^{\infty} \frac{1}{n!} z^n.$

Solution   By the ratio test, the radius of convergence is
$\lim\limits_{n\to\infty} \left| \dfrac{(n+1)!}{n!} \right| = \lim\limits_{n\to\infty} (n+1) = \infty.$ Thus, the series converges for all values of z.

We come now to the main result of this section.

Theorem 4.13 *Suppose the function $f(z) = \sum_{n=0}^{\infty} c_n(z - \alpha)^n$ has radius of convergence $\rho > 0$. Then*

(i) *f is infinitely differentiable for all $z \in D_\rho(\alpha)$, in fact*

(ii) *for all k, $f^{(k)}(z) = \sum_{n=k}^{\infty} n(n - 1) \cdots (n - k + 1)c_n(z - \alpha)^{n-k}$, and*

(iii) *$c_k = \dfrac{f^{(k)}(\alpha)}{k!}$, where $f^{(k)}$ denotes the kth derivative of f. (When $k = 0$, $f^{(k)}$ denotes the function f itself, so that $f^{(0)}(z) = f(z)$ for all z.)*

Proof   If we can establish case (ii) for $k = 1$, the cases for $k = 2, 3, \ldots$ will follow by induction. For instance, the case when $k = 2$ follows by applying the result for $k = 1$ to the series $f'(z) = \sum_{n=1}^{\infty} nc_n(z - \alpha)^{n-1}$.

We begin by defining the following functions:

$$g(z) = \sum_{n=1}^{\infty} nc_n(z - \alpha)^{n-1}, \quad S_j(z) = \sum_{n=0}^{j} c_n(z - \alpha)^n, \quad R_j(z) = \sum_{n=j+1}^{\infty} c_n(z - \alpha)^n.$$

Here $S_j(z)$ is simply the $(j + 1)$st partial sum of the series $f(z)$, and $R_j(z)$ is the sum of the remaining terms of that series. We leave as an exercise that the radius of convergence for $g(z)$ is ρ, the same as that of $f(z)$. For a fixed $z_0 \in D_\rho(\alpha)$, we must prove $f'(z_0) = g(z_0)$, that is, we must prove $\lim_{z \to z_0} \dfrac{f(z) - f(z_0)}{z - z_0} = g(z_0)$. This can be done by showing that for all $\varepsilon > 0$, there exists $\delta > 0$ such that if $z \in D_\rho(\alpha)$ with $0 < |z - z_0| < \delta$ then $\left| \dfrac{f(z) - f(z_0)}{z - z_0} - g(z_0) \right| < \varepsilon$.

Let $z_0 \in D_\rho(\alpha)$ and $\varepsilon > 0$ be given. Choose $r < \rho$ so that $z_0 \in D_r(\alpha)$. Choose δ to be small enough so that $D_\delta(z_0) \subset D_r(\alpha) \subset D_\rho(\alpha)$ (see Figure 4.3), and also small enough to satisfy an additional restriction, which we shall specify in a moment.

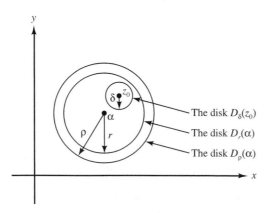

FIGURE 4.3   Choosing δ to prove $f'(z_0) = g(z_0)$.

Since $f(z) = S_j(z) + R_j(z)$, simplifying the right-hand side of the following equation reveals that for all j,

(2) $$\left[\frac{f(z) - f(z_0)}{z - z_0} - g(z_0)\right] = \left[\frac{S_j(z) - S_j(z_0)}{z - z_0} - S_j'(z_0)\right] +$$

$$[S_j'(z_0) - g(z_0)] + \left[\frac{R_j(z) - R_j(z_0)}{z - z_0}\right],$$

where $S_j'(z_0)$ is the derivative of the function S_j evaluated at z_0. Equation (2) has the general form $A = B + C + D$. By the triangle inequality,

$$|A| = |B + C + D| \leq |B| + |C| + |D|,$$

so our proof will be complete if we can show that for a small enough value of δ each of the expressions $|B|$, $|C|$, and $|D|$ can be shown to be less than $\dfrac{\varepsilon}{3}$.

Calculation for $|D|$

$$\left|\frac{R_j(z) - R_j(z_0)}{z - z_0}\right| = \left|\frac{1}{z - z_0}\left(\sum_{n=j+1}^{\infty} c_n[(z - \alpha)^n - (z_0 - \alpha)^n]\right)\right|$$

$$\leq \sum_{n=j+1}^{\infty} |c_n| \left|\frac{(z - \alpha)^n - (z_0 - \alpha)^n}{z - z_0}\right|$$

(Compare with exercise 19 of Section 4.1).

As an exercise, we ask you to establish

(3) $$\left|\frac{(z - \alpha)^n - (z_0 - \alpha)^n}{z - z_0}\right| \leq nr^{n-1}.$$

Assuming this to be the case, we get

(4) $$\left|\frac{R_j(z) - R_j(z_0)}{z - z_0}\right| \leq \sum_{n=j+1}^{\infty} |c_n| nr^{n-1}.$$

Since $r < \rho$, the series $\displaystyle\sum_{n=1}^{\infty} |c_n| nr^{n-1}$ converges (can you explain why?). This means that the tail part of the series, which is the right-hand side of inequality (4), can certainly be made less than $\dfrac{\varepsilon}{3}$ if we choose j large enough, say $j \geq N_1$.

Calculation for $|C|$

Since $S_j'(z_0) = \displaystyle\sum_{n=1}^{j} nc_n(z_0 - \alpha)^{n-1}$, it is clear that $\lim_{j \to \infty} S_j'(z_0) = g(z_0)$. This means there is an integer N_2 such that if $j \geq N_2$, then $\left| S_j'(z_0) - g(z_0) \right| < \dfrac{\varepsilon}{3}$.

Calculation for $|B|$

Define $N = \max\{N_1, N_2\}$. Because $S_N(z)$ is a polynomial, $S'_N(z_0)$ exists. Thus, we can find δ small enough so that it complies with the restriction previously placed on it as well as ensuring

$$\left| \frac{S_N(z) - S_N(z_0)}{z - z_0} - S'_N(z_0) \right| < \frac{\varepsilon}{3} \text{ whenever } z \in D_p(\alpha) \text{ with } 0 < |z - z_0| < \delta.$$

Using this value of N for j in equation (2) together with our chosen δ yields conclusion (ii) of our theorem.

To prove (iii), note that if we set $z = \alpha$ in (ii), all the terms drop out except when $n = k$, giving us $f^{(k)}(\alpha) = k(k-1)\cdots(k-k+1)c_k$. Solving for c_k completes the proof of our theorem.

EXAMPLE 4.18 Show that $\displaystyle\sum_{n=0}^{\infty} (n+1)z^n = \frac{1}{(1-z)^2}$ for all $z \in D_1(0)$.

Solution We know from Theorem 4.8 that $f(z) = \dfrac{1}{1-z} = \displaystyle\sum_{n=0}^{\infty} z^n$ for all $z \in D_1(0)$. If we set $k = 1$ in Theorem 4.13, case (ii), $f'(z) = \dfrac{1}{(1-z)^2} = $

$$\sum_{n=1}^{\infty} nz^{n-1} = \sum_{n=0}^{\infty} (n+1)z^n \text{ for all } z \in D_1(0).$$

EXAMPLE 4.19 The Bessel function of order zero is given by

$$J_0(z) = \sum_{n=0}^{\infty} \frac{(-1)^n}{(n!)^2} \left(\frac{z}{2}\right)^{2n} = 1 - \frac{z^2}{2^2} + \frac{z^4}{2^4 4^4} - \frac{z^6}{2^2 4^2 6^2} + \cdots,$$

and termwise differentiation shows that its derivative is given by

$$J'_0(z) = \sum_{n=0}^{\infty} \frac{(-1)^{n+1}}{n!(n+1)!} \left(\frac{z}{2}\right)^{2n+1} = \frac{-z}{2} + \frac{1}{1!2!} \left(\frac{z}{2}\right)^3 - \frac{1}{2!3!} \left(\frac{z}{2}\right)^5 + \cdots.$$

We leave as an exercise that the radius of convergence of these series is infinity. The Bessel function $J_1(z)$ of order one is known to satisfy the differential equation $J_1(z) = -J'_0(z)$.

EXERCISES FOR SECTION 4.2

1. Prove case (iii) of Theorem 4.12.

2. Consider the following series: $\displaystyle\sum_{n=0}^{\infty} z^n$, $\displaystyle\sum_{n=1}^{\infty} \frac{z^n}{n^2}$, and $\displaystyle\sum_{n=1}^{\infty} \frac{z^n}{n}$.

   (a) Using Theorem 4.12, show that each series has radius of convergence 1.
   (b) Show that the first series converges nowhere on $C_1(0)$.

(c) Show that the second series converges everywhere on $C_1(0)$.

(d) It turns out that the third series converges everywhere on $C_1(0)$ except at the point $z = 1$. This is not easy to prove, but see if you can do so.

3. Show that $\sum_{n=0}^{\infty} (n + 1)^2 z^n = \dfrac{1 + z}{(1 - z)^3}$.

4. Find the radius of convergence of the following.

(a) $g(z) = \sum_{n=0}^{\infty} (-1)^n \dfrac{z^n}{(2n)!}$

(b) $h(z) = \sum_{n=0}^{\infty} n! \, z^n$

(c) $f(z) = \sum_{n=0}^{\infty} \left(\dfrac{4n^2}{2n + 1} - \dfrac{6n^2}{3n + 4} \right) z^n$

(d) $g(z) = \sum_{n=0}^{\infty} \dfrac{(n!)^2}{(2n)!} z^n$

(e) $h(z) = \sum_{n=0}^{\infty} (2 - (-1)^n)^n z^n$

(f) $f(z) = \sum_{n=0}^{\infty} \dfrac{n(n - 1)z^n}{(3 + 4i)^n}$

(g) $g(z) = \sum_{n=0}^{\infty} \left(\dfrac{3n + 7}{4n + 2} \right)^n z^n$

(h) $h(z) = \sum_{n=0}^{\infty} \dfrac{2^n}{1 + 3^n} z^n$

(i) $g(z) = \sum_{n=0}^{\infty} \dfrac{n^n}{n!} z^n$. *Hint:* $\lim_{n \to \infty} [1 + (1/n)]^n = e$.

(j) $g(z) = \sum_{n=0}^{\infty} z^{2n}$

(k) $\sinh(z) + \dfrac{4}{4 - z^2} = 1 + z + \dfrac{z^2}{2^2} + \dfrac{z^3}{3!} + \dfrac{z^4}{2^4} + \dfrac{z^5}{5!} + \dfrac{z^6}{2^6} + \dfrac{z^7}{7!} + \cdots$.

5. Suppose that $\sum_{n=0}^{\infty} c_n z^n$ has radius of convergence R. Show that $\sum_{n=0}^{\infty} c_n^2 z^n$ has radius of convergence R^2.

6. Does there exist a power series $\sum_{n=0}^{\infty} c_n z^n$ that converges at $z_1 = 4 - i$ and diverges at $z_2 = 2 + 3i$? Why?

7. Verify part (ii) of Theorem 4.13 for all k by using mathematical induction.

8. This exercise will establish that the radius of convergence for g given in Theorem 4.13 is ρ, the same as that of the function f.

(a) Explain why the radius of convergence for g is $\dfrac{1}{\lim\sup_{n \to \infty} |nc_n|^{\frac{1}{n-1}}}$.

(b) Show that $\lim\sup_{n \to \infty} |n|^{\frac{1}{n-1}} = 1$. *Hint:* The lim sup equals the limit. Show that

$$\lim_{n \to \infty} \dfrac{\log n}{n - 1} = 0.$$

(c) Assuming that $\lim\sup_{n \to \infty} |c_n|^{\frac{1}{n-1}} = \lim\sup_{n \to \infty} |c_n|^{\frac{1}{n}}$, show that the conclusion for this exercise follows.

(d) Establish the truth of the assumption made in part (c).

9. This exercise will establish the validity of inequality (3) given in the proof of Theorem 4.13.

(a) Show that

$$\left| \frac{s^n - t^n}{s - t} \right| = \left| s^{n-1} + s^{n-2}t + s^{n-3}t^2 + \cdots + st^{n-2} + t^{n-1} \right|$$

$$\leq \left| s^{n-1} \right| + \left| s^{n-2}t \right| + \left| s^{n-3}t^2 \right| + \cdots + \left| st^{n-2} \right| + \left| t^{n-1} \right|,$$

where s and t are arbitrary complex numbers, $s \neq t$.

(b) Argue why in inequality (3) we know that $\left| z - \alpha \right| < r$ and $\left| z_0 - \alpha \right| < r$.

(c) Let $s = z - \alpha$ and $t = z_0 - \alpha$ in part (a) to establish inequality (3).

10. Show that the radius of convergence is infinity of the series for $J_0(z)$ and $J_0'(z)$ given in Example 4.19.

11. Explain what you think might happen if the complex number z is substituted for x in the Maclaurin series for $\sin x$ that is studied in calculus?

12. Write a report on series of complex numbers and/or functions. Include ideas and examples not mentioned in the text. Resources include bibliographical items 10, 83, 116, and 153.

4.3   Julia and Mandelbrot Sets

An impetus for studying complex analysis is the comparison of properties of real numbers and functions with their complex counterparts. In this section we take a look at Newton's method for finding solutions to the equation $f(z) = 0$. We then examine the more general topic of iteration.

Recall from calculus that Newton's method proceeds by starting with a function $f(x)$ and an initial "guess" x_0 as a solution to $f(x) = 0$. We then generate a new guess x_1 by the computation $x_1 = x_0 - \dfrac{f(x_0)}{f'(x_0)}$. Using x_1 in place of x_0, this process is repeated, giving us $x_2 = x_1 - \dfrac{f(x_1)}{f'(x_1)}$. We thus obtain a sequence of points $\{x_k\}$, where $x_{k+1} = x_k - \dfrac{f(x_k)}{f'(x_k)}$. The points $\{x_k\}_{k=0}^{\infty}$ are called the *iterates* of x_0. For functions defined on the real numbers, this method gives remarkably good results, so that the sequence $\{x_k\}$ often converges to a solution of $f(x) = 0$ rather quickly. In the late 1800s the British mathematician Arthur Cayley investigated the question as to whether Newton's method can be applied to complex functions. He wrote a paper giving an analysis for how this method works for quadratic polynomials and indicated his intention to publish a subsequent paper for cubic polynomials. Unfortunately, Cayley died before producing this paper. As you will see, the extension of Newton's method to the complex domain and the more general question of iteration are quite complicated.

EXAMPLE 4.20   Trace out the next five iterates of Newton's method given an initial guess of $z_0 = \frac{1}{4} + \frac{1}{4}i$ as a solution to the equation $f(z) = 0$, where $f(z) = z^2 + 1$.

COLOR PLATE 1 Newton's method applied
to $f(z) = z^3 + 1$.

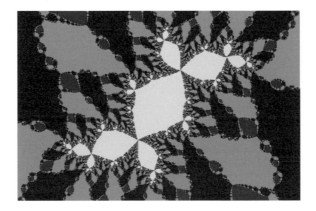

COLOR PLATE 2 The rabbit.

COLOR PLATE 3   A zoom of the rabbit.

COLOR PLATE 4   The Julia set for $f(z) = z^2 - 1.25$.

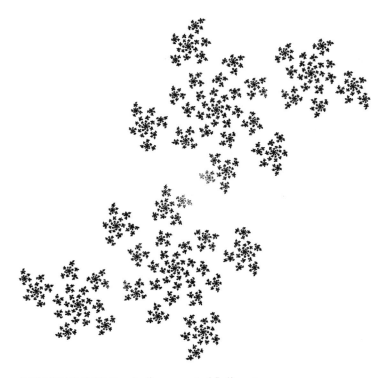

COLOR PLATE 5   A disconnected Julia set.

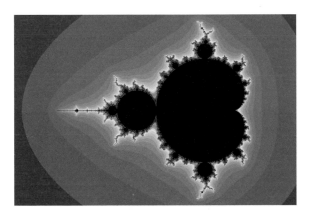

COLOR PLATE 6   The Mandelbrot set (M).

COLOR PLATE 7   A zoom on the upper portion of M.

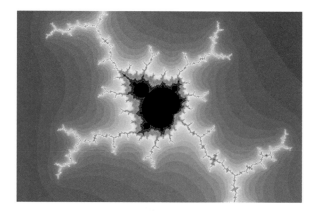

COLOR PLATE 8   A zoom on the upper portion of
color plate 7.

Solution   Given z as an initial guess, our next guess will be $z - \dfrac{f(z)}{f'(z)} = \dfrac{z^2 - 1}{2z}$. With the aid of a computer algebra system, we can easily produce Table 4.1, where values are rounded to five decimal places.

TABLE 4.1   The iterates of $z_0 = \frac{1}{4} + \frac{1}{4}i$ for Newton's method applied to $f(z) = z^2 + 1$.

k	z_k	$f(z_k)$
0	$0.25000 + 0.25000i$	$1.00000 + 0.12500i$
1	$-0.87500 + 1.12500i$	$0.50000 - 1.96875i$
2	$-0.22212 + 0.83942i$	$0.34470 - 0.37290i$
3	$0.03624 + 0.97638i$	$0.04799 + 0.07077i$
4	$-0.00086 + 0.99958i$	$0.00084 - 0.00172i$
5	$0.00000 + 1.00000i$	$0.00000 + 0.00000i$

Figure 4.4 shows the relative positions of these points on the z plane. Notice that the points z_4 and z_5 are so close together that they appear to coincide and that the value for z_5 agrees to five decimal places with the actual solution $z = i$.

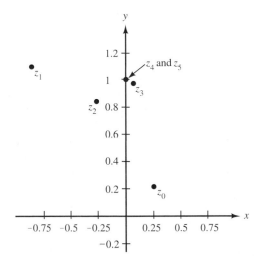

FIGURE 4.4   The iterates of $z_0 = \frac{1}{4} + \frac{1}{4}i$ for Newton's method applied to $f(z) = z^2 + 1$.

The complex version of Newton's method appears also to work quite well. You may recall, however, that with functions defined on the reals, it is not the case that every initial guess produces a sequence that converges to a solution. Example 4.21 shows that the same is true in the complex case.

EXAMPLE 4.21   Show that Newton's method fails for the function $f(z) = z^2 + 1$ if our initial guess is a real number.

Solution   From Example 4.20 we know that given any guess z as a solution of $z^2 + 1 = 0$, the next guess at a solution is $N(z) = z - \dfrac{f(z)}{f'(z)} = \dfrac{z^2 - 1}{2z}$. Let z_0 be any real number, and let $\{z_k\}$ be the sequence of iterations produced by the initial seed z_0. If for any k, $z_k = 0$, the iteration terminates with an undefined result. If all the terms of the sequence $\{z_k\}$ are defined, an easy induction argument shows that all the terms of the sequence will be real. Since the solutions of $z^2 + 1 = 0$ are $\pm i$, the sequence $\{z_k\}$ cannot possibly converge to either solution. In the exercises we ask you to explore in detail what happens when z_0 is in the upper or lower half plane.

The case for cubic polynomials is more complicated than that for quadratics. Fortunately, we can get an idea of what's going on by doing some experimentation with computer graphics. Let us begin with the cubic polynomial $f(z) = z^3 + 1$. $\left(\text{Recall, the roots of this polynomial are at } -1, \dfrac{1}{2} + \dfrac{\sqrt{3}}{2}i, \text{ and } \dfrac{1}{2} - \dfrac{\sqrt{3}}{2}i.\right)$ We associate a color with each root (blue, red, and green, respectively). We form a rectangular region R, which contains the three roots of $f(z)$, and partition this region into equal rectangles R_{ij}. We then choose a point z_{ij} at the center of each rectangle and for each one of these points we apply the following algorithm:

1.   With $N(z) = z - \dfrac{f(z)}{f'(z)}$, compute $N(z_{ij})$. Continue computing successive iterates of this initial point until we either are within a certain preassigned tolerance (say ε) of one of the roots of $f(z) = 0$, or until the number of iterations has exceeded a preassigned maximum.
2.   If step 1 left us within ε of one of the roots of $f(z)$, we color the entire rectangle R_{ij} with the color associated with that root. Otherwise, we assume the initial point z_{ij} does not converge to any root and color the entire rectangle yellow.

Notice that the preceding algorithm does not prove anything. In step 2, there is no a priori reason to justify the assumption mentioned, nor is there any necessity for an initial point z_{ij} to have its sequence of iterates converging to one of the roots of $f(z) = 0$ just because a particular iteration is within ε of that root. Finally, the fact that one point in a rectangle behaves in a certain way does not imply that all the points in that rectangle behave in a like manner. Nevertheless, we can use this algorithm to motivate mathematical explorations. Indeed, computer experiments like the one described have contributed to a lot of exciting mathematics in the past 15 years. Color plate 1 at the end of this section shows the results of applying our algorithm to the cubic polynomial $f(z) = z^3 + 1$. The points in the blue, red, and green regions are those "initial guesses" that will converge to the roots -1, $\frac{1}{2} + \frac{\sqrt{3}}{2}i$, and $\frac{1}{2} - \frac{\sqrt{3}}{2}i$, respectively. (The roots themselves are located in the middle of the three largest colored regions.) The complexity of this picture becomes apparent when we observe that wherever two colors appear to meet, the third color emerges between them. But then, a closer inspection of the area where this third color meets one of the other colors reveals again a different color between them. This process continues with an infinite complexity.

Not all initial guesses will result in a sequence that converges to a solution (let z_0 be any real number, as in Example 4.21). On the other hand, there appear to be no yellow regions with any area in color plate 1, indicating that most initial guesses z_0 at a solution to $z^3 + 1 = 0$ will produce a sequence $\{z_k\}$ which converges to one of the three roots. Color plate 2 illustrates that this is not always the case. It shows the results of applying the preceding algorithm to the polynomial $f(z) = z^3 + (-0.26 + 0.02i)z + (-0.74 + 0.02i)$. The yellow area shown is often referred to as the rabbit. It consists of a main body and two ears. Upon closer inspection (color plate 3) we see that each one of the ears consists of a main body and two ears. Color plate 2 is an example of a fractal image. Mathematicians use the term *fractal* to indicate an object that is self-similar and infinitely replicating.

In 1918 the French mathematicians Gaston Julia and Pierre Fatou noticed this fractal phenomenon when exploring iterations of functions not necessarily connected with Newton's method. Beginning with a function $f(z)$ and a point z_0, they computed the iterates $z_1 = f(z_0), z_2 = f(z_1), \ldots, z_{k+1} = f(z_k), \ldots$, and investigated properties of the sequence $\{z_k\}$. Their findings did not receive a great deal of attention, largely because computer graphics were not available at that time. With the recent proliferation of computers, it is not surprising that these investigations were revived in the 1980s. Detailed studies of Newton's method and the more general topic of iteration were undertaken by a host of mathematicians including Curry, Douady, Garnett, Hubbard, Mandelbrot, Milnor, and Sullivan. We now turn our attention to some of their results by focusing on the iterations produced by quadratics of the form $f_c(z) = z^2 + c$.

EXAMPLE 4.22   Given $f_c(z) = z^2 + c$, analyze all possible iterations for the function $f_0(z) = z^2 + 0$.

Solution   In the exercises we will ask you to verify that if $|z_0| < 1$ the sequence will converge to 0, if $|z_0| > 1$ the sequence will be unbounded, and if $|z_0| = 1$ the sequence will either oscillate chaotically around the unit circle or converge to 1.

Given the function $f_c(z) = z^2 + c$, and an initial seed z_0, the set of iterates given by $z_1 = f_c(z_0)$, $z_2 = f_c(z_1)$, etc., are also called the *orbits* of z_0 generated by $f_c(z)$. Let K_c denote the set of points with bounded orbits for $f_c(z)$. Example 4.22 shows that K_0 is the closed unit disk $\overline{D}_1(0)$. The boundary of K_c is known as the *Julia set* for the function $f_c(z)$. The Julia set for $f_0(z)$ is the unit circle $C_1(0)$. It turns out that K_c is a nice simple set only when $c = 0$ or $c = -2$. Otherwise, K_c is fractal. Color plate 4 shows $K_{-1.25}$. The variation in colors indicate the length of time it takes for points to become "sufficiently unbounded" according to the following algorithm, which uses the same notation as our algorithm for iterations via Newton's method:

1. Compute $f_c(z_{ij})$. Continue computing successive iterates of this initial point until the absolute value of one of the iterations exceeds a certain bound (say L), or until the number of iterations has exceeded a preassigned maximum.

2. If step 1 left us with an iteration whose absolute value exceeds L, we color the entire rectangle R_{ij} with a color indicating the number of iterations needed before this value was attained (the more iterations required, the darker the color). Otherwise, we assume the orbits of the initial point z_{ij} do not diverge to infinity and we color the entire rectangle black.

Notice, again, that this algorithm does not prove anything. It merely guides the direction of our efforts to do rigorous mathematics.

Color plate 5 shows the Julia set for the function $f_c(z)$, where $c = -0.11 - 0.67i$. The boundary of this set is different from that of the other sets we have seen in that it is disconnected. Julia and Fatou independently discovered a simple criterion that can be used to tell when the Julia set for $f_c(z)$ is connected or disconnected. We state their result, but omit the proof, as it is beyond the scope of this text.

Theorem 4.14 *The boundary of K_c is connected if and only if $0 \in K_c$. In other words, the Julia set for $f_c(z)$ is connected if and only if the orbits of 0 are bounded.*

EXAMPLE 4.23 Show that the Julia set for $f_i(z)$ is connected.

Solution We apply Theorem 4.14 and compute the orbits of 0 for $f_i(z) = z^2 + i$. We have $f_i(0) = i$, $f_i(i) = -1 + i$, $f_i(-1 + i) = -i$, $f_i(-i) = -1 + i$, ... Thus, the orbits of 0 are the sequence 0, $-1 + i$, $-i$, $-1 + i$, $-i$, $-1 + i$, $-i$, . . . , which is clearly a bounded sequence. Thus, by Theorem 4.14, the Julia set for $f_i(z)$ is connected.

In 1980, the Polish-born mathematician Benoit Mandelbrot used computer graphics to study the following set:

$$M = \{c: \text{ the Julia set for } f_c(z) \text{ is connected}\}$$
$$= \{c: \text{ the orbits of } 0 \text{ determined by } f_c(z) \text{ are bounded}\}.$$

The set M has come to be known as the Mandelbrot set. Color plate 6 shows its intricate nature. Technically, the Mandelbrot set is not fractal since it is not self-similar (although it may look that way). It is, however, infinitely complex. Color plate 7 shows a zoom over the upper portion of the set shown in color plate 6, and likewise color plate 8 zooms in on the upper portion of color plate 7. Notice that we see in color plate 8 the emergence of another structure very similar to the Mandelbrot set we began with. It is not an exact replica. Nevertheless, if we zoomed in on this set at the right spot (and there are many such choices), we would eventually see yet another "Mandelbrot clone" and so on ad infinitum! The remainder of this section looks at some of the properties of this amazing set.

EXAMPLE 4.24   Show that $\left\{c : |c| \leq \dfrac{1}{4}\right\} \subseteq M$.

Solution   Let $|c| \leq \frac{1}{4}$, and let $\{a_n\}_{n=0}^{\infty}$ be the orbits of 0 generated by $f_c(z) = z^2 + c$. Thus,

$$a_0 = 0,$$
$$a_1 = f_c(a_0) = a_0^2 + c = c,$$
$$a_2 = f_c(a_1) = a_1^2 + c, \quad \text{and in general,}$$
$$a_{n+1} = f_c(a_n) = a_n^2 + c.$$

We shall show that $\{a_n\}$ is bounded. In particular, we shall show that $|a_n| \leq \frac{1}{2}$ for all n by mathematical induction. Clearly $|a_n| \leq \frac{1}{2}$ if $n = 0$ or 1. Assume $|a_n| \leq \frac{1}{2}$ for some value of $n \geq 1$ (our goal is to show $|a_{n+1}| \leq \frac{1}{2}$). Now

$$
\begin{aligned}
|a_{n+1}| &= |a_n^2 + c| \\
&\leq |a_n^2| + |c| \quad \text{by the triangle inequality} \\
&\leq \tfrac{1}{4} + \tfrac{1}{4} \quad \text{by our induction assumption and the fact that } |c| \leq \tfrac{1}{4}.
\end{aligned}
$$

In the exercises, we ask you to show that if $|c| > 2$, then $c \notin M$. Thus, the Mandelbrot set depicted in color plate 6 contains the disk $\overline{D}_{1/4}(0)$, and is contained in the disk $\overline{D}_2(0)$.

There are other methods for determining which points belong to M. To see what they are, we need some additional vocabulary.

Definition 4.4   *The point z_0 is a fixed point for the function $f(z)$ if $f(z_0) = z_0$.*

Definition 4.5   *The point z_0 is an attracting point for the function $f(z)$ if $|f'(z_0)| < 1$.*

The following theorem explains the significance of these terms.

Theorem 4.15   *Suppose z_0 is an attracting fixed point for the function $f(z)$. Then there is a disk $D_r(z_0)$ about z_0 such that the iterates of all the points in $D_r(z_0)$ are drawn towards the point z_0 in the sense that*

if $z \in D_r^*(z_0)$, then $\left| f(z) - z_0 \right| < \left| z - z_0 \right|$.

Proof   Since z_0 is an attracting point for f, we know that $\left| f'(z_0) \right| < 1$. Since f is differentiable at z_0, we know that given any $\varepsilon > 0$ there exists some $r > 0$ such that if z is any point in the disk $D_r^*(z_0)$, then $\left| \dfrac{f(z) - f(z_0)}{z - z_0} - f'(z_0) \right| < \varepsilon$. If we set $\varepsilon = 1 - \left| f'(z_0) \right|$, then we have for all z in $D_r^*(z_0)$ that

$$\left| \frac{f(z) - f(z_0)}{z - z_0} \right| - \left| f'(z_0) \right| \leq \left| \frac{f(z) - f(z_0)}{z - z_0} - f'(z_0) \right| < 1 - \left| f'(z_0) \right|,$$

which gives

$$\left| \frac{f(z) - f(z_0)}{z - z_0} \right| < 1.$$

Thus,

$$\left| f(z) - f(z_0) \right| < \left| z - z_0 \right|.$$

Since z_0 is a fixed point for f, this implies

$$\left| f(z) - z_0 \right| < \left| z - z_0 \right|.$$

In 1905, Fatou showed that if the function $f_c(z)$ has attracting fixed points, the orbits of 0 determined by $f_c(z)$ must converge to one of them. Since a convergent sequence is bounded, this implies that c must belong to M. In the exercises we ask you to show that the main cardiod-shaped body of M in color plate 6 is composed of those points c for which $f_c(z)$ has attracting fixed points. You will find that Theorem 4.16 is a useful characterization of them.

Theorem 4.16   *The points c for which $f_c(z)$ has attracting fixed points satisfy $\left| 1 + (1 - 4c)^{1/2} \right| < 1$ or $\left| 1 - (1 - 4c)^{1/2} \right| < 1$, where the square root is the principal square root function.*

Proof   The point z_0 is a fixed point for $f_c(z)$ if and only if $f_c(z_0) = z_0$. In other words, if and only if $z_0^2 - z_0 + c = 0$. The solutions to this equation are

$$z_0 = \frac{1 + (1 - 4c)^{1/2}}{2} \quad \text{and} \quad z_0 = \frac{1 - (1 - 4c)^{1/2}}{2},$$

where again the fractional exponent is the principal square root function. Now, if z_0 is a fixed point, then $\left| f_c'(z_0) \right| = \left| 2z_0 \right| < 1$. Combining this with the solutions for z_0 gives our desired result.

Definition 4.6   *An n-cycle for a function f is a set* $\{z_0, z_1, \ldots, z_{n-1}\}$ *of n complex numbers such that* $z_k = f(z_{k-1})$ *for* $k = 1, 2, \ldots, n - 1$, *and* $f(z_{n-1}) = z_0$.

Definition 4.7   *An n-cycle* $\{z_0, z_1, \ldots, z_{n-1}\}$ *for a function f is said to be attracting if* $\left| g_n'(z_0) \right| < 1$, *where* g_n *is the composition of f with itself n times. For example, if* $n = 2$, *then* $g_2(z) = f(f(z))$.

EXAMPLE 4.25   Example 4.23 shows that $\{-1 + i, -i\}$ is a 2-cycle for the function $f_i(z)$. It is not an attracting 2-cycle since $g_2(z) = z^4 + 2iz^2 + i - 1$, and $g_2'(z) = 4z^3 + 4iz$. Hence, $\left| g_2'(-1 + i) \right| = \left| 4 + 4i \right|$, so $\left| g_2'(-1 + i) \right| > 1$.

In the exercises, we ask you to establish that if $\{z_0, z_1, \ldots, z_{n-1}\}$ is an attracting *n*-cycle for a function f, then not only does z_0 satisfy $\left| g_n'(z_0) \right| < 1$, but in fact we also have that $\left| g_n'(z_k) \right| < 1$, for $k = 1, 2, \ldots, n - 1$.

It turns out that the large disk to the left of the cardiod in color plate 6 consists of those points c for which $f_c(z)$ has a 2-cycle. The disk to the left of this, as well as the large disks above and below the main cardiod disk, are those points c for which $f_c(z)$ has a 3-cycle.

Continuing with this scheme, we see that the idea of *n*-cycles explains the appearance of the ''buds'' that you see on color plate 6. It does not, however, begin to do justice to the enormous complexity of the entire set. Even color plates 7 and 8 are a mere glimpse into its awesome beauty. In the exercise section, we suggest several references for projects that you may like to pursue for a more detailed study of topics relating to those covered in this section.

EXERCISES FOR SECTION 4.3

1. Prove that Newton's method always works for polynomials of degree 1 (functions of the form $f(z) = az + b$, where $a \neq 0$). How many iterations are necessary before Newton's method produces the solution $z = \dfrac{-b}{a}$ to $f(z) = 0$?

2. Consider the function $f(z) = z^2 + 1$, where $N(z) = z - \dfrac{f(z)}{f'(z)} = \dfrac{z^2 - 1}{2z} = \dfrac{1}{2}\left(z - \dfrac{1}{z} \right)$.

    (a) Show that if $\text{Im}(z_0) > 0$, the sequence $\{z_k\}$ formed by successive iterations of z_0 via $N(z)$ lies entirely within the upper half plane.

    Hint: If $z = r(\cos \theta + i \sin \theta)$, show $N(z) = \dfrac{1}{2}\left(r - \dfrac{1}{r} \right)\cos \theta + i\dfrac{1}{2}\left(r + \dfrac{1}{r} \right) \sin \theta$.

    (b) Show a similar result holds if $\text{Im}(z_0) < 0$.
    (c) Discuss whether $\{z_k\}$ converges to i if $\text{Im}(z_0) > 0$.
    (d) Discuss whether $\{z_k\}$ converges to $-i$ if $\text{Im}(z_0) < 0$.

(e) Use induction to show that if all the terms of the sequence $\{z_k\}$ are defined, then the sequence $\{z_k\}$ is real provided z_0 is real.

(f) For which real numbers will the sequence $\{z_k\}$ be defined?

3. Formulate and solve analogous questions to Exercise 2 for the function $f(z) = z^2 - 1$.

4. Consider the function $f_0(z) = z^2$, and an initial point z_0. Let $\{z_k\}$ be the sequence of iterates of z_0 generated by $f_0(z)$. That is, $z_1 = f_0(z_0)$, $z_2 = f_0(z_1)$, and so forth.

   (a) Show that if $\left| z_0 \right| < 1$, the sequence $\{z_k\}$ converges to 0.

   (b) Show that if $\left| z_0 \right| > 1$, the sequence $\{z_k\}$ is unbounded.

   (c) Show that if $\left| z_0 \right| = 1$, the sequence $\{z_k\}$ either converges to 1 or oscillates around the unit circle. Give a simple criterion that can be applied to z_0 that will dictate which one of these two paths the sequence $\{z_k\}$ takes.

5. Show that the Julia set for $f_{-2}(z)$ is connected.

6. Determine the precise structure of the set K_{-2}.

7. Prove that if $z = c$ is in the Mandelbrot set, then its conjugate $\bar{c}$ is also in the Mandelbrot set. Thus, the Mandelbrot set is symmetric about the x axis. *Hint*: Use mathematical induction.

8. Find a value c in the Mandelbrot set such that its negative, $-c$, is not in the Mandelbrot set.

9. Show that if $\left| c \right| > 2$, then $c \notin M$. *Hint*: Use the inequality $\left| a_n^2 + c \right| \geq \left| a_n^2 \right| - \left| c \right|$.

10. Show that if c is any real number greater than 1/4, then c is not in the Mandelbrot set. *Note*: Combining this with Example 4.23 shows that the cusp in the cardiod section of the Mandelbrot set occurs precisely at $c = 1/4$.

11. Use Theorem 4.16 to show that the point $-\frac{1}{4}\sqrt{3}i$ belongs to the Mandelbrot set.

12. Show that the points c that solve the inequalities of Theorem 4.16 form a cardiod. This cardiod is the main body of the Mandelbrot set shown in color plate 6. *Hint*: It may be helpful to write the inequalities of Theorem 4.16 as

$$\left| \tfrac{1}{2} + (\tfrac{1}{4} - c)^{1/2} \right| < \tfrac{1}{2} \quad \text{or} \quad \left| \tfrac{1}{2} - (\tfrac{1}{4} - c)^{1/2} \right| < \tfrac{1}{2}.$$

13. Suppose that $\{z_0, z_1\}$ is a 2-cycle for f. Show that if z_0 is attracting for $g_2(z)$, then so is the point z_1. *Hint*: Differentiate $g_2(z) = f(f(z))$ using the chain rule, and show that $g_2'(z_0) = g_2'(z_1)$.

14. Generalize Exercise 13 to n-cycles.

The remaining exercises are suggested projects for those wishing to engage in a more detailed study of complex dynamics.

15. Write a report on how complex analysis is used in the study of dynamical systems. Resources include bibliographical items 53, 54, 55, 58, and 143.

16. Write a report on how complex analysis is used in the study of fractals. Resources include bibliographical items 7, 8, 9, 11, 55, 57, 58, 78, 84, 101, 125, 126, 134, 139, 143, 167, 175, and 188.

17. Write a report on how complex analysis is used to study the Julia set. Resources include bibliographical items 144 and 177.

18. Write a report on how complex analysis is used to study the Mandelbrot set. Include ideas and examples that are not mentioned in the text. Resources include bibliographical items 31, 45, 56, 74, 125, 126, and 177.

19. Write a report on how complex functions are used in the study of chaos. Resources include bibliographical items 11, 53, 54, 55, 57, 58, 142, and 168.

5

Elementary Functions

How should complex-valued functions such as e^z, log z, sin z, etc., be defined? Clearly, any responsible definition should satisfy the following criteria: (1) The functions so defined must give the same values as the corresponding functions for real variables when the number z is a real number. (2) As far as possible, the properties of these new functions must correspond with their real counterparts. For example, we would want $e^{z_1+z_2} = e^{z_1}e^{z_2}$ to be valid regardless of whether z were real or complex.

These requirements may seem like a tall order to fill. There is a procedure, however, that offers promising results. It is to put in complex form the expansion of the real functions e^x, sin x, etc., as power series.

5.1   The Complex Exponential Function

A standard result from elementary calculus is that the exponential function can be represented by the power series $e^x = \sum_{n=0}^{\infty} \frac{1}{n!} x^n$. Thus, it is only natural to *define* the complex exponential e^z (also written as exp z) in the following way:

Definition 5.1 $e^z = exp(z) = \sum_{n=0}^{\infty} \frac{1}{n!} z^n.$

Clearly this definition agrees with that of the real exponential function when z is a real number. We now show that this complex exponential has two of the key properties we associate with its real counterpart, and verify the identity $e^{i\theta} = \cos \theta + i \sin \theta$, which we promised to establish back in Chapter 1.

Theorem 5.1   *The function* exp(z) *is an entire function satisfying:*

(i) exp$'(z)$ = exp(z) = e^z. *(In other words,* $\frac{d}{dz} e^z = e^z$.)

(ii) exp($z_1 + z_2$) = exp(z_1)exp(z_2). *(In other words,* $e^{z_1+z_2} = e^{z_1}e^{z_2}$.)

(iii) *If* θ *is a real number, then* $e^{i\theta} = \cos \theta + i \sin \theta$.

Proof   From Example 4.17, we see that the series in Definition 5.1 has an infinite radius of convergence, so exp(z) is entire by Theorem 4.13, (i).

(i) By use of Theorem 4.13, (ii),

$$\exp'(z) = \sum_{n=1}^{\infty} \frac{n}{n!} z^{n-1} = \sum_{n=1}^{\infty} \frac{1}{(n-1)!} z^{n-1} = \sum_{n=0}^{\infty} \frac{1}{n!} z^n = \exp(z).$$

(ii) Let ζ be an arbitrary complex number, and define $g(z)$ by

$$g(z) = \exp(z) \exp(\zeta - z).$$

Using the product rule, chain rule, and part (i), we have

$$g'(z) = \exp(z) \exp(\zeta - z) + \exp(z)[-\exp(\zeta - z)] = 0 \text{ for all } z.$$

This implies that the function g must be constant (see Exercise 17 of Section 3.3). Thus, for all z, $g(z) = g(0)$. Since $\exp(0) = 1$ (verify!), we deduce

$$g(z) = g(0) = \exp(0) \exp(\zeta - 0) = \exp(\zeta). \quad \text{Hence,}$$
$$\exp(z) \exp(\zeta - z) = \exp(\zeta).$$

If we set $z = z_1$, and let $\zeta = z_1 + z_2$, we get $\exp(z_1)\exp(z_1 + z_2 - z_1) = \exp(z_1 + z_2)$, which simplifies to our desired result.

(iii) If θ is a real number, then by Definition 5.1,

$$e^{i\theta} = \exp(i\theta)$$

$$= \sum_{n=0}^{\infty} \frac{1}{n!} (i\theta)^n$$

$$(1) \quad = \sum_{n=0}^{\infty} \left[\frac{1}{(2n)!} (i\theta)^{2n} + \frac{1}{(2n+1)!} (i\theta)^{2n+1} \right] \quad \text{(by separating odd and even exponents)}$$

$$(2) \quad = \sum_{n=0}^{\infty} (-1)^n \frac{\theta^{2n}}{(2n)!} + i \sum_{n=0}^{\infty} (-1)^n \frac{\theta^{2n+1}}{(2n+1)!} \quad \text{(verify!)}$$

$$= \cos \theta + i \sin \theta \quad \text{(by the series representations for the real-valued sine and cosine functions.)}$$

Note that parts (ii) and (iii) of the Theorem 5.1 combine to verify De Moivre's formula, which we gave in Section 1.5 (see identity (3) of that section).

If $z = x + iy$, we also see from conclusions (ii) and (iii) that

$$(3) \quad \exp(z) = e^z = e^{x+iy} = e^x e^{iy} = e^x(\cos y + i \sin y).$$

Some texts start with equation (3) as their definition for $\exp(z)$. In the exercises, we show that this is a natural approach from the standpoint of differential equations.

The notation $\exp(z)$ is preferred over e^z in some situations. For example, $\exp(\frac{1}{5}) = 1.22140275816017 \ldots$ is the value of $\exp(z)$ when $z = 1/5$, and equals the positive fifth root of $e = 2.71828182845904. \ldots$. The notation $e^{1/5}$ is ambiguous and might be interpreted as any of the complex fifth roots of the number e that were discussed in Section 1.5:

$$e^{1/5} \approx 1.22140275816017 \left(\cos \frac{2\pi k}{5} + i \sin \frac{2\pi k}{5} \right), \text{ for } k = 0, 1, \ldots, 4.$$

To prevent this confusion, we often use $\exp(z)$ to denote the single-valued exponential function.

Properties of exp(z)

Since identity (3) involves the periodic functions $\cos y$ and $\sin y$, any two points in the z plane that lie on the same vertical line with their imaginary parts differing by an integral multiple of 2π are mapped onto the same point in the w plane. Thus, the complex exponential function is periodic with period $2\pi i$.

EXAMPLE 5.1   The points

$$z_n = \frac{5}{4} + i \left(\frac{11\pi}{6} + 2\pi n \right), \quad \text{for } n = 0, \pm 1, \pm 2, \ldots$$

in the z plane are mapped onto the single point

$$\begin{aligned} w_0 = \exp(z_n) &= e^{5/4} \left(\cos \frac{11\pi}{6} + i \sin \frac{11\pi}{6} \right) \\ &= \frac{\sqrt{3}}{2} e^{5/4} - \frac{i}{2} e^{5/4} \\ &= 3.022725669 \ldots - i1.745171479 \ldots \end{aligned}$$

in the w plane, as indicated in Figure 5.1.

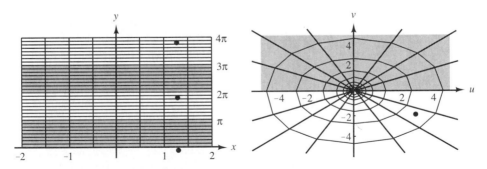

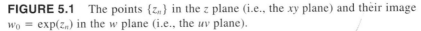

FIGURE 5.1   The points $\{z_n\}$ in the z plane (i.e., the xy plane) and their image $w_0 = \exp(z_n)$ in the w plane (i.e., the uv plane).

Some of the properties of the transformation $w = e^z = \exp(z)$ are:

(4) $e^{z+2n\pi i} = e^z$ for all z, provided n is an integer.

(5) $e^z = 1$ if and only if $z = 2n\pi i$, where n is an integer.

(6) $e^{z_1} = e^{z_2}$ if and only if $z_2 = z_1 + 2n\pi i$, where n is an integer.

Properties (4) through (6) are left for the reader to establish.

Let us investigate the modulus and argument of $w = e^z$. If $z = x + iy$, and we write w in its exponential form as $w = \rho e^{i\phi}$, equation (3) gives us

$$\rho e^{i\phi} = e^x e^{iy}.$$

Using property (4) of Section 1.5, the preceding implies

(7) $\rho = e^x$ and $\phi = y + 2n\pi$, where n is an integer. Therefore,

(8) $\rho = \left| e^z \right| = e^x$ and

(9) $\phi = \arg e^z = \text{Arg } e^z + 2n\pi$, where n is an integer.

If we solve the equations in statement (7) for x and y, we get

(10) $x = \ln \rho$, and $y = \phi + 2n\pi$, where n is an integer.

Thus, for any complex number $w \neq 0$, there are infinitely many complex numbers $z = x + iy$ such that $w = e^z$. From statement (10), we see that the numbers z are

(11) $z = x + iy = \ln \rho + i(\phi + 2n\pi)$
$$= \ln \left| w \right| + i(\text{Arg } w + 2n\pi), \quad \text{where } n \text{ is an integer.}$$

Hence

$$\exp[\ln \left| w \right| + i(\text{Arg } w + 2n\pi)] = w.$$

In summary, the transformation $w = e^z$ maps the complex plane (infinitely often) onto the entire set $S = \{w: w \neq 0\}$ of nonzero complex numbers.

If we restrict the solutions in statement (11) so that only the principal value of the argument, $-\pi < \text{Arg } w \leq \pi$, is used, we can also see that the transformation $w = e^z = e^{x+iy}$ maps the horizontal strip $-\pi < y \leq \pi$ one-to-one and onto the range set $S = \{w: w \neq 0\}$. This strip is called the *fundamental period strip* and is shown in Figure 5.2. Furthermore, the horizontal line $z = t + ib$, for $-\infty < t < \infty$ in the z plane, is mapped onto the ray $w = e^t e^{ib} = e^t(\cos b + i \sin b)$ that is inclined at an angle $\phi = b$ in the w plane. The vertical segment $z = a + i\theta$, for $-\pi < \theta \leq \pi$ in the z plane, is mapped onto the circle centered at the origin with radius e^a in the w plane. That is, $w = e^a e^{i\theta} = e^a(\cos \theta + i \sin \theta)$. In Figure 5.2, the lines r_1, r_2, and r_3 are mapped to the rays r_1^*, r_2^*, and r_3^*, respectively. Likewise, the segments s_1, s_2, and s_3 are mapped to the corresponding circles, s_1^*, s_2^*, and s_3^*, respectively.

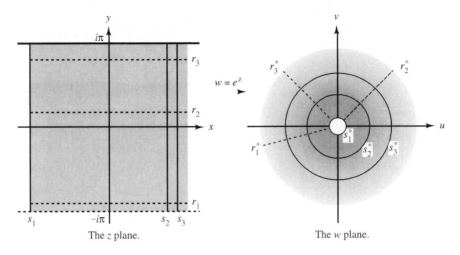

FIGURE 5.2   The fundamental period strip for the mapping $w = \exp(z)$.

EXAMPLE 5.2   Consider a rectangle $R = \{(x, y): a \le x \le b \text{ and } c \le y \le d\}$, where $d - c < 2\pi$. Show that the transformation $w = e^z = e^{x+iy}$ maps R onto a portion of an annular region bounded by two rays.

Solution   The image points in the w plane satisfy the following relationships involving the modulus and argument of w:

$$e^a = \left| e^{a+iy} \right| \le \left| e^{x+iy} \right| \le \left| e^{b+iy} \right| = e^b,$$
$$c = \arg(e^{x+ic}) \le \arg(e^{x+iy}) \le \arg(e^{x+id}) \le d,$$

which is a portion of the annulus $\{\rho e^{i\phi}: e^a \le \rho \le e^b\}$ in the w plane subtended by the rays $\phi = c$ and $\phi = d$. In Figure 5.3, we show the image of the rectangle $R = \left\{(x, y): -1 \le x \le 1 \text{ and } \dfrac{-\pi}{4} \le y \le \dfrac{\pi}{3}\right\}.$

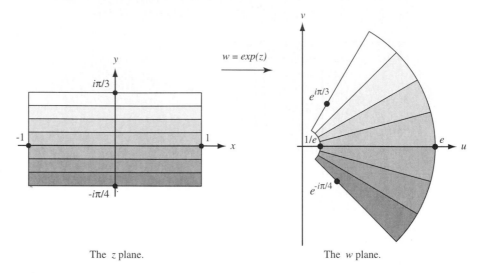

The z plane.   The w plane.

FIGURE 5.3   The image of R under the transformation $w = \exp(z)$.

EXERCISES FOR SECTION 5.1

1. Using Definition 5.1, explain why $\exp(0) = e^0 = 1$.

2. Give a careful justification for statements (1) and (2) in the proof of Theorem 5.1, part (iii).

3. An alternate definition for $\exp(z)$: Many texts take the approach of starting with identity (3) as the definition for $f(z) = \exp(z)$. This exercise shows this is a natural approach from the standpoint of differential equations. Our starting point is that we require $f(z)$ to be the solution to an initial-value problem satisfying the following three properties: (1) f is entire, (2) $f'(z) = f(z)$ for all z, and (3) $f(0) = 1$. Suppose $f(z) = f(x + iy) = u(x, y) + iv(x, y)$ satisfies properties (1), (2), and (3).

   (a) Use the result $f'(z) = u_x(x, y) + iv_x(x, y)$ and the requirement $f'(z) = f(z)$ from property (2) to show that $u_x(x, y) - u(x, y) = 0$.

   (b) Show the result in part (a) implies $\dfrac{\partial}{\partial x}[u(x, y)e^{-x}] = 0$. Note that integrating this last equation partially with respect to x gives $u(x, y)e^{-x} = p(y)$, where $p(y)$ is a function of y alone.

   (c) Using a similar procedure for $v(x, y)$, show we wind up getting the pair of solutions $u(x, y) = p(y)e^x$, and $v(x, y) = q(y)e^x$, where $p(y)$ and $q(y)$ are functions of y alone.

   (d) Now use the Cauchy-Riemann equations to conclude from part (c) that $p(y) = q'(y)$ and $p'(y) = -q(y)$.

   (e) Show the result in part (d) implies $p''(y) + p(y) = 0$ and $q''(y) + q(y) = 0$.

   (f) Given that $f(0) = f(0 + 0i) = u(0, 0) + iv(0, 0) = 1 + 0i$, together with the known solutions to the equations in part (e), conclude that identity (3) follows.

4. Verify properties (4) through (6) given in the text.

5. The questions for this problem relate to Figure 5.2. The shaded portion in the w plane indicates the image of the shaded portion in the z plane, with the darkness of shading indicating the density of corresponding points.

(a) Why is there no shading inside the circle s_1^*?

(b) Explain why the images of r_1, r_2, and r_3 appear to make, respectively, angles of $-7\pi/8$, $\pi/4$, and $3\pi/4$ radians with the positive u axis.

(c) Why does the shading get lighter in the w plane as the distance from the origin increases?

(d) Describe precisely where the images of the points $\pm i\pi$ (in the z plane) should be located in the w plane.

6. Show that $\exp(z + i\pi) = \exp(z - i\pi)$ holds for all z.

7. Find the value of $e^z = u + iv$ for the following values of z.

(a) $\dfrac{-i\pi}{3}$

(b) $\dfrac{1}{2} - \dfrac{i\pi}{4}$

(c) $-4 + i5$

(d) $-1 + \dfrac{i3\pi}{2}$

(e) $1 + \dfrac{i5\pi}{4}$

(f) $\dfrac{\pi}{3} - 2i$

8. Find all values of z for which the following equations hold.

(a) $e^z = -4$

(b) $e^z = 2 + 2i$

(c) $e^z = \sqrt{3} - i$

(d) $e^z = -1 + i\sqrt{3}$

9. Express $\exp(z^2)$ and $\exp(1/z)$ in Cartesian form $u(x, y) + iv(x, y)$.

10. Show that:

(a) $\exp(\bar{z}) = \overline{\exp z}$ holds for all z and that

(b) $\exp(\bar{z})$ is nowhere analytic.

11. Show that $\left| e^{-z} \right| < 1$ if and only if $\operatorname{Re}(z) > 0$.

12. Show that:

(a) $\displaystyle\lim_{z \to 0} \dfrac{e^z - 1}{z} = 1$

(b) $\displaystyle\lim_{z \to i\pi} \dfrac{e^z + 1}{z - i\pi} = -1$

13. Show that $f(z) = ze^z$ is analytic for all z by showing that its real and imaginary parts satisfy the Cauchy-Riemann equations.

14. Find the derivatives of the following:

(a) e^{iz}

(b) $z^4 \exp(z^3)$

(c) $e^{(a+ib)z}$

(d) $\exp(1/z)$

15. Let n be a positive integer. Show that:

(a) $(\exp z)^n = \exp(nz)$

(b) $\dfrac{1}{(\exp z)^n} = \exp(-nz)$

16. Show that the image of the horizontal ray $x > 0$, $y = \pi/3$ under the mapping $w = \exp z$ is a ray.

17. Show that the image of the vertical line segment $x = 2$, $y = t$, where $\pi/6 < t < 7\pi/6$ under the mapping $w = e^z$ is half of a circle.

18. Use the fact that $\exp(z^2)$ is analytic to show that $e^{x^2-y^2} \sin 2xy$ is a harmonic function.

19. Show that the image of the line $x = t$, $y = 2\pi + t$, where $-\infty < t < \infty$ under the mapping $w = \exp z$ is a spiral.

20. Show that the image of the first quadrant $x > 0$, $y > 0$ under the mapping $w = \exp z$ is the region $\left| w \right| > 1$.

21. Let α be a real constant. Show that the mapping $w = \exp z$ maps the horizontal strip $\alpha < y \le \alpha + 2\pi$ one-to-one and onto the range $\left| w \right| > 0$.

22. Explain how the complex function e^z and the function e^x studied in calculus are different. How are they similar?

5.2   Branches of the Complex Logarithm Function

If w is a nonzero complex number, then the equation $w = \exp z$ has infinitely many solutions. Because the function $\exp(z)$ is a many-to-one function, its inverse (the logarithm) is necessarily multivalued. Special consideration must be made to define branches of the logarithm that are one-to-one.

> **Definition 5.2 (Multivalued Logarithm)**   *For $z \neq 0$, we define the function $\log(z)$ as the inverse of the exponential function; that is,*

(1) $\log(z) = w$ *if and only if* $z = \exp(w)$.

> *If we go through the same steps as we did in equations (10) and (11) of Section 5.1, we will see that the values of w that solve equation (1) are*

(2) $\log(z) = \ln|z| + i \arg(z)$ $(z \neq 0)$,

> *where $\arg(z)$ is an argument of z and $\ln|z|$ denotes the natural logarithm of the positive number $|z|$. Using identity (5) in Section 1.4, we can write*

(3) $\log(z) = \ln|z| + i[\mathrm{Arg}(z) + 2\pi n]$,   *where n is an integer.*

We call any one of the values given in equations (2) or (3) a logarithm of z. Notice that the different values of $\log(z)$ all have the same real part and that their imaginary parts differ by the amount $2\pi n$, where n is an integer.

> **Definition 5.3 (Principal Value of the Logarithm)**   *For $z \neq 0$, we define the principal value of the logarithm as follows:*

(4) $\mathrm{Log}(z) = \ln|z| + i\,\mathrm{Arg}(z)$, *where $|z| > 0$ and $-\pi < \mathrm{Arg}(z) \leq \pi$.*

The domain for $f(z) = \mathrm{Log}(z) = w$ is the set of all nonzero complex numbers in the z plane, and its range is the horizontal strip $-\pi < \mathrm{Im}(w) \leq \pi$ in the w plane. Notice that $\mathrm{Log}(z)$ is a single-valued function and corresponds to setting $n = 0$ in formula (3). This is the choice used by popular software programs such as *Mathematica*™, which we used to produce Figure 5.4. Parts (a) and (b) of this figure give the real and imaginary parts (u and v, respectively) for $\mathrm{Log}(z) = \mathrm{Log}(x + iy) = u(x, y) + iv(x, y)$. Figure 5.4(b) illustrates a phenomenon inherent in constructing a logarithm function: It must have a discontinuity! This is the case because as we saw in Chapter 2, any branch we choose for $\arg(z)$ is necessarily a discontinuous function. The principal branch, $\mathrm{Arg}(z)$, is discontinuous at each point along the negative x axis.

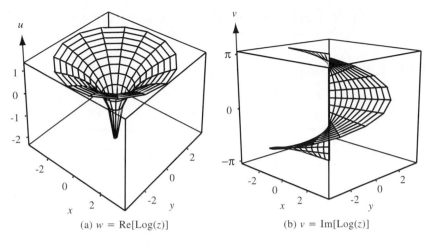

(a) $w = \text{Re}[\text{Log}(z)]$ (b) $v = \text{Im}[\text{Log}(z)]$

FIGURE 5.4   The real and imaginary parts of $\text{Log}(z)$.

EXAMPLE 5.3   By standard computations, we have

$$\log(1 + i) = \ln|1 + i| + i\arg(1 + i) = \frac{\ln 2}{2} + i\left(\frac{\pi}{4} + 2\pi n\right),$$

where n is an integer; the principal value is

$$\text{Log}(1 + i) = \frac{\ln 2}{2} + i\frac{\pi}{4}.$$

Properties of log(z) and Log(z)

From equations (1) and (4), it follows that

(5) $\exp[\text{Log}(z)] = z$ for all $z \neq 0$,

and that the mapping $w = \text{Log}\, z$ is one-to-one from the domain $D = \{z: |z| > 0\}$ in the z plane onto the horizontal strip $-\pi < \text{Im}(w) \leq \pi$ in the w plane. The mapping $w = \text{Log}(z)$ and its branch cut are shown in Figure 5.5.

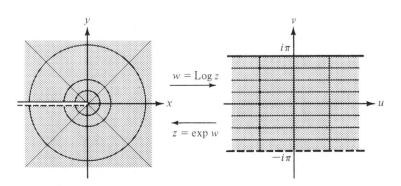

FIGURE 5.5   The single-valued mapping $w = \text{Log}(z)$.

Even though Log(z) is not continuous along the negative real axis, it is still possible to compute the logarithm of negative numbers.

EXAMPLE 5.4   By using equation (4) we see that

(a) $\text{Log}(-e) = \ln|-e| + i\,\text{Arg}(-e) = 1 + i\pi$,   and

(b) $\text{Log}(-1) = \ln|-1| + i\,\text{Arg}(-1) = i\pi$.

When $z = x + i0$, where x is a positive real number, the principal value of the complex logarithm of z is

(6) $\text{Log}(x + i0) = \ln x$,   where $x > 0$.

Hence Log(z) is an extension of the real function $\ln x$ to the complex case. To find the derivative of Log(z), we use polar coordinates $z = re^{i\theta}$, and formula (4) becomes

(7) $\text{Log}(z) = \ln r + i\theta$,   where $-\pi < \theta \le \pi$ and $r > 0$.

In equation (7) we have $u(r, \theta) = \ln r$ and $v(r, \theta) = \theta$. The polar form of the Cauchy-Riemann equations are

(8) $u_r = \dfrac{1}{r} v_\theta = \dfrac{1}{r}$ and $v_r = \dfrac{-1}{r} u_\theta = 0$

and appear to hold for all $z \ne 0$. But since Log(z) is discontinuous along the negative x axis where $\theta = \pi$, the Cauchy-Riemann equations (8) are valid only for $-\pi < \theta < \pi$. The derivative of Log(z) is found by using the results of equations (8) and identity (15) of Section 3.2, and we find that

(9) $\dfrac{d}{dz}\text{Log}(z) = e^{-i\theta}(u_r + iv_r) = \dfrac{1}{r}e^{-i\theta} = \dfrac{1}{z}$,   where $-\pi < \theta < \pi$ and $r > 0$.

Let z_1 and z_2 be nonzero complex numbers. The multivalued function log z obeys the familiar properties of logarithms:

(10) $\log(z_1 z_2) = \log(z_1) + \log(z_2)$,

(11) $\log(z_1/z_2) = \log(z_1) - \log(z_2)$,   and

(12) $\log(1/z) = -\log(z)$.

Identity (10) is easy to establish. Using identity (12) in Section 1.4 concerning the argument of a product, we write

$$\begin{aligned}
\log(z_1 z_2) &= \ln|z_1||z_2| + i\,\text{arg}(z_1 z_2) \\
&= \ln|z_1| + \ln|z_2| + i\,\text{arg}(z_1) + i\,\text{arg}(z_2) \\
&= [\ln|z_1| + i\,\text{arg}(z_1)] + [\ln|z_2| + i\,\text{arg}(z_2)] = \log(z_1) + \log(z_2).
\end{aligned}$$

Identities (11) and (12) are easy to verify and are left as exercises.

It should be noted that identities (10)–(12) do not in general hold true when log(z) is replaced everywhere by Log(z). For example, if we make the specific choices $z_1 = -\sqrt{3} + i$ and $z_2 = -1 + i\sqrt{3}$, then their product is $z_1 z_2 = -4i$. Computing the principal value of the logarithms, we find that

$$\text{Log}(z_1 z_2) = \text{Log}(-4i) = \ln 4 - \frac{i\pi}{2}.$$

The sum of the logarithms is given by

$$\text{Log}(z_1) + \text{Log}(z_2) = \text{Log}(-\sqrt{3} + i) + \text{Log}(-1 + i\sqrt{3})$$

$$= \ln 2 + \frac{i5\pi}{6} + \ln 2 + \frac{i2\pi}{3} = \ln 4 + \frac{i3\pi}{2},$$

and identity (10) does not hold for these principal values of the logarithm.

We can construct many different branches of the multivalued logarithm function. Let α denote a fixed real number, and choose the value of $\theta = \arg(z)$ that lies in the range $\alpha < \theta \le \alpha + 2\pi$. Then the function

(13) $f(z) = \ln r + i\theta,$ where $r > 0$ and $\alpha < \theta \le \alpha + 2\pi$

is a single-valued branch of the logarithm function. The branch cut for f is the ray $\theta = \alpha$, and each point along this ray is a point of discontinuity of f. Since $\exp[f(z)] = z$, we conclude that the mapping $w = f(z)$ is a one-to-one mapping of the domain $|z| > 0$ onto the horizontal strip $\alpha < \text{Im}(w) \le \alpha + 2\pi$. If $\alpha < c < d < \alpha + 2\pi$, then the function $w = f(z)$ maps the set $D = \{re^{i\theta}: a < r < b, c < \theta < d\}$ one-to-one and onto the rectangle $R = \{u + iv: \ln a < u < \ln b, c < v < d\}$. The mapping $w = f(z)$, its branch cut $\theta = \alpha$, and the set D and its image R are shown in Figure 5.6.

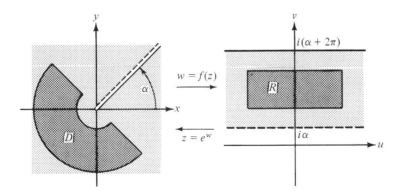

FIGURE 5.6   The branch $w = f(z)$ of the logarithm.

Any branch of the logarithm function is discontinuous along its branch cut. In particular, $f(z) = \text{Log}(z)$ is discontinuous at $z = 0$ and at points on the negative x axis. Hence the derivative of $\text{Log}(z)$ does not exist at $z = 0$ and at points on the negative x axis. However, if we focus our attention on the multivalued function $w = \log(z)$, then implicit differentiation will permit us to use the formula $dw/dz = 1/z$. This can be justified by starting with the second formula in equations (1) and differentiating implicitly with respect to z to get

$$(14) \quad 1 = \exp(w)\,\frac{dw}{dz}.$$

We can substitute $\exp(w) = z$ in equation (14) and obtain

$$\frac{dw}{dz} = \frac{1}{\exp(w)} = \frac{1}{z}.$$

Therefore we have shown that

$$(15) \quad \frac{d}{dz}\log(z) = \frac{1}{z}$$

holds for all $z \neq 0$.

It is appropriate to consider the Riemann surface for the multivalued function $w = \log(z)$. This requires infinitely many copies of the z plane cut along the negative x axis, which we will label S_k for $k = \ldots, -n, \ldots, -1, 0, 1, \ldots, n, \ldots$. Now stack these cut planes directly upon each other so that the corresponding points have the same position. Join the sheet S_k to S_{k+1} as follows. For each integer k the edge of the sheet S_k in the upper half plane is joined to the edge of the sheet S_{k+1} in the lower half plane. The Riemann surface for the domain of $\log(z)$ looks like a spiral staircase that extends upward on the sheets $S_1, S_2, \ldots$ and downward on the sheets $S_{-1}, S_{-2}, \ldots$, as shown in Figure 5.7. Polar coordinates are used for z on each sheet:

$$(16) \quad \text{on } S_k, \text{ use } \quad z = r(\cos\theta + i\sin\theta), \quad \text{where}$$
$$r = |z| \text{ and } 2\pi k - \pi < \theta \le \pi + 2\pi k.$$

The correct branch of $\log(z)$ on each sheet is

$$(17) \quad \text{on } S_k, \text{ use } \quad \log(z) = \ln r + i\theta, \quad \text{where}$$
$$r = |z| \text{ and } 2\pi k - \pi < \theta \le \pi + 2\pi k.$$

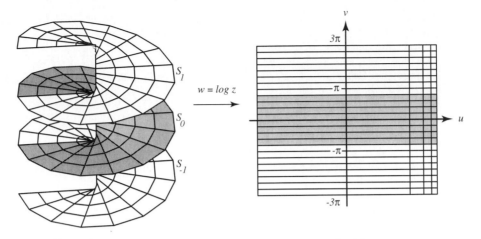

FIGURE 5.7   The Riemann surface for the mapping $w = \log(z)$.

EXERCISES FOR SECTION 5.2

1. Find the principal value $\text{Log}(z) = u + iv$ for the following:
   (a) $\text{Log}(ie^2)$
   (b) $\text{Log}(\sqrt{3} - i)$
   (c) $\text{Log}(i\sqrt{2} - \sqrt{2})$
   (d) $\text{Log}[(1 + i)^4]$

2. Find *all* the values of $\log(z)$ for the following:
   (a) $\log(-3)$
   (b) $\log(4i)$
   (c) $\log 8$
   (d) $\log(-\sqrt{3} - i)$

3. Find *all* the values of z for which the following equations hold:

   (a) $\text{Log}(z) = 1 - \dfrac{i\pi}{4}$
   (b) $\text{Log}(z - 1) = \dfrac{i\pi}{2}$

   (c) $\exp(z) = -ie$
   (d) $\exp(z + 1) = i$

4. Use properties of $\arg(z)$ in Section 1.4 to establish identity (11).

5. Use properties of $\arg(z)$ in Section 1.4 to establish identity (12).

6. Show that $\log(z) = \text{Log}(z) + i2\pi n$ where n is an integer.

7. Let $w = \log(f(z))$. Use implicit differentiation to find dw/dz at points where $f(z)$ is analytic and nonzero.

8. Use implicit differentiation to find dw/dz for the following:
   (a) $w = \log(z^2 - z + 2)$
   (b) $w = z \log(z)$

9. Show that $f(z) = \text{Log}(iz)$ is analytic everywhere except at points on the ray given by $y \geq 0$, $x = 0$. Find $f'(z)$.

10. Show that $f(z) = [\text{Log}(z + 5)]/(z^2 + 3z + 2)$ is analytic everywhere except at the points -1, -2 and at points on the ray $x \leq -5$, $y = 0$.

11. Show that if $\text{Re}(z) > 0$, then $\text{Log } z = \frac{1}{2} \ln(x^2 + y^2) + i \arctan(y/x)$, where the principal branch of the real function $\arctan t$ is used; that is, $-\pi/2 < \arctan t < \pi/2$.

12. Show that:
    (a) $\ln(x^2 + y^2)$ (b) $\arctan(y/x)$
    are harmonic functions in the right half plane $\operatorname{Re}(z) > 0$.
13. Show that $z^n = \exp[n \log(z)]$ where n is an integer.
14. (a) Show that $\operatorname{Log}(z_1 z_2) = \operatorname{Log}(z_1) + \operatorname{Log}(z_2)$ holds true provided that $\operatorname{Re}(z_1) > 0$ and $\operatorname{Re}(z_2) > 0$.
    (b) Does $\operatorname{Log}[(-1 + i)^2] = 2 \operatorname{Log}(-1 + i)$?
15. (a) Is it always true that $\operatorname{Log}(1/z) = -\operatorname{Log}(z)$? Why?
    (b) Is it always true that $\dfrac{d}{dz} \operatorname{Log}(z) = \dfrac{1}{z}$? Why?
16. Construct branches of $\log(z + 2)$ that are analytic at all points in the plane except at points on the following rays:
    (a) $x \geq -2,\ y = 0$ (b) $x = -2,\ y \geq 0$ (c) $x = -2,\ y \leq 0$
17. Construct a branch of $\log(z + 4)$ that is analytic at the point $z = -5$ and takes on the value $7\pi i$ there.
18. Using the polar coordinate notation $z = re^{i\theta}$, discuss the possible interpretations of the function $f(z) = \ln r + i\theta$.
19. Show that the mapping $w = \operatorname{Log}(z)$ maps the ray $r > 0$, $\theta = \pi/3$ one-to-one and onto the horizontal line $v = \pi/3$.
20. Show that the mapping $w = \operatorname{Log}(z)$ maps the semicircle $r = 2$, $-\pi/2 \leq \theta \leq \pi/2$ one-to-one and onto the vertical line segment $u = \ln 2$, $-\pi/2 \leq v \leq \pi/2$.
21. Find specific values of z_1 and z_2 so that $\operatorname{Log}(z_1/z_2) \neq \operatorname{Log}(z_1) - \operatorname{Log}(z_2)$.
22. Show that $\log[\exp(z)] = z + i2\pi n$, where n is an integer.
23. Show why the solutions to equation (1) are given by those in equation (2). *Hint*: Mimic the process used in obtaining identities (10) and (11) of Section 5.1.
24. Explain why the function $\operatorname{Log}(z)$ is not defined when $z = 0$.

5.3   Complex Exponents

In Section 1.5 we indicated that the complex numbers are complete in the sense that it is possible to make sense out of expressions such as $\sqrt{1 + i}$ or $(-1)^i$ without having to appeal to a number system beyond the framework of complex numbers. We are now in a position to come up with a meaningful definition of what is meant by a complex number raised to a complex power.

> **Definition 5.4**   *Let c be a complex number. We define z^c by the equation*

(1) $z^c = \exp[c \log(z)]$.

This definition makes sense, since if both z and c are real numbers with $z > 0$, equation (1) gives the familiar (real) definition for z^c.

Since $\log(z)$ is multivalued, the function z^c will in general be multivalued. The function f given by

(2) $f(z) = \exp[c \operatorname{Log}(z)]$

is called the *principal branch* of the multivalued function z^c. Note that the principal branch of z^c is obtained from equation (1) by replacing $\log(z)$ with the principal branch of the logarithm.

EXAMPLE 5.5   Find the principal values of $\sqrt{1 + i}$ and $(-1)^i$.

Solution   From Examples 5.3 and 5.4,

$$\mathrm{Log}(1 + i) = \frac{\ln 2}{2} + i\frac{\pi}{4} = \ln 2^{1/2} + i\frac{\pi}{4}, \quad \text{and}$$

$$\mathrm{Log}(-1) = i\pi.$$

By using equation (1), therefore, we see that the principal values of $\sqrt{1 + i}$ and $(-1)^i$ are

$$\sqrt{1 + i} = (1 + i)^{1/2}$$

$$= \exp\left[\frac{1}{2}\left(\ln 2^{1/2} + i\frac{\pi}{4}\right)\right]$$

$$= \exp\left(\ln 2^{1/4} + i\frac{\pi}{8}\right)$$

$$= 2^{1/4}\left(\cos\frac{\pi}{8} + i\sin\frac{\pi}{8}\right)$$

$$\approx 1.09684 + 0.45509i, \quad \text{and}$$

$$(-1)^i = \exp(i(i\pi))$$

$$= \exp(-\pi)$$

$$\approx 0.04321.$$

It is interesting that the result of raising a real number to a complex power may be a real number in a nontrivial way.

Let us now consider the various possibilities that may arise when we apply formula (1):

Case (i): Let us suppose $c = k$, where k is an integer. Then if $z = re^{i\theta}$,

$$k \log(z) = k \ln(r) + ik(\theta + 2\pi n),$$

where n is an integer. Recalling that the complex exponential function has period $2\pi i$, we have

$$(3) \quad z^k = \exp[k \log(z)]$$

$$= \exp[k \ln(r) + ik(\theta + 2\pi n)]$$

$$= r^k \exp(ik\theta) = r^k(\cos k\theta + i\sin k\theta),$$

which is the single-valued kth power of z that was studied in Section 1.5.

Case (ii): If $c = 1/k$, where k is an integer, and $z = re^{i\theta}$, then

$$\frac{1}{k}\log z = \frac{1}{k}\ln r + \frac{i(\theta + 2\pi n)}{k},$$

where n is an integer. Hence, formula (1) becomes

(4) $z^{1/k} = \exp\left[\dfrac{1}{k}\log(z)\right]$

$\quad = \exp\left[\dfrac{1}{k}\ln(r) + i\,\dfrac{\theta + 2\pi n}{k}\right]$

$\quad = r^{1/k}\exp\left(i\,\dfrac{\theta + 2\pi n}{k}\right) = r^{1/k}\left[\cos\left(\dfrac{\theta + 2\pi n}{k}\right) + i\sin\left(\dfrac{\theta + 2\pi n}{k}\right)\right].$

Using again the periodicity of the complex exponential function, we see that equation (4) gives k distinct values corresponding to $n = 0, 1, \ldots, k - 1$. Therefore, the fractional power $z^{1/k}$ is the multivalued kth root function.

Case (iii): If j and k are positive integers that have no common factors, and $c = j/k$, then formula (1) becomes

(5) $z^{j/k} = r^{j/k}\exp\left[i\,\dfrac{(\theta + 2\pi n)j}{k}\right] = r^{j/k}\left[\cos\left(\dfrac{(\theta + 2\pi n)j}{k}\right) + i\sin\left(\dfrac{(\theta + 2\pi n)j}{k}\right)\right],$

and again there are k distinct values corresponding to $n = 0, 1, \ldots, k - 1$.

Case (iv): If c is not a rational number, then there are infinitely many values for z^c.

EXAMPLE 5.6   The values of $2^{1/9 + i/50}$ are given by

$2^{1/9 + i/50} = \exp\left[\left(\dfrac{1}{9} + \dfrac{i}{50}\right)(\ln 2 + i2\pi n)\right]$

$\quad = \exp\left[\dfrac{\ln 2}{9} - \dfrac{n\pi}{25} + i\left(\dfrac{\ln 2}{50} + \dfrac{2\pi n}{9}\right)\right]$

$\quad = 2^{1/9}e^{-n\pi/25}\left[\cos\left(\dfrac{\ln 2}{50} + \dfrac{2\pi n}{9}\right) + i\sin\left(\dfrac{\ln 2}{50} + \dfrac{2\pi n}{9}\right)\right],$

where n is an integer, and the principal value of $2^{1/9 + i/50}$ is given by

$2^{1/9 + i/50} = 2^{1/9}\left[\cos\left(\dfrac{\ln 2}{50}\right) + i\sin\left(\dfrac{\ln 2}{50}\right)\right] \approx 1.079956 + 0.014972i.$

Terms for the multivalued expression with $n = -9, -8, \ldots, 8, 9$ are shown in Figure 5.8, and exhibit a spiral pattern that is often present in complex powers.

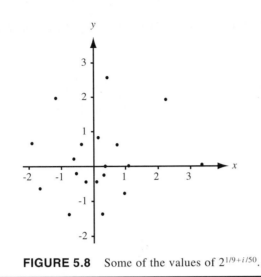

FIGURE 5.8　Some of the values of $2^{1/9+i/50}$.

Some of the rules for exponents carry over from the real case. If c and d are complex numbers and $z \neq 0$, then

(6)　　$z^{-c} = \dfrac{1}{z^c}$;

(7)　　$z^c z^d = z^{c+d}$;

(8)　　$\dfrac{z^c}{z^d} = z^{c-d}$;

(9)　　$(z^c)^n = z^{cn}$,　where n is an integer.

Identity (9) does not hold if n is replaced with an arbitrary complex value.

EXAMPLE 5.7

$(i^2)^i = \exp[i \log(-1)] = e^{-(1+2n)\pi}$,　where n is an integer and
$(i)^{2i} = \exp(2i \log i) = e^{-(1+4n)\pi}$,　where n is an integer.

Since these sets of solutions are not equal, identity (9) does not always hold.

The derivative of $f(z) = \exp[c \, \text{Log}(z)]$ can be found by using the chain rule, and we see that

(10) $f'(z) = \dfrac{c}{z} \exp[c \, \text{Log}(z)].$

If we use $z^c = \exp[c \, \text{Log}(z)]$, then equation (10) can be written in the familiar form we learn in calculus. That is,

(11) $\dfrac{d}{dz} z^c = \dfrac{c}{z} z^c,$

which holds true when z^c is the principal value, when z lies in the domain $r > 0$, $-\pi < \theta < \pi$, and when c is a complex number.

We can use definition (1) to define the exponential function with base b where $b \neq 0$ is a complex number, and we write

(12) $b^z = \exp[z \, \log(b)].$

If a value of $\log(b)$ is specified, then b^z in equation (12) can be made single-valued, and the rules of differentiation can be used to show that the resulting branch of b^z is an entire function. The derivative of b^z is then given by the familiar rule

(13) $\dfrac{d}{dz} b^z = b^z \log(b).$

EXERCISES FOR SECTION 5.3

1. Find the principal value of the following:
   (a) 4^i (b) $(1 + i)^{\pi i}$ (c) $(-1)^{1/\pi}$ (d) $(1 + i\sqrt{3})^{i/2}$
2. Find *all* values of the following:
   (a) i^i (b) $(-1)^{\sqrt{2}}$ (c) $(i)^{2/\pi}$ (d) $(1 + i)^{2-i}$
3. Show that if $z \neq 0$, then z^0 has a unique value.
4. Find *all* values of $(-1)^{3/4}$ and $(i)^{2/3}$.
5. Use polar coordinates $z = re^{i\theta}$, and show that the principal branch of z^i is given by the equation

   $$z^i = e^{-\theta}[\cos(\ln r) + i \sin(\ln r)], \quad \text{where } r > 0 \text{ and } -\pi < \theta \leq \pi.$$

6. Let α be a real number. Show that the principal branch of z^α is given by the equation

   $$z^\alpha = r^\alpha \cos \alpha\theta + ir^\alpha \sin \alpha\theta, \quad \text{where } -\pi < \theta \leq \pi.$$

   Find $(d/dz)z^\alpha$.
7. Establish identity (13); that is, $(d/dz)b^z = b^z \log(b)$.
8. Let $z_n = (1 + i)^n$ for $n = 1, 2, \ldots$. Show that the sequence $\{z_n\}$ is a solution to the linear difference equation

   $$z_n = 2z_{n-1} - 2z_{n-2} \quad \text{for } n = 3, 4, \ldots .$$

   Hint: Show that the equation holds true when the values z_n, z_{n-1}, and z_{n-2} are substituted.
9. Verify identity (6).   10. Verify identity (7).
11. Verify identity (8).   12. Verify identity (9).

13. Is 1 raised to any power always equal to 1? Why?
14. Construct an example that shows that the principal value of $(z_1z_2)^{1/3}$ need not be equal to the product of the principal values of $z_1^{1/3} \, z_2^{1/3}$.

5.4   Trigonometric and Hyperbolic Functions

Given the success we have had in using power series to define the complex exponential, we have reason to believe this approach will be fruitful for other elementary functions as well. The power series expansions for the real-valued sine and cosine functions are

$$\sin x = \sum_{n=0}^{\infty} (-1)^n \frac{x^{2n+1}}{(2n+1)!} \quad \text{and} \quad \cos x = \sum_{n=0}^{\infty} (-1)^n \frac{x^{2n}}{(2n)!}.$$

Thus, it is natural to make the following definitions.

Definition 5.5

$$\sin z = \sum_{n=0}^{\infty} (-1)^n \frac{z^{2n+1}}{(2n+1)!}, \quad \cos z = \sum_{n=0}^{\infty} (-1)^n \frac{z^{2n}}{(2n)!}.$$

Clearly, these definitions agree with their real counterparts when z is real. Additionally, it is easy to show that $\cos z$ and $\sin z$ are entire functions. (We leave the argument as an exercise.) Their derivatives can be computed by appealing to Theorem 4.13, part (ii):

$$(1) \quad \frac{d}{dz} \sin z = \frac{d}{dz} \left[\sum_{n=0}^{\infty} (-1)^n \frac{z^{2n+1}}{(2n+1)!} \right]$$

$$= \sum_{n=0}^{\infty} (-1)^n \frac{(2n+1)z^{2n}}{(2n+1)!} \quad \text{(Exercise: Explain why the index } n \text{ stays at 0 here.)}$$

$$= \sum_{n=0}^{\infty} (-1)^n \frac{z^{2n}}{(2n)!}$$

$$= \cos z.$$

It is left as an exercise to show that

$$(2) \quad \frac{d}{dz} \cos z = -\sin z.$$

We also ask you to establish that for all complex values z,

$$(3) \quad \sin^2 z + \cos^2 z = 1.$$

The other trigonometric functions are defined in terms of the sine and cosine functions and are given by

$$(4) \quad \tan z = \frac{\sin z}{\cos z}, \quad \cot z = \frac{\cos z}{\sin z};$$

(5) $\sec z = \dfrac{1}{\cos z}$, $\csc z = \dfrac{1}{\sin z}$.

The rules for differentiating a quotient can now be used in equations (4) and (5) together with identity (3) to establish the differentiation formulas

(6) $\dfrac{d}{dz} \tan z = \sec^2 z$, $\dfrac{d}{dz} \cot z = -\csc^2 z$;

(7) $\dfrac{d}{dz} \sec z = \sec z \tan z$, $\dfrac{d}{dz} \csc z = -\csc z \cot z$.

Many algebraic properties of the sine and cosine extend to the complex domain. It is easy to show by appealing to Definition 5.5 that

(8) $\sin(-z) = -\sin z$ and $\cos(-z) = \cos z$ for all z.

To establish other properties, it will be useful to have formulas to compute $\cos z$ and $\sin z$ that are of the form $u + iv$. (Additionally, the applications in Chapters 9 and 10 will use these formulas.) We begin by observing that the argument given to prove part (iii) in Theorem 5.1 easily generalizes to the complex case with the aid of Definition 5.5. That is,

(9) $e^{iz} = \cos z + i \sin z$

for all z, whether z is real or complex. Hence,

(10) $e^{-iz} = \cos(-z) + i \sin(-z) = \cos z - i \sin z$.

Subtracting equation (10) from equation (9) and solving for $\sin z$ gives

(11) $\sin z = \dfrac{1}{2i} (e^{iz} - e^{-iz}) = \dfrac{1}{2i} (e^{-y+ix} - e^{y-ix})$.

Now we appeal to Theorem 5.1, parts (ii) and (iii), to get

$$\sin z = \frac{1}{2i} [e^{-y}(\cos x + i \sin x) - e^{y}(\cos x - i \sin x)]$$

$$= \sin x \, \frac{e^y + e^{-y}}{2} + i \cos x \, \frac{e^y - e^{-y}}{2}.$$

Similarly

(12) $\cos z = \tfrac{1}{2}(e^{iz} + e^{-iz}) = \tfrac{1}{2}(e^{-y+ix} + e^{y-ix})$

$$= \cos x \left(\frac{e^y + e^{-y}}{2} \right) - i \sin x \left(\frac{e^y - e^{-y}}{2} \right).$$

You may recall that the hyperbolic cosine and hyperbolic sine of the real variable y are

(13) $\cosh y = \dfrac{e^y + e^{-y}}{2}$ and $\sinh y = \dfrac{e^y - e^{-y}}{2}$.

Substituting identities (13) into the proper places of equation (12) gives

(14) $\sin z = \sin x \cosh y + i \cos x \sinh y$.

A similar derivation leads to the formula for $\cos z$:

(15) $\cos z = \cos x \cosh y - i \sin x \sinh y$.

A graph of the mapping $w = \sin z = \sin(x + iy) = \sin x \cosh y + i \cos x \sinh y$ can be obtained parametrically. Consider the vertical line segments in the z plane obtained by successively setting $x = \dfrac{-\pi}{2} + \dfrac{k\pi}{12}$ for $k = 0, 1, \dots, 12$, and for each x value letting y vary continuously, $-3 \le y \le 3$. The image of these vertical segments are hyperbolas in the uv plane, as Figure 5.9 illustrates. We will give a more detailed investigation of the mapping $w = \sin z$ in Chapter 9.

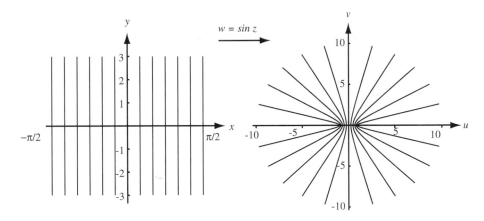

FIGURE 5.9   Vertical segments are mapped onto hyperbolas by $w = \sin(z)$.

Identities (14) and (15) can be used to investigate the periodic character of the trigonometric functions, and we have

(16) $\cos(z + 2\pi) = \cos z$ and $\sin(z + 2\pi) = \sin z$

and

(17) $\cos(z + \pi) = -\cos z$ and $\sin(z + \pi) = -\sin z$.

Identities (17) can be used to show that

(18) $\tan(z + \pi) = \tan z$ and $\cot(z + \pi) = \cot z$.

A solution of the equation $f(z) = 0$ is called a *zero* of the given function f. The zeros of the sine and cosine function are real, and we find that

(19) $\sin z = 0$ if and only if $z = n\pi$,

where $n = 0, \pm 1, \pm 2, \dots$, and

(20) $\cos z = 0$ if and only if $z = (n + \frac{1}{2})\pi$ where $n = 0, \pm 1, \pm 2, \dots$.

EXAMPLE 5.8   Let us verify identity (20). We start with equation (15) and write

(21) $0 = \cos x \cosh y - i \sin x \sinh y.$

Equating the real and imaginary parts of equation (21), we obtain

(22) $0 = \cos x \cosh y$ and $0 = \sin x \sinh y.$

Since the real-valued function $\cosh y$ is never zero, the first equation in (22) implies that $0 = \cos x$, from which we obtain $x = (n + \frac{1}{2})\pi$ for $n = 0, \pm 1, \pm 2, \ldots$. By using these values for x the second equation in (22) becomes

$$0 = \sin[(n + \tfrac{1}{2})\pi] \sinh y = (-1)^n \sinh y.$$

Hence $y = 0$, and the zeros of $\cos z$ are the values $z = (n + \frac{1}{2})\pi$ where $n = 0, \pm 1, \pm 2, \ldots$.

The standard trigonometric identities are valid for complex variables:

(23) $\sin(z_1 + z_2) = \sin z_1 \cos z_2 + \cos z_1 \sin z_2$ and

(24) $\cos(z_1 + z_2) = \cos z_1 \cos z_2 - \sin z_1 \sin z_2.$

When $z_1 = z_2$, identities (23) and (24) become

(25) $\sin 2z = 2 \sin z \cos z$ and $\cos 2z = \cos^2 z - \sin^2 z.$

Other useful identities are

(26) $\sin(-z) = -\sin z$ and $\cos(-z) = \cos z,$

(27) $\sin\left(\dfrac{\pi}{2} + z\right) = \sin\left(\dfrac{\pi}{2} - z\right)$ and $\sin\left(\dfrac{\pi}{2} - z\right) = \cos z.$

EXAMPLE 5.9   Let us show how identity (24) is proven in the complex case. We start with the definitions (3) and (4) and the right side of identity (24). Then we write

(28) $\cos z_1 \cos z_2 = \frac{1}{4}\left[e^{i(z_1+z_2)} + e^{i(z_1-z_2)} + e^{i(z_2-z_1)} + e^{-i(z_1+z_2)}\right]$ and

$-\sin z_1 \sin z_2 = \frac{1}{4}\left[e^{i(z_1+z_2)} - e^{i(z_1-z_2)} - e^{i(z_2-z_1)} + e^{-i(z_1+z_2)}\right].$

When these expressions are added, we obtain

$$\cos z_1 \cos z_2 - \sin z_1 \sin z_2 = \frac{1}{2}\left[e^{i(z_1+z_2)} + e^{-i(z_1+z_2)}\right] = \cos(z_1 + z_2),$$

and identity (24) is established.

Identities involving moduli of cosine and sine are also important. If we start with identity (14) and compute the square of the modulus of sin z, the result is

$$
\begin{aligned}
\left| \sin z \right|^2 &= \left| \sin x \cosh y + i \cos x \sinh y \right|^2 \\
&= \sin^2 x \cosh^2 y + \cos^2 x \sinh^2 y \\
&= \sin^2 x (\cosh^2 y - \sinh^2 y) + \sinh^2 y (\cos^2 x + \sin^2 x).
\end{aligned}
$$

Using the hyperbolic identity $\cosh^2 y - \sinh^2 y = 1$ yields

$$
(29) \quad \left| \sin z \right|^2 = \sin^2 x + \sinh^2 y.
$$

A similar derivation shows that

$$
(30) \quad \left| \cos z \right|^2 = \cos^2 x + \sinh^2 y.
$$

If we set $z = x_0 + iy$ in equation (29) and let $y \to \infty$, then the result is

$$
\lim_{y \to \infty} \left| \sin(x_0 + iy) \right|^2 = \sin^2 x_0 + \lim_{y \to \infty} \sinh^2 y = \infty.
$$

This shows that sin z is not a bounded function, and it is also evident that cos z is not a bounded function. This is one of the important differences between the real and complex cases of the functions sine and cosine.

From the periodic character of the trigonometric functions it is apparent that any point $w = \cos z$ in the range of cos z is actually the image of an infinite number of points.

EXAMPLE 5.10   Let us find all the values of z for which the equation $\cos z = \cosh 2$ holds true. Starting with identity (15), we write

$$
(31) \quad \cos x \cosh y - i \sin x \sinh y = \cosh 2.
$$

Equating the real and imaginary parts in equation (31) results in

$$
(32) \quad \cos x \cosh y = \cosh 2 \quad \text{and} \quad \sin x \sinh y = 0.
$$

The second equation in (32) implies either that $x = n\pi$ where n is an integer or that $y = 0$. Using the latter choice $y = 0$ and the first equation in (32) leads to the impossible situation $\cos x = (\cosh 2 / \cosh 0) = \cosh 2 > 1$. Therefore $x = n\pi$ where n is an integer. Since $\cosh y \geq 1$ for all values of y, we see that the term $\cos x$ in the first equation in (32) must also be positive. For this reason we eliminate the odd values of n and see that $x = 2\pi k$ where k is an integer.

We now solve the equation $\cosh y \cos 2\pi k = \cosh y = \cosh 2$ and use the fact that $\cosh y$ is an even function to conclude that $y = \pm 2$. Therefore the solutions to the equation $\cos z = \cosh 2$ are $z = 2\pi k \pm 2i$ where k is an integer.

The hyperbolic functions also find practical use in putting the tangent function into $u + iv$ form. By earlier results, we have

$$
\tan z = \tan(x + iy) = \frac{\sin(x + iy)}{\cos(x + iy)} = \frac{\sin x \cosh y + i \cos x \sinh y}{\cos x \cosh y - i \sin x \sinh y}.
$$

If we multiply each term on the right by the conjugate of the denominator, the simplified result will be

$$(33) \quad \tan z = \frac{\cos x \sin x + i \cosh y \sinh y}{\cos^2 x \cosh^2 y + \sin^2 x \sinh^2 y}.$$

We leave as an exercise that the identities $\cosh^2 y - \sinh^2 y = 1$, and $\sinh 2y = \cosh y \sinh y$ can be used in simplifying equation (33) to get

$$(34) \quad \tan z = \frac{\sin 2x}{\cos 2x + \cosh 2y} + i\frac{\sinh 2y}{\cos 2x + \cosh 2y}.$$

As was with the case with $\sin z$, a graph of the mapping $w = \tan z$ can be obtained parametrically. Consider the vertical line segments in the z plane obtained by successively setting $x = \dfrac{-\pi}{4} + \dfrac{k\pi}{16}$ for $k = 0, 1, \ldots, 8$, and for each x value letting y vary continuously, $-3 \le y \le 3$. The image of these vertical segments are circular arcs in the uv plane, as Figure 5.10 shows. We will give a more detailed investigation of the mapping $w = \tan z$ in Chapter 9.

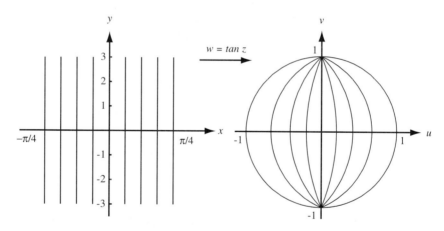

FIGURE 5.10   Vertical segments are mapped onto circular arcs by $w = \tan z$.

The hyperbolic cosine and hyperbolic sine of a complex variable are defined by the equations

$$(35) \quad \cosh z = \tfrac{1}{2}(e^z + e^{-z}) \quad \text{and}$$

$$(36) \quad \sinh z = \tfrac{1}{2}(e^z - e^{-z}).$$

The other hyperbolic functions are given by the formulas

$$(37) \quad \tanh z = \frac{\sinh z}{\cosh z} \quad \text{and} \quad \coth z = \frac{\cosh z}{\sinh z};$$

(38) $\operatorname{sech} z = \dfrac{1}{\cosh z}$ and $\operatorname{csch} z = \dfrac{1}{\sinh z}$.

The derivatives of the hyperbolic functions follow the same rules as in calculus:

(39) $\dfrac{d}{dz} \cosh z = \sinh z$ and $\dfrac{d}{dz} \sinh z = \cosh z$;

(40) $\dfrac{d}{dz} \tanh z = \operatorname{sech}^2 z$ and $\dfrac{d}{dz} \coth z = -\operatorname{csch}^2 z$;

(41) $\dfrac{d}{dz} \operatorname{sech} z = -\operatorname{sech} z \tanh z$ and $\dfrac{d}{dz} \operatorname{csch} z = -\operatorname{csch} z \coth z$.

The hyperbolic cosine and hyperbolic sine can be expressed as

(42) $\cosh z = \cosh x \cos y + i \sinh x \sin y$ and

(43) $\sinh z = \sinh x \cos y + i \cosh x \sin y$.

The trigonometric and hyperbolic functions are all defined in terms of the exponential function, and they can easily be shown to be related by the following identities:

(44) $\cosh(iz) = \cos z$ and $\sinh(iz) = i \sin z$;

(45) $\sin(iz) = i \sinh z$ and $\cos(iz) = \cosh z$.

Some of the identities involving the hyperbolic functions are

(46) $\cosh^2 z - \sinh^2 z = 1$,

(47) $\sinh(z_1 + z_2) = \sinh z_1 \cosh z_2 + \cosh z_1 \sinh z_2$,

(48) $\cosh(z_1 + z_2) = \cosh z_1 \cosh z_2 + \sinh z_1 \sinh z_2$,

(49) $\cosh(z + 2\pi i) = \cosh z$,

(50) $\sinh(z + 2\pi i) = \sinh z$,

(51) $\cosh(-z) = \cosh z$,   and

(52) $\sinh(-z) = -\sinh z$.

We conclude this section with an example from electronics. In the theory of electric circuits it is shown that the voltage drop E_R across a resistance R obeys Ohm's law:

(53) $E_R = IR$,

where I is the current flowing through the resistor. It is also known that the current and voltage drop across an inductor L obey the equation

(54) $E_L = L \dfrac{dI}{dt}$.

The current and voltage drop across a capacitor C are related by

(55) $E_C = \dfrac{1}{C} \displaystyle\int_{t_0}^{t} I(\tau)\, d\tau.$

The voltages E_L, E_R, and E_C and the impressed voltage $E(t)$ in Figure 5.11 satisfy the equation

(56) $E_L + E_R + E_C = E(t).$

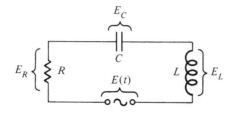

FIGURE 5.11   An LRC circuit.

Suppose that the current $I(t)$ in the circuit is given by

(57) $I(t) = I_0 \sin \omega t.$

Using equations (53), (54), and (57), we obtain

(58) $E_R = RI_0 \sin \omega t$ and

(59) $E_L = \omega L I_0 \cos \omega t,$

and we can set $t_0 = \pi/2$ in equation (55) to obtain

(60) $E_C = -\dfrac{1}{\omega C} I_0 \cos \omega t.$

If we write equation (57) as a complex current

(61) $I^* = I_0 e^{i\omega t}$

and use the understanding that the actual physical current I is the imaginary part of I^*, then equations (58)–(60) can be written

(62) $E_R^* = RI_0 e^{i\omega t} = RI^*,$

(63) $E_L^* = i\omega L I_0 e^{i\omega t} = i\omega L I^*,$ and

(64) $E_C^* = \dfrac{1}{i\omega C} I_0 e^{i\omega t} = \dfrac{1}{i\omega C} I^*.$

Substituting equations (62)–(64) into equation (56) results in

(65) $E^* = E_R^* + E_L^* + E_C^* = \left[R + i\left(\omega L - \dfrac{1}{\omega C} \right) \right] I^*,$

and the complex quantity Z defined by

(66) $Z = R + i\left(\omega L - \dfrac{1}{\omega C}\right)$

is called the *complex impedance*. Using equation (66), we can write

(67) $E^* = ZI^*$,

which is the complex extension of Ohm's law.

EXERCISES FOR SECTION 5.4

1. By making use of appropriate theorems in Section 4.2, show that $\sin z$ and $\cos z$ are entire functions.

2. Establish that $\dfrac{d}{dz} \cos z = -\sin z$.

3. This exercise will demonstrate that for all z, $\sin^2 z + \cos^2 z = 1$.
   (a) Define the function $g(z) = \sin^2 z + \cos^2 z$. Explain why g is entire.
   (b) Show that g is constant. *Hint*: look at $g'(z)$.
   (c) Use part (b) to establish that for all z, $\sin^2 z + \cos^2 z = 1$.

4. Show by appealing to Definition 5.4 that $\sin(-z) = -\sin z$, and $\cos(-z) = \cos z$ for all z.

5. Verify identity (9). *Hint*: Use a similar argument to the one used in the proof of Theorem 4.14, part (iii).

6. Show that equation (33) simplifies to equation (34). *Hint*: Use the facts that $\cosh^2 y - \sinh^2 y = 1$, and $\sinh 2y = \cosh y \sinh y$.

7. Explain why the pictures in Figures 5.6 and 5.7 came out the way they did.

8. Show that:
   (a) $\sin(\pi - z) = \sin z$
   (b) $\sin\left(\dfrac{\pi}{2} - z\right) = \cos z$

9. Express the following quantities in $u + iv$ form:
   (a) $\cos(1 + i)$
   (b) $\sin\left(\dfrac{\pi + 4i}{4}\right)$
   (c) $\sin 2i$
   (d) $\cos(-2 + i)$
   (e) $\tan\left(\dfrac{\pi + 2i}{4}\right)$
   (f) $\tan\left(\dfrac{\pi + i}{2}\right)$

10. Find the derivatives of the following:
    (a) $\sin(1/z)$
    (b) $z \tan z$
    (c) $\sec z^2$
    (d) $z \csc^2 z$

11. Establish identity (15).

12. Show that:
    (a) $\overline{\sin z} = \sin \bar{z}$ holds for all z and that
    (b) $\sin \bar{z}$ is nowhere analytic.

13. Show that:
    (a) $\lim\limits_{z \to 0} \dfrac{\cos z - 1}{z} = 0$ and that
    (b) $\lim\limits_{y \to +\infty} \tan(x_0 + iy) = i$, where x_0 is any fixed real number.

14. Find all values of z for which the following equations hold:
    (a) $\sin z = \cosh 4$
    (b) $\cos z = 2$
    (c) $\sin z = i \sinh 1$

15. Show that the zeros of $\sin z$ are $z = n\pi$ where n is an integer.

16. Establish identity (23).   17. Establish identity (30).
18. Establish the following relation: $|\sinh y| \le |\sin z| \le \cosh y$.
19. Use the result of Exercise 18 to help establish the inequality $|\cos z|^2 + |\sin z|^2 \ge 1$, and show that equality holds if and only if z is a real number.
20. Show that the mapping $w = \sin z$ maps the y axis one-to-one and onto the v axis.
21. Use the fact that $\sin iz$ is analytic to show that $\sinh x \cos y$ is a harmonic function.
22. Show that the transformation $w = \sin z$ maps the ray $x = \pi/2$, $y > 0$ one-to-one and onto the ray $u > 1$, $v = 0$.
23. Express the following quantities in $u + iv$ form.

   (a) $\sinh(1 + i\pi)$ (b) $\cosh\dfrac{i\pi}{2}$ (c) $\cosh\left(\dfrac{4 - i\pi}{4}\right)$

24. Establish identity (46).
25. Show that:
   (a) $\sinh(z + i\pi) = -\sinh z$ (b) $\tanh(z + i\pi) = \tanh z$
26. Find all values of z for which the following equations hold:
   (a) $\sinh z = i/2$ (b) $\cosh z = 1$
27. Find the derivatives of the following:
   (a) $z \sinh z$ (b) $\cosh z^2$ (c) $z \tanh z$
28. Show that:
   (a) $\sin iz = i \sinh z$ (b) $\cosh(iz) = \cos z$
29. Establish identity (42).
30. Show that:
   (a) $\cosh \bar{z} = \overline{\cosh z}$ and that   (b) $\cosh \bar{z}$ is nowhere analytic.
31. Establish identity (48).
32. Find the complex impedance Z if $R = 10$, $L = 10$, $C = 0.05$, and $\omega = 2$.
33. Find the complex impedance Z if $R = 15$, $L = 10$, $C = 0.05$, and $\omega = 4$.
34. Explain how $\sin z$ and the function $\sin x$ studied in calculus are different. How are they similar?
35. Look up the article on trigonometry and discuss what you found. Use bibliographical item 80.

5.5   Inverse Trigonometric and Hyperbolic Functions

The trigonometric and hyperbolic functions were expressed in Section 5.4 in terms of the exponential function. When we solve for their inverses, we will obtain formulas that involve the logarithm. Since the trigonometric and hyperbolic functions are all periodic, they are many-to-one. Hence their inverses are necessarily multivalued. The formulas for the inverse trigonometric functions are given by

(1) $\arcsin z = -i \log[iz + (1 - z^2)^{1/2}]$,

(2) $\arccos z = -i \log[z + i(1 - z^2)^{1/2}]$,   and

(3) $\arctan z = \dfrac{i}{2} \log\left(\dfrac{i + z}{i - z}\right)$.

The derivatives of the functions in formulas (1)–(3) can be found by implicit differentiation and are given by the formulas:

(4) $\quad \dfrac{d}{dz} \arcsin z = \dfrac{1}{(1 - z^2)^{1/2}}$,

(5) $\quad \dfrac{d}{dz} \arccos z = \dfrac{-1}{(1 - z^2)^{1/2}}$,   and

(6) $\quad \dfrac{d}{dz} \arctan z = \dfrac{1}{1 + z^2}$.

We shall establish equations (1) and (4) and leave the others as exercises. Starting with $w = \arcsin z$, we write

$$z = \sin w = \frac{1}{2i} (e^{iw} - e^{-iw}),$$

which also can be written as

(7) $\quad e^{iw} - 2iz - e^{-iw} = 0$.

If each term in equation (7) is multiplied by e^{iw}, the result is

(8) $\quad (e^{iw})^2 - 2ize^{iw} - 1 = 0$,

which is a quadratic equation in terms of e^{iw}. Using the quadratic equation to solve for e^{iw} in equation (8), we obtain

(9) $\quad e^{iw} = \dfrac{2iz + (4 - 4z^2)^{1/2}}{2} = iz + (1 - z^2)^{1/2}$,

where the square root is a multivalued function. Taking the logarithm of both sides of equation (9) leads to the desired equation

$$w = \arcsin z = -i \log[iz + (1 - z^2)^{1/2}],$$

where the multivalued logarithm is used. To construct a specific branch of arcsin z, we must first select a branch of the square root and then select a branch of the logarithm.

The derivative of $w = \arcsin z$ is found by starting with the equation $\sin w = z$ and using implicit differentiation to obtain

$$\frac{dw}{dz} = \frac{1}{\cos w}.$$

When the principal value is used, $w = \operatorname{Arcsin} z = -i \operatorname{Log}[iz + (1 - z^2)^{1/2}]$ maps the upper half plane, $\operatorname{Im}(z) > 0$, onto a portion of the upper half plane $\operatorname{Im}(w) > 0$, that lies in the vertical strip $\dfrac{-\pi}{2} < \operatorname{Re}(w) < \dfrac{\pi}{2}$. The image of a rectangular grid in the z plane is a ''spider web'' in the w plane, as Figure 5.12 shows.

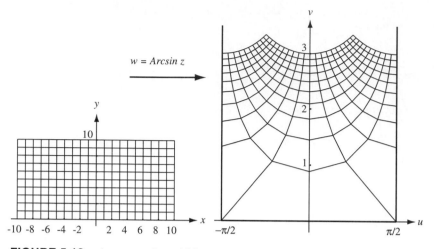

$w = Arcsin\ z$

FIGURE 5.12   A rectangular grid is mapped onto a spider web by $w = $ Arcsin z.

EXAMPLE 5.11   The values of arcsin $\sqrt{2}$ are given by

(10)   arcsin $\sqrt{2} = -i \log[i\sqrt{2} + (1 - (\sqrt{2})^2)^{1/2}] = -i \log (i\sqrt{2} \pm i)$.

Using straightforward techniques, we simplify equation (10) and obtain

$$\text{arcsin } \sqrt{2} = -i \log[(\sqrt{2} \pm 1)i]$$
$$= i\left[\ln(\sqrt{2} \pm 1) + i\left(\frac{\pi}{2} + 2\pi n\right)\right]$$
$$= \frac{\pi}{2} + 2\pi n - i \ln(\sqrt{2} \pm 1), \quad \text{where } n \text{ is an integer.}$$

If we observe that

$$\ln(\sqrt{2} - 1) = \ln \frac{(\sqrt{2} - 1)(\sqrt{2} + 1)}{\sqrt{2} + 1} = \ln \frac{1}{\sqrt{2} + 1} = -\ln(\sqrt{2} + 1),$$

then we can write

$$\text{arcsin } \sqrt{2} = \frac{\pi}{2} + 2\pi n \pm i \ln(\sqrt{2} + 1), \quad \text{where } n \text{ is an integer.}$$

EXAMPLE 5.12   Suppose that we make specific choices in equation (10). We select $+i$ as the value of the square root $[1 - (\sqrt{2})^2]^{1/2}$ and use the principal value of the logarithm. The result will be

$$f(\sqrt{2}) = \text{arcsin } \sqrt{2} = -i \text{ Log}(i\sqrt{2} + i) = \frac{\pi}{2} - i \ln(\sqrt{2} + 1),$$

and the corresponding value of the derivative is given by

$$f'(\sqrt{2}) = \frac{1}{[1 - (\sqrt{2})^2]^{1/2}} = \frac{1}{i} = -i.$$

The inverse hyperbolic functions are given by the equations:

(11) $\text{arcsinh } z = \log[z + (z^2 + 1)^{1/2}],$

(12) $\text{arccosh } z = \log[z + (z^2 - 1)^{1/2}],$ and

(13) $\text{arctanh } z = \frac{1}{2} \log\left(\frac{1 + z}{1 - z}\right).$

The derivatives of the inverse hyperbolic functions are given by

(14) $\dfrac{d}{dz} \text{arcsinh } z = \dfrac{1}{(z^2 + 1)^{1/2}},$

(15) $\dfrac{d}{dz} \text{arccosh } z = \dfrac{1}{(z^2 - 1)^{1/2}},$ and

(16) $\dfrac{d}{dz} \text{arctanh } z = \dfrac{1}{1 - z^2}.$

To establish identity (13), we start with $w = \text{arctanh } z$ and obtain

$$z = \tanh w = \frac{e^w - e^{-w}}{e^w + e^{-w}} = \frac{e^{2w} - 1}{e^{2w} + 1},$$

which can be solved for e^{2w} to yield $e^{2w} = (1 + z)/(1 - z)$. After taking the logarithms of both sides, we obtain the result

$$w = \text{arctanh } z = \frac{1}{2} \log\left(\frac{1 + z}{1 - z}\right),$$

and identity (13) is established.

EXAMPLE 5.13   Calculation reveals that

$$\text{arctanh}(1 + 2i) = \frac{1}{2} \log \frac{1 + 1 + 2i}{1 - 1 - 2i} = \frac{1}{2} \log(-1 + i)$$

$$= \frac{1}{4} \ln 2 + i\left(\frac{3}{8} + n\right) \pi, \quad \text{where } n \text{ is an integer.}$$

EXERCISES FOR SECTION 5.5

1. Find *all* values of the following:
   (a) $\arcsin \frac{5}{4}$ (b) $\arccos \frac{5}{3}$ (c) $\arcsin 3$
   (d) $\arccos 3i$ (e) $\arctan 2i$ (f) $\arctan i$

2. Find *all* values of the following:
   (a) $\text{arcsinh } i$ (b) $\text{arcsinh } \frac{3}{4}$ (c) $\text{arccosh } i$
   (d) $\text{arccosh } \frac{1}{2}$ (e) $\text{arctanh } i$ (f) $\text{arctanh } i\sqrt{3}$

3. Establish equations (2) and (5).
4. Establish equations (3) and (6).
5. Establish the identity $\arcsin z + \arccos z = (\pi/2) + 2\pi n$ where n is an integer.
6. Establish equation (16).
7. Establish equations (11) and (14).
8. Establish equations (12) and (15).

⑥

Complex Integration

6.1 Complex Integrals

In Chapter 3 we saw how the derivative of a complex function is defined. We now turn our attention to the problem of integrating complex functions. We will find that integrals of analytic functions are well behaved and that many properties from calculus carry over to the complex case. To introduce the integral of a complex function, we start by defining what is meant by the integral of a complex-valued function of a real variable. Let

$$f(t) = u(t) + iv(t) \quad \text{for } a \le t \le b,$$

where $u(t)$ and $v(t)$ are real-valued functions of the real variable t. If u and v are continuous functions on the interval, then from calculus we know that u and v are integrable functions of t. Therefore we make the following definition for the definite integral of f:

$$(1) \qquad \int_a^b f(t) \, dt = \int_a^b u(t) \, dt + i \int_a^b v(t) \, dt.$$

Integrals of this type can be evaluated by finding the antiderivatives of u and v and evaluating the definite integrals on the right side of equation (1). That is, if $U'(t) = u(t)$ and $V'(t) = v(t)$, then we write

$$(2) \qquad \int_a^b f(t) \, dt = U(b) - U(a) + i[V(b) - V(a)].$$

EXAMPLE 6.1 Let us show that

$$(3) \qquad \int_0^1 (t - i)^3 \, dt = \frac{-5}{4}.$$

Since the complex integral is defined in terms of real integrals, we write the integrand in equation (3) in terms of its real and imaginary parts: $f(t) = (t - i)^3 = t^3 - 3t + i(-3t^2 + 1)$. Here we see that u and v are given by $u(t) = t^3 - 3t$ and $v(t) = -3t^2 + 1$. The integrals of u and v are easy to compute, and we find that

$$\int_0^1 (t^3 - 3t) \, dt = \frac{-5}{4} \quad \text{and} \quad \int_0^1 (-3t^2 + 1) \, dt = 0.$$

Hence definition (1) can be used to conclude that

$$\int_0^1 (t - i)^3 \, dt = \int_0^1 u(t) \, dt + i \int_0^1 v(t) \, dt = \frac{-5}{4}.$$

Our knowledge about the elementary functions can be used to find their integrals.

EXAMPLE 6.2   Let us show that

$$\int_0^{\pi/2} \exp(t + it) \, dt = \frac{1}{2} (e^{\pi/2} - 1) + \frac{i}{2} (e^{\pi/2} + 1).$$

Using the method suggested by equations (1) and (2), we obtain

$$\int_0^{\pi/2} \exp(t + it) \, dt = \int_0^{\pi/2} e^t \cos t \, dt + i \int_0^{\pi/2} e^t \sin t \, dt.$$

The integrals can be evaluated via integration by parts, and we have

$$\int_0^{\pi/2} \exp(t + it) \, dt = \frac{1}{2} e^t(\cos t + \sin t) + \frac{i}{2} e^t(\sin t - \cos t) \Big|_{t=0}^{t=\pi/2}$$

$$= \frac{1}{2} (e^{\pi/2} - 1) + \frac{i}{2} (e^{\pi/2} + 1).$$

Complex integrals have properties that are similar to those of real integrals. Let $f(t) = u(t) + iv(t)$ and $g(t) = p(t) + iq(t)$ be continuous on $a \le t \le b$. Then the integral of their sum is the sum of their integrals; so we can write

$$(4) \qquad \int_a^b [f(t) + g(t)] \, dt = \int_a^b f(t) \, dt + \int_a^b g(t) \, dt.$$

It is convenient to divide the interval $a \le t \le b$ into $a \le t \le c$ and $c \le t \le b$ and integrate $f(t)$ over these subintervals. Hence we obtain the formula

$$(5) \qquad \int_a^b f(t) \, dt = \int_a^c f(t) \, dt + \int_c^b f(t) \, dt.$$

Constant multiples are dealt with in the same manner as in calculus. If $c + id$ denotes a complex constant, then

$$(6) \qquad \int_a^b (c + id)f(t) \, dt = (c + id) \int_a^b f(t) \, dt.$$

If the limits of integration are reversed, then

$$(7) \qquad \int_b^a f(t) \, dt = -\int_a^b f(t) \, dt.$$

Let us emphasize that we are dealing with complex integrals. We write the integral of the product as follows:

$$(8) \quad \int_a^b f(t)g(t) \, dt = \int_a^b [u(t)p(t) - v(t)q(t)] \, dt +$$
$$i \int_a^b [u(t)q(t) + v(t)p(t)] \, dt.$$

EXAMPLE 6.3 Let us prove equation (6). We start by writing

$$(c + id)f(t) = cu(t) - dv(t) + i[cv(t) + du(t)].$$

Using definition (1), the left side of equation (6) can be written as

$$(9) \quad c \int_a^b u(t) \, dt - d \int_a^b v(t) \, dt + ic \int_a^b v(t) \, dt + id \int_a^b u(t) \, dt,$$

which is easily seen to be equivalent to the product

$$(10) \quad (c + id)\left[\int_a^b u(t) \, dt + i \int_a^b v(t) \, dt \right].$$

It is worthwhile to point out the similarity between equation (2) and its counterpart in calculus. Suppose that U and V are differentiable on $a < t < b$ and $F(t) = U(t) + iV(t)$, then $F'(t)$ is defined to be

$$F'(t) = U'(t) + iV'(t),$$

and equation (2) takes on the familiar form

$$(11) \quad \int_a^b f(t) \, dt = F(b) - F(a), \quad \text{where } F'(t) = f(t).$$

This can be viewed as an extension of the fundamental theorem of calculus. In Section 6.5 we will see how the extension is made to the case of analytic functions of a complex variable. For now, note that we have the following important case of equation (11):

$$(12) \quad \int_a^b f'(t) \, dt = f(b) - f(a).$$

EXAMPLE 6.4 Let us use equation (11) to show that $\int_0^\pi \exp(it) \, dt = 2i$.

Solution If we let $F(t) = -i \exp(it) = \sin t - i \cos t$ and $f(t) = \exp(it) = \cos t + i \sin t$, then $F'(t) = f(t)$, and from equation (11) we obtain

$$\int_0^\pi \exp(it)dt = \int_0^\pi f(t) \, dt = F(\pi) - F(0) = -ie^{i\pi} + ie^0 = 2i.$$

EXERCISES FOR SECTION 6.1

For Exercises 1–4, use equations (1) and (2) to find the following definite integrals.

1. $\int_0^1 (3t - i)^2 \, dt$ **2.** $\int_0^1 (t + 2i)^3 \, dt$ **3.** $\int_0^{\pi/2} \cosh(it) \, dt$ **4.** $\int_0^2 \frac{t}{t + i} \, dt$

5. Find $\int_0^{\pi/4} t \exp(it) \, dt$.

6. Let m and n be integers. Show that

$$\int_0^{2\pi} e^{imt} e^{-int} \, dt = \begin{cases} 0 & \text{when } m \neq n, \\ 2\pi & \text{when } m = n. \end{cases}$$

7. Show that $\int_0^\infty e^{-zt} \, dt = 1/z$ provided that $\text{Re}(z) > 0$.

8. Let $f(t) = u(t) + iv(t)$ where u and v are differentiable. Show that $\int_a^b f(t) f'(t) \, dt = \frac{1}{2}[f(b)]^2 - \frac{1}{2}[f(a)]^2$.

9. Establish identity (4).   **10.** Establish identity (5).

11. Establish identity (7).   **12.** Establish identity (8).

6.2   Contours and Contour Integrals

In Section 6.1 we learned how to evaluate integrals of the form $\int_a^b f(t) \, dt$, where f was complex-valued and $[a, b]$ was an interval on the real axis (so that t was real, with $t \in [a, b]$). In this section we shall define and evaluate integrals of the form $\int_C f(z) \, dz$, where f is complex-valued and C is a contour in the plane (so that z is complex, with $z \in C$). Our main result is Theorem 6.1, which will show how to transform the latter type of integral into the kind we investigated in Section 6.1.

We will use concepts first introduced in Section 1.6, where we defined the concept of a curve in the plane. Recall that to represent a curve C we used the parametric notation

(1) $C: z(t) = x(t) + iy(t)$ for $a \leq t \leq b$,

where $x(t)$ and $y(t)$ are continuous functions. We now want to place a few more restrictions on the type of curve that we will be studying. The following discussion will lead to the concept of a contour, which is a type of curve that is adequate for the study of integration.

Recall that C is said to be *simple* if it does not cross itself, which is expressed by requiring that $z(t_1) \neq z(t_2)$ whenever $t_1 \neq t_2$. A curve C with the property that $z(b) = z(a)$ is said to be a *closed curve*. If $z(b) = z(a)$ is the only point of intersection, then we say that C is a *simple closed curve*. As the parameter t increases from the value a to the value b, the point $z(t)$ starts at the *initial point* $z(a)$, moves along the curve C, and ends up at the *terminal points* $z(b)$. If C is simple, then $z(t)$ moves continuously from $z(a)$ to $z(b)$ as t increases, and the curve is given an *orientation*, which we indicate by drawing arrows along the curve. Figure 6.1 illustrates how the terms "simple" and "closed" can be used to describe a curve.

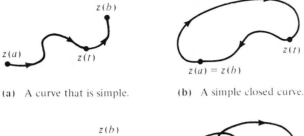

(a) A curve that is simple.   (b) A simple closed curve.

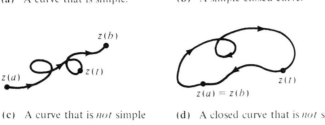

(c) A curve that is *not* simple   (d) A closed curve that is *not* simple.
     and *not* closed.

FIGURE 6.1   The terms ''simple'' and ''closed'' used to describe curves.

The complex-valued function $z(t)$ in equation (1) is said to be *differentiable* if both $x(t)$ and $y(t)$ are differentiable for $a \le t \le b$. Here the one-sided derivatives* of $x(t)$ and $y(t)$ are required to exist at the endpoints of the interval. The derivative $z'(t)$ with respect to t is defined by the equation

(2) $z'(t) = x'(t) + iy'(t)$ for $a \le t \le b$.

The curve C defined by equation (1) is said to be *smooth* if $z'(t)$, given by equation (2), is continuous and nonzero on the interval. If C is a smooth curve, then C has a nonzero tangent vector at each point $z(t)$, which is given by the vector $z'(t)$. If $x'(t_0) = 0$, then the tangent vector $z'(t_0) = iy'(t_0)$ is vertical. If $x'(t_0) \ne 0$, then the slope dy/dx of the tangent line to C at the point $z(t_0)$ is given by $y'(t_0)/x'(t_0)$. Hence the angle of inclination $\theta(t)$ of the tangent vector $z'(t)$ is defined for all values of t and is the continuous function given by

$$\theta(t) = \arg[z'(t)] = \arg[(x'(t) + iy'(t)].$$

Therefore a smooth curve has a continuously turning tangent vector. A smooth curve has no corners or cusps. Figure 6.2 illustrates this concept.

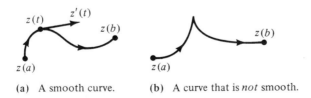

(a) A smooth curve.   (b) A curve that is *not* smooth.

FIGURE 6.2   The term ''smooth'' used to describe curves.

---

*The derivative on the right $x'(a^+)$ and on the left $x'(b^-)$ are defined by the following limits:

$$x'(a^+) = \lim_{t \to a^+} \frac{x(t) - x(a)}{t - a} \quad \text{and} \quad x'(b^-) = \lim_{t \to b^-} \frac{x(t) - x(b)}{t - b}.$$

If C is a smooth curve, then ds, the differential of arc length, is given by

(3) $ds = \sqrt{[x'(t)]^2 + [y'(t)]^2}\, dt = |z'(t)|\, dt.$

Since $x'(t)$ and $y'(t)$ are continuous functions, then so is the function $\sqrt{[x'(t)]^2 + [y'(t)]^2}$, and the length L of the curve C is given by the definite integral

(4) $L = \displaystyle\int_a^b \sqrt{[x'(t)]^2 + [y'(t)]^2}\, dt = \int_a^b |z'(t)|\, dt.$

Now consider C to be a curve with parameterization

$C\colon z_1(t) = x(t) + iy(t)$ for $a \le t \le b.$

The *opposite curve* $-C$ traces out the same set of points in the plane but in the reverse order, and it has the parameterization

$-C\colon z_2(t) = x(-t) + iy(-t)$ for $-b \le t \le -a.$

Since $z_2(t) = z_1(-t)$, it is easy to see that $-C$ is merely C traversed in the opposite sense. This is illustrated in Figure 6.3.

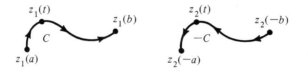

FIGURE 6.3   The curve C and its opposite curve $-C$.

A curve C that is constructed by joining finitely many smooth curves end to end is called a *contour*. Let $C_1, C_2, \ldots, C_n$ denote n smooth curves such that the terminal point of C_k coincides with the initial point of C_{k+1} for $k = 1, 2, \ldots, n - 1$. Then the contour C is expressed by the equation

(5) $C = C_1 + C_2 + \cdots + C_n.$

A synonym for contour is *path*.

EXAMPLE 6.5   Let us find a parameterization of the polygonal path C from $-1 + i$ to $3 - i$, which is shown in Figure 6.4.

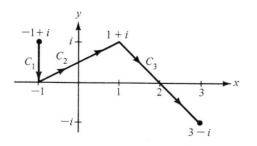

FIGURE 6.4   The polygonal path $C = C_1 + C_2 + C_3$ from $-1 + i$ to $3 - i$.

Solution   The contour is conveniently expressed as three smooth curves $C = C_1 + C_2 + C_3$. A formula for the straight line segment joining two points was given by equation (2) in Section 1.6. If we set $z_0 = -1 + i$ and $z_1 = -1$, then the segment C_1 joining z_0 to z_1 is given by

$$C_1: z_1(t) = z_0 + t(z_1 - z_0) = (-1 + i) + t[-1 - (-1 + i)],$$

which can be simplified to obtain

$$C_1: z_1(t) = -1 + i(1 - t) \quad \text{for } 0 \le t \le 1.$$

In a similar fashion the segments C_2 and C_3 are given by

$$C_2: z_2(t) = (-1 + 2t) + it \qquad \text{for } 0 \le t \le 1 \quad \text{and}$$
$$C_3: z_3(t) = (1 + 2t) + i(1 - 2t) \quad \text{for } 0 \le t \le 1.$$

We are now ready to define the integral of a complex function along a contour C in the plane with initial point A and terminal point B. Our approach is to mimic what is done in calculus. We create a partition $P_n = \{z_0 = A, z_1, z_2, \ldots, z_n = B\}$ of points that proceed along C from A to B and form the differences $\Delta z_k = z_k - z_{k-1}$ for $k = 1, 2, \ldots, n$. Between each pair of partition points z_{k-1} and z_k we select a point c_k on C, where the function $f(c_k)$ is evaluated (see Figure 6.5). These values are used to make a Riemann sum for the partition:

$$(6) \quad S(P_n) = \sum_{k=1}^{n} f(c_k)(z_k - z_{k-1}) = \sum_{k=1}^{n} f(c_k)\Delta z_k.$$

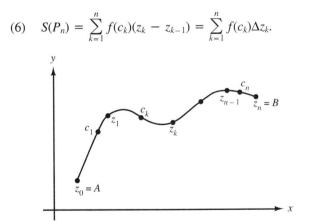

FIGURE 6.5   Partition points $\{z_k\}$, and function evaluation points $\{c_k\}$ for a Riemann sum along the contour C from $z = A$ to $z = B$.

Assume now that there exists a unique complex number L that is the limit of every sequence $\{S(P_n)\}$ of Riemann sums given in equation (6), where the maximum of $|\Delta z_k|$ tends toward 0, for the sequence of partitions. We define the number L as the value of the integral of $f(z)$ taken along the contour C. We thus have the following.

Definition 6.1 *Let C be a contour, then* $\displaystyle\int_C f(z)\,dz = \lim_{n\to\infty}\sum_{k=1}^{n} f(c_k)\Delta z_k$,

provided the limit exists in the sense previously discussed.

You will notice that in this definition, the value of the integral depends on the contour. In Section 6.3 the Cauchy-Goursat Theorem will establish the remarkable property that *if $f(z)$ is analytic*, then $\displaystyle\int_C f(z)\,dz$ is *independent* of the contour.

EXAMPLE 6.6 Use a Riemann sum to construct an approximation for the contour integral $\int_C \exp z\,dz$, where C is the line segment joining the point $A = 0$ to $B = 2 + i\dfrac{\pi}{4}$.

Solution Set $n = 8$ in equation (6) and form the partition P_8: $z_k = \dfrac{k}{4} + i\dfrac{\pi k}{32}$ for $k = 0, 1, 2, \ldots, 8$. For this situation, we have a uniform increment $\Delta z_k = \dfrac{1}{4} + i\dfrac{\pi}{32}$. For convenience we select $c_k = \dfrac{z_{k-1} + z_k}{2} = \dfrac{2k - 1}{8} + i\dfrac{\pi(2k - 1)}{64}$ for $k = 1, 2, \ldots, 8$. The points $\{z_k\}$ and $\{c_k\}$ are shown in Figure 6.6.

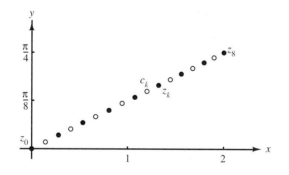

FIGURE 6.6 Partition and evaluation points for the Riemann sum $S(P_8)$.

One possible Riemann sum, then, is

$$S(P_8) = \sum_{k=1}^{8} f(c_k)\,\Delta z_k = \sum_{k=1}^{8} \exp\left[\frac{2k - 1}{8} + i\frac{\pi(2k - 1)}{64}\right]\left(\frac{1}{4} + i\frac{\pi}{32}\right).$$

By rounding the terms in this Riemann sum to two decimal digits, we obtain an approximation for the integral:

$$S(P_8) \approx (0.28 + 0.13i) + (0.33 + 0.19i) + (0.41 + 0.29i) + (0.49 + 0.42i)$$
$$+ (0.57 + 0.6i) + (0.65 + 0.84i) + (0.72 + 1.16i) + (0.78 + 1.57i),$$
$$S(P_8) \approx 4.23 + 5.20i.$$

This compares favorably with the precise value of the integral, which you will soon see equals $\exp\left(2 + i\dfrac{\pi}{4}\right) - 1 = -1 + e^2\dfrac{\sqrt{2}}{2} + ie^2\dfrac{\sqrt{2}}{2} \approx 4.22485 + 5.22485i.$

In general, obtaining an exact value for an integral given by Definition 6.1 is a daunting task. Fortunately, there is a beautiful theory that allows for an easy computation of many contour integrals. Suppose we have a parameterization of the contour C given by the function $z(t)$ for $a \leq t \leq b$. That is, C is the range of the function $z(t)$ over the interval $[a, b]$, as Figure 6.7 shows.

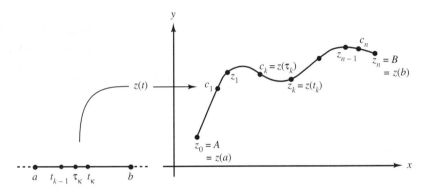

FIGURE 6.7   A parameterization of the contour C by $z(t)$ for $a \leq t \leq b$.

It follows that

$$\lim_{n\to\infty} \sum_{k=1}^{n} f(c_k)\Delta z_k = \lim_{n\to\infty} \sum_{k=1}^{n} f(c_k)(z_k - z_{k-1})$$

$$= \lim_{n\to\infty} \sum_{k=1}^{n} f(z(\tau_k))[z(t_k) - z(t_{k-1})],$$

where the τ_k and t_k are those points contained in the interval $[a, b]$ with the property that $c_k = z(\tau_k)$ and $z_k = z(t_k)$, as is also shown in Figure 6.7. If for all k we multiply the kth term in the last sum by $\dfrac{t_k - t_{k-1}}{t_k - t_{k-1}}$, we get

$$\lim_{n\to\infty} \sum_{k=1}^{n} f(z(\tau_k))\left[\frac{z(t_k) - z(t_{k-1})}{t_k - t_{k-1}}\right](t_k - t_{k-1}) = \lim_{n\to\infty} \sum_{k=1}^{n} f(z(\tau_k))\left[\frac{z(t_k) - z(t_{k-1})}{t_k - t_{k-1}}\right]\Delta t_k.$$

The quotient inside the last summation looks suspiciously like a derivative, and the entire quantity looks like a Riemann sum. Assuming no difficulties, this last expression should equal

$$\int_a^b f(z(t))z'(t)\, dt, \text{ as defined in Section 6.1.}$$

It would be especially nice if we were to get the same limit *regardless of how we parameterize the contour C.* As the following theorem states, this is indeed the case.

> **Theorem 6.1** *Suppose $f(z)$ is a continuous complex-valued function defined on a set containing the contour C. Let $z(t)$ be any parameterization of C for $a \le t \le b$. Then*

$$(7) \quad \int_C f(z)\, dz = \int_a^b f(z(t)) z'(t)\, dt.$$

Proof   We omit the proof of this theorem since it involves ideas (such as the theory of the Riemann-Stieltjes integral) that are beyond the scope of this book. A more rigorous development of the contour integral based on Riemann sums is found in advanced texts such as L. V. Ahlfors, *Complex Analysis*, 3rd ed. (New York: McGraw-Hill, 1979).

There are two important facets of Theorem 6.1 that are worth mentioning. First, the theorem makes the problem of evaluating complex-valued functions along contours easy since it reduces our task to one that requires the evaluation complex-valued functions over real intervals—a procedure we studied in Section 6.1. Second, according to the theorem this transformation yields the same answer regardless of the parameterization we choose for C, a truly remarkable fact.

EXAMPLE 6.7

Let us give an exact calculation of the integral in Example 6.6. That is, we want $\int_C \exp z\, dz$, where C is the line segment joining $A = 0$ to $B = 2 + i\dfrac{\pi}{4}$. According to equation (2) of Section 1.6, we can parameterize C by $z(t) = \left(2 + i\dfrac{\pi}{4}\right)t$, for $0 \le t \le 1$. Since $z'(t) = \left(2 + i\dfrac{\pi}{4}\right)$, according to Theorem 6.1 we have that

$$\int_C \exp z\, dz = \int_0^1 \exp\left[\left(2 + i\frac{\pi}{4}\right)t\right]\left(2 + i\frac{\pi}{4}\right) dt$$

$$= \left(2 + i\frac{\pi}{4}\right) \int_0^1 e^{2t} e^{i\pi t/4}\, dt$$

$$= \left(2 + i\frac{\pi}{4}\right) \int_0^1 e^{2t}[\cos(\pi t/4) + i\,\sin(\pi t/4)]\, dt$$

$$= \left(2 + i\frac{\pi}{4}\right)\left[\int_0^1 e^{2t}\cos(\pi t/4)\, dt + i\int_0^1 e^{2t}\sin(\pi t/4)\, dt\right].$$

Each integral in the last expression can be done using integration by parts. We leave as an exercise that the final answer simplifies to $\exp\left(2 + i\frac{\pi}{4}\right) - 1$, as claimed in Example 6.6.

EXAMPLE 6.8 Evaluate $\int_C \frac{1}{z - 2}\, dz$, where C is the upper semicircle with radius 1 centered at $x = 2$ oriented in a position (i.e., counterclockwise) direction.

    Solution   The function $z(t) = 2 + e^{it}$, for $0 \le t \le \pi$ is a parameterization for C. We apply Theorem 6.1 with $f(z) = \frac{1}{z - 2}$. (Note: $f(z(t)) = \frac{1}{z(t) - 2}$, and $z'(t) = ie^{it}$.) Hence,

$$\int_C \frac{1}{z - 2}\, dz = \int_0^\pi \frac{1}{(2 + e^{it}) - 2}\, ie^{it}\, dt = \int_0^\pi i\, dt = i\pi.$$

To help convince yourself that the value of the integral is independent of the parameterization chosen for the given contour, try working through this example with $z(t) = 2 + e^{i\pi t}$, for $0 \le t \le 1$.

    There is a convenient bookkeeping device that helps us remember how to apply Theorem 6.1. Since $\int_C f(z)\, dz = \int_a^b f(z(t))\, z'(t)\, dt$, we can symbolically equate z with $z(t)$ and dz with $z'(t)\, dt$. This should be easy to remember because z is supposed to be a point on the contour C parameterized by $z(t)$, and $\frac{dz}{dt} = z'(t)$ according to the Leibniz notation for the derivative.

    If $z(t) = x(t) + iy(t)$, then by the preceding paragraph we have

(8) $dz = z'(t)\, dt = [x'(t) + iy'(t)]\, dt = dx + i\, dy,$

where dx and dy are the differentials for $x(t)$ and $y(t)$, respectively. (That is, dx is equated with $x'(t)\, dt$ and dy with $y'(t)\, dt$.) The expression dz is often called the *complex differential* of z. Just as dx and dy are intuitively considered to be small segments along the x and y axes in real variables, we can think of dz as representing a very tiny piece of the contour C. Moreover, if we write

(9) $\left| dz \right| = \left| [x'(t) + iy'(t)]\, dt \right| = \left| [x'(t) + iy'(t)] \right|\, dt = \sqrt{[x(t)]^2 + [y(t)]^2}\, dt,$

then we know from calculus that the length of the curve C, $L(C)$, is given by

(10) $L(C) = \int_a^b \sqrt{[x(t)]^2 + [y(t)]^2}\, dt = \int_C \left| dz \right|,$

so we can think of $\left| dz \right|$ as representing the length of dz.

Suppose $f(z) = u(z) + iv(z)$, and $z(t) = x(t) + iy(t)$ is a parameterization for the contour C. Then

$$(11)\quad \int_C f(z)\, dz = \int_a^b f(z(t))z'(t)\, dt$$

$$= \int_a^b [u(z(t)) + iv(z(t))][x'(t) + iy'(t)]\, dt$$

$$= \int_a^b [u(z(t))x'(t) - v(z(t))y'(t)]\, dt$$

$$+ i \int_a^b [v(z(t))x'(t) + u(z(t))y'(t)]\, dt$$

$$= \int_a^b (ux' - vy')\, dt + i \int_a^b (vx' + uy')\, dt,$$

where we are equating u with $u(z(t))$, x' with $x'(t)$, etc.

If we use the differentials given in equation (8), then equation (11) can be written in terms of line integrals of the real-valued functions u and v, giving

$$(12)\quad \int_C f(z)\, dz = \int_C u\, dx - v\, dy + i \int_C v\, dx + u\, dy,$$

which is easy to remember if we recall that symbolically

$$f(z)\, dz = (u + iv)(dx + i\, dy).$$

We emphasize that equation (12) is merely a notational device for applying equation (7) in Theorem 6.1. We recommend you carefully apply the theorem as illustrated in Examples 6.7 and 6.8 before using any shortcuts suggested by equation (12).

EXAMPLE 6.9   Let us show that

$$\int_{C_1} z\, dz = \int_{C_2} z\, dz = 4 + 2i,$$

where C_1 is the line segment from $-1 - i$ to $3 + i$ and C_2 is the portion of the parabola $x = y^2 + 2y$ joining $-1 - i$ to $3 + i$, as indicated in Figure 6.8.

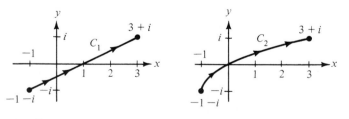

(a) The line segment.   (b) The portion of the parabola.

FIGURE 6.8   The two contours C_1 and C_2 joining $-1 - i$ to $3 + i$.

The line segment joining $(-1, -1)$ to $(3, 1)$ is given by the slope intercept formula $y = \frac{1}{2}x - \frac{1}{2}$, which can be written as $x = 2y + 1$. It is convenient to choose the parameterization $y = t$ and $x = 2t + 1$. Then the segment C_1 can be given by

$$C_1: z(t) = 2t + 1 + it \quad \text{and} \quad dz = (2 + i)\, dt \quad \text{for } -1 \le t \le 1.$$

Along C_1, we have $f(z(t)) = 2t + 1 + it$. Computing the value of the integral in equation (7), we obtain

$$\int_{C_1} z\, dz = \int_{-1}^{1} (2t + 1 + it)(2 + i)\, dt,$$

which can be evaluated by using straightforward techniques to obtain

$$\int_{C_1} z\, dz = \int_{-1}^{1} (3t + 2)\, dt + i \int_{-1}^{1} (4t + 1)\, dt = 4 + 2i.$$

Similarly, for the portion of the parabola $x = y^2 + 2y$ joining $(-1, -1)$ to $(3, 1)$, it is convenient to choose the parameterization $y = t$ and $x = t^2 + 2t$. Then C_2 can be given by

$$C_2: z(t) = t^2 + 2t + it \quad \text{and} \quad dz = (2t + 2 + i)\, dt \quad \text{for } -1 \le t \le 1.$$

Along C_2 we have $f(z(t)) = t^2 + 2t + it$. Computing the value of the integral in equation (7), we obtain

$$\int_{C_2} z\, dz = \int_{-1}^{1} (t^2 + 2t + it)(2t + 2 + i)\, dt$$

$$= \int_{-1}^{1} (2t^3 + 6t^2 + 3t)\, dt + i \int_{-1}^{1} (3t^2 + 4t)\, dt = 4 + 2i.$$

In this example, the value of the two integrals is the same. This does not hold in general, as is shown in Example 6.10.

EXAMPLE 6.10   Let us show that

$$\int_{C_1} \bar{z}\, dz = -\pi i, \quad \text{but} \quad \int_{C_2} \bar{z}\, dz = -4i,$$

where C_1 is the semicircular path from -1 to 1 and C_2 is the polygonal path from -1 to 1, respectively, that are shown in Figure 6.9.

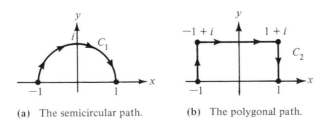

(a) The semicircular path.   (b) The polygonal path.

FIGURE 6.9   The two contours C_1 and C_2 joining -1 to 1.

Solution   The semicircle C_1 can be parameterized by

$$C_1: z(t) = -\cos t + i \sin t \quad \text{and} \quad dz = (\sin t + i \cos t)\, dt \quad \text{for } 0 \le t \le \pi.$$

Along C_1 we have $f(z(t)) = -\cos t - i \sin t$. Computing the value of the integral equation in (7), we obtain

$$\int_{C_1} \bar{z}\, dz = \int_0^\pi (-\cos t - i \sin t)(\sin t + i \cos t)\, dt$$

$$= -i \int_0^\pi (\cos^2 t + \sin^2 t)\, dt = -\pi i.$$

The polygonal path C_2 must be parameterized in three parts, one for each line segment:

$$
\begin{aligned}
z_1(t) &= -1 + it, & dz_1 &= i\, dt, & f(z_1(t)) &= -1 - it, \\
z_2(t) &= -1 + 2t + i, & dz_2 &= 2\, dt, & f(z_2(t)) &= -1 + 2t - i, \\
z_3(t) &= 1 + i(1 - t), & dz_3 &= -i\, dt, & f(z_3(t)) &= 1 - i(1 - t),
\end{aligned}
$$

where all of the parameters t are to be taken on the interval $0 \le t \le 1$. The value of the integral in equation (7) is obtained by adding the three integrals along the above three segments, and the result is

$$\int_0^1 (-1 - it)i\, dt + \int_0^1 (-1 + 2t - i)2\, dt + \int_0^1 [1 - i(1 - t)](-i)\, dt.$$

A straightforward calculation now shows that

$$\int_{C_2} \bar{z}\, dz = \int_0^1 (6t - 3)\, dt + i \int_0^1 (-4)\, dt = -4i.$$

We remark that the value of the contour integral along C_1 is *not* the same as the value of the contour integral along C_2, although both integrals have the same initial and terminal points.

Contour integrals have properties that are similar to those of integrals of a complex function of a real variable, which were studied in Section 6.1. If C is given by equation (1), then the contour integral for the opposite contour $-C$ is given by

$$(13) \quad \int_{-C} f(z)\, dz = \int_{-b}^{-a} f(z(-\tau))[-z'(-\tau)]\, d\tau.$$

Using the change of variable $t = -\tau$ in equation (13) and identity (7) of Section 6.1, we obtain

$$(14) \quad \int_{-C} f(z)\, dz = -\int_C f(z)\, dz.$$

If two functions f and g can be integrated over the same path of integration C, then their sum can be integrated over C, and we have the familiar result

(15) $\displaystyle \int_C [f(z) + g(z)] \, dz = \int_C f(z) \, dz + \int_C g(z) \, dz.$

Constant multiples are dealt with in the same manner as in identity (6) in Section 6.1:

(16) $\displaystyle \int_C (c + id) f(z) \, dz = (c + id) \int_C f(z) \, dz.$

If two contours C_1 and C_2 are placed end to end so that the terminal point of C_1 coincides with the initial point of C_2, then the contour $C = C_1 + C_2$ is a *continuation* of C_1, and we have the property

(17) $\displaystyle \int_{C_1+C_2} f(z) \, dz = \int_{C_1} f(z) \, dz + \int_{C_2} f(z) \, dz.$

If the contour C has two parameterizations

$$C\colon z_1(t) = x_1(t) + iy_1(t) \qquad \text{for } a \le t \le b \quad \text{and}$$
$$C\colon z_2(\tau) = x_2(\tau) + iy_2(\tau) \qquad \text{for } \alpha \le \tau \le \beta,$$

and there exists a differentiable function $\tau = \phi(t)$ such that

(18) $\alpha = \phi(a), \quad \beta = \phi(b), \quad \text{and} \quad \phi'(t) > 0 \quad \text{for } a < t < b,$

then we say that $z_2(\tau)$ is a *reparameterization* of the contour C. If f is continuous on C, then we have

(19) $\displaystyle \int_a^b f(z_1(t)) z_1'(t) \, dt = \int_\alpha^\beta f(z_2(\tau)) z_2'(\tau) \, d\tau.$

Identity (19) shows that the value of a contour integral is invariant under a change in the parametric representation of its contour if the reparameterization satisfies equations (18).

There are a few important inequalities relating to complex integrals, which we now state.

Lemma 6.1 (Integral Triangle Inequality) *If $f(t) = u(t) + iv(t)$ is a continuous function of the real parameter t, then*

(20) $\displaystyle \left| \int_a^b f(t) \, dt \right| \le \int_a^b |f(t)| \, dt.$

Proof Write the value of the integral in polar form:

(21) $\displaystyle r_0 e^{i\theta_0} = \int_a^b f(t) \, dt \quad \text{and} \quad r_0 = \int_a^b e^{-i\theta_0} f(t) \, dt.$

Taking the real part of the second integral in equations (21), we write

$$r_0 = \int_a^b \text{Re}[e^{-i\theta_0}f(t)] \, dt.$$

Using equation (2) of Section 1.3, we obtain the relation

$$\text{Re}[e^{-i\theta_0}f(t)] \le \left| e^{-i\theta_0}f(t) \right| \le \left| f(t) \right|.$$

The left and right sides can be used as integrands, and then familiar results from calculus can be used to obtain

$$r_0 = \int_a^b \text{Re}[e^{-i\theta_0}f(t)] \, dt \le \int_a^b \left| f(t) \right| \, dt.$$

Since

$$r_0 = \left| \int_a^b f(t) \, dt \right|,$$

we have established inequality (20).

Lemma 6.2 (ML Inequality)   *If $f(z) = u(x, y) + iv(x, y)$ is continuous on the contour C, then*

(22) $$\left| \int_C f(z) \, dz \right| \le ML,$$

> *where L is the length of the contour C and M is an upper bound for the modulus $\left| f(z) \right|$ on C.*

Proof   When inequality (20) is used with Theorem 6.1, we get

(23) $$\left| \int_C f(z) \, dz \right| = \left| \int_a^b f(z(t))z'(t) \, dt \right| \le \int_a^b \left| f(z(t))z'(t) \right| \, dt.$$

Let M be the positive real constant such that

$$\left| f(z) \right| \le M \quad \text{for all } z \text{ on } C.$$

Then equation (9) and inequality (23) imply that

$$\left| \int_C f(z) \, dz \right| \le \int_a^b M \left| z'(t) \right| \, dt = ML.$$

Therefore inequality (22) is proved.

EXAMPLE 6.11   Let us use inequality (22) to show that

$$\left| \int_C \frac{1}{z^2 + 1} \, dz \right| \leq \frac{1}{2\sqrt{5}},$$

where C is the straight line segment from 2 to $2 + i$. Here $\left| z^2 + 1 \right| = \left| z - i \right| \times \left| z + i \right|$, and the terms $\left| z - i \right|$ and $\left| z + i \right|$ represent the distance from the point z to the points i and $-i$, respectively. We refer to Figure 6.10 and use a geometric argument to see that

$$\left| z - i \right| \geq 2 \quad \text{and} \quad \left| z + i \right| \geq \sqrt{5} \quad \text{for } z \text{ on } C.$$

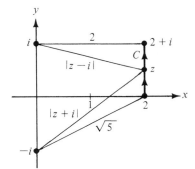

FIGURE 6.10   The distances $\left| z - i \right|$ and $\left| z + i \right|$ for z on C.

Here we have

$$\left| f(z) \right| = \frac{1}{\left| z - i \right| \left| z + i \right|} \leq \frac{1}{2\sqrt{5}} = M,$$

and $L = 1$, so inequality (22) implies that

$$\left| \int_C \frac{1}{z^2 + 1} \, dz \right| \leq ML = \frac{1}{2\sqrt{5}}.$$

EXERCISES FOR SECTION 6.2

1. Sketch the following curves.
   (a) $z(t) = t^2 - 1 + i(t + 4)$ for $1 \leq t \leq 3$
   (b) $z(t) = \sin t + i \cos 2t$ for $-\pi/2 \leq t \leq \pi/2$
   (c) $z(t) = 5 \cos t - i3 \sin t$ for $\pi/2 \leq t \leq 2\pi$
2. Give a parameterization of the contour $C = C_1 + C_2$ indicated in Figure 6.11.
3. Give a parameterization of the contour $C = C_1 + C_2 + C_3$ indicated in Figure 6.12.

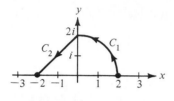

FIGURE 6.11   Accompanies
Exercise 2.

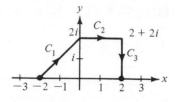

FIGURE 6.12   Accompanies
Exercise 3.

4. Consider the integral $\int_{C_1^+(0)} z^2\, dz$ (see Section 1.6 for the meaning of $C_1^+(0)$).

   (a) Given a Riemann sum approximation for the above integral by selecting $n = 4$, and the following points: $z_k = e^{ik\pi(t/4)}$; $c_k = \dfrac{z_k + z_{k-1}}{2}$ for appropriate values of k.

   (b) Compute the integral exactly by selecting a parameterization for $C_1^+(0)$ and applying Theorem 6.1.

5. Show that the integral in Example 6.7 simplifies to $\exp\left(2 + i\dfrac{\pi}{4}\right) - 1$.

6. Evaluate $\int_C y\, dz$ for $-i$ to i along the following contours as shown in Figures 6.13(a) and 6.13(b).

   (a) The polygonal path C with vertices $-i$, $-1 - i$, -1, and i.

   (b) The contour C that is the left half of the circle $|z| = 1$.

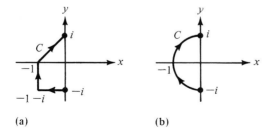

(a)   (b)

FIGURE 6.13   Accompanies Exercise 6.

7. Evaluate $\int_C x\, dz$ from -4 to 4 along the following contours as shown in Figures 6.14(a) and 6.14(b).

   (a) The polygonal path C with vertices -4, $-4 + 4i$, $4 + 4i$, and 4.

   (b) The contour C that is the upper half of the circle $|z| = 4$.

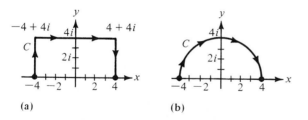

(a)   (b)

FIGURE 6.14   Accompanies Exercise 7.

8. Evaluate $\int_C z \, dz$, where C is the circle $|z| = 4$ taken with the counterclockwise orientation. *Hint:* Let C: $z(t) = 4 \cos t + i4 \sin t$ for $0 \le t \le 2\pi$.

9. Evaluate $\int_C \bar{z} \, dz$, where C is the circle $|z| = 4$ taken with the counterclockwise orientation.

10. Evaluate $\int_C (z + 1) \, dz$, where C given by C: $z(t) = \cos t + i \sin t$ for $0 \le t \le \pi/2$.

11. Evaluate $\int_C z \, dz$, where C is the line segment from i to 1 and $z(t) = t + (1 - t)i$ for $0 \le t \le 1$.

12. Evaluate $\int_C z^2 \, dz$, where C is the line segment from 1 to $1 + i$ and $z(t) = 1 + it$ for $0 \le t \le 1$.

13. Evaluate $\int_C (x^2 - iy^2) \, dz$, where C is the upper semicircle C: $z(t) = \cos t + i \sin t$ for $0 \le t \le \pi$.

14. Evaluate $\int_C |z^2| \, dz$, where C given by C: $z(t) = t + it^2$ for $0 \le t \le 1$.

15. Evaluate $\int_C |z - 1|^2 \, dz$, where C is the upper half of the circle $|z| = 1$ taken with the counterclockwise orientation.

16. Evaluate $\int_C (1/z) \, dz$, where C is the circle $|z| = 2$ taken with the clockwise orientation. *Hint:* C: $z(t) = 2 \cos t - i2 \sin t$ for $0 \le t \le 2\pi$.

17. Evaluate $\int_C (1/\bar{z}) \, dz$, where C is the circle $|z| = 2$ taken with the clockwise orientation.

18. Evaluate $\int_C \exp z \, dz$, where C is the straight line segment joining 1 to $1 + i\pi$.

19. Show that $\int_C \cos z \, dz = \sin(1 + i)$, where C is the polygonal path from 0 to $1 + i$ that consists of the line segments from 0 to 1 and 1 to $1 + i$.

20. Show that $\int_C \exp z \, dz = \exp(1 + i) - 1$, where C is the straight line segment joining 0 to $1 + i$.

21. Evaluate $\int_C \bar{z} \exp z \, dz$, where C is the square with vertices 0, 1, $1 + i$, and i taken with the counterclockwise orientation.

22. Let $z(t) = x(t) + iy(t)$ for $a \le t \le b$ be a smooth curve. Give a meaning for each of the following expressions.

   (a) $z'(t)$ (b) $|z'(t)| \, dt$ (c) $\int_a^b z'(t) \, dt$ (d) $\int_a^b |z'(t)| \, dt$

23. Let f be a continuous function on the circle $|z - z_0| = R$. Let the circle C have the parameterization C: $z(\theta) = z_0 + Re^{i\theta}$ for $0 \le \theta \le 2\pi$. Show that

$$\int_C f(z) \, dz = iR \int_0^{2\pi} f(z_0 + Re^{i\theta}) e^{i\theta} \, d\theta.$$

24. Use the results of Exercise 23 to show that

   (a) $\displaystyle \int_C \frac{1}{z - z_0} \, dz = 2\pi i$ and

   (b) $\displaystyle \int_C \frac{1}{(z - z_0)^n} \, dz = 0$, where $n \ne 1$ is an integer,

   where the contour C is the circle $|z - z_0| = R$ taken with the counterclockwise orientation.

25. Explain how contour integrals studied in complex analysis and line integrals studied in calculus are different. How are they similar?

26. Write a report on contour integrals. Include some of the more complicated techniques in your discussion. Resources include bibliographical items 5, 16, 81, 82, and 157.

6.3   The Cauchy-Goursat Theorem

The Cauchy-Goursat theorem states that within certain domains the integral of an analytic function over a simple closed contour is zero. An extension of this theorem will allow us to replace integrals over certain complicated contours with integrals over contours that are easy to evaluate. We will show how to use the technique of partial fractions together with the Cauchy-Goursat theorem to evaluate certain integrals. In Section 6.4 we will see that the Cauchy-Goursat theorem implies that an analytic function has an antiderivative. To start with, we need to introduce a few new concepts.

We saw in Section 1.6 that with each simple closed contour C there are associated two disjoint domains, each of which has C as its boundary. The contour C divides the plane into two domains. One domain is bounded and is called the *interior* of C, and the other domain is unbounded and is called the *exterior* of C. Figure 6.15 illustrates this concept. This result is known as the Jordan Curve Theorem.

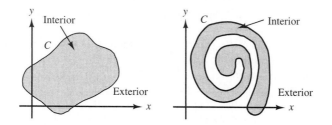

FIGURE 6.15   The interior and exterior of simple closed contours.

In Section 1.6 we saw that a domain D is an open connected set. In particular, if z_1 and z_2 are any pair of points in D, then they can be joined by a curve that lies entirely in D. A domain D is said to be *simply connected* if it has the property that any simple closed contour C contained in D has its interior contained in D. In other words, there are no ''holes'' in a simply connected domain. A domain that is not simply connected is said to be a *multiply connected domain*. Figure 6.16 illustrates the use of the terms ''simply connected'' and ''multiply connected.''

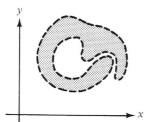

(a)   A simply connected domain.

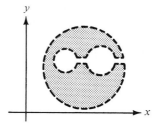

(b)   A simply connected domain.

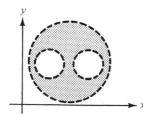

(c)   A multiply connected domain.

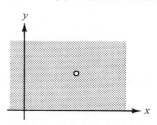

(d)   A multiply connected domain.

FIGURE 6.16   Simply connected and multiply connected domains.

Let the simple closed contour C have the parameterization C: $z(t) = x(t) + iy(t)$ for $a \le t \le b$. If C is parameterized so that the interior of C is kept on the left as $z(t)$ moves around C, then we say that C is oriented in the *positive* (counterclockwise) sense; otherwise, C is oriented *negatively*. If C is positively oriented, then $-C$ is negatively oriented. Figure 6.17 illustrates the concept of positive and negative orientation.

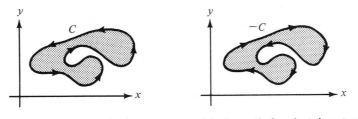

(a)   A positively oriented contour.   (b)   A negatively oriented contour.

FIGURE 6.17   Simple closed contours that are positively and negatively oriented.

An important result from the calculus of real variables is known as Green's theorem and is concerned with the line integral of real-valued functions.

Theorem 6.2 (Green's Theorem)   *Let C be a simple closed contour with positive orientation, and let R be the domain that forms the interior of C. If P and Q are continuous and have continuous partial derivatives P_x, P_y, Q_x, and Q_y at all points on C and R, then*

(1) $$\int_C P(x, y)\, dx + Q(x, y)\, dy = \iint_R [Q_x(x, y) - P_y(x, y)]\, dx\, dy.$$

Proof for a Standard Region*   If R is a standard region, then there exist functions $y = g_1(x)$ and $y = g_2(x)$ for $a \le x \le b$ whose graphs form the lower and upper portions of C, respectively, as indicated in Figure 6.18. Since C is to be given the positive (counterclockwise) orientation, these functions can be used to express C as the sum of two contours C_1 and C_2 where

C_1: $z_1(t) = t + ig_1(t)$ for $a \le t \le b$ and
C_2: $z_2(t) = -t + ig_2(-t)$ for $-b \le t \le -a$.

We now use the functions $g_1(x)$ and $g_2(x)$ to express the double integral of $-P_y(x, y)$ over R as an iterated integral, first with respect to y and second with respect to x, as follows:

(2) $$-\iint_R P_y(x, y)\, dx\, dy = -\int_a^b \left[\int_{g_1(x)}^{g_2(x)} P_y(x, y)\, dy \right] dx.$$

---

*A standard region is bounded by a contour C, which can be expressed in the two forms $C = C_1 + C_2$ and $C = C_3 + C_4$ that are used in the proof.

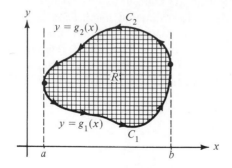

FIGURE 6.18   Integration over a standard region where $C = C_1 + C_2$.

Computing the first iterated integral on the right side of equation (2), we obtain

(3) $$-\iint_R P_y(x, y)\, dx\, dy = \int_a^b P(x, g_1(x))\, dx - \int_a^b P(x, g_2(x))\, dx.$$

In the second integral on the right side of equation (3) we can use the change of variable $x = -t$ and manipulate the integral to obtain

(4) $$-\iint_R P_y(x, y)\, dx\, dy = \int_a^b P(x, g_1(x))\, dx + \int_{-b}^{-a} P(-t, g_2(-t))(-1)\, dt.$$

When the two integrals on the right side of equation (4) are interpreted as contour integrals along C_1 and C_2, respectively, we see that

(5) $$-\iint_R P_y(x, y)\, dx\, dy = \int_{C_1} P(x, y)\, dx + \int_{C_2} P(x, y)\, dx = \int_C P(x, y)\, dx.$$

To complete the proof, we rely on the fact that for a standard region, there exist functions $x = h_1(y)$ and $x = h_2(y)$ for $c \le y \le d$ whose graphs form the left and right portions of C, respectively, as indicated in Figure 6.19. Since C has the positive orientation, it can be expressed as the sum of two contours C_3 and C_4, where

C_3: $z_3(t) = h_1(-t) - it$ for $-d \le t \le -c$ and
C_4: $z_4(t) = h_2(t) + it$ for $c \le t \le d$.

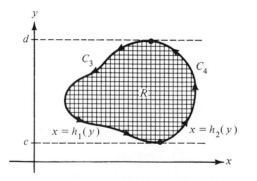

FIGURE 6.19   Integration over a standard region where $C = C_3 + C_4$.

Using the functions $h_1(y)$ and $h_2(y)$, we express the double integral of $Q_x(x, y)$ over R as an iterated integral:

(6) $$\iint_R Q_x\,(x, y)\,dx\,dy = \int_c^d \left[\int_{h_1(y)}^{h_2(y)} Q_x(x, y)\,dx \right] dy.$$

A similar derivation will show that equation (6) is equivalent to

(7) $$\iint_R Q_x(x, y)\,dx\,dy = \int_C Q(x, y)\,dy.$$

When equations (5) and (7) are added, the result is equation (1), and the proof is complete.

We are now ready to state our main result in this section.

Theorem 6.3 (Cauchy-Goursat Theorem)   *Let f be analytic in a simply connected domain D. If C is a simple closed contour that lies in D, then*

(8) $$\int_C f(z)\,dz = 0.$$

Proof   If we add the additional hypothesis that the derivative $f'(z)$ is also continuous, the proof is more intuitive. It was Augustin Cauchy who first proved this theorem under the hypothesis that $f'(z)$ is continuous. His proof, which we will now state, used Green's theorem.

Proof Using Green's Theorem   We assume that C is oriented in the positive sense and use equation (12) in Section 6.2 to write

(9) $$\int_C f(z)\,dz = \int_C u\,dx - v\,dy + i \int_C v\,dx + u\,dy.$$

If we use Green's theorem on the real part of the right side of equation (9) with $P = u$ and $Q = -v$, then we obtain

(10) $$\int_C u\,dx - v\,dy = \iint_R (-v_x - u_y)\,dx\,dy,$$

where R is the region that is the interior of C. If we use Green's theorem on the imaginary part, the result will be

(11) $$\int_C v\,dx + u\,dy = \iint_R (u_x - v_y)\,dx\,dy.$$

The Cauchy-Riemann equations $u_x = v_y$ and $u_y = -v_x$ can be used in equations (10) and (11) to see that the value of equation (9) is given by

$$\int_C f(z)\,dz = \int\int_R 0\,dx\,dy + i\int\int_R 0\,dx\,dy = 0,$$

and the proof is complete.

A proof that does not require the continuity of $f'(z)$ was devised by Edward Goursat (1858–1936) in 1883.

Goursat's Proof of Theorem 6.3   We first establish the result for a triangular contour C with positive orientation. Construct four positively oriented contours C^1, C^2, C^3, and C^4 that are the triangles obtained by joining the midpoints of the sides of C as shown in Figure 6.20.

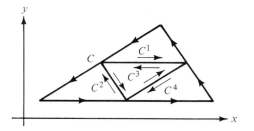

FIGURE 6.20   The triangular contours C and C^1, C^2, C^3, C^4.

Since each contour is positively oriented, if we sum the integrals along the four triangular contours, then the integrals along the segments interior to C cancel out in pairs. The result is

(12) $$\int_C f(z)\,dz = \sum_{k=1}^{4}\int_{C^k} f(z)\,dz.$$

Let C_1 be selected from C^1, C^2, C^3, and C^4 so that the following relation holds true:

(13) $$\left|\int_C f(z)\,dz\right| \le \sum_{k=1}^{4}\left|\int_{C^k} f(z)\,dz\right| \le 4\left|\int_{C_1} f(z)\,dz\right|.$$

We can proceed inductively and carry out a similar subdivision process to obtain a sequence of triangular contours $\{C_n\}$, where the interior of C_{n+1} lies in the interior of C_n and the following inequality holds:

(14) $$\left|\int_{C_n} f(z)\,dz\right| \le 4\left|\int_{C_{n+1}} f(z)\,dz\right| \qquad \text{for } n = 1, 2, \ldots .$$

Let T_n denote the closed region that consists of C_n and its interior. Since the length of the sides of C_n go to zero as $n \to \infty$, there exists a unique point z_0 that belongs to all of the closed triangular regions T_n. Since f is analytic at the point z_0, there exists a function $\eta(z)$ with

$$(15) \quad f(z) = f(z_0) + f'(z_0)(z - z_0) + \eta(z)(z - z_0).$$

Using equation (15) and integrating f along C_n, we find that

$$(16) \quad \int_{C_n} f(z)\, dz = \int_{C_n} f(z_0)\, dz + \int_{C_n} f'(z_0)(z - z_0)\, dz$$
$$+ \int_{C_n} \eta(z)(z - z_0)\, dz$$
$$= [f(z_0) - f'(z_0)z_0] \int_{C_n} 1\, dz \;+\; f'(z_0) \int_{C_n} z\, dz$$
$$+ \int_{C_n} \eta(z)(z - z_0)\, dz$$
$$= \int_{C_n} \eta(z)(z - z_0)\, dz.$$

If $\varepsilon > 0$ is given, then a $\delta > 0$ can be found such that

$$(17) \quad |z - z_0| < \delta \quad \text{implies that } |\eta(z)| < \frac{\varepsilon}{L^2},$$

where L is the length of the original contour C. An integer n can now be chosen so that C_n lies in the neighborhood $|z - z_0| < \delta$, as shown in Figure 6.21.

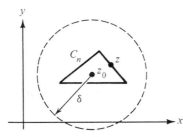

FIGURE 6.21   The contour C_n that lies in the neighborhood $|z - z_0| < \delta$.

Since the distance between a point z on a triangle and a point z_0 interior to the triangle is no greater than half the perimeter of the triangle, it follows that

$$(18) \quad |z - z_0| < \tfrac{1}{2} L_n \quad \text{for all } z \text{ on } C_n,$$

where L_n is the length of the triangle C_n. From the preceding construction process, it follows that

$$(19) \quad L_n = (\tfrac{1}{2})^n L \quad \text{and} \quad |z - z_0| < (\tfrac{1}{2})^{n+1} L \quad \text{for } z \text{ on } C_n.$$

We can use equations (14), (17), and (19) of this section and equation (23) of Section 6.2 to obtain the following estimate:

$$\left| \int_C f(z)\, dz \right| \le 4^n \int_{C_n} |\eta(z)(z - z_0)| \, |dz|$$

$$\le 4^n \int_{C_n} \frac{\varepsilon}{L^2} \left(\frac{1}{2} \right)^{n+1} L \, |dz|$$

$$= \frac{2^{n-1}\varepsilon}{L} \int_{C_n} |dz|$$

$$= \frac{2^{n-1}\varepsilon}{L} \left(\frac{1}{2} \right)^n L = \frac{\varepsilon}{2}.$$

Since ε was arbitrary, it follows that equation (12) holds true for the triangular contour C. If C is a polygonal contour, then interior edges can be added until the interior is subdivided into a finite number of triangles. The integral around each triangle is zero, and the sum of all these integrals is equal to the integral around the polygonal contour C. Therefore equation (12) holds true for polygonal contours. The proof for an arbitrary simple closed contour is established by approximating the contour "sufficiently close" with a polygonal contour.

EXAMPLE 6.12   Let us recall that exp z, cos z, and z^n, where n is a positive integer are all entire functions and have continuous derivatives. The Cauchy-Goursat theorem implies that for any simple closed contour we have

$$\int_C \exp z \, dz = 0, \quad \int_C \cos z \, dz = 0, \quad \int_C z^n \, dz = 0.$$

EXAMPLE 6.13   If C is a simple closed contour such that the origin does not lie interior to C, then there is a simply connected domain D that contains C in which $f(z) = 1/z^n$ is analytic, as is indicated in Figure 6.22. The Cauchy-Goursat theorem implies that

$$\int_C \frac{1}{z^n} \, dz = 0 \quad \text{provided that the origin does not lie interior to } C.$$

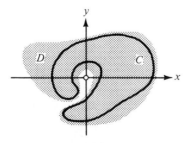

FIGURE 6.22   A simple connected domain D containing the simple closed contour C that does not contain the origin.

It is desirable to be able to replace integrals over certain complicated contours with integrals that are easy to evaluate. If C_1 is a simple closed contour that can be continuously deformed into another simple closed contour C_2 without passing through a point where f is not analytic, then the value of the contour integral of f over C_1 is the same as the value of the integral of f over C_2. To be precise, we state the following result.

Theorem 6.4 (Deformation of Contour)   *Let C_1 and C_2 be two simple closed positively oriented contours such that C_1 lies interior to C_2. If f is analytic in a domain D that contains both C_1 and C_2 and the region between them, as shown in Figure 6.23, then*

$$\int_{C_1} f(z) \, dz = \int_{C_2} f(z) \, dz.$$

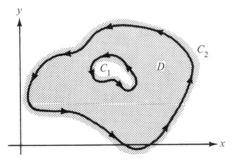

FIGURE 6.23   The domain D that contains the simple closed contours C_1 and C_2 and the region between them.

Proof   Assume that both C_1 and C_2 have positive (counterclockwise) orientation. We construct two disjoint contours or *cuts* L_1 and L_2 that join C_1 to C_2. Hence the contour C_1 will be cut into two contours C_1^* and C_1^{**}, and the contour C_2 will be cut into C_2^* and C_2^{**}. We now form two new contours:

$$K_1 = -C_1^* + L_1 + C_2^* - L_2 \quad \text{and} \quad K_2 = -C_1^{**} + L_2 + C_2^{**} - L_1,$$

which are shown in Figure 6.24. The function f will be analytic on a simply connected domain D_1 that contains K_1, and f will be analytic on the simply connected domain D_2 that contains K_2, as is illustrated in Figure 6.24.

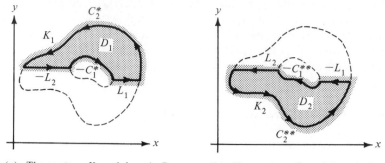

(a) The contour K_1 and domain D_1.   (b) The contour K_2 and domain D_2.

FIGURE 6.24   The cuts L_1 and L_2 and the contours K_1 and K_2 used to prove the Deformation Theorem.

The Cauchy-Goursat theorem can be applied to the contours K_1 and K_2, and the result is

(20) $\displaystyle\int_{K_1} f(z)\,dz = 0$ and $\displaystyle\int_{K_2} f(z)\,dz = 0.$

Adding contours, we observe that

(21) $K_1 + K_2 = -C_1^* + L_1 + C_2^* - L_2 - C_1^{**} + L_2 + C_2^{**} - L_1$
$$= C_2^* + C_2^{**} - C_1^* - C_1^{**} = C_2 - C_1.$$

We can use identities (14) and (17) of Section 6.2 and equations (20) and (21) given in this section to conclude that

$$\int_{C_2} f(z)\,dz - \int_{C_1} f(z)\,dz = \int_{K_1} f(z)\,dz + \int_{K_2} f(z)\,dz = 0,$$

which completes the proof of Theorem 6.4.

We now state an important result that is proven by the deformation theorem. This result will occur several times in the theory to be developed and is an important tool for computations.

EXAMPLE 6.14   Let z_0 denote a fixed complex value. If C is a simple closed contour with positive orientation such that z_0 lies interior to C, then

(22) $\displaystyle\int_C \frac{dz}{z - z_0} = 2\pi i$ and

$\displaystyle\int_C \frac{dz}{(z - z_0)^n} = 0$ where $n \ne 1$ is an integer.

Solution   Since z_0 lies interior to C, we can choose R so that the circle C_R will center z_0 and radius R lies interior to C. Hence $f(z) = 1/(z - z_0)^n$ is analytic in a domain D that contains both C and C_R and the region between them, as shown in Figure 6.25. Let C_R have the parameterization

$$C_R: z(\theta) = z_0 + Re^{i\theta} \quad \text{and} \quad dz = i\, Re^{i\theta}\, d\theta \quad \text{for } 0 \le \theta \le 2\pi.$$

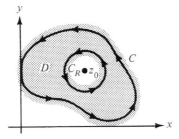

FIGURE 6.25   The domain D that contains both C and C_R.

The deformation theorem implies that the integral of f over C_R has the same value as the integral of f over C, and we obtain

$$\int_C \frac{dz}{z - z_0} = \int_{C_R} \frac{dz}{z - z_0} = \int_0^{2\pi} \frac{i\, Re^{i\theta}}{Re^{i\theta}}\, d\theta = i \int_0^{2\pi} d\theta = 2\pi i$$

and

$$\int_C \frac{dz}{(z - z_0)^n} = \int_{C_R} \frac{dz}{(z - z_0)^n} = \int_0^{2\pi} \frac{i\, Re^{i\theta}}{R^n e^{in\theta}}\, d\theta = iR^{1-n} \int_0^{2\pi} e^{i(1-n)\theta}\, d\theta$$

$$= \left. \frac{R^{1-n}}{1 - n} e^{i(1-n)\theta} \right|_{\theta=0}^{\theta=2\pi} = \frac{R^{1-n}}{1 - n} - \frac{R^{1-n}}{1 - n} = 0.$$

The deformation theorem is an extension of the Cauchy-Goursat theorem to a doubly connected domain in the following sense. Let D be a domain that contains C_1 and C_2 and the region between them, as shown in Figure 6.25. Then the contour $C = C_2 - C_1$ is a parameterization of the boundary of the region R that lies between C_1 and C_2 so that the points of R lie to the left of C as a point $z(t)$ moves around C. Hence C is a positive orientation of the boundary of R, and Theorem 6.4 implies that

$$\int_C f(z)\, dz = 0.$$

We can extend Theorem 6.4 to multiply connected domains with more than one "hole." The proof, which is left for the reader, involves the introduction of several cuts and is similar to the proof of Theorem 6.4.

Theorem 6.5 (Extended Cauchy-Goursat Theorem)

Let $C, C_1, C_2, \ldots, C_n$ be simple closed positively oriented contours with the property that C_k lies interior to C for $k = 1, 2, \ldots, n$, and the set interior to C_k has no points in common with the set interior to C_j if $k \neq j$. Let f be analytic on a domain D that contains all the contours and the region between C and $C_1 + C_2 + \cdots + C_n$, which is shown in Figure 6.26. Then

$$(23) \quad \int_C f(z)\, dz = \sum_{k=1}^{n} \int_{C_k} f(z)\, dz.$$

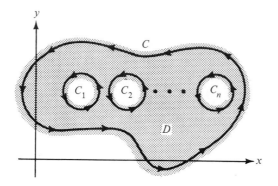

FIGURE 6.26   The multiply connected domain D and the contours C and $C_1, C_2, \ldots, C_n$ in the statement of the Extended Cauchy-Goursat Theorem.

EXAMPLE 6.15   If C is the circle $|z| = 2$ taken with positive orientation, then

$$(24) \quad \int_C \frac{2z\, dz}{z^2 + 2} = 4\pi i.$$

Solution   Using partial fractions, the integral in equation (24) can be written as

$$(25) \quad \int_C \frac{2z\, dz}{z^2 + 2} = \int_C \frac{dz}{z + i\sqrt{2}} + \int_C \frac{dz}{z - i\sqrt{2}}.$$

Since the points $z = \pm i\sqrt{2}$ lie interior to C, Example 6.14 implies that

$$(26) \quad \int_C \frac{dz}{z \pm i\sqrt{2}} = 2\pi i.$$

The results in (26) can be used in (25) to conclude that

$$\int_C \frac{2z\, dz}{z^2 + 2} = 2\pi i + 2\pi i = 4\pi i.$$

EXAMPLE 6.16   If C is the circle $|z - i| = 1$ taken with positive orientation, then

$$(27) \quad \int_C \frac{2z \, dz}{z^2 + 2} = 2\pi i.$$

Solution   Using partial fractions, the integral in equation (27) can be written as

$$(28) \quad \int_C \frac{2z \, dz}{z^2 + 2} = \int_C \frac{dz}{z + i\sqrt{2}} + \int_C \frac{dz}{z - i\sqrt{2}}.$$

In this case, only the point $z = i\sqrt{2}$ lies interior to C, so the second integral on the right side of equation (28) has the value $2\pi i$. The function $f(z) = 1/(z + i\sqrt{2})$ is analytic on a simply connected domain that contains C. Hence by the Cauchy-Goursat theorem the first integral on the right side of equation (28) is zero (see Figure 6.27). Therefore

$$\int_C \frac{2z \, dz}{z^2 + 2} = 0 + 2\pi i = 2\pi i.$$

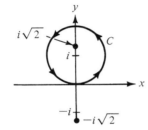

FIGURE 6.27   The circle $|z - i| = 1$ and the points $z = \pm i\sqrt{2}$.

EXAMPLE 6.17   Show that

$$\int_C \frac{z - 2}{z^2 - z} \, dz = -6\pi i$$

where C is the "figure eight" contour shown in Figure 6.28(a).

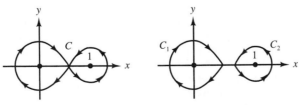

(a) The figure eight contour C.   (b) The contours C_1 and C_2.

FIGURE 6.28   The contour $C = C_1 + C_2$.

Solution   Partial fractions can be used to express the integral as

$$(29) \quad \int_C \frac{z-2}{z^2-z} \, dz = 2 \int_C \frac{1}{z} \, dz - \int_C \frac{1}{z-1} \, dz.$$

Using the Cauchy-Goursat theorem and property (14) of Section 6.2 together with Example 6.13, we compute the value of the first integral on the right side of equation (29):

$$(30) \quad 2 \int_C \frac{1}{z} \, dz = 2 \int_{C_1} \frac{1}{z} \, dz + 2 \int_{C_2} \frac{1}{z} \, dz$$

$$= -2 \int_{-C_1} \frac{1}{z} \, dz + 0 = -2(2\pi i) = -4\pi i.$$

In a similar fashion we find that

$$(31) \quad -\int_C \frac{dz}{z-1} = -\int_{C_1} \frac{dz}{z-1} - \int_{C_2} \frac{dz}{z-1} = 0 - 2\pi i = -2\pi i.$$

The results of equations (30) and (31) can be used in equation (29) to conclude that

$$\int_C \frac{z-2}{z^2-z} \, dz = -4\pi i - 2\pi i = -6\pi i.$$

EXERCISES FOR SECTION 6.3

1. Determine the domain of analyticity for the following functions, and conclude that $\int_C f(z) \, dz = 0$, where C is the circle $|z| = 1$ with positive orientation.

   (a) $f(z) = \dfrac{z}{z^2 + 2}$ (b) $f(z) = \dfrac{1}{z^2 + 2z + 2}$

   (c) $f(z) = \tan z$ (d) $f(z) = \text{Log}(z + 5)$

2. Show that $\int_C z^{-1} \, dz = 2\pi i$, where C is the square with vertices $1 \pm i$, $-1 \pm i$ with positive orientation.

3. Show that $\int_C (4z^2 - 4z + 5)^{-1} \, dz = 0$, where C is the unit circle $|z| = 1$ with positive orientation.

4. Find $\int_C (z^2 - z)^{-1} \, dz$ for the following contours.

   (a) The circle $|z - 1| = 2$ with positive orientation.

   (b) The circle $|z - 1| = \frac{1}{2}$ with positive orientation.

5. Find $\int_C (2z - 1)(z^2 - z)^{-1} \, dz$ for the following contours.

   (a) The circle $|z| = 2$ with positive orientation.

   (b) The circle $|z| = \frac{1}{2}$ with positive orientation.

6. Evaluate $\int_C (z^2 - z)^{-1} \, dz$, where C is the figure eight contour shown in Figure 6.28(a).

7. Evaluate $\int_C (2z - 1)(z^2 - z)^{-1} \, dz$, where C is the figure eight contour shown in Figure 6.28(a).

8. Evaluate $\int_C (4z^2 + 4z - 3)^{-1} \, dz = \int_C (2z - 1)^{-1}(2z + 3)^{-1} \, dz$ for the following contours.

   (a) The circle $|z| = 1$ with positive orientation.

   (b) The circle $\left|z + \frac{3}{2}\right| = 1$ with positive orientation.

   (c) The circle $|z| = 3$ with positive orientation.

9. Evaluate $\int_C (z^2 - 1)^{-1} \, dz$ for the contours given in Figure 6.29.

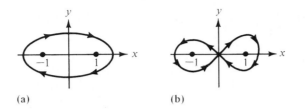

(a)   (b)

FIGURE 6.29   Accompanies Exercise 9.

10. Let C be the triangle with vertice 0, 1, and i with positive orientation. Parameterize C and show that

$$\int_C 1 \, dz = 0 \quad \text{and} \quad \int_C z \, dz = 0.$$

11. Let the circle $|z| = 1$ be given the parameterization

$$C: z(t) = \cos t + i \sin t \quad \text{for } -\pi \le t \le \pi.$$

Use the principal branch of the square root function:

$$z^{1/2} = r^{1/2}\cos \frac{\theta}{2} + ir^{1/2}\sin \frac{\theta}{2} \quad \text{for } -\pi < \theta \le \pi$$

and find $\int_C z^{1/2} \, dz$.

12. Evaluate $\int_C |z|^2 \exp z \, dz$, where C is the unit circle $|z| = 1$ with positive orientation.

13. Let $f(z) = u(r, \theta) + iv(r, \theta)$ be analytic for all values of $z = re^{i\theta}$. Show that

$$\int_0^{2\pi} [u(r, \theta) \cos \theta - v(r, \theta) \sin \theta] \, d\theta = 0.$$

Hint: Integrate f around the circle $|z| = 1$.

14. Show by using Green's theorem that the area enclosed by a simple closed contour C is $\frac{1}{2} \int_C x \, dy - y \, dx$.

15. Compare the various methods for evaluating contour integrals. What are the limitations of each method?

6.4   The Fundamental Theorems of Integration

Let f be analytic in the simply connected domain D. The theorems in this section show that an antiderivative F can be constructed by contour integration. A consequence will be the fact that in a simply connected domain, the integral of an analytic function f along any contour joining z_1 to z_2 is the same, and its value is given by $F(z_2) - F(z_1)$. Hence we will be able to use the antiderivative formulas from calculus to compute the value of definite integrals.

> **Theorem 6.6 (Indefinite Integrals or Antiderivatives)**   *Let f be analytic in the simply connected domain D. If z_0 is a fixed value in D and if C is any contour in D with initial point z_0 and terminal point z, then the function given by*

(1) $F(z) = \int_C f(\xi)\, d\xi = \int_{z_0}^{z} f(\xi)\, d\xi$

is analytic in D and

(2) $F'(z) = f(z)$.

Proof   We first establish that the integral is independent of the path of integration. Hence we will need to keep track only of the endpoints, and we can use the notation

$$\int_C f(\xi)\, d\xi = \int_{z_0}^{z} f(\xi)\, d(\xi).$$

Let C_1 and C_2 be two contours in D, both with the initial point z_0 and the terminal point z, as shown in Figure 6.30. Then $C = C_1 - C_2$ is a simple closed contour, and the Cauchy-Goursat theorem implies that

$$\int_{C_1} f(\xi)\, d\xi - \int_{C_2} f(\xi)\, d\xi = \int_{C_1 - C_2} f(\xi)\, d\xi = 0.$$

Therefore the contour integral in equation (1) is independent of path. Here we have taken the liberty of drawing contours that intersect only at the endpoints. A slight modification of the foregoing proof will show that a finite number of other points of intersection are permitted.

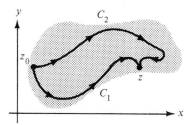

FIGURE 6.30   The contours C_1 and C_2 joining z_0 to z.

We now show that $F'(z) = f(z)$. Let z be held fixed, and let Δz be chosen small enough so that the point $z + \Delta z$ also lies in the domain D. Since z is held fixed, $f(z) = K$ where K is a constant, and equation (12) of Section 6.1 implies that

(3) $\displaystyle \int_{z}^{z+\Delta z} f(z)\, d\xi = \int_{z}^{z+\Delta z} K\, d\xi = K\, \Delta z = f(z)\, \Delta z.$

Using the additive property of contours and the definition of F given in equation (1), it follows that

(4) $\displaystyle F(z + \Delta z) - F(z) = \int_{z_0}^{z+\Delta z} f(\xi)\, d\xi - \int_{z_0}^{z} f(\xi)\, d\xi$

$\displaystyle \qquad\qquad = \int_{C_2} f(\xi)\, d\xi - \int_{C_1} f(\xi)\, d\xi = \int_C f(\xi)\, d\xi,$

where the contour C is the straight line segment joining z to $z + \Delta z$ and C_1 and C_2 join z_0 to z and z_0 to $z + \Delta z$, respectively, as shown in Figure 6.31.

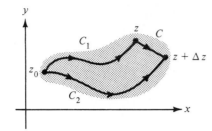

FIGURE 6.31   The contours C_1 and C_2 and the line segment $C = -C_1 + C_2$.

Since f is continuous at z, then if $\varepsilon > 0$, there is a $\delta > 0$ so that

$$(5) \quad |f(\xi) - f(z)| < \varepsilon \quad \text{whenever } |\xi - z| < \delta.$$

If we require that $|\Delta z| < \delta$, then using equations (3) and (4), inequality (5), and inequality (22) of Section 6.2, we obtain the following estimate:

$$(6) \quad \left| \frac{F(z + \Delta z) - F(z)}{\Delta z} - f(z) \right| = \frac{1}{|\Delta z|} \left| \int_C f(\xi) \, d\xi - \int_C f(z) \, d\xi \right|$$

$$\leq \frac{1}{|\Delta z|} \int_C |f(\xi) - f(z)| \, |d\xi|$$

$$< \frac{1}{|\Delta z|} \varepsilon |\Delta z| = \varepsilon.$$

Consequently, the left side of equation (6) tends to 0 as $\Delta z \to 0$; that is, $F'(z) = f(z)$, and the theorem is proven.

It is important to notice that the line integral of an analytic function is independent of path. An easy calculation shows

$$\int_{C_1} z \, dz = \int_{C_2} z \, dz = 4 + 2i,$$

where C_1 and C_2 were contours joining $-1 - i$ to $3 + i$. Since the integrand $f(z) = z$ is an analytic function, Theorem 6.6 implies that the value of the two integrals is the same; hence one calculation would suffice.

If we set $z = z_1$ in Theorem 6.6, then we obtain the following familiar result for evaluating a definite integral of an analytic function.

Theorem 6.7 (Definite Integrals)   *Let f be analytic in a simply connected domain D. If z_0 and z_1 are two points in D, then*

$$(7) \quad \int_{z_0}^{z_1} f(z) \, dz = F(z_1) - F(z_0)$$

where F is any antiderivative of f.

Proof   If F is chosen to be the function in equation (1), then equation (7) holds true. If G is any other antiderivative of f, then $H(z) = G(z) - F(z)$ is analytic, and $H'(z) = 0$ for all points z in D. Hence $H(z) = K$ where K is a constant, and $G(z) = F(z) + K$. Therefore $G(z_1) - G(z_0) = F(z_1) - F(z_0)$, and Theorem 6.7 is proven.

Theorem 6.7 is an important method for evaluating definite integrals when the integrand is an analytic function. In essence, it permits us to use all the rules of integration that were introduced in calculus. For analytic integrands, application of Theorem 6.7 is easier to use than the method of parameterization of a contour.

EXAMPLE 6.18   Show that $\int_1^i \cos z \, dz = -\sin 1 + i \sinh 1$.

Solution   An antiderivative of $f(z) = \cos z$ is $F(z) = \sin z$. Hence

$$\int_1^i \cos z \, dz = \sin i - \sin 1 = -\sin 1 + i \sinh 1.$$

EXAMPLE 6.19   Evaluate $\left(2 + \dfrac{i\pi}{4} \right) \displaystyle\int_0^1 e^{2t} e^{i\pi t/4} \, dt$.

Solution   In Example 6.7, we broke the integrand up into its real and imaginary parts, which then required integration by parts. Using Theorem 6.7, however, we see that

$$\left(2 + \frac{i\pi}{4} \right) \int_0^1 e^{2t} e^{i\pi t/4} = \left(2 + \frac{i\pi}{4} \right) \int_0^1 e^{t(2+i\pi/4)} \, dt$$

$$= \left(2 + \frac{i\pi}{4} \right) \left(\frac{1}{2 + \dfrac{i\pi}{4}} \right) e^{t(2+i\pi/4)} \Bigg|_0^1$$

$$= e^{(2+i\pi/4)} - e^0$$

$$= e^{(2+i\pi/4)} - 1.$$

EXAMPLE 6.20   Show that

$$\int_4^{8+6i} \frac{dz}{2z^{1/2}} = 1 + i,$$

where $z^{1/2}$ is the principal branch of the square root function and the integral is to be taken along the line segment joining 4 to $8 + 6i$.

Solution   Example 3.8 showed that if $F(z) = z^{1/2}$, then $F'(z) = 1/(2z^{1/2})$, where the principal branch of the square root function is used in both the formulas for F and F'. Hence

$$\int_4^{8+6i} \frac{dz}{2z^{1/2}} = (8 + 6i)^{1/2} - 4^{1/2} = 3 + i - 2 = 1 + i.$$

EXAMPLE 6.21   Let $D = \{z = re^{i\theta}: r > 0 \text{ and } -\pi < \theta < \pi\}$ be the simply connected domain shown in Figure 6.32. Then $F(z) = \text{Log } z$ is analytic in D, and its derivative is $F'(z) = 1/z$. If C is a contour in D that joins the point z_1 to the point z_2, then Theorem 6.7 implies that

$$\int_{z_1}^{z_2} \frac{dz}{z} = \int_C \frac{dz}{z} = \text{Log } z_2 - \text{Log } z_1.$$

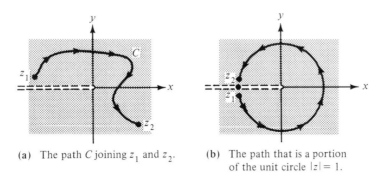

(a)   The path C joining z_1 and z_2.

(b)   The path that is a portion of the unit circle $|z| = 1$.

FIGURE 6.32   The simply connected domain D in Examples 6.21 and 6.22.

EXAMPLE 6.22   As a consequence of Example 6.21, let us show that

$$\int_C \frac{dz}{z} = 2\pi i, \quad \text{where } C \text{ is the unit circle } |z| = 1,$$

taken with positive orientation.

Solution   If we let z_2 approach -1 through the upper half plane and z_1 approaches -1 through the lower half plane, then we can integrate around the portion of the circle shown in Figure 6.32(b) and take limits to obtain

$$\int_C \frac{dz}{z} = \lim_{\substack{z_1 \to -1 \\ z_2 \to -1}} \int_{z_1}^{z_2} \frac{dz}{z} = \lim_{\substack{z_2 \to -1 \\ Im(z_2)>0}} \mathrm{Log}\, z_2 - \lim_{\substack{z_1 \to -1 \\ Im(z_1)<0}} \mathrm{Log}\, z_1$$

$$= i\pi - (-i\pi) = 2\pi i.$$

EXERCISES FOR SECTION 6.4

For Exercises 1–14, use antiderivatives to find the value of the definite integral.

1. $\int_{1+i}^{2+i} z^2 \, dz$

2. $\int_1^i \frac{1+z}{z} \, dz$ (use Log z)

3. $\int_2^{i\pi/2} \exp z \, dz$

4. $\int_i^{1+i} (z^2 + z^{-2}) \, dz$

5. $\int_{-i}^{1+i} \cos z \, dz$

6. $\int_0^{\pi-2i} \sin \frac{z}{2} \, dz$

7. $\int_{-1-i\pi/2}^{2+\pi i} z \exp z \, dz$

8. $\int_{1-2i}^{1+2i} z \exp(z^2) \, dz$

9. $\int_0^i z \cos z \, dz$

10. $\int_0^i \sin^2 z \, dz$

11. $\int_1^{1+i} \mathrm{Log}\, z \, dz$

12. $\int_2^{2+i} \frac{dz}{z^2 - z}$

13. $\int_2^{2+i} \frac{2z-1}{z^2 - z} \, dz$

14. $\int_2^{2+i} \frac{z-2}{z^2 - z} \, dz$

15. Show that $\int_{z_1}^{z_2} 1 \, dz = z_2 - z_1$ by parameterizing the line segment from z_1 to z_2.

16. Let z_1 and z_2 be points in the right half plane. Show that

$$\int_{z_1}^{z_2} \frac{dz}{z^2} = \frac{1}{z_1} - \frac{1}{z_2}.$$

17. Find

$$\int_9^{3+4i} \frac{dz}{2z^{1/2}},$$

where $z^{1/2}$ is the principal branch of the square root function and the integral is to be taken along the line segment from 9 to $3 + 4i$.

18. Find $\int_{-2i}^{2i} z^{1/2} \, dz$, where $z^{1/2}$ is the principal branch of the square root function and the integral is to be taken along the right half of the circle $|z| = 2$.

19. Using the equation

$$\frac{1}{z^2 + 1} = \frac{i}{2} \frac{1}{z + i} - \frac{i}{2} \frac{1}{z - i},$$

show that if z lies in the right half plane, then

$$\int_0^z \frac{d\xi}{\xi^2 + 1} = \arctan z = \frac{i}{2} \mathrm{Log}(z + i) - \frac{i}{2} \mathrm{Log}(z - i).$$

20. Let f' and g' be analytic for all z. Show that

$$\int_{z_1}^{z_2} f(z)g'(z) \, dz = f(z_2)g(z_2) - f(z_1)g(z_1) - \int_{z_1}^{z_2} f'(z)g(z) \, dz.$$

21. Compare the various methods for evaluating contour integrals. What are the limitations of each method?

22. Explain how the fundamental theorem of calculus studied in complex analysis and the fundamental theorem of calculus studied in calculus are different. How are they similar?

6.5   Integral Representations for Analytic Functions

We now present some major results in the theory of functions of a complex variable. The first result is known as Cauchy's integral formula and shows that the value of an analytic function f can be represented by a certain contour integral. The nth derivative, $f^{(n)}(z)$, will have a similar representation. In Chapter 7 we will show how the Cauchy integral formulae are used to prove Taylor's theorem, and we will establish the power series representation for analytic functions. The Cauchy integral formulae will also be a convenient tool for evaluating certain contour integrals.

> **Theorem 6.8 (Cauchy's Integral Formula)** *Let f be analytic in the simply connected domain D, and let C be a simple closed positively oriented contour that lies in D. If z_0 is a point that lies interior to C, then*

$$(1) \quad f(z_0) = \frac{1}{2\pi i} \int_C \frac{f(z)}{z - z_0}\, dz.$$

Proof   Since f is continuous at z_0, if $\varepsilon > 0$ is given, there is a $\delta > 0$ such that

$$(2) \quad \left| f(z) - f(z_0) \right| < \varepsilon \quad \text{whenever} \quad \left| z - z_0 \right| < \delta.$$

Also the circle C_0: $\left| z - z_0 \right| = \frac{1}{2}\delta$ lies interior to C as shown in Figure 6.33.

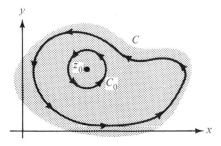

FIGURE 6.33   The contours C and C_0 in the proof of Cauchy's integral formula.

Since $f(z_0)$ is a fixed value, we can use the result of Exercise 24 of Section 6.2 to conclude that

$$(3) \quad f(z_0) = \frac{f(z_0)}{2\pi i} \int_{C_0} \frac{dz}{z - z_0} = \frac{1}{2\pi i} \int_{C_0} \frac{f(z_0)}{z - z_0}\, dz.$$

Using the deformation theorem we see that

$$(4) \quad \frac{1}{2\pi i} \int_C \frac{f(z)}{z - z_0}\, dz = \frac{1}{2\pi i} \int_{C_0} \frac{f(z)}{z - z_0}\, dz.$$

Using inequality (2), equations (3) and (4), and inequality (22) of Section 6.2, we obtain the following estimate:

$$(5) \quad \left| \frac{1}{2\pi i} \int_C \frac{f(z)\, dz}{z - z_0} - f(z_0) \right| = \left| \frac{1}{2\pi i} \int_{C_0} \frac{f(z)\, dz}{z - z_0} - \frac{1}{2\pi i} \int_{C_0} \frac{f(z_0)\, dz}{z - z_0} \right|$$

$$\leq \frac{1}{2\pi} \int_{C_0} \frac{|f(z) - f(z_0)|}{|z - z_0|} \, |dz|$$

$$\leq \frac{1}{2\pi} \frac{\varepsilon}{(1/2)\delta} \pi\delta = \varepsilon.$$

Since ε can be made arbitrarily small, the theorem is proven.

EXAMPLE 6.23   Show that

$$\int_C \frac{\exp z}{z - 1} \, dz = i2\pi e,$$

where C is the circle $|z| = 2$ with positive orientation.

   Solution   Here we have $f(z) = \exp z$ and $f(1) = e$. The point $z_0 = 1$ lies interior to C, so Cauchy's integral formula implies that

$$e = f(1) = \frac{1}{2\pi i} \int_C \frac{\exp z}{z - 1} \, dz,$$

and multiplication by $2\pi i$ will establish the desired result.

EXAMPLE 6.24   Show that

$$\int_C \frac{\sin z}{4z + \pi} \, dz = \frac{-\sqrt{2}\pi i}{4},$$

where C is the circle $|z| = 1$ with positive orientation.

   Solution   Here we have $f(z) = \sin z$. We can manipulate the integral and use Cauchy's integral formula to obtain

$$\int_C \frac{\sin z}{4z + \pi} \, dz = \frac{1}{4} \int_C \frac{\sin z}{z + (\pi/4)} \, dz = \frac{1}{4} \int_C \frac{f(z)}{z - (-\pi/4)} \, dz$$

$$= \frac{1}{4} (2\pi i) f\left(\frac{-\pi}{4} \right) = \frac{\pi i}{2} \sin\left(\frac{-\pi}{4} \right) = \frac{-\sqrt{2}\pi i}{4}.$$

   We now state a general result that shows how differentiation under the integral sign can be accomplished. The proof can be found in some advanced texts. See, for instance, Rolf Nevanlinna and V. Paatero, *Introduction to Complex Analysis* (Reading, Massachusetts: Addison-Wesley Publishing Company, 1969), Section 9.7.

Theorem 6.9 (Leibniz's Rule)   *Let D be a simply connected domain, and let I: $a \le t \le b$ be an interval of real numbers. Let $f(z, t)$ and its partial derivative $f_z(z, t)$ with respect to z be continuous functions for all z in D and all t in I. Then*

(6) $$F(z) = \int_a^b f(z, t)\, dt$$

is analytic for z in D, and

$$F'(z) = \int_a^b f_z(z, t)\, dt.$$

We now show how Theorem 6.8 can be generalized to give an integral representation for the *n*th derivative, $f^{(n)}(z)$. Leibniz's rule will be used in the proof, and we shall see that this method of proof will be a mnemonic device for remembering how the denominator is written.

Theorem 6.10 (Cauchy's Integral Formulae for Derivatives)   *Let f be analytic in the simply connected domain D, and let C be a simple closed positively oriented contour that lies in D. If z is a point that lies interior to C, then*

(7) $$f^{(n)}(z) = \frac{n!}{2\pi i} \int_C \frac{f(\xi)}{(\xi - z)^{n+1}}\, d\xi.$$

Proof   We will establish the theorem for the case $n = 1$. We start by using the parameterization

$$C: \xi = \xi(t) \quad \text{and} \quad d\xi = \xi'(t)\, dt \quad \text{for } a \le t \le b.$$

We use Theorem 6.8 and write

(8) $$f(z) = \frac{1}{2\pi i} \int_C \frac{f(\xi)}{\xi - z}\, d\xi = \frac{1}{2\pi i} \int_a^b \frac{f(\xi(t))\xi'(t)\, dt}{\xi(t) - z}.$$

The integrand on the right side of equation (8) can be considered as a function $f(z, t)$ of the two variables z and t, where

(9) $$f(z, t) = \frac{f(\xi(t))\xi'(t)}{\xi(t) - z} \quad \text{and} \quad f_z(z, t) = \frac{f(\xi(t))\xi'(t)}{(\xi(t) - z)^2}.$$

Using equations (9) and Leibniz's rule, we see that $f'(z)$ is given by

$$f'(z) = \frac{1}{2\pi i} \int_a^b \frac{f(\xi(t))\xi'(t)\, dt}{(\xi(t) - z)^2} = \frac{1}{2\pi i} \int_C \frac{f(\xi)\, d\xi}{(\xi - z)^2},$$

and the proof for the case $n = 1$ is complete. We can apply the same argument to the analytic function f' and show that its derivative f'' has representation (7) with $n = 2$. The principle of mathematical induction will establish the theorem for any value of n.

EXAMPLE 6.25   Let z_0 denote a fixed complex value. If C is a simple closed positively oriented contour such that z_0 lies interior to C, then

(10) $$\int_C \frac{dz}{z - z_0} = 2\pi i \quad \text{and} \quad \int_C \frac{dz}{(z - z_0)^{n+1}} = 0,$$

where $n \geq 1$ is a positive integer.

Solution   Here we have $f(z) = 1$ and the nth derivative is $f^{(n)}(z) = 0$. Theorem 6.8 implies that the value of the first integral in equations (10) is given by

$$\int_C \frac{dz}{z - z_0} = 2\pi i f(z_0) = 2\pi i,$$

and Theorem 6.10 can be used to conclude that

$$\int_C \frac{dz}{(z - z_0)^{n+1}} = \frac{2\pi i}{n!} f^{(n)}(z_0) = 0.$$

We remark that this is the same result that was proven earlier in Example 6.14. It should be obvious that the technique of using Theorems 6.8 and 6.10 is easier.

EXAMPLE 6.26   Show that

$$\int_C \frac{\exp z^2}{(z - i)^4} \, dz = \frac{-4\pi}{3e},$$

where C is the circle $|z| = 2$ with positive orientation.

Solution   Here we have $f(z) = \exp z^2$, and a straightforward calculation shows that $f^{(3)}(z) = (12z + 8z^3) \exp z^2$. Using Cauchy's integral formulae with $n = 3$, we conclude that

$$\int_C \frac{\exp z^2}{(z - i)^4} \, dz = \frac{2\pi i}{3!} f^{(3)}(i) = \frac{2\pi i}{6} \frac{4i}{e} = \frac{-4\pi}{3e}.$$

EXAMPLE 6.27   Show that

$$\int_C \frac{\exp(i\pi z) \, dz}{2z^2 - 5z + 2} = \frac{2\pi}{3},$$

where C is the circle $|z| = 1$ with positive orientation.

Solution   By factoring the denominator we obtain $2z^2 - 5z + 2 = (2z - 1)(z - 2)$. Only the root $z_0 = \frac{1}{2}$ lies interior to C. Now we set $f(z) = [\exp(i\pi z)]/(z - 2)$ and use Theorem 6.8 to conclude that

$$\int_C \frac{\exp(i\pi z)\, dz}{2z^2 - 5z + 2} = \frac{1}{2} \int_C \frac{f(z)\, dz}{z - (1/2)} = \frac{1}{2}(2\pi i)f(\tfrac{1}{2}) = \pi i\, \frac{\exp(i\pi/2)}{(1/2) - 2}$$

$$= \frac{2\pi}{3}.$$

We now state two important corollaries to Theorem 6.10.

Corollary 6.1   *If f is analytic in the domain D, then all derivatives f', f'', . . . , $f^{(n)}$, . . . exist and are analytic in D.*

Proof   For each point z_0 in D, there exists a closed disk $|z - z_0| \le R$ that is contained in D. The circle C: $|z - z_0| = R$ can be used in Theorem 6.10 to show that $f^{(n)}(z_0)$ exists for all n.

This result is interesting, since the definition of analytic function means that the derivative f' exists at all points in D. Here we find something more, that the derivatives of all orders exist!

Corollary 6.2   *If u is a harmonic function at each point (x, y) in the domain D, then all partial derivatives u_x, u_y, u_{xx}, u_{xy}, and u_{yy} exist and are harmonic functions.*

Proof   For each point (x_0, y_0) in D there exists a closed disk $|z - z_0| \le R$ that is contained in D. A conjugate harmonic function v exists in this disk, so the function $f(z) = u + iv$ is an analytic function. We use the Cauchy-Riemann equations and see that $f'(z) = u_x + iv_x = v_y - iu_y$. Since f' is analytic, the functions u_x and u_y are harmonic. Again, we can use the Cauchy-Riemann equations to see that

$$f''(z) = u_{xx} + iv_{xx} = v_{yx} - iu_{yx} = -u_{yy} - iv_{yy}.$$

Since f'' is analytic, the functions u_{xx}, u_{xy}, and u_{yy} are harmonic.

EXERCISES FOR SECTION 6.5

For Exercises 1–15, assume that the contour C has positive orientation.

1. Find $\int_C (\exp z + \cos z)z^{-1}\, dz$, where C is the circle $|z| = 1$.
2. Find $\int_C (z + 1)^{-1}(z - 1)^{-1}\, dz$, where C is the circle $|z - 1| = 1$.
3. Find $\int_C (z + 1)^{-1}(z - 1)^{-2}\, dz$, where C is the circle $|z - 1| = 1$.
4. Find $\int_C (z^3 - 1)^{-1}\, dz$, where C is the circle $|z - 1| = 1$.

5. Find $\int_C (z \cos z)^{-1} \, dz$, where C is the circle $|z| = 1$.
6. Find $\int_C z^{-4} \sin z \, dz$, where C is the circle $|z| = 1$.
7. Find $\int_C z^{-3} \sinh(z^2) \, dz$, where C is the circle $|z| = 1$.
8. Find $\int_C z^{-2} \sin z \, dz$ along the following contours:
   (a) The circle $|z - (\pi/2)| = 1$.   (b) The circle $|z - (\pi/4)| = 1$.
9. Find $\int_C z^{-n} \exp z \, dz$, where C is the circle $|z| = 1$ and n is a positive integer.
10. Find $\int_C z^{-2}(z^2 - 16)^{-1} \exp z \, dz$ along the following contours:
    (a) The circle $|z| = 1$.   (b) The circle $|z - 4| = 1$.
11. Find $\int_C (z^4 + 4)^{-1} \, dz$, where C is the circle $|z - 1 - i| = 1$.
12. Find $\int_C (z^2 + 1)^{-1} \, dz$ along the following contours:
    (a) The circle $|z - i| = 1$.   (b) The circle $|z + i| = 1$.
13. Find $\int_C (z^2 + 1)^{-1} \sin z \, dz$ along the following contours:
    (a) The circle $|z - i| = 1$.   (b) The circle $|z + i| = 1$.
14. Find $\int_C (z^2 + 1)^{-2} \, dz$, where C is the circle $|z - i| = 1$.
15. Find $\int_C z^{-1}(z - 1)^{-1} \exp z \, dz$ along the following contours:
    (a) The circle $|z| = 1/2$.   (b) The circle $|z| = 2$.

For Exercises 16–19, assume that the contour C has positive orientation.

16. Let $P(z) = a_0 + a_1 z + a_2 z^2 + a_3 z^3$ be a cubic polynomial. Find $\int_C P(z) z^{-n} \, dz$, where C is the circle $|z| = 1$ and n is a positive integer.
17. Let f be analytic in the simply connected domain D, and let C be a simple closed contour in D. Suppose that z_0 lies exterior to C. Find $\int_C f(z)(z - z_0)^{-1} \, dz$.
18. Let z_1 and z_2 be two complex numbers that lie interior to the simple closed contour C. Show that $\int_C (z - z_1)^{-1}(z - z_2)^{-1} \, dz = 0$.
19. Let f be analytic in the simply connected domain D, and let z_1 and z_2 be two complex numbers that lie interior to the simple closed contour C that lies in D. Show that

$$\frac{f(z_2) - f(z_1)}{z_2 - z_1} = \frac{1}{2\pi i} \int_C \frac{f(z) \, dz}{(z - z_1)(z - z_2)}.$$

State what happens when $z_2 \to z_1$.

20. The *Legendre polynomial* $P_n(z)$ is defined by

$$P_n(z) = \frac{1}{2^n n!} \frac{d^n}{dz^n} [(z^2 - 1)^n].$$

Use Cauchy's integral formula to show that

$$P_n(z) = \frac{1}{2\pi i} \int_C \frac{(\xi^2 - 1)^n \, d\xi}{2^n (\xi - z)^{n+1}}.$$

where z lies inside C.

21. Discuss the importance of being able to define an analytic function $f(z)$ with the contour integral in formula (1). How does this differ from other definitions of a function that you have learned?

22. Write a report on Cauchy integral formula. Include examples of complicated examples discussed in the literature. Resources include bibliographical items 13, 59, 107, 110, 118, 119, and 187.

6.6   The Theorems of Morera and Liouville and Some Applications

In this section we investigate some of the qualitative properties of analytic and harmonic functions. Our first result shows that the existence of an antiderivative for a continuous function is equivalent to the statement that the integral of f is independent of the path of integration. This result is stated in a form that will serve as a converse to the Cauchy-Goursat theorem.

Theorem 6.11 (Morera's Theorem)   *Let f be a continuous function in a simply connected domain D. If*

$$\int_C f(z)\, dz = 0$$

for every closed contour in D, then f is analytic in D.

Proof   Select a point z_0 in D and define $F(z)$ by the following integral:

$$F(z) = \int_{z_0}^{z} f(\xi)\, d\xi.$$

The function $F(z)$ is uniquely defined because if C_1 and C_2 are two contours in D, both with initial point z_0 and terminal point z, then $C = C_1 - C_2$ is a closed contour in D, and

$$0 = \int_C f(\xi)\, d\xi = \int_{C_1} f(\xi)\, d\xi - \int_{C_2} f(\xi)\, d\xi.$$

Since $f(z)$ is continuous, then if $\varepsilon > 0$, there exists a $\delta > 0$ such that $|\xi - z| < \delta$ implies that $|f(\xi) - f(z)| < \varepsilon$. Now we can use the identical steps to those in the proof of Theorem 6.6 to show that $F'(z) = f(z)$. Hence $F(z)$ is analytic on D, and Corollary 6.1 implies that $F'(z)$ and $F''(z)$ are also analytic. Therefore $f'(z) = F''(z)$ exists for all z in D, and we have proven that $f(z)$ is analytic on D.

Cauchy's integral formula shows how the value $f(z_0)$ can be represented by a certain contour integral. If we choose the contour of integration C to be a circle with center z_0, then we can show that the value $f(z_0)$ is the integral average of the values of $f(z)$ at points z on the circle C.

Theorem 6.12 (Gauss's Mean Value Theorem)   *If f is analytic in a simply connected domain D that contains the circle C: $|z - z_0| = R$, then*

(1) $f(z_0) = \dfrac{1}{2\pi} \displaystyle\int_0^{2\pi} f(z_0 + Re^{i\theta})\, d\theta.$

Proof   The circle C can be given the parameterization

(2) $C\colon z(\theta) = z_0 + Re^{i\theta}$ and $dz = i\,Re^{i\theta}\,d\theta$ for $0 \le \theta \le 2\pi$.

We can use the parameterization (2) and Cauchy's integral formula to obtain

$$f(z_0) = \frac{1}{2\pi i}\int_0^{2\pi}\frac{f(z_0 + Re^{i\theta})i\,Re^{i\theta}\,d\theta}{Re^{i\theta}} = \frac{1}{2\pi}\int_0^{2\pi}f(z_0 + Re^{i\theta})\,d\theta,$$

and Theorem 6.12 is proven.

We now prove an important result concerning the modulus of an analytic function.

> **Theorem 6.13 (Maximum Modulus Principle)**   *Let f be analytic and nonconstant in the domain D. Then $\left|f(z)\right|$ does not attain a maximum value at any point z_0 in D.*

Proof by Contradiction   Assume the contrary, and suppose that there exists a point z_0 in D such that

(3) $\left|f(z)\right| \le \left|f(z_0)\right|$ holds for all z in D.

If $C_0\colon \left|z - z_0\right| = R$ is any circle contained in D, then we can use identity (1) and property (22) of Section 6.2 to obtain

(4) $\left|f(z_0)\right| = \left|\dfrac{1}{2\pi}\displaystyle\int_0^{2\pi}f(z_0 + re^{i\theta})\,d\theta\right| \le \dfrac{1}{2\pi}\displaystyle\int_0^{2\pi}\left|f(z_0 + re^{i\theta})\right|\,d\theta$ for $0 \le r \le R$.

But in view of inequality (3), we can treat $\left|f(z)\right| = \left|f(z_0 + re^{i\theta})\right|$ as a real-valued function of the real variable θ and obtain

(5) $\dfrac{1}{2\pi}\displaystyle\int_0^{2\pi}\left|f(z_0 + re^{i\theta})\right|\,d\theta \le \dfrac{1}{2\pi}\displaystyle\int_0^{2\pi}\left|f(z_0)\right|\,d\theta = \left|f(z_0)\right|$ for $0 \le r \le R$.

If we combine inequalities (4) and (5), the result is the equation

$$\left|f(z_0)\right| = \frac{1}{2\pi}\int_0^{2\pi}\left|f(z_0 + re^{i\theta})\right|\,d\theta,$$

which can be written as

(6) $\displaystyle\int_0^{2\pi}\left(\left|f(z_0)\right| - \left|f(z_0 + re^{i\theta})\right|\right)\,d\theta = 0$ for $0 \le r \le R$.

A theorem from the calculus of real-valued functions states that if the integral of a nonnegative continuous function taken over an interval is zero, then that function must be identically zero. Since the integrand in equation (6) is a nonnegative real-valued function, we conclude that it is identically zero; that is,

(7) $\left|f(z_0)\right| = \left|f(z_0 + re^{i\theta})\right|$ for $0 \le r \le R$ and $0 \le \theta \le 2\pi$.

If the modulus of an analytic function is constant, then the results of Example 3.13 show that the function is constant. Therefore identity (7) implies that

(8) $f(z) = f(z_0)$ for all z in the disk D_0: $|z - z_0| \le R$.

Now let Z denote an arbitrary point in D, and let C be a contour in D that joins z_0 to Z. Let $2d$ denote the minimum distance from C to the boundary of D. Then we can find consecutive points $z_0, z_1, z_2, \ldots, z_n = Z$ along C with $|z_{k+1} - z_k| \le d$, such that the disks D_k: $|z - z_k| \le d$ for $k = 0, 1, \ldots, n$ are contained in D and cover C, as shown in Figure 6.34.

Since each disk D_k contains the center z_{k+1} of the next disk D_{k+1}, it follows that z_1 lies in D_0, and from equation (8) we see that $f(z_1) = f(z_0)$. Hence $|f(z)|$ also assumes its maximum value at z_1. An argument identical to the one given above will show that

(9) $f(z) = f(z_1) = f(z_0)$ for all z in the disk D_1.

We can proceed inductively and show that

(10) $f(z) = f(z_{k+1}) = f(z_k)$ for all z in the disk D_{k+1}.

By using equations (8), (9), and (10) it follows that $f(Z) = f(z_0)$. Therefore f is constant in D. With this contradiction the proof of the theorem is complete.

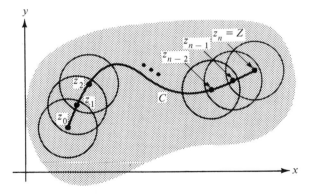

FIGURE 6.34   The "chain of disks" $D_0, D_1, \ldots, D_n$ that cover C.

The maximum modulus principle is sometimes stated in the following weaker form.

Theorem 6.13* (Maximum Modulus Principle)   *Let f be analytic and nonconstant in the bounded domain D. If f is continuous on the closed region R that consists of D and all of its boundary points B, then $|f(z)|$ assumes its maximum value, and does so only at point(s) z_0 on the boundary B.*

EXAMPLE 6.28   Let $f(z) = az + b$, where the domain is the disk $D = \{z: |z| < 1\}$. Then f is continuous on the closed region $R = \{z: |z| \leq 1\}$. Prove that

$$\max_{|z| \leq 1} |f(z)| = |a| + |b|$$

and that this value is assumed by f at a point $z_0 = e^{i\theta_0}$ on the boundary of D.

Solution   From the triangle inequality and the fact that $|z| \leq 1$ it follows that

$$|f(z)| = |az + b| \leq |az| + |b| \leq |a| + |b|.$$

If we choose $z_0 = e^{i\theta_0}$, where $\theta_0 = \arg b - \arg a$, then

$$\arg az_0 = \arg a + (\arg b - \arg a) = \arg b,$$

so the vectors az_0 and b lie on the same ray through the origin. Hence $|az_0 + b| = |az_0| + |b| = |a| + |b|$, and the result is established.

Theorem 6.14 (Cauchy's Inequalities)   *Let f be analytic in the simply connected domain D that contains the circle C: $|z - z_0| = R$. If $|f(z)| \leq M$ holds for all points z on C, then*

$$(11) \quad |f^{(n)}(z_0)| \leq \frac{n!M}{R^n} \quad for \ n = 1, 2, \dots .$$

Proof   Let C have the parameterization

$$C: z(\theta) = z_0 + Re^{i\theta} \quad \text{and} \quad dz = i\,Re^{i\theta}\,d\theta \quad \text{for } 0 \leq \theta \leq 2\pi.$$

We can use Cauchy's integral formulae and write

$$(12) \quad f^{(n)}(z_0) = \frac{n!}{2\pi i} \int_C \frac{f(z)\,dz}{(z - z_0)^{n+1}} = \frac{n!}{2\pi i} \int_0^{2\pi} \frac{f(z_0 + Re^{i\theta})i\,Re^{i\theta}\,d\theta}{R^{n+1}e^{i(n+1)\theta}}.$$

Using equation (12) and property (22) of Section 6.2, we obtain

$$|f^{(n)}(z_0)| \leq \frac{n!}{2\pi R^n} \int_0^{2\pi} |f(z_0 + Re^{i\theta})|\,d\theta$$

$$\leq \frac{n!}{2\pi R^n} \int_0^{2\pi} M\,d\theta = \frac{n!}{2\pi R^n} M2\pi = \frac{n!M}{R^n},$$

and Theorem 6.14 is established.

The next result shows that a nonconstant entire function cannot be a bounded function.

Theorem 6.15 (Liouville's Theorem)   *If f is an entire function and is bounded for all values of z in the complex plane, then f is constant.*

Proof   Suppose that $|f(z)| \leq M$ holds for all values of z. Let z_0 denote an arbitrary point. Then we can use the circle $C: |z - z_0| = R$, and Cauchy's inequality with $n = 1$ implies that

$$(13) \quad |f'(z_0)| \leq \frac{M}{R}.$$

If we let $R \rightarrow \infty$ in inequality (13), then we see that $f'(z_0) = 0$. Hence $f'(z) = 0$ for all z. If the derivative of an analytic function is zero for all z, then it must be a constant function. Therefore f is constant, and the theorem is proven.

EXAMPLE 6.29   The function $\sin z$ is *not* a bounded function.

Solution   One way to see this is to observe that $\sin z$ is a nonconstant entire function, and therefore Liouville's theorem implies that $\sin z$ cannot be bounded. Another way is to investigate the behavior of real and imaginary parts of $\sin z$. If we fix $x = \pi/2$ and let $y \rightarrow \infty$, then we see that

$$\lim_{y \to +\infty} \sin\left(\frac{\pi}{2} + iy\right) = \lim_{y \to +\infty} \sin\frac{\pi}{2} \cosh y + i \cos\frac{\pi}{2} \sinh y$$

$$= \lim_{y \to +\infty} \cosh y = +\infty.$$

Liouville's theorem can be used to establish an important theorem of elementary algebra.

Theorem 6.16 (The Fundamental Theorem of Algebra)   *If P(z) is a polynomial of degree n, then P has at least one zero.*

Proof by Contradiction   Assume the contrary and suppose that $P(z) \neq 0$ for all z. Then the function $f(z) = 1/P(z)$ is an entire function. We show that f is bounded as follows. First we write $P(z) = a_n z^n + a_{n-1} z^{n-1} + \cdots + a_1 z + a_0$ and consider the equation

$$(14) \quad |f(z)| = \frac{1}{|P(z)|} = \frac{1}{|z|^n} \frac{1}{\left| a_n + \dfrac{a_{n-1}}{z} + \dfrac{a_{n-2}}{z^2} + \cdots + \dfrac{a_1}{z^{n-1}} + \dfrac{a_0}{z^n} \right|}.$$

Since $\left|a_k\right|/\left|z^{n-k}\right| = \left|a_k\right|/r^{n-k} \to 0$ as $\left|z\right| = r \to \infty$, it follows that

(15) $\quad a_n + \dfrac{a_{n-1}}{z} + \dfrac{a_{n-2}}{z^2} + \cdots + \dfrac{a_0}{z^n} \to a_n \quad$ as $\left|z\right| \to \infty$.

If we use statement (15) in equation (14), then the result is

$$\left|f(z)\right| \to 0 \quad \text{as } \left|z\right| \to \infty.$$

In particular, we can find a value of R such that

(16) $\quad \left|f(z)\right| \le 1 \quad$ for all $\left|z\right| \ge R$.

Consider

$$\left|f(z)\right| = ([u(x, y)]^2 + [v(x, y)]^2)^{1/2},$$

which is a continuous function of the two real variables x and y. A result from calculus regarding real functions says that a continuous function on a closed and bounded set is bounded. Hence $\left|f(z)\right|$ is a bounded function on the closed disk

$$x^2 + y^2 \le R^2;$$

that is, there exists a positive real number K such that

(17) $\quad \left|f(z)\right| \le K \quad$ for all $\left|z\right| \le R$.

Combining inequalities (16) and (17), it follows that $\left|f(z)\right| \le M = \max\{K, 1\}$ holds for all z. Liouville's theorem can now be used to conclude that f is constant. With this contradiction the proof of the theorem is complete.

Corollary 6.3 *Let P be a polynomial of degree n. Then P can be expressed as the product of linear factors. That is,*

$$P(z) = A(z - z_1)(z - z_2) \cdots (z - z_n)$$

where $z_1, z_2, \ldots, z_n$ are the zeros of P counted according to multiplicity and A is a constant.

EXERCISES FOR SECTION 6.6

For Exercises 1–4, express the given polynomial as a product of linear factors.

1. Factor $P(z) = z^4 + 4$.
2. Factor $P(z) = z^2 + (1 + i)z + 5i$.
3. Factor $P(z) = z^4 - 4z^3 + 6z^2 - 4z + 5$.
4. Factor $P(z) = z^3 - (3 + 3i)z^2 + (-1 + 6i)z + 3 - i$. *Hint*: Show that $P(i) = 0$.
5. Let $f(z) = az^n + b$, where the region is the disk $R = \{z: \left|z\right| \le 1\}$. Show that

$$\max_{|z| \le 1} \left|f(z)\right| = \left|a\right| + \left|b\right|.$$

6. Show that $\cos z$ is *not* a bounded function.

7. Let $f(z) = z^2$, where the region is the rectangle $R = \{z = x + iy: 2 \le x \le 3$ and $1 \le y \le 3\}$.
   Find the following:
   (a) $\max_R |f(z)|$ (b) $\min_R |f(z)|$
   (c) $\max_R \operatorname{Re} [f(z)]$ (d) $\min_R \operatorname{Im} [f(z)]$

   Hint for (a) *and* (b): $|z|$ is the distance from 0 to z.

8. Let $F(z) = \sin z$, where the region is the rectangle

$$R = \left\{z = x + iy: 0 \le x \le \frac{\pi}{2} \text{ and } 0 \le y \le 2\right\}.$$

   Find $\max_R |f(z)|$. *Hint*: $|\sin z|^2 = \sin^2 x + \sinh^2 y$.

9. Let f be analytic in the disk $|z| < 5$, and suppose that $|f(\xi)| \le 10$ for values of ξ on the circle $|\xi - 1| = 3$. Find a bound for $|f^{(3)}(1)|$.

10. Let f be analytic in the disk $|z| < 5$, and suppose that $|f(\xi)| \le 10$ for values of ξ on the circle $|\xi - 1| = 3$. Find a bound for $|f^{(3)}(0)|$.

11. Let f be an entire function such that $|f(z)| \le M|z|$ holds for all z.
    (a) Show that $f''(z) = 0$ for all z, and   (b) conclude that $f(z) = az + b$.

12. Establish the following *minimum modulus principle*. Let f be analytic and nonconstant in the domain D. If $|f(z)| \ge m$, where $m > 0$ holds for all z in D, then $|f(z)|$ does *not* attain a minimum value at any point z_0 in D.

13. Let $u(x, y)$ be harmonic for all (x, y). Show that

$$u(x_0, y_0) = \frac{1}{2\pi} \int_0^{2\pi} u(x_0 + R\cos\theta, y_0 + R\sin\theta)\, d\theta, \quad \text{where } R > 0.$$

    Hint: Consider $f(z) = u(x, y) + iv(x, y)$.

14. Establish the following *maximum principle for harmonic functions*. Let $u(x, y)$ be harmonic and nonconstant in the simply connected domain D. Then u does not take on a maximum value at any point (x_0, y_0) in D. *Hint*: Let $f(z) = u(x, y) + iv(x, y)$ be analytic in D, and consider $F(z) = \exp[f(z)]$ where $|F(z)| = e^{u(x,y)}$.

15. Let f be an entire function that has the property $|f(z)| \ge 1$ for all z. Show that f is constant.

16. Let f be a nonconstant analytic function in the closed disk $R = \{z: |z| \le 1\}$. Suppose that $|f(z)| = K$ for all z on the circle $|z| = 1$. Show that f has a zero in D. *Hint*: Use both the maximum and minimum modulus principles.

17. Why is it important to study the fundamental theorem of algebra in a complex analysis course?

18. Look up the article on Morera's theorem and discuss what you found. Use bibliographical item 163.

19. Look up the article on Liouville's theorem and discuss what you found. Use bibliographical item 117.

20. Write a report on the fundamental theorem of algebra. Discuss ideas that you found in the literature that are not mentioned in the text. Resources include bibliographical items 6, 18, 29, 38, 60, 66, 150, 151, 170, and 184.

21. Write a report on zeros of polynomials and/or complex functions. Resources include bibliographical items 50, 65, 67, 102, 109, 120, 121, 122, 140, 152, 162, 171, 174, and 178.

7

Taylor and Laurent Series

Throughout this text we have compared and contrasted properties of complex functions with functions whose domain and range lie entirely within the reals. There are many similarities, such as the standard differentiation formulas. On the other hand, there are some surprises, and in this chapter we will encounter one of the hallmarks distinguishing complex functions from their real counterparts.

It is possible for a function defined on the real numbers to be differentiable everywhere and yet not be expressible as a power series (see Exercise 27 at the end of Section 7.2). In the complex case, however, things are much simpler! It turns out that if a complex function is analytic in the disk $D_r(\alpha)$, its Taylor series about α will converge to the function at every point in this disk. Thus, analytic functions are locally nothing more than glorified polynomials.

We shall also see that complex functions are the key to unlocking many of the mysteries encountered when power series are first introduced in a calculus course. We begin by discussing an important property associated with power series—uniform convergence.

7.1   Uniform Convergence

Recall that if we have a function $f(z)$ defined on a set T, the sequence of functions $\{S_n(z)\}$ converges to the function f at the point $z = z_0 \in T$ provided $\lim_{n \to \infty} S_n(z_0) = f(z_0)$. Thus, for the particular point z_0, this means that for each $\varepsilon > 0$, there exists a positive integer N_{ε, z_0} (which depends on both ε and z_0) such that

(1)   if $n \geq N_{\varepsilon, z_0}$, then $\left| S_n(z_0) - f(z_0) \right| < \varepsilon$.

If $S_n(z)$ is the nth partial sum of the series $\sum_{k=0}^{\infty} c_k(z - \alpha)^k$, statement (1) becomes

(2)   if $n \geq N_{\varepsilon, z_0}$, then $\left| \sum_{k=0}^{n-1} c_k(z_0 - \alpha)^k - f(z_0) \right| < \varepsilon$.

It is important to stress that for a given value of ε, the integer N_{ε, z_0} we need to satisfy statement (1) will often depend on our choice of z_0. This is not the case if the sequence $\{S_n(z)\}$ converges *uniformly*. For a uniformly convergent sequence, it is possible to find an integer N_ε (which depends *only* on ε) that guarantees statement (1) no matter which value for $z_0 \in T$ we pick. In other words, if n is large enough, the function $S_n(z)$ is *uniformly close* to $f(z)$. Formally, we have the following definition.

> **DEFINITION 7.1**   *The sequence $\{S_n(z)\}$ converges uniformly to $f(z)$ on the set T if for every $\varepsilon > 0$, there exists a positive integer N_ε (which depends* only *on ε) such that*

(3)   *if $n \geq N_\varepsilon$, then $\left| S_n(z) - f(z) \right| < \varepsilon$ for all $z \in T$.*

If in the preceding, $S_n(z)$ is the nth partial sum of a series $\sum_{k=0}^{\infty} c_k(z - \alpha)^k$, we say that

the series $\sum_{k=0}^{\infty} c_k(z - \alpha)^k$ converges uniformly to $f(z)$ on the set T.

EXAMPLE 7.1   The sequence $\{S_n(z)\} = \left\{ e^z + \dfrac{1}{n} \right\}$ converges uniformly to $f(z) = e^z$ on the entire complex plane because for any $\varepsilon > 0$, statement (3) is satisfied for all z if N_ε is any integer greater than $1/\varepsilon$. We leave the details for showing this as an exercise.

A good example of a sequence of functions that does not converge uniformly is the sequence of partial sums comprising the geometric series. Recall that the geometric series has $S_n(z) = \sum_{k=0}^{n-1} z^k = \dfrac{1 - z^n}{1 - z}$ converging to $f(z) = \dfrac{1}{1 - z}$ for all $z \in D_1(0)$. Since the real numbers are a subset of the complex numbers, we can show statement (3) is not satisfied by demonstrating it does not hold when we restrict our attention to the real numbers. In that context, $D_1(0)$ becomes the open interval $(-1, 1)$, and the inequality $\left| S_n(z) - f(z) \right| < \varepsilon$ becomes $\left| S_n(x) - f(x) \right| < \varepsilon$, which for real variables is equivalent to $f(x) - \varepsilon < S_n(x) < f(x) + \varepsilon$. If statement (3) were to be satisfied, then given $\varepsilon > 0$, $S_n(x)$ should (for large enough values of n) be within an ε-bandwidth of the function $f(x)$ *for all x* in the interval $(-1, 1)$. Figure 7.1 illustrates that there is an ε such that no matter how large n is, we can find $x_0 \in (-1, 1)$ such that $S_n(x_0)$ is outside this bandwidth. In other words, this figure illustrates the negation of statement (3), which in technical terms is

(4)   there exists $\varepsilon > 0$ such that for all positive integers N, there is some $n \geq N$ and some $z_0 \in T$ such that $\left| S_n(z_0) - f(z_0) \right| \geq \varepsilon$.

We leave the verification of statement (4) when applied to the partial sums of a geometric series as an exercise.

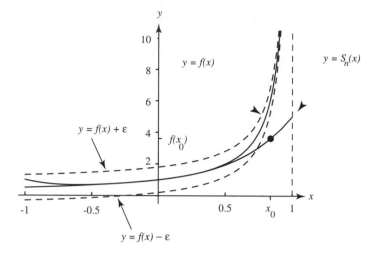

FIGURE 7.1 The geometric series does not converge uniformly on $(-1, 1)$.

There is a useful procedure known as the Weierstrass M-test, which can help determine whether an infinite series is uniformly convergent.

Theorem 7.1 (Weierstrass M-test): *Suppose the infinite series* $\displaystyle\sum_{k=0}^{\infty} u_k(z)$

has the property that for each k, $\left| u_k(z) \right| \le M_k$ *for all $z \in T$. If* $\displaystyle\sum_{k=0}^{\infty} M_k$ *converges,*

then $\displaystyle\sum_{k=0}^{\infty} u_k(z)$ *converges uniformly on T.*

Proof   Let $S_n(z) = \displaystyle\sum_{k=0}^{n-1} u_k(z)$ be the nth partial sum of the series. If $n > m$,

$$\left| S_n(z) - S_m(z) \right| = \left| u_m(z) + u_{m+1}(z) + \cdots + u_{n-1}(z) \right| \le \sum_{k=m}^{n-1} M_k.$$

Since the series $\displaystyle\sum_{k=0}^{\infty} M_k$ converges, the last expression can be made as small as we wish by choosing m large enough. Thus, given $\varepsilon > 0$, there is a positive integer N_ε such that if $n, m > N_\varepsilon$, then $\left| S_n(z) - S_m(z) \right| < \varepsilon$. But this means that for all $z \in T$, $\{S_n(z)\}$ is a Cauchy sequence. According to Theorem 4.2, this sequence must converge to a number which we might as well designate by $f(z)$. That is,

$$f(z) = \lim_{n \to \infty} S_n(z) = \sum_{k=0}^{\infty} u_k(z).$$

This gives us a function to which the series $\sum_{k=0}^{\infty} u_k(z)$ converges; it remains to be shown that the convergence is uniform. Let $\varepsilon > 0$ be given. Again, since $\sum_{k=0}^{\infty} M_k$ converges, there exists N_ε such that if $n \geq N_\varepsilon$, then $\sum_{k=n}^{\infty} M_k < \varepsilon$. Thus, if $n \geq N_\varepsilon$, we have for all $z \in T$ that

$$\left| f(z) - S_n(z) \right| = \left| \sum_{k=0}^{\infty} u_k(z) - \sum_{k=0}^{n-1} u_k(z) \right|$$

$$= \left| \sum_{k=n}^{\infty} u_k(z) \right|$$

$$\leq \sum_{k=n}^{\infty} M_k$$

$$< \varepsilon, \quad \text{which completes the argument.}$$

As an application of the Weierstrass M-test, we have the following.

Theorem 7.2   *Suppose the power series $\sum_{k=0}^{\infty} c_k(z - \alpha)^k$ has radius of convergence $\rho > 0$. Then for each r, $0 < r < \rho$, the series converges uniformly on the closed disk $\overline{D}_r(\alpha) = \{z: \left| z - \alpha \right| \leq r\}$.*

Proof   Given r with $0 < r < \rho$, choose $z_0 \in D_\rho(\alpha)$ such that $\left| z_0 - \alpha \right| = r$. Since $\sum_{k=0}^{\infty} c_k(z - \alpha)^k$ converges absolutely for $z \in D_\rho(\alpha)$ (Theorems 4.9 and 4.11, part ii), we know that $\sum_{k=0}^{\infty} \left| c_k(z_0 - \alpha)^k \right| = \sum_{k=0}^{\infty} \left| c_k \right| r^k$ converges. For all $z \in \overline{D}_r(\alpha)$ it is clear that

$$\left| c_k(z - \alpha)^k \right| = \left| c_k \right| \left| z - \alpha \right|^k \leq \left| c_k \right| r^k.$$

The conclusion now follows from the Weierstrass M-test with $M_k = \left| c_k \right| r^k$.

Corollary 7.1   *For each r, $0 < r < 1$, the geometric series converges uniformly on the closed disk $\overline{D}_r(0)$.*

The following theorem gives important properties of uniformly convergent sequences.

Theorem 7.3   *Suppose $\{S_k\}$ is a sequence of continuous functions defined on a set T containing the contour C. If $\{S_k\}$ converges uniformly to f on the set T, then*

(i) f is continuous on T, and

(ii) $\lim_{k \to \infty} \int_C S_k(z)\,dz = \int_C \lim_{k \to \infty} S_k(z)\,dz = \int_C f(z)\,dz.$

Proof (i) Given $z_0 \in T$, we must prove $\lim_{z \to z_0} f(z) = f(z_0)$. Let $\varepsilon > 0$ be given. Since S_k converges uniformly on T, there exists a positive integer N_ε such that for all $z \in T$, $\left| f(z) - S_k(z) \right| < \dfrac{\varepsilon}{3}$ whenever $k \geq N_\varepsilon$. Since S_{N_ε} is continuous at z_0, there exists $\delta > 0$ such that if $\left| z - z_0 \right| < \delta$, then $\left| S_{N_\varepsilon}(z) - S_{N_\varepsilon}(z_0) \right| < \dfrac{\varepsilon}{3}$. Hence, if $\left| z - z_0 \right| < \delta$, we have

$$
\begin{aligned}
\left| f(z) - f(z_0) \right| &= \left| f(z) - S_{N_\varepsilon}(z) + S_{N_\varepsilon}(z) - S_{N_\varepsilon}(z_0) + S_{N_\varepsilon}(z_0) - f(z_0) \right| \\
&\leq \left| f(z) - S_{N_\varepsilon}(z) \right| + \left| S_{N_\varepsilon}(z) - S_{N_\varepsilon}(z_0) \right| + \left| S_{N_\varepsilon}(z_0) - f(z_0) \right| \\
&< \frac{\varepsilon}{3} + \frac{\varepsilon}{3} + \frac{\varepsilon}{3} = \varepsilon.
\end{aligned}
$$

(ii) Let $\varepsilon > 0$ be given, and let L be the length of the contour C. Since $\{S_k\}$ converges uniformly to f on T, there exists a positive integer N_ε such that if $k \geq N_\varepsilon$, then $\left| S_k(z) - f(z) \right| < \dfrac{\varepsilon}{L}$ for all $z \in T$. Since $C \subset T$, $\max_{z \in C} \left| S_k(z) - f(z) \right| < \dfrac{\varepsilon}{L}$, so if $k \geq N_\varepsilon$,

$$
\begin{aligned}
\left| \int_C S_k(z)\,dz - \int_C f(z)\,dz \right| &= \left| \int_C [S_k(z) - f(z)]\,dz \right| \\
&\leq \max_{z \in C} \left| S_k(z) - f(z) \right| L \quad \text{(by Lemma 6.2)} \\
&< \left(\frac{\varepsilon}{L} \right) L = \varepsilon.
\end{aligned}
$$

Corollary 7.2 *If the series* $\displaystyle\sum_{n=0}^{\infty} c_n(z - \alpha)^n$ *converges uniformly to $f(z)$ on the set T, and C is a contour contained in T, then*

$$
\sum_{n=0}^{\infty} \int_C c_n(z - \alpha)^n\,dz = \int_C \sum_{n=0}^{\infty} c_n(z - \alpha)^n\,dz = \int_C f(z)\,dz.
$$

EXAMPLE 7.2 Show that $\mathrm{Log}(1 - z) = \displaystyle\sum_{n=1}^{\infty} \frac{1}{n} z^n$ for all $z \in D_1(0)$.

Solution Given $z_0 \in D_1(0)$, choose r such that $0 \leq \left| z_0 \right| < r < 1$, thus ensuring that $z_0 \in \overline{D}_r(0)$. By Corollary 7.1, the geometric series $\displaystyle\sum_{n=0}^{\infty} z^n$ converges uniformly to $\dfrac{1}{1-z}$ on $\overline{D}_r(0)$. If C is any contour contained in $\overline{D}_r(0)$, Corollary 7.2 gives

$$\int_C \frac{1}{1-z}\, dz = \sum_{n=0}^{\infty} \int_C z^n\, dz.$$

Now, the composite function $\mathrm{Log}(1 - z)$ is an antiderivative for $\dfrac{1}{1-z}$ on $\overline{D}_r(0)$, where Log is the principal branch of the logarithm. Clearly, $\dfrac{1}{n+1}\,z^{n+1}$ is an anti-derivative for the function z^n. Hence, if C is the straight line segment joining 0 to z_0, Theorem 6.7 gives

$$\mathrm{Log}(1 - z)\,\Big|_0^{z_0} = \sum_{n=0}^{\infty} \frac{1}{n+1}\,z^{n+1}\,\Big|_0^{z_0},$$

which becomes

$$\mathrm{Log}(1 - z_0) = \sum_{n=0}^{\infty} \frac{1}{n+1}\,z_0^{n+1} = \sum_{n=1}^{\infty} \frac{1}{n}\,z_0^{n}.$$

Since $z_0 \in D_1(0)$ was arbitrary, we are done.

EXERCISES FOR SECTION 7.1

1. This question relates to Figure 7.1.
   (a) For x near -1, is the graph of $S_n(x)$ above or below $f(x)$? Explain.
   (b) Is the index n in $S_n(x)$ odd or even? Explain.
   (c) Assuming the graph is accurate to scale, what is the value of n in $S_n(x)$?
2. Complete the details to verify the claim of Example 7.1.
3. Show that statement (4) holds if $S_n(z) = \displaystyle\sum_{k=0}^{n-1} z^k = \dfrac{1-z^n}{1-z}$, $f(z) = \dfrac{1}{1-z}$, and $T = D_1(0)$.

   Hint: Given $\varepsilon > 0$, and a positive integer n, consider $z_n = \varepsilon^{1/n}$.
4. Prove that the following series converge uniformly on the sets indicated.

   (a) $\displaystyle\sum_{k=0}^{\infty} \frac{1}{k^2}\,z^k$ on $\overline{D}_1(0) = \{z\colon |z| \le 1\}$

   (b) $\displaystyle\sum_{k=0}^{\infty} \frac{1}{(z^2 - 1)^k}$ on $\{z\colon |z| \ge 2\}$

   (c) $\displaystyle\sum_{k=0}^{\infty} \frac{z+i}{z^2 + k^2}$ on $\overline{D}_R(0)$, where $0 < R < \infty$

   (d) $\displaystyle\sum_{k=0}^{\infty} \frac{z^k}{z^{2k} + 1}$ on $\overline{D}_r(0)$, where $0 < r < 1$.

5. Why can't we use the arguments of Theorem 7.2 to prove that the geometric series converges uniformly on *all* of $D_1(0)$?
6. By starting with the series for $\cos z$ given in Section 5.4, choose an appropriate contour and use the methodology of Example 7.2 to obtain the series for $\sin z$.

7. Suppose that $\{f_n(z)\}$ and $\{g_n(z)\}$ converge uniformly on the set T.
   (a) Show that the sequence $\{f_n(z) + g_n(z)\}$ converges uniformly on the set T.
   (b) Show by example that it is not necessarily the case that $\{f_n(z)g_n(z)\}$ converges uniformly on the set T.
8. On what portion of $D_1(0)$ does the sequence $\{nz^n\}_{n=1}^{\infty}$ converge, and on what portion does it converge uniformly?

7.2   Taylor Series Representations

In Section 4.2 we saw that functions defined by power series have derivatives of all orders. In Section 6.5 we saw that analytic functions also have derivatives of all orders. It seems natural, therefore, that there would be some connection between analytic functions and power series. As you might guess, the connection exists via the Taylor and Maclaurin series of analytic functions.

Definition 7.2   *If $f(z)$ is analytic at $z = \alpha$, the series*

$$f(\alpha) + f'(\alpha)(z - \alpha) + \frac{f^{(2)}(\alpha)}{2!}(z - \alpha)^2 + \frac{f^{(3)}(\alpha)}{3!}(z - \alpha)^3$$

$$+ \cdots = \sum_{k=0}^{\infty} \frac{f^{(k)}(\alpha)}{k!}(z - \alpha)^k$$

is called the Taylor series for f centered around α. When the center is $\alpha = 0$, the series is called the Maclaurin series for f.

To investigate when the preceding series converges we will need the following lemma.

Lemma 7.1   *If z, z_0, and α are complex numbers with $z \neq z_0$, and $z \neq \alpha$, then*

(1)
$$\frac{1}{z - z_0} = \frac{1}{z - \alpha} + \frac{z_0 - \alpha}{(z - \alpha)^2} + \frac{(z_0 - \alpha)^2}{(z - \alpha)^3} +$$

$$\cdots + \frac{(z_0 - \alpha)^n}{(z - \alpha)^{n+1}} + \frac{1}{z - z_0}\frac{(z_0 - \alpha)^{n+1}}{(z - \alpha)^{n+1}},$$

where n is a positive integer.

Proof
$$\frac{1}{z - z_0} = \frac{1}{(z - \alpha) - (z_0 - \alpha)}$$

$$= \frac{1}{z - \alpha}\frac{1}{1 - (z_0 - \alpha)/(z - \alpha)}.$$

The result now follows from Corollary 4.3 if in that corollary we replace z by $\frac{z_0 - \alpha}{z - \alpha}$. We leave the verification details as an exercise.

Theorem 7.4 (Taylor's Theorem)   *Suppose f is analytic in a domain G, and that $D_R(\alpha)$ is contained in G. Then the Taylor series for f converges to $f(z)$ for all z in $D_R(\alpha)$; that is,*

(2) $f(z) = \sum_{k=0}^{\infty} \dfrac{f^{(k)}(\alpha)}{k!} (z - \alpha)^k$ *for all $z \in D_R(\alpha)$.*

Furthermore, this representation is valid in the largest disk with center α that is contained in G, and the convergence is uniform on any closed subdisk $\overline{D}_r(\alpha) = \{z: |z - \alpha| \leq r\}$ for $0 < r < R$.

Proof   We observe first that if we can establish equation (2), the uniform convergence on $\overline{D}_r(\alpha)$ for $0 < r < R$ will follow immediately from Theorem 7.2 by equating c_k of that theorem with $\dfrac{f^{(k)}(\alpha)}{k!}$.

Let $z_0 \in D_R(\alpha)$ be given, and let r designate the distance between z_0 and α, so that $|z_0 - \alpha| = r$. We note that $0 \leq r < R$, since z_0 belongs to the open disk $D_R(\alpha)$. Choose ρ such that

(3) $0 \leq r < \rho < R$,

and let $C = C_\rho^+(\alpha)$ be the positively oriented circle centered at α with radius ρ as shown in Figure 7.2.

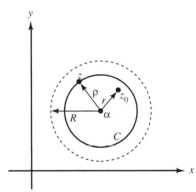

FIGURE 7.2   The constructions for Taylor's theorem.

Since C is contained in G, we can use the Cauchy integral formula to get

$$f(z_0) = \frac{1}{2\pi i} \int_C \frac{1}{z - z_0} f(z) \, dz.$$

Replacing $\dfrac{1}{z - z_0}$ in the integrand by its equivalent expression in Lemma 7.1 gives

$$(4) \quad f(z_0) = \frac{1}{2\pi i} \int_C \left[\frac{1}{z - \alpha} + \frac{z_0 - \alpha}{(z - \alpha)^2} + \cdots + \frac{(z_0 - \alpha)^n}{(z - \alpha)^{n+1}} \right.$$
$$\left. + \frac{1}{z - z_0} \frac{(z_0 - \alpha)^{n+1}}{(z - \alpha)^{n+1}} \right] f(z) \, dz$$
$$= \frac{1}{2\pi i} \int_C \frac{f(z) \, dz}{z - \alpha} + \frac{z_0 - \alpha}{2\pi i} \int_C \frac{f(z) \, dz}{(z - \alpha)^2} + \cdots + \frac{(z_0 - \alpha)^n}{2\pi i} \int_C \frac{f(z) \, dz}{(z - \alpha)^{n+1}}$$
$$+ \frac{(z_0 - \alpha)^{n+1}}{2\pi i} \int_C \frac{f(z) \, dz}{(z - z_0)(z - \alpha)^{n+1}} \, ,$$

where n is a positive integer. The last term in this expression can be put in the form

$$(5) \quad E_n(z_0) = \frac{1}{2\pi i} \int_C \frac{(z_0 - \alpha)^{n+1} f(z) \, dz}{(z - z_0)(z - \alpha)^{n+1}} \, .$$

Recall also by the Cauchy integral formula that for $k = 0, 1, 2, \ldots$,

$$(6) \quad \frac{2\pi i}{k!} f^{(k)}(\alpha) = \int_C \frac{f(z) \, dz}{(z - \alpha)^{k+1}} \, .$$

Substituting equations (5) and (6) into equation (4) now gives

$$(7) \quad f(z_0) = \sum_{k=0}^{n} \frac{f^{(k)}(\alpha)}{k!} (z_0 - \alpha)^k + E_n(z_0).$$

The summation on the right-hand side of equation (7) is the first $n + 1$ terms of the Taylor series. Our proof will be complete if we can show that the remainder, $E_n(z_0)$, can be made as small as we please by taking n to be sufficiently large.

We will use the ML inequality (Lemma 6.2) to get a bound for $|E_n(z_0)|$. According to our constructions shown in Figure 7.2, we have

$$(8) \quad |z_0 - \alpha| = r \quad \text{and} \quad |z - \alpha| = \rho.$$

By inequality (6) of Section 1.3, we also have

$$(9) \quad \begin{aligned} |z - z_0| &= |(z - \alpha) - (z_0 - \alpha)| \\ &\geq |z - \alpha| - |z_0 - \alpha| \\ &= \rho - r. \end{aligned}$$

If we set $M = \max |f(z)|$ for z on C, equations (8) and (9) allow us to conclude

$$(10) \quad \left| \frac{(z_0 - \alpha)^{n+1} f(z)}{(z - z_0)(z - \alpha)^{n+1}} \right| \leq \left(\frac{r}{\rho} \right)^{n+1} \left(\frac{1}{\rho - r} \right) M \quad \text{for } z \in C.$$

The length of the circle C is $2\pi\rho$, so using the ML inequality in conjunction with equations (5) and (10) gives

$$(11) \quad |E_n(z_0)| \leq \frac{1}{2\pi} \left(\frac{r}{\rho} \right)^{n+1} \left(\frac{1}{\rho - r} \right) M(2\pi\rho).$$

According to equation (3), the fraction $\dfrac{r}{\rho}$ is less than 1, so $\left(\dfrac{r}{\rho}\right)^{n+1}$ (and hence the right side of equation (11)) goes to zero as n goes to infinity. Thus, given any $\varepsilon > 0$, we can find an integer N_ε such that $\left|E_n(z_0)\right| < \varepsilon$ for $n \geq N_\varepsilon$, and our proof is complete.

If f is an entire function, then f has no singular points, and equation (2) holds for all complex numbers z, making the radius of convergence of the Taylor series equal to infinity. The fact that equation (2) is valid in the largest disk with center α that is contained in G is made clear by the following corollary.

Corollary 7.3 *Suppose that f is analytic in the domain G that contains the point $z = \alpha$. Let z_0 be a singular point of minimum distance to α. If $\left|z_0 - \alpha\right| = R$, then*

(i) *the Taylor series (2) converges to $f(z)$ on all of $D_R(\alpha)$, and*

(ii) *if $S > R$, the Taylor series (2) does not converge to $f(z)$ on all of $D_S(\alpha)$.*

Proof
(i)   The argument for (i) is identical to the proof of Theorem 7.2.
(ii)   If $\left|z_0 - \alpha\right| = R$, then $z_0 \in D_S(\alpha)$ whenever $S > R$. If for some $S > R$, the Taylor series converged to $f(z)$ on all of $D_S(\alpha)$, then according to Theorem 4.13, f would be differentiable at z_0, contradicting the fact that z_0 is a singular point.

EXAMPLE 7.3 Show that $\dfrac{1}{(1 - z)^2} = \displaystyle\sum_{n=0}^{\infty} (n + 1)z^n$ is valid for $z \in D_1(0)$.

Solution   We established this identity with the use of Theorem 4.13 in Example 4.18. We will now do so via Theorem 7.4. If $f(z) = \dfrac{1}{(1 - z)^2}$, a standard induction argument (which we leave as an exercise) will show that $f^{(n)}(z) = \dfrac{(n + 1)!}{(1 - z)^{n+2}}$. Thus, $f^{(n)}(0) = (n + 1)!$, and identity (2) gives (with n taking the role of k)

$$(12) \quad f(z) = \frac{1}{(1 - z)^2} = \sum_{n=0}^{\infty} \frac{f^{(n)}(0)}{n!} z^n = \sum_{n=0}^{\infty} \frac{(n + 1)!}{n!} z^n = \sum_{n=0}^{\infty} (n + 1)z^n.$$

Furthermore, since the point $z_0 = 1$ is the closest singularity to the point $\alpha = 0$, Corollary 7.2 assures us that equation (12) must be valid for all $z \in D_1(0)$.

EXAMPLE 7.4   Show that for $z \in D_1(0)$,

(13) $$\frac{1}{1 - z^2} = \sum_{n=0}^{\infty} z^{2n} \quad \text{and} \quad \frac{1}{1 + z^2} = \sum_{n=0}^{\infty} (-1)^n z^{2n}.$$

Solution   We know that for $z \in D_1(0)$,

(14) $$\frac{1}{1 - z} = \sum_{n=0}^{\infty} z^n.$$

If we let z^2 take the role of z in equation (14), we get that $\dfrac{1}{1 - z^2} = \sum_{n=0}^{\infty} (z^2)^n = \sum_{n=0}^{\infty} z^{2n}$ for $z^2 \in D_1(0)$. But $z^2 \in D_1(0)$ if and only if $z \in D_1(0)$. Letting $-z^2$ take the role of z in (14) gives the second part of equation (13).

Corollary 7.3 clears up what often seems to be a mystery when series are first introduced in calculus. The calculus analog of equation (13) is

(15) $$\frac{1}{1 - x^2} = \sum_{n=0}^{\infty} x^{2n} \quad \text{and} \quad \frac{1}{1 + x^2} = \sum_{n=0}^{\infty} (-1)^n x^{2n} \quad \text{for } x \in (-1, 1).$$

For many students, it makes sense that the first series in equation (15) converges only on the interval $(-1, 1)$ because $\dfrac{1}{1 - x^2}$ is undefined at the points $x = \pm 1$. It seems unclear as to why this should also be the case for the series representing $\dfrac{1}{1 + x^2}$, since the real-valued function $f(x) = \dfrac{1}{1 + x^2}$ is defined everywhere. The explanation, of course, comes from the complex domain. The complex function $f(z) = \dfrac{1}{1 + z^2}$ is *not* defined everywhere. In fact, the singularities of f are at the points $\pm i$, and the distance between them and the point $\alpha = 0$ equals 1. According to Corollary 7.3, therefore, equation (13) is valid only for $z \in D_1(0)$, and thus equation (15) is valid only for $x \in (-1, 1)$.

Alas, there is a potential fly in this ointment. Corollary 7.3 applies to Taylor series. To form the Taylor series of a function, we must compute its derivatives. Since we did not get the series in equation (13) by computing derivatives, how do we know they are indeed the Taylor series centered about $\alpha = 0$? Perhaps the Taylor series would give completely different expressions than the ones given by equation (13). Fortunately, the following theorem removes this possibility.

Theorem 7.5 (Uniqueness of Power Series)   *Suppose that in some disk $D_r(\alpha)$ we have*

$$f(z) = \sum_{n=0}^{\infty} a_n (z - \alpha)^n = \sum_{n=0}^{\infty} b_n (z - \alpha)^n.$$

Then $a_n = b_n$ for $n = 0, 1, 2, \ldots$.

Proof   By Theorem 4.12, $a_n = \dfrac{f^{(n)}(\alpha)}{n!} = b_n$ for $n = 0, 1, 2, \ldots$.

Thus, any power series representation of $f(z)$ is automatically the Taylor series.

EXAMPLE 7.5   Find the Maclaurin series of $f(z) = \sin^3 z$.

Solution   Computing derivatives for $f(z)$ would be an onerous task. Fortunately, we can make use of the trigonometric identity

$$\sin^3 z = \frac{3}{4}\sin z - \frac{1}{4}\sin 3z.$$

Recall that the series for $\sin z$ (valid for all z) is

$$\sin z = \sum_{n=0}^{\infty} (-1)^n \frac{z^{2n+1}}{(2n+1)!} .$$

Using this identity, we obtain

$$
\begin{aligned}
\sin^3 z &= \frac{3}{4}\sum_{n=0}^{\infty} (-1)^n \frac{z^{2n+1}}{(2n+1)!} - \frac{1}{4}\sum_{n=0}^{\infty} (-1)^n \frac{(3z)^{2n+1}}{(2n+1)!} \\
&= \frac{3}{4}\sum_{n=0}^{\infty} (-1)^n \frac{z^{2n+1}}{(2n+1)!} - \frac{3}{4}\sum_{n=0}^{\infty} (-1)^n \frac{9^n z^{2n+1}}{(2n+1)!} \\
&= \sum_{n=0}^{\infty} (-1)^n \frac{3(1 - 9^n)z^{2n+1}}{4(2n+1)!} = \sum_{n=1}^{\infty} (-1)^n \frac{3(1 - 9^n)z^{2n+1}}{4(2n+1)!} .
\end{aligned}
$$

By the uniqueness of power series, this last expression is the Maclaurin series for $\sin^3 z$.

The preceding argument used some obvious results of power series representations that we have not yet formally stated. The requisite results are part of the following.

Theorem 7.6   *Let f and g have the power series representations*

$$f(z) = \sum_{n=0}^{\infty} a_n(z - \alpha)^n \quad \textit{for } z \in D_{r_1}(\alpha) \quad \textit{and}$$

$$g(z) = \sum_{n=0}^{\infty} b_n(z - \alpha)^n \quad \textit{for } z \in D_{r_2}(\alpha).$$

If $r = \min\{r_1, r_2\}$, and β is any complex constant, then

(16) $\beta f(z) = \displaystyle\sum_{n=0}^{\infty} \beta a_n(z - \alpha)^n \quad \textit{for } z \in D_{r_1}(\alpha)$

(17) $f(z) + g(z) = \displaystyle\sum_{n=0}^{\infty} (a_n + b_n)(z - \alpha)^n \quad \textit{for } z \in D_r(\alpha) \quad \textit{and}$

(18) $f(z)g(z) = \sum\limits_{n=0}^{\infty} c_n(z - \alpha)^n \quad for \; z \in D_r(\alpha), \quad where \; c_n = \sum\limits_{k=0}^{n} a_k b_{n-k}.$

Identity (18) *is known as the* Cauchy product *of the series for f(z) and g(z).*

Proof   We leave the details for (16) and (17) as an exercise. To establish (18) we observe that the function $h(z) = f(z)g(z)$ is analytic in $D_r(\alpha)$. Thus, for $z \in D_r(\alpha)$,

$$h'(z) = f(z)g'(z) + f'(z)g(z), \quad h''(z) = f''(z)g(z) + 2f'(z)g'(z) + f(z)g''(z).$$

By mathematical induction, the preceding pattern can be generalized to the nth derivative, giving Leibniz's formula for the derivative of a product of functions:

(19) $h^{(n)}(z) = \sum\limits_{k=0}^{n} \dfrac{n!}{k!(n - k)!} \; f^k(z)g^{(n-k)}(z).$

(We will ask you to show this as an exercise.)

By Theorem 4.13, we know that

$$\frac{f^k(\alpha)}{k!} = a_k, \quad and \quad \frac{g^{(n-k)}(\alpha)}{(n - k)!} = b_{n-k},$$

so equation (19) becomes

(20) $\dfrac{h^{(n)}(\alpha)}{n!} = \sum\limits_{k=0}^{n} \dfrac{f^k(\alpha)}{k!} \dfrac{g^{(n-k)}(\alpha)}{(n - k)!} = \sum\limits_{k=0}^{\infty} a_k b_{n-k}.$

Now, according to equation (2), we know that

(21) $h(z) = \sum\limits_{k=0}^{\infty} \dfrac{h^{(n)}(\alpha)}{n!} (z - \alpha)^n.$

Substituting equation (20) into equation (21) gives equation (18) because of the uniqueness of power series.

EXAMPLE 7.6   Use the Cauchy product of series to show that

$$\frac{1}{(1 - z)^2} = \sum\limits_{n=0}^{\infty} (n + 1)z^n \quad for \; z \in D_1(0).$$

Solution   Let $f(z) = g(z) = \dfrac{1}{1 - z} = \sum\limits_{n=0}^{\infty} z^n$ for $z \in D_1(0)$. In terms of Theorem 7.6, we have $a_n = b_n = 1$ for all n, and thus equation (18) gives

$$\frac{1}{(1 - z)^2} = h(z) = f(z)g(z) = \sum\limits_{n=0}^{\infty} \left(\sum\limits_{k=0}^{n} a_k b_{n-k} \right) z^n = \sum\limits_{n=0}^{\infty} (n + 1)z^n, \quad as \; required.$$

EXERCISES FOR SECTION 7.2

1. Show that $\sinh z = \sum\limits_{n=0}^{\infty} \dfrac{z^{2n+1}}{(2n+1)!}$ for all z.

2. Show that $\cosh z = \sum\limits_{n=0}^{\infty} \dfrac{z^{2n}}{(2n)!}$ for all z.

3. Show that $\mathrm{Log}(1+z) = \sum\limits_{n=1}^{\infty} \dfrac{(-1)^{n-1}}{n} z^n$ for all $z \in D_1(0)$.

4. Find the Maclaurin series for $\arctan z$. *Hint*: Choose an appropriate contour and integrate the appropriate series given in Example 7.4.

5. Find the Maclaurin series for $\cos^3 z$. *Hint*: Use the trigonometric identity $4 \cos^3 z = \cos 3z + 3 \cos z$.

6. Find the Taylor series for $f(z) = \dfrac{1-z}{z-2}$ centered about $\alpha = 1$. Where does this series converge? *Hint*: $\dfrac{1-z}{z-2} = \dfrac{z-1}{1-(z-1)} = (z-1)\dfrac{1}{1-(z-1)}$. Expand the last expression using a geometric series.

7. Find the Taylor series for $f(z) = \dfrac{1-z}{z-3}$ centered about $\alpha = 1$. Where does this series converge? *Hint*: $\dfrac{1-z}{z-3} = \left(\dfrac{1}{2}\right)\dfrac{z-1}{1-[(z-1)/2]}$.

8. Let $f(z) = \dfrac{\sin z}{z}$, and set $f(0) = 0$.

    (a) Explain why f is analytic at $z = 0$.
    (b) Find the Maclaurin series for $f(z)$.

    (c) Find the Maclaurin series for $g(z) = \int_0^z f(\zeta)\, d\zeta$.

9. Find the Maclaurin series for $f(z) = (z^2 + 1)\sin z$.

10. Let β denote a fixed complex number, and let $f(z) = (1+z)^\beta = \exp[\beta \, \mathrm{Log}(1+z)]$ be the principal branch of $(1+z)^\beta$. Establish the binomial expansion

$$(1+z)^\beta = 1 + \beta z + \frac{\beta(\beta-1)}{2!}z^2 + \frac{\beta(\beta-1)(\beta-2)}{3!}z^3 + \cdots$$

$$= 1 + \sum_{n=1}^{\infty} \frac{\beta(\beta-1)(\beta-2)\cdots(\beta-n+1)}{n!} z^n \text{ for all } z \in D_1(0).$$

11. Show that $f(z) = \dfrac{1}{1-z}$ has its Taylor series representation about the point $\alpha = i$ given by

$$f(z) = \sum_{n=0}^{\infty} \frac{(z-i)^n}{(1-i)^{n+1}} \text{ for all } z \in \{z \colon |z-i| < \sqrt{2}\}$$

12. Use the identity $\cos z = \dfrac{1}{2}(e^{iz} + e^{-iz})$ to find the Maclaurin series for $f(z) = e^z \cos z = \dfrac{1}{2}e^{(1+i)z} + \dfrac{1}{2}e^{(1-i)z}$.

13. Suppose that $f(z) = \sum_{n=0}^{\infty} c_n z^n$ is an entire function.

    (a) Find a series representation for $\overline{f(z)}$ using powers of $\bar{z}$.
    (b) Show that $\overline{f(\bar{z})}$ is an entire function.
    (c) Does $\overline{f(\bar{z})} = f(z)$? Why?

14. Find $f^{(3)}(0)$ for the following functions.

    (a) $f(z) = \sum_{n=0}^{\infty} (3 + (-1)^n)^n z^n$

    (b) $g(z) = \sum_{n=0}^{\infty} \frac{(1 + i)^n}{n} z^n$

    (c) $h(z) = \sum_{n=0}^{\infty} \frac{z^n}{(\sqrt{3} + i)^n}$

15. Let $f(z) = \sum_{n=0}^{\infty} c_n z^n = 1 + z + 2z^2 + 3z^3 + 5z^4 + 8z^5 + 13z^6 + \cdots$, where the coefficients c_n are the Fibonacci numbers defined by $c_0 = 1$, $c_1 = 1$, and $c_n = c_{n-1} + c_{n-2}$ for $n \geq 2$.

    (a) Show that $f(z) = \dfrac{1}{1 - z - z^2}$ for all z in the disk $D_R(0)$ for some number R. *Hint*: Show that f satisfies the equation $f(z) = 1 + z f(z) + z^2 f(z)$.
    (b) Find the value of R in part a for which the series representation is valid. *Hint*: Find the singularities of $f(z)$.

16. Complete the details in the verification of Lemma 7.1.

17. We used Lemma 7.1 in establishing identity (4). However, Lemma 7.1 is valid provided $z \neq z_0$ and $z \neq \alpha$. Explain why in identity (4) this is indeed the case.

18. Prove by mathematical induction that $f^{(n)}(z) = \dfrac{(n + 1)!}{(1 - z)^{n+2}}$ in Example 7.3.

19. Establish identities (16) and (17).

20. Use Maclaurin series and the Cauchy product in identity (18) to verify the identity $\sin 2z = 2 \cos z \sin z$ up to terms involving z^5.

21. The Fresnel integrals $C(z)$ and $S(z)$ are defined by

$$C(z) = \int_0^z \cos(\xi^2)\, d\xi \quad \text{and} \quad S(z) = \int_0^z \sin(\xi^2)\, d\xi,$$

    and $F(z)$ is defined by the equation $F(z) = C(z) + iS(z)$.
    (a) Establish the identity

$$F(z) = \int_0^z \exp(i\xi^2)\, d\xi.$$

    (b) Integrate the power series for $\exp(i\xi^2)$ and obtain the power series for $F(z)$.
    (c) Use the partial sum involving terms up to z^9 to find approximations to $C(1.0)$ and $S(1.0)$.

22. Compute the Taylor series for the principal logarithm $f(z) = \text{Log } z$ expanded about the center $z_0 = -1 + i$. *Hint*: Use $f'(z) = [z - (-1 + i) + (-1 + i)]^{-1}$ and expand $f'(z)$ in powers of $[z - (-1 + i)]$, then apply Corollary 7.2.

23. Let f be defined in a domain that contains the origin. The function f is said to be even if $f(-z) = f(z)$, and it is called odd if $f(-z) = -f(z)$.
    (a) Show that the derivative of an odd function is an even function.
    (b) Show that the derivative of an even function is an odd function.
    Hint: Use limits.

24. (a) If $f(z)$ is even, show that all the coefficients of the odd powers of z in the Maclaurin series are zero.

(b) If $f(z)$ is odd, show that all the coefficients of the even powers of z in the Maclaurin series are zero.

25. Establish identity (19) by using mathematical induction.

26. Consider the following function:

$$f(z) = \begin{cases} \dfrac{1}{1-z} & \text{when } z \neq \dfrac{1}{2}, \\ 0 & \text{when } z = \dfrac{1}{2}. \end{cases}$$

(a) Use Theorem 7.4 to show that the Maclaurin series for $f(z)$ equals $\displaystyle\sum_{n=0}^{\infty} z^n$.

(b) Obviously, the radius of convergence of this series equals 1 (ratio test). However, the distance between 0 and the nearest singularity of f equals $\frac{1}{2}$. Explain why this does not contradict Corollary 7.3.

27. Consider the *real-valued* function f defined on the *real numbers* as follows:

$$f(x) = \begin{cases} e^{-1/x^2} & \text{when } x \neq 0, \\ 0 & \text{when } x = 0. \end{cases}$$

(a) Show that for all $n > 0$, $f^{(n)}(0) = 0$, where $f^{(n)}$ is the nth derivative of f. *Hint*: Use the limit definition for the derivative to establish the case for $n = 1$, then use mathematical induction to complete your argument.

(b) Explain why this gives an example of a function that, although differentiable everywhere on the real line, is not expressible as a Taylor series about 0. *Hint:* Evaluate the Taylor series representation for $f(x)$ when $x \neq 0$, and show that the series does not equal $f(x)$.

(c) Explain why a similar argument could not be made for the *complex-valued* function g defined on the *complex numbers* as follows:

$$g(z) = \begin{cases} e^{-1/z^2} & \text{when } z \neq 0, \\ 0 & \text{when } z = 0. \end{cases}$$

Hint: Show that $g(z)$ is not even continuous at $z = 0$ by taking limits along the real and imaginary axes.

28. Explain how Laurent series and series solutions for differential equations studied in calculus are different. How are they similar?

29. Write a report on series of complex numbers and/or functions. Include ideas and examples not mentioned in the text. Resources include bibliographical items 10, 83, 116, and 153.

30. Write a report on the topic of analytic continuation. Be sure to discuss the chain of power series and disks of convergence. Resources include bibliographical items 4, 19, 46, 51, 52, 93, 106, 128, 129, 141, 145, and 166.

7.3   Laurent Series Representations

Suppose $f(z)$ is not analytic in $D_R(\alpha)$, but is analytic in $D_R^*(\alpha) = \{z \colon 0 < |z - \alpha| < R\}$. For example, the function $f(z) = \dfrac{1}{z^3}\, e^z$ is not analytic when $z = 0$ but is analytic for $|z| > 0$. Clearly, this function does not have a Maclaurin series representation. If we use the Maclaurin series for $g(z) = e^z$, however, and formally divide each term in that series by z^3, we obtain the representation

$$f(z) = \frac{1}{z^3} e^z = \frac{1}{z^3} + \frac{1}{z^2} + \frac{1}{2!z} + \frac{1}{3!} + \frac{z}{4!} + \frac{z^2}{5!} + \frac{z^3}{6!} + \cdots$$

that is valid for all z such that $|z| > 0$.

This example raises the question as to whether it might be possible to generalize the Taylor series method to functions analytic in an annulus

$$A = \{z: r < |z - \alpha| < R\}.$$

Perhaps we can represent these functions with a series that employs negative powers of z in some way as we did with $f(z) = \dfrac{1}{z^3} e^z$. As you will see shortly, this is indeed the case. We begin by defining a series that allows for negative powers of z.

Definition 7.3 *Let c_n be a collection of complex numbers for $n = 0, \pm 1,$ $\pm 2, \pm 3, \ldots$. The doubly infinite series $\displaystyle\sum_{n=-\infty}^{\infty} c_n(z - \alpha)^n$, also called a Laurent series, is defined by*

$$(1) \qquad \sum_{n=-\infty}^{\infty} c_n(z - \alpha)^n = \sum_{n=1}^{\infty} c_{-n}(z - \alpha)^{-n} + \sum_{n=0}^{\infty} c_n(z - \alpha)^n,$$

provided the series on the right-hand side of this equation converge.

Note: *You may recall that by $\displaystyle\sum_{n=0}^{\infty} c_n(z - \alpha)^n$ we really mean*

$c_0 + \displaystyle\sum_{n=1}^{\infty} c_n(z - \alpha)^n$. *At times it will be convenient to write $\displaystyle\sum_{n=-\infty}^{\infty} c_n(z - \alpha)^n$ as*

$$\sum_{n=-\infty}^{\infty} c_n(z - \alpha)^n = \sum_{n=-\infty}^{-1} c_n(z - \alpha)^n + \sum_{n=0}^{\infty} c_n(z - \alpha)^n$$

rather than using the expression given in equation (1).

Definition 7.4 *Given $0 \le r < R$, we define the annulus centered at α with radii r and R by*

$$A(r, R, \alpha) = \{z: r < |z - \alpha| < R\}.$$

The closed annulus centered at α with radii r and R is denoted by

$$\overline{A}(r, R, \alpha) = (z: r \le |z - \alpha| \le R).$$

Figure 7.3 illustrates these terms.

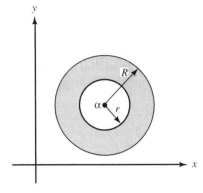

FIGURE 7.3   The closed annulus $\overline{A}(r, R, \alpha)$. The shaded portion is the open annulus $A(r, R, \alpha)$.

Theorem 7.7   *Suppose the Laurent series* $\sum\limits_{n=-\infty}^{\infty} c_n(z - \alpha)^n$ *converges on an annulus* $A(r, R, \alpha)$. *Then the series converges uniformly on any subannulus* $\overline{A}(s, t, \alpha)$, *where* $r < s < t < R$.

Proof   According to Definition 7.3,

$$\sum_{n=-\infty}^{\infty} c_n(z - \alpha)^n = \sum_{n=1}^{\infty} c_{-n}(z - \alpha)^{-n} + \sum_{n=0}^{\infty} c_n(z - \alpha)^n.$$

By Theorem 7.2, the series $\sum\limits_{n=0}^{\infty} c_n(z - \alpha)^n$ must converge uniformly on $\overline{D}_t(\alpha)$.

By the Weierstrass M-test, it is easy to show that the series $\sum\limits_{n=1}^{\infty} c_{-n}(z - \alpha)^{-n}$ converges uniformly on $\{z: |z| \geq s\}$ (we leave the details as an exercise). Combining these two facts yields the required result.

The main result of this section specifies how functions analytic in an annulus can be expanded in a Laurent series. In it, we will use symbols of the form $C_\rho^+(\alpha)$, which we remind you designate the positively oriented circle with radius ρ and center α. That is, $C_\rho^+(\alpha) = \{z: |z - \alpha| < \rho\}$, oriented in the positive direction.

Theorem 7.8 (Laurent's Theorem)   *Suppose* $0 \leq r < R$, *and that* f *is analytic in the annulus* $A = A(r, R, \alpha) = \{z: r < |z - \alpha| < R\}$ *shown in Figure 7.3. If* ρ *is any number such that* $r < \rho < R$, *then for all* $z_0 \in A$, f *has the Laurent series representation*

(2) $f(z_0) = \sum\limits_{n=-\infty}^{\infty} c_n(z_0 - \alpha)^n = \sum\limits_{n=1}^{\infty} c_{-n}(z_0 - \alpha)^{-n} + \sum\limits_{n=0}^{\infty} c_n(z_0 - \alpha)^n$

where for $n = 0, 1, 2, \ldots$, *the coefficients* c_{-n} *and* c_n *are given by*

(3) $c_{-n} = \dfrac{1}{2\pi i} \displaystyle\int_{C_\rho^+(\alpha)} \dfrac{f(z)}{(z-\alpha)^{-n+1}}\, dz$ and $c_n = \dfrac{1}{2\pi i} \displaystyle\int_{C_\rho^+(\alpha)} \dfrac{f(z)}{(z-\alpha)^{n+1}}\, dz.$

Moreover, the convergence in equation (2) is uniform on any closed subannulus $\overline{A}(s, t, \alpha) = \{z\colon s \le |z-\alpha| \le t\}$, where $r < s < t < R$.

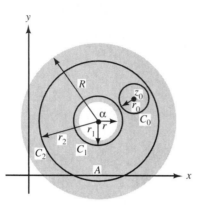

FIGURE 7.4   The annulus A (shaded) and in its interior the circles C_0, C_1, and C_2.

Proof   If we can establish equation (2), the uniform convergence on $\overline{A}(s, t, \alpha)$ will follow from Theorem 7.7. Let z_0 be an arbitrary point of A. Choose r_0 small enough so that the circle $C_0 = C_{r_0}^+(z_0)$ is contained in A. Since f is analytic in $D_{r_0}(z_0)$, the Cauchy integral formula gives

(4) $f(z_0) = \dfrac{1}{2\pi i} \displaystyle\int_{C_0} \dfrac{f(z)}{(z-z_0)}\, dz.$

Let $C_1 = C_{r_1}^+(\alpha)$ and $C_2 = C_{r_2}^+(\alpha)$, where we choose r_1 and r_2 so that C_0 lies in the region between C_1 and C_2, and $r < r_1 < r_2 < R$, as shown in Figure 7.4. Let D be the domain consisting of the annulus A except for the point z_0. The domain D includes the contours C_0, C_1, and C_2, as well as the region between C_2 and $C_0 + C_1$. In addition, since z_0 does not belong to D, the function $\dfrac{f(z)}{(z-z_0)}$ is analytic on D, so by the extended Cauchy-Goursat theorem we obtain

(5) $\dfrac{1}{2\pi i} \displaystyle\int_{C_2} \dfrac{f(z)}{(z-z_0)}\, dz = \dfrac{1}{2\pi i} \displaystyle\int_{C_0} \dfrac{f(z)}{(z-z_0)}\, dz + \dfrac{1}{2\pi i} \displaystyle\int_{C_1} \dfrac{f(z)}{(z-z_0)}\, dz.$

Combining equation (5) with equation (4) and rearranging terms gives

(6) $f(z_0) = \dfrac{1}{2\pi i} \displaystyle\int_{C_2} \dfrac{f(z)}{(z-z_0)}\, dz - \dfrac{1}{2\pi i} \displaystyle\int_{C_1} \dfrac{f(z)}{(z-z_0)}\, dz.$

If $z \in C_2$, then $|z_0 - \alpha|$, so by use of the geometric series (Theorem 4.8) we have

(7) $\quad \dfrac{1}{z - z_0} = \dfrac{1}{(z - \alpha) - (z_0 - \alpha)}$

$$= \dfrac{1}{(z - \alpha)\left(1 - \dfrac{z_0 - \alpha}{z - \alpha}\right)}$$

$$= \sum_{n=0}^{\infty} \dfrac{(z_0 - \alpha)^n}{(z - \alpha)^{n+1}} \quad \text{(provided } z \in C_2\text{)}.$$

Moreover, by the Weierstrass M-test, it is possible to show that the preceding series converges uniformly for $z \in C_2$. We leave the details as an exercise.

Likewise, if $z \in C_1$, we leave as an exercise for you to show by Corollary 4.2 that

(8) $\quad \dfrac{1}{z - z_0} = -\sum_{n=0}^{\infty} \dfrac{(z - \alpha)^n}{(z_0 - \alpha)^{n+1}},$

and that the convergence is uniform for $z \in C_1$.

Substituting the series for equations (7) and (8) into equation (6) yields

(9) $\quad f(z_0) = \dfrac{1}{2\pi i} \int_{C_2} \sum_{n=0}^{\infty} \dfrac{(z_0 - \alpha)^n}{(z - \alpha)^{n+1}} f(z) \, dz + \dfrac{1}{2\pi i} \int_{C_1} \sum_{n=0}^{\infty} \dfrac{(z - \alpha)^n}{(z_0 - \alpha)^{n+1}} f(z) \, dz.$

Since the series in equation (9) converge uniformly on C_2 and C_1, respectively, we can interchange the summations and the integrals in accord with Corollary 7.2 to obtain

(10) $\quad f(z_0) = \displaystyle\sum_{n=0}^{\infty} \left[\dfrac{1}{2\pi i} \int_{C_2} \dfrac{f(z) \, dz}{(z - \alpha)^{n+1}} \right] (z_0 - \alpha)^n +$

$$\sum_{n=0}^{\infty} \left[\dfrac{1}{2\pi i} \int_{C_1} f(z)(z - \alpha)^n \, dz \right] \dfrac{1}{(z_0 - \alpha)^{n+1}}.$$

If we move some terms around in the second series of equation (10) and reindex, we get

(11) $\quad f(z_0) = \displaystyle\sum_{n=0}^{\infty} \left[\dfrac{1}{2\pi i} \int_{C_2} \dfrac{f(z) \, dz}{(z - \alpha)^{n+1}} \right] (z_0 - \alpha)^n +$

$$\sum_{n=1}^{\infty} \left[\dfrac{1}{2\pi i} \int_{C_1} \dfrac{f(z)}{(z - \alpha)^{-n+1}} \, dz \right] (z_0 - \alpha)^{-n}.$$

We apply the extended Cauchy-Goursat theorem once more to conclude that the integrals taken over C_2 and C_1 in equation (11) give the same result if they are taken over the contour $C_\rho^+(\alpha)$, where ρ is any number such that $r < \rho < R$. This yields

(12) $\quad f(z_0) = \displaystyle\sum_{n=0}^{\infty} \left[\dfrac{1}{2\pi i} \int_{C_\rho^+(\alpha)} \dfrac{f(z) \, dz}{(z - \alpha)^{n+1}} \right] (z_0 - \alpha)^n +$

$$\sum_{n=1}^{\infty} \left[\dfrac{1}{2\pi i} \int_{C_\rho^+(\alpha)} \dfrac{f(z)}{(z - \alpha)^{-n+1}} \, dz \right] (z_0 - \alpha)^{-n}.$$

Finally, writing the second series first in equation (12) gives

$$(13) \quad f(z_0) = \sum_{n=1}^{\infty} \left[\frac{1}{2\pi i} \int_{C_\rho^+(\alpha)} \frac{f(z)}{(z-\alpha)^{-n+1}} \, dz \right] (z_0 - \alpha)^{-n} + \sum_{n=0}^{\infty} \left[\frac{1}{2\pi i} \int_{C_\rho^+(\alpha)} \frac{f(z) \, dz}{(z-\alpha)^{n+1}} \right] (z_0 - \alpha)^n.$$

Since $z_0 \in A$ was arbitrary, this establishes equations (2) and (3), and our proof of Theorem 7.8 is complete.

What happens to the Laurent series if f is analytic in the disk $D_R(\alpha)$? If we look at equation (11), we see that the coefficient for the positive power $(z_0 - \alpha)^n$ equals $f^{(n)}(z_0)/n!$ by using Cauchy's integral formula for derivatives. Hence, the series in equation (2) involving the positive powers of $(z_0 - \alpha)$ is actually the Taylor series for f. The Cauchy-Goursat theorem shows us that the coefficients for the negative powers of $(z_0 - \alpha)$ equal zero. In this case, therefore, there are no negative powers involved, and the Laurent series reduces to the Taylor series.

Our next theorem delineates two important aspects of the Laurent series.

Theorem 7.9   *Suppose that f is analytic in the annulus $A(r, R, \alpha)$, and has Laurent series $f(z) = \sum_{n=-\infty}^{\infty} c_n(z - \alpha)^n$ for all $z \in A(r, R, \alpha)$.*

(i) If $f(z) = \sum_{n=-\infty}^{\infty} b_n(z - \alpha)^n$ for all $z \in A(r, R, \alpha)$, then $b_n = c_n$ for all n. (In other words, the Laurent series for f in a given annulus is unique.)

(ii) For all $z \in A(r, R, \alpha)$, the derivatives for $f(z)$ may be obtained by termwise differentiation of its Laurent series.

Proof   We will prove part (i) only, since the proof for part (ii) involves no new ideas beyond what you have already seen in the proof of Theorem 4.13. Since the series $\sum_{n=-\infty}^{\infty} b_n(z - \alpha)^n$ converges pointwise on $A(r, R, \alpha)$, Theorem 7.7 guarantees that this series converges uniformly on $C_\rho^+(\alpha)$ for $0 \le r < \rho < R$. By Laurent's theorem,

$$c_n = \frac{1}{2\pi i} \int_{C_\rho^+(\alpha)} \frac{f(z)}{(z - \alpha)^{n+1}} \, dz$$

$$= \frac{1}{2\pi i} \int_{C_\rho^+(\alpha)} (z - \alpha)^{-n-1} \sum_{m=-\infty}^{\infty} b_m(z - \alpha)^m \, dz$$

$$= \sum_{m=-\infty}^{\infty} \frac{b_m}{2\pi i} \int_{C_\rho^+(\alpha)} (z - \alpha)^{m-n-1} \, dz \quad \text{(since the convergence is uniform).}$$

Since $(z - \alpha)^{m-n-1}$ has an antiderivative for all z except when $m = n$, all the terms in the preceding expression drop out except when $m = n$, giving us

$$c_n = \frac{b_n}{2\pi i} \int_{C_\rho^+(\alpha)} (z - \alpha)^{-1} \, dz = b_n.$$

The uniqueness of the Laurent series is an important property because the coefficients in the Laurent expansion of a function are seldom found by using equation (3). The following examples illustrate some methods for finding the Laurent series coefficients.

EXAMPLE 7.7   Find three different Laurent series representations for $f(z) = 3/(2 + z - z^2)$ involving powers of z.

Solution   The function f has poles to $z = -1, 2$, and is analytic in the disk $D: |z| < 1$, in the annulus $A: 1 < |z| < 2$, and in the region $R: |z| > 2$. We will find a different Laurent series for f in each of the three domains D, A, and R. We start by writing f in its partial fraction form:

$$(14) \quad f(z) = \frac{3}{(1 + z)(2 - z)} = \frac{1}{1 + z} + \frac{1}{2}\frac{1}{1 - (z/2)}.$$

We can use Theorem 4.8 and Corollary 4.2 to obtain the following representations for the terms on the right side of equation (14):

$$(15) \quad \frac{1}{1 + z} = \sum_{n=0}^{\infty} (-1)^n z^n \quad \text{valid for } |z| < 1,$$

$$(16) \quad \frac{1}{1 + z} = \sum_{n=1}^{\infty} \frac{(-1)^{n+1}}{z^n} \quad \text{valid for } |z| > 1,$$

$$(17) \quad \frac{1/2}{1 - (z/2)} = \sum_{n=0}^{\infty} \frac{z^n}{2^{n+1}} \quad \text{valid for } |z| < 2, \quad \text{and}$$

$$(18) \quad \frac{1/2}{1 - (z/2)} = \sum_{n=1}^{\infty} \frac{-2^{n-1}}{z^n} \quad \text{valid for } |z| > 2.$$

Representations (15) and (17) are both valid in the disk D, and thus we have

$$(19) \quad f(z) = \sum_{n=0}^{\infty} \left[(-1)^n + \frac{1}{2^{n+1}}\right] z^n \quad \text{valid for } |z| < 1,$$

which is a Laurent series that reduces to a Maclaurin series. In the annulus A, representations (16) and (17) are valid; hence we get

$$(20) \quad f(z) = \sum_{n=1}^{\infty} \frac{(-1)^{n+1}}{z^n} + \sum_{n=0}^{\infty} \frac{z^n}{2^{n+1}} \quad \text{valid for } 1 < |z| < 2.$$

Finally, in the region R we can use representations (16) and (18) to obtain

$$(21) \quad f(z) = \sum_{n=1}^{\infty} \frac{(-1)^{n+1} - 2^{n-1}}{z^n} \quad \text{valid for } |z| > 2.$$

EXAMPLE 7.8   Find the Laurent series representation for $f(z) = (\cos z - 1)/z^4$ that involves powers of z.

Solution We can use the Maclaurin series for cos $z - 1$ to write

$$f(z) = \frac{\dfrac{-1}{2!} z^2 + \dfrac{1}{4!} z^4 - \dfrac{1}{6!} z^6 + \cdots}{z^4}.$$

We formally divide each term by z^4 to obtain the Laurent series

$$f(z) = \frac{-1}{2z^2} + \frac{1}{24} - \frac{z^2}{720} + \cdots \quad \text{valid for } z \neq 0.$$

EXAMPLE 7.9 Find the Laurent series for $\exp(-1/z^2)$ centered at $z_0 = 0$.

Solution The Maclaurin series for $\exp Z$ is given by

(22) $\quad \exp Z = \sum_{n=0}^{\infty} \frac{Z^n}{n!} \quad$ valid for all Z.

Using the substitution $Z = -z^{-2}$ in equation (22), we obtain

$$\exp\left(\frac{-1}{z^2}\right) = \sum_{n=0}^{\infty} \frac{(-1)^n}{n! z^{2n}} \quad \text{valid for } |z| > 0.$$

EXERCISES FOR SECTION 7.3

1. Find two Laurent series expansions for $f(z) = 1/(z^3 - z^4)$ that involve powers of z.
2. Find the Laurent series for $f(z) = (\sin 2z)/z^4$ that involves powers of z.
3. Show that

$$f(z) = \frac{1}{1-z} = \frac{1}{1-i} \frac{1}{1 - \dfrac{z-i}{1-i}}$$

has a Laurent series representation about the point $z_0 = i$ given by

$$f(z) = \frac{1}{1-z} = -\sum_{n=1}^{\infty} \frac{(1-i)^{n-1}}{(z-i)^n} \quad \text{valid for } |z - i| > \sqrt{2}.$$

4. Show that

$$\frac{1-z}{z-2} = -\sum_{n=0}^{\infty} \frac{1}{(z-1)^n}$$

is valid for $|z - 1| > 1$. *Hint*: Use the hint for Exercise 6 in Section 7.2.
5. Show that

$$\frac{1-z}{z-3} = -\sum_{n=0}^{\infty} \frac{2^n}{(z-1)^n}$$

is valid for $|z - 1| > 2$. *Hint*: Use the hint for Exercise 7 in Section 7.2.

6. Find the Laurent series for $\sin(1/z)$ centered at $\alpha = 0$.

7. Find the Laurent series for $f(z) = (\cosh z - \cos z)/z^5$ that involves powers of z.

8. Find the Laurent series for $f(z) = 1/[z^4(1 - z)^2]$ that involves powers of z and is valid for $|z| > 1$. *Hint*: $1/[1 - (1/z)]^2 = z^2/(1 - z)^2$.

9. Find two Laurent series for $z^{-1}(4 - z)^{-2}$ involving powers of z. *Hint*: Use the result of Example 7.6.

10. Find three Laurent series for $(z^2 - 5z + 6)^{-1}$ centered at $\alpha = 0$.

11. Let a and b be positive real numbers with $b > a > 1$. Show that

$$\text{Log}\,\frac{z - a}{z - b} = \sum_{n=1}^{\infty} \frac{b^n - a^n}{nz^n}$$

holds for $|z| > b$. *Hint*: $\text{Log}(z - a)/(z - b) = \text{Log}[1 - (a/z)] - \text{Log}[1 - (b/z)]$.

12. Use the Maclaurin series for $\sin z$ and then long division to show that the Laurent series for $\csc z$ with $\alpha = 0$ is

$$\csc z = \frac{1}{z} + \frac{z}{6} + \frac{7z^3}{360} + \cdots .$$

13. Can $\text{Log}\,z$ be represented by a Maclaurin series or a Laurent series about the point $\alpha = 0$? Give a reason for your answer.

14. Show that $\cosh[z + (1/z)] = \sum_{n=-\infty}^{\infty} a_n z^n$, where

$$a_n = \frac{1}{2\pi} \int_0^{2\pi} \cos n\theta \cosh(2 \cos \theta)\, d\theta.$$

Hint: Let the path of integration be the circle $C: |z| = 1$.

15. The *Bessel function* $J_n(z)$ is sometimes defined by the generating function

$$\exp\left[\frac{z}{2}\left(t - \frac{1}{t}\right)\right] = \sum_{n=-\infty}^{\infty} J_n(z)t^n.$$

Use the circle $C: |z| = 1$ as the contour of integration, and show that

$$J_n(z) = \frac{1}{\pi} \int_0^{\pi} \cos(n\theta - z \sin \theta)\, d\theta.$$

16. Consider the real-valued function $u(\theta) = 1/(5 - 4 \cos \theta)$.
    (a) Use the substitution $\cos \theta = (1/2)(z + 1/z)$ and obtain

$$u(\theta) = f(z) = \frac{-z}{(z - 2)(2z - 1)} = \frac{1}{3}\frac{1}{1 - z/2} - \frac{1}{3}\frac{1}{1 - 2z}.$$

    (b) Expand the function $f(z)$ in part (a) in a Laurent series that is valid in the annulus $1/2 < |z| < 2$ and get

$$f(z) = \frac{1}{3} + \frac{1}{3}\sum_{n=1}^{\infty} 2^{-n}(z^n + z^{-n}).$$

    (c) Use the substitutions $\cos(n\theta) = (1/2)(z^n + z^{-n})$ in part (b) and obtain the Fourier series for $u(\theta)$:

$$u(\theta) = \frac{1}{3} + \frac{1}{3}\sum_{n=1}^{\infty} 2^{-n+1} \cos(n\theta).$$

17. Suppose that the Laurent expansion $f(z) = \sum_{n=-\infty}^{\infty} a_n z^n$ converges in the annulus $r_1 < |z| < r_2$ where $r_1 < 1$ and $1 < r_2$. Consider the real-valued function $u(\theta) = f(e^{i\theta})$. Show that $u(\theta)$ has the Fourier series expansion

$$u(\theta) = f(e^{i\theta}) = \sum_{n=-\infty}^{\infty} a_n e^{i\theta}, \quad \text{where } a_n = \frac{1}{2\pi} \int_0^{2\pi} e^{-in\phi} f(e^{i\phi}) \, d\phi.$$

18. *The Z-transform.* Let $\{a_n\}$ be a sequence of complex numbers satisfying the growth condition $|a_n| \le MR^n$ for $n = 0, 1, \ldots$ for some fixed positive values M and R. Then the Z-transform of the sequence $\{a_n\}$ is the function $F(z)$ defined by

$$Z(\{a_n\}) = F(z) = \sum_{n=0}^{\infty} a_n z^{-n}.$$

Prove that $F(z)$ converges for $|z| > R$.

19. With reference to Exercise 18, find $Z(\{a_n\})$ for the following sequences:

   (a) $a_n = 2$ (b) $a_n = 1/n!$ (c) $a_n = 1/(n + 1)$

   (d) $a_n = 1$ when n is even, $a_n = 0$ when n is odd

20. With reference to Exercise 18, prove the shifting property for the Z-transform:

$$Z(\{a_{n+1}\}) = z[Z(\{a_n\}) - a_0].$$

21. Use the Weierstrass M-test to show that the series $\sum_{n=1}^{\infty} c_{-n}(z - \alpha)^{-n}$ of Theorem 7.7 converges uniformly on the set $(z: |z| \ge s)$ as claimed.

22. Show that the series in equation (7) converges uniformly for $z \in C_2$.

23. Establish the validity of equation (8) by appealing to Corollary 4.2.

24. Show that the series in equation (8) converges uniformly for $z \in C_1$.

7.4   Singularities, Zeros, and Poles

The point α is called a *singular point*, or *singularity*, of the complex function f if $f(z)$ is not analytic at $z = \alpha$, but every neighborhood $D_R(\alpha)$ of α contains at least one point at which f is analytic. For example, the function $f(z) = 1/(1 - z)$ is not analytic at $z = 1$, but is analytic for all other values of z. Thus, the point $z = 1$ is a singular point of $f(z)$. As another example, consider $g(z) = \text{Log } z$. We saw in Section 5.2 that g is analytic for all z except at the origin and at the points on the negative real axis. Thus, the origin and each point on the negative real axis is a singularity of g.

    The point α is called an *isolated singularity* of a complex function f if f is not analytic at α, but there exists a real number $R > 0$ such that f is analytic everywhere in the punctured disk $D_R^*(\alpha)$. Our function $f(z) = 1/(1 - z)$ has an isolated singularity at $z = 1$, but the singularity at $z = 0$ (or at any point of the negative real axis) is not isolated for $g(z) = \text{Log } z$. Functions with isolated singularities have a Laurent series, since the punctured disk $D_R^*(\alpha)$ is the same as the annulus $A(0, R, \alpha)$. We now look at this special case of Laurent's theorem in order to classify three types of isolated singularities.

Definition 7.5   *Let f have an isolated singularity at α with Laurent series expansion*

$$f(z) = \sum_{n=-\infty}^{\infty} c_n(z - \alpha)^n \quad valid \ for \ all \ z \ \epsilon \ A(0, R, \alpha).$$

Then we distinguish the following types of singularities at α:

(i) *If $c_n = 0$ for $n = -1, -2, -3, \ldots$, then we say that f has a* removable singularity *at α.*

(ii) *If k is a positive integer such that $c_{-k} \neq 0$ and $c_n = 0$ for $n = -k - 1$, $-k - 2, -k - 3, \ldots$, then we say that f has a* pole of order k *at α.*

(iii) *If $c_n \neq 0$ for infinitely many negative integers n, then we say that f has an* essential singularity *at α.*

Let us investigate some examples in the three cases that arise.

(i) If f has a removable singularity at α, then it has a Laurent series

(1) $$f(z) = \sum_{n=0}^{\infty} c_n(z - \alpha)^n \quad \text{valid for all } z \ \epsilon \ A(0, R, \alpha).$$

By Theorem 4.13, we see that the power series in equation (1) defines an analytic function in the disk $D_R(\alpha)$. If we use this series to define $f(\alpha) = c_0$, then the function f becomes analytic at $z = \alpha$, and the singularity is "removed." As an example, consider the function $f(z) = \dfrac{\sin z}{z}$. It is undefined at $z = 0$, and has an isolated singularity at $z = 0$. The Laurent series for f is given by

$$f(z) = \frac{\sin z}{z} = \frac{1}{z}\left(z - \frac{z^3}{3!} + \frac{z^5}{5!} - \frac{z^7}{7!} + \cdots\right)$$

$$= 1 - \frac{z^2}{3!} + \frac{z^4}{5!} - \frac{z^6}{7!} + \cdots \quad \text{valid for } |z| > 0.$$

We can "remove" this singularity if we define $f(0) = 1$, for then f will be analytic at 0 in accordance with Theorem 4.13. Another example is $g(z) = \dfrac{\cos z - 1}{z^2}$, which has an isolated singularity at $z = 0$. The Laurent series for f is given by

$$g(z) = \frac{1}{z^2}\left(-\frac{z^2}{2!} + \frac{z^4}{4!} - \frac{z^6}{6!} + \cdots\right)$$

$$= -\frac{1}{2} + \frac{z^2}{4!} - \frac{z^4}{6!} + \cdots \quad \text{valid for } |z| > 0.$$

If we choose to define $g(0) = -\frac{1}{2}$, then g will be analytic for all z.

(ii) If f has a pole of order k at α, the Laurent series for f is given by

(2) $$f(z) = \sum_{n=-k}^{\infty} c_n(z - \alpha)^n \text{ valid for all } z \ \epsilon \ A(0, R, \alpha), \text{ where } c_{-k} \neq 0.$$

For example, $f(z) = \dfrac{\sin z}{z^3} = \dfrac{1}{z^2} - \dfrac{1}{3!} + \dfrac{z^2}{5!} - \dfrac{z^4}{7!} + \cdots$

has a pole of order 2 at $z = 0$.

If f has a pole of order 1 at α, then we say that f has a *simple pole* at α. An example is

$$g(z) = \dfrac{1}{z}\, e^z = \dfrac{1}{z} + 1 + \dfrac{z}{2!} + \dfrac{z^2}{3!} + \cdots ,$$

which has a simple pole at $z = 0$.

(iii) If an infinite number of negative powers of $(z - \alpha)$ occur in the Laurent series, then f has an essential singularity. For example

$$f(z) = z^2 \sin \dfrac{1}{z} = z - \dfrac{1}{3!z} + \dfrac{1}{5!z^3} - \dfrac{1}{7!z^5} + \cdots$$

has an essential singularity at the origin.

> **Definition 7.6**   *A function $f(z)$ analytic in $D_R(\alpha)$ is said to have a* zero of order k *at the point $z = \alpha$ if and only if*

(3) $f^{(n)}(\alpha) = 0$ *for $n = 0, 1, \ldots, k - 1$, and $f^{(k)}(\alpha) \neq 0$.*

> *A zero of order one is sometimes called a* simple zero.

> **Theorem 7.10**   *A function $f(z)$ analytic in $D_R(\alpha)$ has a zero of order k at the point $z = \alpha$ if and only if its Taylor series given by $f(z) = \sum\limits_{n=0}^{\infty} c_n (z - \alpha)^n$ has*

(4) $c_0 = c_1 = \cdots = c_{k-1} = 0$ and $c_k \neq 0.$

> **Proof**   Statement (4) follows immediately from equation (3) since we have $c_n = \dfrac{f^{(n)}(\alpha)}{n!}$ according to Taylor's theorem.

EXAMPLE 7.10   We see from Theorem 7.10 that the function

$$f(z) = z \sin z^2 = z^3 - \dfrac{z^7}{3!} + \dfrac{z^{11}}{5!} - \dfrac{z^{15}}{7!} + \cdots$$

has a zero of order 3 at $z = 0$. Furthermore, Definition 7.6 confirms this because

$$
\begin{aligned}
f'(z) &= 2z^2\cos z^2 + \sin z^2, \\
f''(z) &= -4z^3\sin z^2 + 4z^2\cos z^2 + 2z \cos z^2, \\
f'''(z) &= -8z^4\cos z^2 - 12z^2\sin z^2 - 8z^3\cos z^2 \\
&\quad + 8z \cos z^2 - 4z^2\sin z^2 + 2 \cos z^2.
\end{aligned}
$$

Clearly we have $f(0) = f'(0) = f''(0) = 0$, but $0 \neq f'''(0) = 2$.

Theorem 7.11 *Suppose f is analytic in $D_R(\alpha)$. Then $f(z)$ has a zero of order k at $z = \alpha$ if and only if f can be expressed in the form*

(5) $f(z) = (z - \alpha)^k g(z),$

where g is analytic at $z = \alpha$, and $g(\alpha) \neq 0$.

Proof Suppose that $f(z)$ has a zero of order k at α. Using equation (4) of Theorem 7.10 we can write $f(z)$ as

$$
\begin{aligned}
(6) \quad f(z) &= \sum_{n=0}^{\infty} c_n (z - \alpha)^n \\
&= \sum_{n=k}^{\infty} c_n (z - \alpha)^n \quad \text{(by equations (4))} \\
&= \sum_{n=0}^{\infty} c_{n+k} (z - \alpha)^{n+k} \quad \text{(by reindexing)} \\
&= (z - \alpha)^k \sum_{n=0}^{\infty} c_{n+k} (z - \alpha)^n, \quad \text{where } c_k \neq 0.
\end{aligned}
$$

The series on the right side of equation (6) defines a function which we shall denote by $g(z)$. That is,

$$
g(z) = \sum_{n=0}^{\infty} c_{n+k} (z - \alpha)^n = c_k + \sum_{n=1}^{\infty} c_{n+k} (z - \alpha)^n \quad \text{valid for all } z \text{ in } D_R(\alpha).
$$

By Theorem 4.13, g is analytic in $D_R(\alpha)$. Also, $g(\alpha) = c_k \neq 0$.

Conversely, suppose that f has the form given by equation (5). Since $g(z)$ is analytic at $z = \alpha$, it has the power series representation

(7) $g(z) = \sum_{n=0}^{\infty} b_n (z - \alpha)^n, \quad \text{where } g(\alpha) = b_0 \neq 0 \text{ by assumption.}$

If we multiply both sides of equation (7) by $(z - \alpha)^k$, we obtain the following power series representation for f:

$$
f(z) = g(z)(z - \alpha)^k = \sum_{n=0}^{\infty} b_n (z - \alpha)^{n+k} = \sum_{n=k}^{\infty} b_{n-k} (z - \alpha)^n.
$$

By Theorem 7.10, $f(z)$ has a zero of order k at $z = \alpha$, and our proof is complete.

An immediate consequence of Theorem 7.11 is the following result. The proof is left as an exercise.

Corollary 7.4 *If $f(z)$ and $g(z)$ are analytic at $z = \alpha$, and have zeros of orders m and n, respectively, at $z = \alpha$, then their product $h(z) = f(z)g(z)$ has a zero of order $m + n$ at $z = \alpha$.*

EXAMPLE 7.11   Let $f(z) = z^3 \sin z$. Then $f(z)$ can be factored as the product of z^3 and $\sin z$, which have zeros of order $m = 3$, and $n = 1$ at $z = 0$. Hence, $z = 0$ is a zero of order 4 of $f(z)$.

Our next result gives a useful way to characterize a pole.

Theorem 7.12   *A function $f(z)$ analytic in the punctured disk $D_R^*(\alpha)$ has a pole of order k at $z = \alpha$ if and only if f can be expressed in the form*

(8) $$f(z) = \frac{h(z)}{(z - \alpha)^k},$$

where $h(z)$ is analytic at $z = \alpha$, and $h(\alpha) \neq 0$.

Proof   Suppose that $f(z)$ has a pole of order k at $z = \alpha$. The Laurent series for f can then be written as

(9) $$f(z) = \frac{1}{(z - \alpha)^k} \sum_{n=0}^{\infty} c_{n-k}(z - \alpha)^n, \quad \text{where } c_{-k} \neq 0.$$

The series on the right side of equation (9) defines a function which we shall denote by $h(z)$. That is,

(10) $$h(z) = \sum_{n=0}^{\infty} c_{n-k}(z - \alpha)^n \quad \text{for all } z \text{ in } D_R^*(\alpha) = \{z : 0 < |z - \alpha| < R\}.$$

If we specify that $h(\alpha) = c_{-k}$, then h is analytic in all of $D_R(\alpha)$, with $h(\alpha) \neq 0$.

Conversely, suppose that f has the form given by equation (8). Since $h(z)$ is analytic at $z = \alpha$ with $h(\alpha) \neq 0$, it has a power series representation

(11) $$h(z) = \sum_{n=0}^{\infty} b_n(z - \alpha)^n, \quad \text{where } b_0 \neq 0.$$

If we divide both sides of equation (11) by $(z - \alpha)^k$, we obtain the following Laurent series representation for f:

$$f(z) = \sum_{n=0}^{\infty} b_n(z - \alpha)^{n-k} = \sum_{n=-k}^{\infty} b_{n+k}(z - \alpha)^n = \sum_{n=-k}^{\infty} c_n(z - \alpha)^n, \text{ where } c_n = b_{n+k}.$$

Since $c_{-k} = b_0 \neq 0$, $f(z)$ has a pole of order k at $z = \alpha$. This completes the proof.

The following results will be useful in determining the order of a zero or a pole. The proofs follow easily from Theorems 7.11 and 7.12 and are left as exercises.

Theorem 7.13
 (i) *If $f(z)$ is analytic and has a zero of order k at $z = \alpha$, then $g(z) = 1/f(z)$ has a pole of order k at $z = \alpha$.*

(ii) *If $f(z)$ has a pole of order k at $z = \alpha$, then $g(z) = 1/f(z)$ has a removable singularity at $z = \alpha$. If we define $h(\alpha) = 0$, then $h(z)$ has a zero of order k at $z = \alpha$.*

Corollary 7.5   *If $f(z)$ and $g(z)$ have poles of orders m and n, respectively, at $z = \alpha$, then their product $h(z) = f(z)g(z)$ has a pole of order $m + n$ at $z = \alpha$.*

Corollary 7.6   *Let $f(z)$ and $g(z)$ be analytic with zeros of orders m and n, respectively, at $z = \alpha$. Then their quotient $h(z) = f(z)/g(z)$ has the following behavior:*
   (i) *If $m > n$, then $h(z)$ has a removable singularity at $z = \alpha$. If we define $h(\alpha) = 0$, then $h(z)$ has a zero of order $m - n$ at $z = \alpha$.*
  (ii) *If $m < n$, then $h(z)$ has a pole of order $n - m$ at $z = \alpha$.*
 (iii) *If $m = n$, then $h(z)$ has a removable singularity at $z = \alpha$, and can be defined so that $h(z)$ is analytic at $z = \alpha$ by $h(\alpha) = \lim\limits_{z \to \alpha} h(z)$.*

EXAMPLE 7.12   Locate the zeros and poles of $h(z) = (\tan z)/z$ and determine their order.

Solution   In Section 5.4 we saw that the zeros of $f(z) = \sin z$ occur at the points $z = n\pi$, where n is an integer. Since $f'(n\pi) = \cos n\pi \neq 0$, the zeros of f are simple. In a similar fashion it can be shown that the function $g(z) = z \cos z$ has simple zeros at the points $z = 0$ and $z = (n + \frac{1}{2})\pi$ where n is an integer. From the information given we find that $h(z) = f(z)/g(z)$ has the following behavior.
   (i) h has simple zeros at $z = n\pi$ where $n = \pm 1, \pm 2, \ldots$.
  (ii) h has simple poles at $z = (n + \frac{1}{2})\pi$ where n is an integer.
 (iii) h is analytic at 0 and $\lim\limits_{z \to 0} h(z) \neq 0$.

EXAMPLE 7.13   Locate the poles of $g(z) = 1/(5z^4 + 26z^2 + 5)$, and specify their order.

Solution   The roots of the quadratic equation $5Z^2 + 26Z + 5 = 0$ occur at the points $Z_1 = -5$ and $Z_2 = -1/5$. If we use the substitution $Z = z^2$, then we see that the function $f(z) = 5z^4 + 26z^2 + 5$ has simple zeros at the points $\pm i\sqrt{5}$ and $\pm i/\sqrt{5}$. Theorem 7.13 implies that g has simple poles at $\pm i\sqrt{5}$ and $\pm i/\sqrt{5}$.

EXAMPLE 7.14   Locate the poles of $g(z) = (\pi \cot \pi z)/z^2$, and specify their order.

Solution   The functions $z^2 \sin \pi z$ and $f(z) = (z^2 \sin \pi z)/(\pi \cos \pi z)$ have a zero of order 3 at $z = 0$ and simple zeros at the points $z = \pm 1, \pm 2, \ldots$. Using Theorem 7.13, we see that g has a pole of order 3 at $z = 0$ and simple poles at the points $z = \pm 1, \pm 2, \ldots$.

EXERCISES FOR SECTION 7.4

Locate the zeros of the functions in Exercises 1 and 2, and determine their order.

1. **(a)** $(1 + z^2)^4$ **(b)** $\sin^2 z$ **(c)** $z^2 + 2z + 2$
   (d) $\sin z^2$ **(e)** $z^4 + 10z^2 + 9$ **(f)** $1 + \exp z$
2. **(a)** $z^6 + 1$ **(b)** $z^3 \exp(z - 1)$ **(c)** $z^6 + 2z^3 + 1$
   (d) $z^3 \cos^2 z$ **(e)** $z^8 + z^4$ **(f)** $z^2 \cosh z$

Locate the poles of the functions in Exercises 3 and 4, and determine their order.

3. **(a)** $(z^2 + 1)^{-3}(z - 1)^{-4}$ **(b)** $z^{-1}(z^2 - 2z + 2)^{-2}$
   (c) $(z^6 + 1)^{-1}$ **(d)** $(z^4 + z^3 - 2z^2)^{-1}$
   (e) $(3z^4 + 10z^2 + 3)^{-1}$ **(f)** $(i + 2/z)^{-1}(3 + 4/z)^{-1}$
4. **(a)** $z \cot z$ **(b)** $z^{-5}\sin z$ **(c)** $(z^2 \sin z)^{-1}$
   (d) $z^{-1}\csc z$ **(e)** $(1 - \exp z)^{-1}$ **(f)** $z^{-5}\sinh z$

Locate the singularities of the functions in Exercises 5 and 6, and determine their type.

5. **(a)** $z^{-2}(z - \sin z)$ **(b)** $\sin(1/z)$
   (c) $z \exp(1/z)$ **(d)** $\tan z$
6. **(a)** $(z^2 + z)^{-1}\sin z$ **(b)** $z/\sin z$
   (c) $(\exp z - 1)/z$ **(d)** $(\cos z - \cos 2z)/z^4$

For Exercises 7–10, use L'Hôpital's rule to find the limit.

7. $\displaystyle \lim_{z \to 1+i} \frac{z - 1 - i}{z^4 + 4}$ 8. $\displaystyle \lim_{z \to i} \frac{z^2 - 2iz - 1}{z^4 + 2z^2 + 1}$

9. $\displaystyle \lim_{z \to i} \frac{1 + z^6}{1 + z^2}$ 10. $\displaystyle \lim_{z \to 0} \frac{\sin z + \sinh z - 2z}{z^5}$

11. Let f be analytic and have a zero of order k at z_0. Show that $f'(z)$ has a zero of order $k - 1$ at z_0.
12. Let f and g be analytic at z_0 and have zeros of order m and n, respectively, at z_0. What can you say about the zero of $f + g$ at z_0?
13. Let f and g have poles of order m and n, respectively, at z_0. Show that $f + g$ has either a pole or removable singularity at z_0.
14. Let f be analytic and have a zero of order k at z_0. Show that $f'(z)/f(z)$ has a simple pole at z_0.
15. Let f have a pole of order k at z_0. Show that $f'(z)$ has a pole of order $k + 1$ at z_0.
16. Establish Corollary 7.4.   17. Establish Corollary 7.5.
18. Establish Corollary 7.6.   19. Find the singularities of $\cot z - 1/z$.

20. Find the singularities of $\text{Log } z^2$.   21. Find the singularities of $\dfrac{1}{\sin(1/z)}$.

22. If $f(z)$ has a removable singularity at z_0, then prove that $1/f(z)$ has either a removable singularity or a pole at z_0.
23. How are the definitions of singularity in complex analysis and asymptote studied in calculus different? How are they similar?

7.5   Applications of Taylor and Laurent Series

In this section we show how Taylor and Laurent series can be used to derive important properties of analytic functions. We begin by showing that the zeros of an analytic function must be "isolated" unless the function is identically zero. A point α of a set T is called *isolated* if there exists a disk $D_r(\alpha)$ about α that does not contain any other points of T.

> **Theorem 7.14**   *Suppose f is analytic at α and that $f(\alpha) = 0$. If f is not identically zero, then there exists a punctured disk $D_r^*(\alpha)$ in which f has no zeros.*

Proof   By Taylor's theorem, there exists some disk $D_R(\alpha)$ about α such that

$$f(z) = \sum_{n=0}^{\infty} \frac{f^{(n)}(\alpha)}{n!} (z - \alpha)^n \quad \text{for all } z \in D_R(\alpha).$$

Now, if all the Taylor coefficients $f^{(n)}(\alpha)/n!$ of f were zero, then f would be identically zero on $D_R(\alpha)$. A proof similar to the proof of the maximum modulus principle given in Section 6.6 would then show that f is identically zero, contradicting our assumption about f.

Thus, not all the Taylor coefficients of f are zero, and we may select the smallest integer k such that $f^{(k)}(\alpha)/k! \neq 0$. According to the results of the previous section, f has a zero of order k at α and can be written in the form

$$f(z) = (z - \alpha)^k g(z),$$

where g is analytic at α and $g(\alpha) \neq 0$. Since g is a continuous function, there exists a disk $D_r(\alpha)$ throughout which g is nonzero. Therefore, $f(z) \neq 0$ in the punctured disk $D_r(\alpha)$.

The following corollaries are given as exercises.

> **Corollary 7.7**   *Suppose that f is analytic in the domain D, and that $\alpha \in D$. If there exists a sequence of points $\{z_n\}$ in D such that $z_n \to \alpha$, and $f(z_n) = 0$, then $f(z) = 0$ for all $z \in D$.*

> **Corollary 7.8**   *Suppose f and g are analytic in the domain D, where $\alpha \in D$. If there exists a sequence $\{z_n\}$ in D such that $z_n \to \alpha$, and $f(z_n) = g(z_n)$ for all n, then $f(z) = g(z)$ for all $z \in D$.*

Theorem 7.14 allows us to give a simple argument for one version of L'Hôpital's rule.

> **Corollary 7.9 (L'Hôpital's Rule)**   *Suppose f and g are analytic at α. If $f(\alpha) = 0$ and $g(\alpha) = 0$, but $g'(\alpha) \neq 0$, then*
>
> $$\lim_{z \to \alpha} \frac{f(z)}{g(z)} = \frac{f'(\alpha)}{g'(\alpha)}.$$

Proof   Since $g'(\alpha) \neq 0$, g is not identically zero, so by Theorem 7.14, there is a punctured disk $D_r^*(\alpha)$ in which $g(z) \neq 0$. Thus, the quotient $\dfrac{f(z)}{g(z)} = \dfrac{f(z) - f(\alpha)}{g(z) - g(\alpha)}$ is defined for all $z \in D_r^*(\alpha)$, and we can write

$$\lim_{z \to \alpha} \frac{f(z)}{g(z)} = \lim_{z \to \alpha} \frac{f(z) - f(\alpha)}{g(z) - g(\alpha)} = \lim_{z \to \alpha} \frac{[f(z) - f(\alpha)]/(z - \alpha)}{[g(z) - g(\alpha)]/(z - \alpha)} = \frac{f'(\alpha)}{g'(\alpha)}.$$

The following theorem can be used to get Taylor series for quotients of analytic functions. Its proof involves ideas from Section 7.2, and we leave it as an exercise.

Theorem 7.15 (Division of Power Series)   *Suppose f and g are analytic at α with power series representations*

$$f(z) = \sum_{n=0}^{\infty} a_n(z - \alpha)^n \quad and \quad g(z) = \sum_{n=0}^{\infty} b_n(z - \alpha)^n \quad for \ all \ z \in D_R(\alpha).$$

If $g(\alpha) \neq 0$, then the quotient f/g has the power series representation

$$\frac{f(z)}{g(z)} = \sum_{n=0}^{\infty} c_n(z - \alpha)^n,$$

where the coefficients satisfy the equations

$$a_n = b_0 c_n + b_1 c_{n-1} + \cdots + b_{n-1} c_1 + b_n c_0.$$

In other words, the series for the quotient $\dfrac{f(z)}{g(z)}$ can be obtained by the familiar process of dividing the series for f by the series for g using the standard long division algorithm.

EXAMPLE 7.15   Find the first few terms of the Maclaurin series for the function $f(z) = \sec z$ if $|z| < \dfrac{\pi}{2}$, and compute $f^{(4)}(0)$.

Solution   Using long division, we see that

$$\sec z = \frac{1}{\cos z} = \frac{1}{1 - \dfrac{z^2}{2!} + \dfrac{z^4}{4!} - \dfrac{z^6}{6!} + \cdots} = 1 + \frac{1}{2}z^2 + \frac{5}{24}z^4 + \cdots.$$

Moreover, using Taylor's theorem, we see that if $f(z) = \sec z$, then $\dfrac{f^{(4)}(0)}{4!} = \dfrac{5}{24}$, so $f^{(4)}(0) = 5$.

We close this section with some results concerning the behavior of complex functions at points near the different types of isolated singularities. Our first theorem is due to the German mathematician G. F. Bernhard Riemann (1826–1866).

Theorem 7.16 (Riemann)   *Suppose that f is analytic in $D_r^*(\alpha)$. If f is bounded in $D_r^*(\alpha)$, then either f is analytic at α or f has a removable singularity at α.*

Proof   Consider the function g defined as follows:

$$(1) \quad g(z) = \begin{cases} (z - \alpha)^2 f(z) & \text{when } z \neq \alpha, \\ 0 & \text{when } z = \alpha. \end{cases}$$

Clearly, g is analytic in at least $D_r^*(\alpha)$. By straightforward calculation,

$$g'(\alpha) = \lim_{z \to \alpha} \frac{g(z) - g(\alpha)}{z - \alpha} = \lim_{z \to \alpha} (z - \alpha) f(z) = 0.$$

The last equation follows because f is bounded. Thus, g is also analytic at α, with $g'(\alpha) = 0$.

By Taylor's theorem, g has the representation

$$(2) \quad g(z) = \sum_{n=2}^{\infty} \frac{g^{(n)}(\alpha)}{n!} (z - \alpha)^n \quad \text{for all } z \in D_r(\alpha)$$

We can divide both sides of equation (2) by $(z - \alpha)^2$ and use equation (1) to obtain the following power series representation for f:

$$f(z) = \sum_{n=2}^{\infty} \frac{g^{(n)}(\alpha)}{n!} (z - \alpha)^{n-2} = \sum_{n=0}^{\infty} \frac{g^{(n+2)}(\alpha)}{(n+2)!} (z - \alpha)^n.$$

By Theorem 4.13, f is analytic at α if we define $f(\alpha) = \dfrac{g^{(2)}(\alpha)}{2!}$. This completes the proof.

The following corollary is given as an exercise.

Corollary 7.10   *Suppose that f is analytic in $D_r^*(\alpha)$. Then f can be defined to be analytic at α if and only if $\lim_{z \to \alpha} f(z)$ exists (and is finite).*

Theorem 7.17   *The function f has a pole of order k at α, if and only if $\lim_{z \to \alpha} |f(z)| = \infty$.*

Proof   Suppose, first, that f has a pole of order k at α. Using Theorem 7.12, we can say that

$$f(z) = \frac{h(z)}{(z - \alpha)^k},$$

where h is analytic at α, and $h(\alpha) \neq 0$. Since

$$\lim_{z \to \alpha} |h(z)| = |h(\alpha)| \neq 0 \quad \text{and} \quad \lim_{z \to \alpha} |(z - \alpha)^k| = 0,$$

we conclude that

$$\lim_{z \to \alpha} |f(z)| = \lim_{z \to \alpha} |h(z)| \lim_{z \to \alpha} \frac{1}{|(z - \alpha)^k|} = \infty.$$

Conversely, suppose that $\lim_{z \to \alpha} |f(z)| = \infty$. By the definition of a limit, there must

be some $\delta > 0$ such that $|f(z)| > 1$ if $z \in D_\delta^*(\alpha)$. Thus, the function $g(z) = \dfrac{1}{f(z)}$ is

analytic and bounded, by equation 1, in $D_\delta^*(\alpha)$. By Theorem 7.16, we may define g at α so that g is analytic in all of $D_\delta(\alpha)$. In fact,

$$|g(\alpha)| = \lim_{z \to \alpha} \frac{1}{|f(z)|} = 0,$$

so α is a zero of g. We claim that α must be of finite order, for otherwise we would

have $g^{(n)}(\alpha) = 0$ for all n, and hence $g(z) = \displaystyle\sum_{n=0}^{\infty} \frac{g^{(n)}(\alpha)}{n!} (z - \alpha)^n = 0$ for all $z \in D_\delta(\alpha)$.

Since $g(z) = \dfrac{1}{f(z)}$ is analytic in $D_\delta^*(\alpha)$, this is impossible, so we can let k be the order

of the zero of g at α. By Theorem 7.13 it follows that f has a pole of order k, and this completes our proof.

Theorem 7.18   *The function f has an essential singularity at α if and only if $\lim_{z \to \alpha} |f(z)|$ does not exist.*

Proof   We see from Corollary 7.10 and Theorem 7.17 that the conclusion of Theorem 7.18 is the only option possible.

EXAMPLE 7.16   Show that the function g defined by

$$g(z) = \begin{cases} e^{-1/z^2} & \text{when } z \neq 0, \\ 0 & \text{when } z = 0, \end{cases}$$

is not continuous at $z = 0$.

Solution   In Exercise 27 of Section 7.2, we asked you to show this by computing limits along the real and imaginary axes. Note, however, that the Laurent series for $g(z)$ in the annulus $D_r^*(0)$ is

$$g(z) = 1 + \sum_{n=1}^{\infty} (-1)^n \frac{1}{z^{2n}},$$

so that 0 is an essential singularity for g. According to Theorem 7.18, $\lim_{z \to 0} |g(z)|$

does not exist, so g is not continuous at 0.

EXERCISES FOR SECTION 7.5

1. Consider the function $f(z) = z \sin(1/z)$.
   (a) Show that there is a sequence (z_n) of points converging to $z = 0$ such that $f(z_n) = 0$ for $n = 1, 2, 3, \ldots$.
   (b) Does this contradict Corollary 7.7?

2. Determine whether there exists a function $f(z)$ that is analytic at $z = 0$ such that

$$f\left(\frac{1}{2n}\right) = 0 \quad \text{and} \quad f\left(\frac{1}{2n-1}\right) = 1 \quad \text{for } n = 1, 2, \ldots.$$

3. Determine whether there exists a function $f(z)$ that is analytic at $z = 0$ such that

$$f\left(\frac{1}{n}\right) = f\left(\frac{-1}{n}\right) = \frac{1}{n^2} \quad \text{for } n = 1, 2, \ldots.$$

4. Determine whether there exists a function $f(z)$ that is analytic at $z = 0$ such that

$$f\left(\frac{1}{n}\right) = f\left(\frac{-1}{n}\right) = \frac{1}{n^3} \quad \text{for } n = 1, 2, \ldots.$$

5. Prove Corollaries 7.7 and 7.8.

6. Prove Theorem 7.15.

7. Let $f(z) = \tan z$.
   (a) Use Theorem 7.15 to find the first few terms of the Maclaurin series for $f(z)$ if
   $$|z| < \frac{\pi}{2}.$$
   (b) What are the values of $f^{(6)}(0)$ and $f^{(7)}(0)$?

8. Prove Corollary 7.10.

9. Show that the real function $f(x)$ defined by

$$f(x) = \begin{cases} x \sin \dfrac{1}{x} & \text{when } x \neq 0, \\ 0 & \text{when } x = 0, \end{cases}$$

   is continuous at $x = 0$, but that the corresponding function $g(z)$ defined by

$$g(z) = \begin{cases} z \sin \dfrac{1}{z} & \text{when } z = 0, \\ 0 & \text{when } z \neq 0, \end{cases}$$

   is *not* continuous at $z = 0$.

10. Write a report on analytic functions. Include a discussion of the Cauchy-Riemann equations and the other conditions that guarantee that $f(z)$ is analytic. Resources include bibliographical items 21, 39, 62, 72, 86, 155, and 161.

11. Write a report on infinite products of complex numbers and/or functions. Resources include bibliographical items 4, 19, 51, 129, 145, and 181.

12. Write a report on the Bieberbach conjecture. Your report should be more of a narrative about the conjecture and its eventual proof. Resources include bibliographical items 49, 73, 108, 148, and 189.

8

Residue Theory

8.1 The Residue Theorem

The Cauchy integral formulae in Section 6.5 are useful in evaluating contour integrals over a simple closed contour C where the integrand has the form $f(z)/(z - z_0)^k$ and f is an analytic function. In this case, the singularity of the integrand is at worst a pole of order k at z_0. In this section we extend this result to integrals that have a finite number of isolated singularities and lie inside the contour C. This new method can be used in cases where the integrand has an essential singularity at z_0 and is an important extension of the previous method.

Let f have a nonremovable isolated singularity at the point z_0. Then f has the Laurent series representation

$$(1) \quad f(z) = \sum_{n=-\infty}^{\infty} a_n(z - z_0)^n \quad \text{valid for } 0 < |z - z_0| < R.$$

The coefficient a_{-1} of $1/(z - z_0)$ is called the *residue* of f at z_0, and we use the notation

$$(2) \quad \text{Res}[f, z_0] = a_{-1}.$$

EXAMPLE 8.1   If $f(z) = \exp(2/z)$, then the Laurent series (1) has the form

$$f(z) = \exp\left(\frac{2}{z}\right) = 1 + \frac{2}{z} + \frac{2^2}{2!z^2} + \frac{2^3}{3!z^3} + \cdots,$$

and we see that $\text{Res}[f, 0] = 2$.

EXAMPLE 8.2   If $g(z) = \dfrac{3}{2z + z^2 - z^3}$, show that $\text{Res}[g, 0] = \dfrac{3}{2}$.

Solution   Using Example 7.7, we find that g has three Laurent series representations involving powers of z. The Laurent series of the form (1) is given by

$$g(z) = \sum_{n=0}^{\infty} \left[(-1)^n + \frac{1}{2^{n+1}} \right] z^{n-1} \quad \text{valid for } 0 < |z| < 1.$$

Computing the first few coefficients, we obtain

$$g(z) = \frac{3}{2}\frac{1}{z} - \frac{3}{4} + \frac{9}{8} z - \frac{15}{16} z^2 + \cdots .$$

Therefore $\text{Res}[g, 0] = \frac{3}{2}$.

Let us recall that the Laurent series coefficients in equation (1) are given by

$$(3) \quad a_n = \frac{1}{2\pi i} \int_C \frac{f(\xi)\, d\xi}{(\xi - z_0)^{n+1}} \quad \text{for } n = 0, \pm 1, \pm 2, \ldots$$

where $C = C_r^+(z_0) = \{z\colon |z - z_0| = r\}$ is any positively oriented circle with $r < R$. This gives us an important fact concerning $\text{Res}[f, z_0]$. If we set $n = -1$ in equation (3), then we obtain

$$(4) \quad \int_C f(\xi)\, d\xi = 2\pi i a_{-1} = 2\pi i \, \text{Res}[f, z_0]$$

where z_0 is the only singularity of f that lies inside C. If we are able to find the Laurent series expansion for f given in equation (1), then equation (4) gives us an important tool for evaluating contour integrals.

EXAMPLE 8.3   Evaluate $\int_c \exp(2/z)\, dz$, where C is the unit circle $|z| = 1$ taken with positive orientation.

Solution   We have seen that the residue of $f(z) = \exp(2/z)$ at $z_0 = 0$ is $\text{Res}[f, 0] = 2$. Using equation (4), we find that

$$\int_C \exp\left(\frac{2}{z}\right) dz = 2\pi i \, \text{Res}[f, 0] = 4\pi i.$$

Theorem 8.1 (Cauchy's Residue Theorem)   *Let D be a simply connected domain, and let C be a simple closed positively oriented contour that lies in D. If f is analytic inside C and on C, except at the points $z_1, z_2, \ldots, z_n$ that lie inside C, then*

$$(5) \quad \int_C f(z)\, dz = 2\pi i \sum_{k=1}^{n} \text{Res}[f, z_k].$$

The situation is illustrated in Figure 8.1.

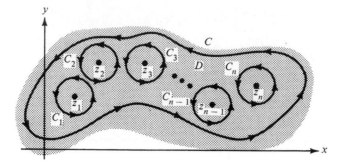

FIGURE 8.1   The domain D and contour C and the singular points $z_1, z_2, \ldots, z_n$ in the statement of Cauchy's residue theorem.

Proof   Since there are a finite number of singular points inside C, there exists an $r > 0$ such that the positively oriented circles $C_k = C_r^+(z_k)$ (for $k = 1, 2, \ldots, n$) are mutually disjoint and all lie inside C. Using Theorem 6.5, the extended Cauchy-Goursat theorem, it follows that

(6) $$\int_C f(z)\, dz = \sum_{k=1}^{n} \int_{C_k} f(z)\, dz.$$

Since f is analytic in a punctured disk with center z_k that contains the circle C_k, equation (4) can be used to obtain

(7) $$\int_{C_k} f(z)\, dz = 2\pi i \operatorname{Res}[f, z_k] \quad \text{for } k = 1, 2, \ldots, n.$$

Using equation (7) in equation (6) results in

$$\int_C f(z)\, dz = 2\pi i \sum_{k=1}^{n} \operatorname{Res}[f, z_k],$$

and the theorem is proven.

8.2   Calculation of Residues

The calculation of a Laurent series expansion is tedious in most circumstances. Since the residue at z_0 involves only the coefficient a_{-1} in the Laurent expansion, we seek a method to calculate the residue from special information about the nature of the singularity at z_0.

If f has a removable singularity at z_0, then $a_{-n} = 0$ for $n = 1, 2, \ldots$. Therefore if z_0 is a removable singularity, then $\operatorname{Res}[f, z_0] = 0$.

Theorem 8.2 (Residues at Poles)

(i) *If f has a simple pole at z_0, then*

(1) $\operatorname{Res}[f, z_0] = \lim_{z \to z_0} (z - z_0) f(z).$

(ii) *If f has a pole of order 2 at z_0, then*

(2) $\operatorname{Res}[f, z_0] = \lim_{z \to z_0} \dfrac{d}{dz} (z - z_0)^2 f(z).$

(iii) *If f has a pole of order k at z_0, then*

(3) $\operatorname{Res}[f, z_0] = \dfrac{1}{(k - 1)!} \lim_{z \to z_0} \dfrac{d^{k-1}}{dz^{k-1}} [(z - z_0)^k f(z)].$

Proof   If f has a simple pole at z_0, then we write

(4) $f(z) = \dfrac{a_{-1}}{z - z_0} + a_0 + a_1(z - z_0) + a_2(z - z_0)^2 + \cdots.$

If we multiply both sides of equation (4) by $(z - z_0)$ and take the limit as $z \to z_0$, then we obtain

$$\lim_{z \to z_0} (z - z_0) f(z) = \lim_{z \to z_0} [a_{-1} + a_0(z - z_0) + a_1(z - z_0)^2 + \cdots]$$
$$= a_{-1} = \operatorname{Res}[f, z_0],$$

and equation (1) is established.

Since equation (2) is a special case of equation (3), let us suppose that f has a pole of order k at z_0. Then f can be written as

(5) $f(z) = \dfrac{a_{-k}}{(z - z_0)^k} + \dfrac{a_{-k+1}}{(z - z_0)^{k-1}} + \cdots + \dfrac{a_{-1}}{z - z_0} + a_0 + a_1(z - z_0) + \cdots.$

If we multiply both sides of equation (5) by $(z - z_0)^k$, then the result is

(6) $(z - z_0)^k f(z) = a_{-k} + \cdots + a_{-1}(z - z_0)^{k-1} + a_0(z - z_0)^k + \cdots.$

We can differentiate both sides of equation (6) $k - 1$ times to obtain

(7) $\dfrac{d^{k-1}}{dz^{k-1}} [(z - z_0)^k f(z)] = (k - 1)! a_{-1} + k! a_0(z - z_0)$

$$+ \dfrac{(k + 1)!}{2} a_1(z - z_0)^2 + \cdots.$$

If we let $z \to z_0$ in equation (7), then

$$\lim_{z \to z_0} \dfrac{d^{k-1}}{dz^{k-1}} [(z - z_0)^k f(z)] = (k - 1)! a_{-1} = (k - 1)! \operatorname{Res}[f, z_0],$$

and equation (3) is established.

EXAMPLE 8.4   Find the residue of $f(z) = \dfrac{\pi \cot \pi z}{z^2}$ at $z_0 = 0$.

Solution   We can write $f(z) = (\pi \cos \pi z)/(z^2 \sin \pi z)$. Since $z^2 \sin \pi z$ has a zero of order 3 at $z_0 = 0$, we see that f has a pole of order 3 at $z_0 = 0$. Therefore using equation (3), we find that

$$\operatorname{Res}[f, 0] = \frac{1}{2!} \lim_{z \to 0} \frac{d^2}{dz^2} \pi z \cot \pi z$$

$$= \frac{1}{2} \lim_{z \to 0} \frac{d}{dz} (\pi \cot \pi z - \pi^2 z \csc^2 \pi z)$$

$$= \pi^2 \lim_{z \to 0} (\pi z \cot \pi z - 1) \csc^2 \pi z$$

$$= \pi^2 \lim_{z \to 0} \frac{\pi z \cos \pi z - \sin \pi z}{\sin^3 \pi z}.$$

This last limit involves an indeterminate form and can be evaluated by using L'Hôpital's rule:

$$\operatorname{Res}[f, 0] = \pi^2 \lim_{z \to 0} \frac{-\pi^2 z \sin \pi z}{3\pi \sin^2 \pi z \cos \pi z}$$

$$= \frac{-\pi^2}{3} \lim_{z \to 0} \frac{\pi z}{\sin \pi z} \lim_{z \to 0} \frac{1}{\cos \pi z} = \frac{-\pi^2}{3}.$$

EXAMPLE 8.5   Find $\int_C [dz/(z^4 + z^3 - 2z^2)]$, where C is the circle $|z| = 3$ taken with the positive orientation.

Solution   The integrand can be written as $f(z) = 1/[z^2(z + 2)(z - 1)]$. The singularities of f that lie inside C are simple poles at the points 1 and -2 and a pole of order 2 at the origin. We compute the residues as follows:

$$\operatorname{Res}[f, 0] = \lim_{z \to 0} \frac{d}{dz} [z^2 f(z)] = \lim_{z \to 0} \frac{-2z - 1}{(z^2 + z - 2)^2} = \frac{-1}{4},$$

$$\operatorname{Res}[f, 1] = \lim_{z \to 1} (z - 1)f(z) = \lim_{z \to 1} \frac{1}{z^2(z + 2)} = \frac{1}{3},$$

$$\operatorname{Res}[f, -2] = \lim_{z \to -2} (z + 2)f(z) = \lim_{z \to -2} \frac{1}{z^2(z - 1)} = \frac{-1}{12}.$$

The value of the integral is now found by using the residue theorem.

$$\int_C \frac{dz}{z^4 + z^3 - 2z^2} = 2\pi i \left[\frac{-1}{4} + \frac{1}{3} - \frac{1}{12} \right] = 0.$$

The value 0 for the integral is not an obvious answer, and all of the preceding calculations are required to find it.

EXAMPLE 8.6 Find $\int_C (z^4 + 4)^{-1} \, dz$, where C is the circle $\left| z - 1 \right| = 2$ taken with the positive orientation.

Solution The singularities of the integrand $f(z) = 1/(z^4 + 4)$ that lie inside C are simple poles that occur at the points $1 \pm i$. (The points $-1 \pm i$ lie outside C.) It is tedious to factor the denominator, so we use a different approach. If z_0 is any one of the singularities of f, then L'Hôpital's rule can be used to compute Res $[f, z_0]$ as follows:

$$\mathrm{Res}[f, z_0] = \lim_{z \to z_0} \frac{z - z_0}{z^4 + 4} = \lim_{z \to z_0} \frac{1}{4z^3} = \frac{1}{4z_0^3}.$$

Since $z_0^4 = -4$, this can be further simplified to yield $\mathrm{Res}[f, z_0] = (-1/16)z_0$. Hence $\mathrm{Res}[f, 1 + i] = (-1 - i)/16$, and $\mathrm{Res}[f, 1 - i] = (-1 + i)/16$. The residue theorem can now be used to obtain

$$\int_C \frac{dz}{z^4 + 4} = 2\pi i \left(\frac{-1 - i}{16} + \frac{-1 + i}{16} \right) = \frac{-\pi i}{4}.$$

The theory of residues can be used to expand the quotient of two polynomials into its *partial fraction* representation.

> **Lemma 8.1** *Let $P(z)$ be a polynomial of degree at most 2. If a, b, and c are distinct complex numbers, then*

$$(8) \quad f(z) = \frac{P(z)}{(z - a)(z - b)(z - c)} = \frac{A}{z - a} + \frac{B}{z - b} + \frac{C}{z - c}$$

where

$$A = \mathrm{Res}[f, a] = \frac{P(a)}{(a - b)(a - c)},$$

$$B = \mathrm{Res}[f, b] = \frac{P(b)}{(b - a)(b - c)},$$

$$C = \mathrm{Res}[f, c] = \frac{P(c)}{(c - a)(c - b)}.$$

Proof It will suffice to prove that $A = \mathrm{Res}[f, a]$. We can expand f in its Laurent series about the point $z = a$ by expanding the three terms on the right side of equation (8) in their Laurent series about $z = a$ and adding them. The term $A/(z - a)$ is itself a one-term Laurent series. The term $B/(z - b)$ is analytic at $z = a$, and its Laurent series is actually a Taylor series,

$$(9) \quad \frac{B}{z - b} = \frac{-B}{b - a} \frac{1}{1 - \dfrac{z - a}{b - a}} = -\sum_{n=0}^{\infty} \frac{B}{(b - a)^{n+1}} (z - a)^n.$$

The expansion for the term $C/(z - c)$ is given by

(10) $\dfrac{C}{z - c} = -\displaystyle\sum_{n=0}^{\infty} \dfrac{C}{(c - a)^{n+1}} (z - a)^n.$

We can substitute equations (9) and (10) into equation (8) to obtain

$$f(z) = \dfrac{A}{z - a} - \sum_{n=0}^{\infty} \left[\dfrac{B}{(b - a)^{n+1}} + \dfrac{C}{(c - a)^{n+1}} \right] (z - a)^n.$$

Therefore $A = \text{Res}[f, a]$, and calculation reveals that

$$\text{Res}[f, a] = \lim_{z \to a} \dfrac{P(z)}{(z - b)(z - c)} = \dfrac{P(a)}{(a - b)(a - c)}.$$

EXAMPLE 8.7   Express $f(z) = \dfrac{3z + 2}{z(z - 1)(z - 2)}$ in partial fractions.

Solution   Computing the residues, we obtain

$$\text{Res}[f, 0] = 1, \quad \text{Res}[f, 1] = -5, \quad \text{Res}[f, 2] = 4.$$

Therefore

$$\dfrac{3z + 2}{z(z - 1)(z - 2)} = \dfrac{1}{z} - \dfrac{5}{z - 1} + \dfrac{4}{z - 2}.$$

If a repeated root occurs, then the process is similar.

Lemma 8.2   *If $P(z)$ has degree at most 2, then*

(11) $f(z) = \dfrac{P(z)}{(z - a)^2(z - b)} = \dfrac{A}{(z - a)^2} + \dfrac{B}{z - a} + \dfrac{C}{z - b}$

where $A = \text{Res}[(z - a)f(z), a]$, $B = \text{Res}[f, a]$, and $C = \text{Res}[f, b]$.

EXAMPLE 8.8   Express $f(z) = \dfrac{z^2 + 3z + 2}{z^2(z - 1)}$ in partial fractions.

Solution   Calculating the residues we find that

$$\text{Res}[zf(z), 0] = \lim_{z \to 0} \dfrac{z^2 + 3z + 2}{z - 1} = -2,$$

$$\text{Res}[f, 0] = \lim_{z \to 0} \dfrac{d}{dz} \dfrac{z^2 + 3z + 2}{z - 1}$$

$$= \lim_{z \to 0} \dfrac{(2z + 3)(z - 1) - (z^2 + 3z + 2)}{(z - 1)^2} = -5,$$

$$\text{Res}[f, 1] = \lim_{z \to 1} \dfrac{z^2 + 3z + 2}{z^2} = 6.$$

Hence $A = -2$, $B = -5$, and $C = 6$, and we can use equation (11) to obtain

$$\frac{z^2 + 3z + 2}{z^2(z - 1)} = \frac{-2}{z^2} - \frac{5}{z} + \frac{6}{z - 1}.$$

EXERCISES FOR SECTION 8.2

Find Res$[f,0]$ for the functions in Exercises 1–4.

1. (a) $z^{-1}\exp z$ (b) $z^{-3}\cosh 4z$
    (c) csc z (d) $(z^2 + 4z + 5)/(z^2 + z)$

2. (a) cot z (b) $z^{-3}\cos z$ (c) $z^{-1}\sin z$ (d) $(z^2 + 4z + 5)/z^3$

3. (a) $\exp(1 + 1/z)$ (b) $z^4\sin(1/z)$ (c) z^{-1}csc z

4. (a) z^{-2}csc z (b) $(\exp 4z - 1)/\sin^2 z$ (c) z^{-1}csc$^2 z$

For Exercises 5–15, assume that the contour C has positive orientation.

5. Find $\displaystyle\int_C \frac{dz}{z^4 + 4}$, where C is the circle $|z + 1 - i| = 1$.

6. Find $\displaystyle\int_C \frac{dz}{z(z^2 - 2z + 2)}$, where C is the circle $|z - i| = 2$.

7. Find $\displaystyle\int_C \frac{\exp z\, dz}{z^3 + z}$, where C is the circle $|z| = 2$.

8. Find $\displaystyle\int_C \frac{\sin z\, dz}{4z^2 - \pi^2}$, where C is the circle $|z| = 2$.

9. Find $\displaystyle\int_C \frac{\sin z\, dz}{z^2 + 1}$, where C is the circle $|z| = 2$.

10. Find $\int_C (z - 1)^{-2}(z^2 + 4)^{-1}\, dz$ along the following contours:
    (a) the circle $|z| = 4$ (b) the circle $|z - 1| = 1$

11. Find $\int_C (z^6 + 1)^{-1}dz$ along the following contours:
    (a) the circle $|z - i| = \frac{1}{2}$ (b) the circle $|z - (1 + i)/2| = 1$
    Hint: If z_0 is a singularity of $f(z) = 1/(z^6 + 1)$, then show that Res$[f, z_0] = (-1/6)z_0$.

12. Find $\int_C (3z^4 + 10z^2 + 3)^{-1}\, dz$ along the following contours:
    (a) the circle $|z - i\sqrt{3}| = 1$ (b) the circle $|z - i/\sqrt{3}| = 1$

13. Find $\int_C (z^4 - z^3 - 2z^2)^{-1}\, dz$ along the following contours:
    (a) the circle $|z| = \frac{1}{2}$ (b) the circle $|z| = \frac{3}{2}$

14. Find $\displaystyle\int_C \frac{dz}{z^2\sin z}$, where C is the circle $|z| = 1$.

15. Find $\displaystyle\int_C \frac{dz}{z\sin^2 z}$, where C is the circle $|z| = 1$.

16. Let f and g have an isolated singularity at z_0. Show that Res$[f + g, z_0] = $ Res$[f, z_0] + $ Res$[g, z_0]$.

17. Let f and g be analytic at z_0. If $f(z_0) \neq 0$ and g has a simple zero at z_0, then show that

$$\text{Res}\left[\frac{f}{g}, z_0\right] = \frac{f(z_0)}{g'(z_0)}.$$

18. Use residues to find the partial fraction representations of the following functions.

(a) $\dfrac{1}{z^2 + 3z + 2}$

(b) $\dfrac{3z - 3}{z^2 - z - 2}$

(c) $\dfrac{z^2 - 7z + 4)}{z^2(z + 4)}$

(d) $\dfrac{10z}{(z^2 + 4)(z^2 + 9)}$

(e) $\dfrac{2z^2 - 3z - 1}{(z - 1)^3}$

(f) $\dfrac{z^3 + 3z^2 - z + 1}{z(z + 1)^2(z^2 + 1)}$

19. Let f be analytic in a simply connected domain D, and let C be a simply closed positively oriented contour in D. If z_0 is the only zero of f in D and z_0 lies interior to C, then show that

$$\frac{1}{2\pi i} \int_C \frac{f'(z)}{f(z)}\, dz = k$$

where k is the order of the zero at z_0.

20. Let f be analytic at the points $z = 0, \pm 1, \pm 2, \ldots\ldots$. If $g(z) = \pi f(z) \cot \pi z$, then show that

$$\text{Res}[\, g, n] = f(n) \quad \text{for } n = 0, \pm 1, \pm 2, \ldots\ldots$$

21. Write a report on how complex analysis is used in the study of partial fractions. Resources include bibliographical items 10 and 63.

22. Write a report on residue theorem. Include ideas and examples that are not mentioned in the text. Resources include bibliographical items 22, 116, and 153.

8.3   Trigonometric Integrals

The evaluation of certain definite integrals can be accomplished with the aid of the residue theorem. If the definite integral can be interpreted as the parametric form of a contour integral of an analytic function along a simple closed contour, then the residue theorem can be used to evaluate the equivalent complex integral.

The method in this section can be used to evaluate integrals of the form

(1) $\displaystyle\int_0^{2\pi} F(\cos\theta, \sin\theta)\, d\theta,$

where $F(u, v)$ is a function of the two real variables u and v. Let us consider the contour C that consists of the unit circle $|z| = 1$, taken with the parameterization

(2) $C\colon z = \cos\theta + i\sin\theta, \quad dz = (-\sin\theta + i\cos\theta)\, d\theta \quad \text{for } 0 \le \theta \le 2\pi.$

Using $1/z = \cos\theta - i\sin\theta$ and (2), we can obtain

(3) $\cos\theta = \dfrac{1}{2}\left(z + \dfrac{1}{z}\right), \quad \sin\theta = \dfrac{1}{2i}\left(z - \dfrac{1}{z}\right), \quad \text{and} \quad d\theta = \dfrac{dz}{iz}.$

If we use the substitutions (3) in expression (1), then the definite integral is transformed into a contour integral

(4) $\displaystyle\int_0^{2\pi} F(\cos\theta, \sin\theta)\, d\theta = \int_C f(z)\, dz,$

where the new integrand is

(5) $f(z) = \dfrac{F\left(\dfrac{1}{2}\left(z + \dfrac{1}{z}\right), \dfrac{1}{2i}\left(z - \dfrac{1}{z}\right)\right)}{iz}.$

Suppose that f is analytic for $|z| \le 1$, except at the points $z_1, z_2, \ldots, z_n$ that lie interior to C. Then the residue theorem can be used to conclude that

(6) $\displaystyle\int_0^{2\pi} F(\cos\theta, \sin\theta)\, d\theta = 2\pi i \sum_{k=1}^{n} \text{Res}[f, z_k].$

The situation is illustrated in Figure 8.2.

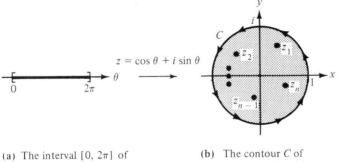

(a) The interval $[0, 2\pi]$ of
     integration for $F(\cos\theta, \sin\theta)$.

(b) The contour C of
     integration for $f(z)$.

FIGURE 8.2   The change of variables from a definite integral on $[0, 2\pi]$ to a contour integral around C.

EXAMPLE 8.9   Show that $\displaystyle\int_0^{2\pi} \dfrac{d\theta}{1 + 3\cos^2\theta} = \pi.$

Solution   The complex integrand f of equation (5) is given by

$$f(z) = \frac{1}{iz[1 + \frac{3}{4}(z + z^{-1})^2]} = \frac{-i4z}{3z^4 + 10z^2 + 3}.$$

The singularities of f are poles that are located at the points where $3(z^2)^2 + 10(z^2) + 3 = 0$. The quadratic formula can be used to see that the singular points of f satisfy the relation $z^2 = (-10 \pm \sqrt{100 - 36})/6 = (-5 \pm 4)/3$. Hence the only singularities of f that lie inside the circle C: $|z| = 1$ are simple poles located at the two points $z_1 = i/\sqrt{3}$ and $z_2 = -i/\sqrt{3}$. Theorem 8.2 with the aid of L'Hôpital's rule can be used to calculate the residues of f at z_k (for $k = 1, 2$) as follows:

$$\begin{aligned}
\text{Res}[f, z_k] &= \lim_{z \to z_k} \frac{-i4z(z - z_k)}{3z^4 + 10z^2 + 3} \\
&= \lim_{z \to z_k} \frac{-i4(2z - z_k)}{12z^3 + 20z} \\
&= \frac{-i4z_k}{12z_k^3 + 20z_k} = \frac{-i}{3z_k^2 + 5}.
\end{aligned}$$

Since $z_k = \pm i/\sqrt{3}$ and $z_k^2 = -1/3$, we see that the residues are given by $\text{Res}[f, z_k] = -i/(3(-1/3) + 5) = -i/4$. Equation (6) can now be used to compute the value of the integral

$$\int_0^{2\pi} \frac{d\theta}{1 + 3\cos^2\theta} = 2\pi i\left(\frac{-i}{4} + \frac{-i}{4}\right) = \pi.$$

EXAMPLE 8.9* Show that $\displaystyle\int_0^{2\pi} \frac{dt}{1 + 3\cos^2 t} = \pi.$

Solution   Using a Computer Algebra System
The indefinite integral, or antiderivative can be obtained by using software such as Mathematica or MAPLE. It is

$$g(t) = \frac{-\arctan(2\cot t)}{2}.$$

Since $\cot 0$ and $\cot 2\pi$ are not defined, the computations for both $g(0)$ and $g(2\pi)$ are indeterminate. The graph $s = g(t)$ is shown in Figure 8.3 and reveals another problem: $g(t)$ has a discontinuity at $t = \pi$. This is a violation of the fundamental theorem of calculus, which asserts that the integral of a continuous function over $(0, 2\pi)$ must be continuous. The integration algorithm used by computer algebra systems (the Risch-Norman algorithm) gives the preceding antiderivative $g(t)$ and all mathematicians should beware.

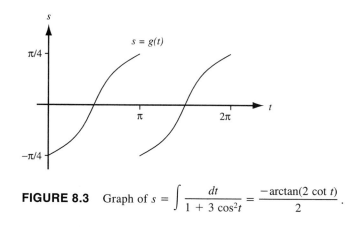

FIGURE 8.3   Graph of $s = \displaystyle\int \frac{dt}{1 + 3\cos^2 t} = \frac{-\arctan(2\cot t)}{2}.$

The proper value for the definite integral can be obtained by using $g(t)$ on the open subintervals $(0, \pi)$ and $(\pi, 2\pi)$ where it is continuous. Limits must be used over $(0, \pi)$ and $(\pi, 2\pi)$. Therefore, the value of the definite integral is

$$\int_0^{2\pi} \frac{dt}{1 + 3\cos^2 t} = \int_0^{\pi} \frac{dt}{1 + 3\cos^2 t} + \int_{\pi}^{2\pi} \frac{dt}{1 + 3\cos^2 t}$$

$$= \lim_{t \to \pi^-} g(t) - \lim_{t \to 0^+} g(t) + \lim_{t \to 2\pi^-} g(t) - \lim_{t \to \pi^+} g(t)$$

$$= \frac{\pi}{4} - \frac{-\pi}{4} + \frac{\pi}{4} - \frac{-\pi}{4} = \pi.$$

EXAMPLE 8.10   Show that $\displaystyle\int_0^{2\pi} \frac{\cos 2\theta \, d\theta}{5 - 4\cos\theta} = \frac{\pi}{6}$.

Solution   For values of z that lie on the circle C: $|z| = 1$, we have

$$z^2 = \cos 2\theta + i\sin 2\theta \quad \text{and} \quad z^{-2} = \cos 2\theta - i\sin 2\theta.$$

We can solve for $\cos 2\theta$ and $\sin 2\theta$ to obtain the substitutions

(7) $\cos 2\theta = \dfrac{1}{2}(z^2 + z^{-2})$ and $\sin 2\theta = \dfrac{1}{2i}(z^2 - z^{-2})$.

Using the substitutions in equations (3) and (7), we find that the complex integrand f in equation (5) can be written as

$$f(z) = \frac{\frac{1}{2}(z^2 + z^{-2})}{iz[5 - 2(z + z^{-1})]} = \frac{i(z^4 + 1)}{2z^2(z - 2)(2z - 1)}.$$

The singularities of f that lie inside C are poles that are located at the points $z_1 = 0$ and $x_2 = \frac{1}{2}$. Using Theorem 8.2 to calculate the residues results in

$$\text{Res}[f, 0] = \lim_{z \to 0} \frac{d}{dz} z^2 f(z) = \lim_{z \to 0} \frac{d}{dz} \frac{i(z^4 + 1)}{2(2z^2 - 5z + 2)}$$

$$= \lim_{z \to 0} i \frac{4z^3(2z^2 - 5z + 2) - (4z - 5)(z^4 + 1)}{2(2z^2 - 5z + 2)^2} = \frac{5i}{8}$$

and

$$\text{Res}[f, \tfrac{1}{2}] = \lim_{z \to 1/2} (z - \tfrac{1}{2})f(z) = \lim_{z \to 1/2} \frac{i(z^4 + 1)}{4z^2(z - 2)} = \frac{-17i}{24}.$$

Therefore using equation (6), we conclude that

$$\int_0^{2\pi} \frac{\cos 2\theta \, d\theta}{5 - 4\cos\theta} = 2\pi i \left(\frac{5i}{8} - \frac{17i}{24} \right) = \frac{\pi}{6}.$$

EXERCISES FOR SECTION 8.3

Use residues to find the following:

1. $\int_0^{2\pi} \dfrac{d\theta}{3 \cos \theta + 5} = \dfrac{\pi}{2}$

2. $\int_0^{2\pi} \dfrac{d\theta}{4 \sin \theta + 5}$

3. $\int_0^{2\pi} \dfrac{d\theta}{15 \sin^2\theta + 1} = \dfrac{\pi}{2}$

4. $\int_0^{2\pi} \dfrac{d\theta}{5 \cos^2\theta + 4}$

5. $\int_0^{2\pi} \dfrac{\sin^2\theta \, d\theta}{5 + 4 \cos \theta} = \dfrac{\pi}{4}$

6. $\int_0^{2\pi} \dfrac{\sin^2\theta \, d\theta}{5 - 3 \cos \theta}$

7. $\int_0^{2\pi} \dfrac{d\theta}{(5 + 3 \cos \theta)^2} = \dfrac{5\pi}{32}$

8. $\int_0^{2\pi} \dfrac{d\theta}{(5 + 4 \cos \theta)^2}$

9. $\int_0^{2\pi} \dfrac{\cos 2\theta \, d\theta}{5 + 3 \cos \theta} = \dfrac{\pi}{18}$

10. $\int_0^{2\pi} \dfrac{\cos 2\theta \, d\theta}{13 - 12 \cos \theta}$

11. $\int_0^{2\pi} \dfrac{d\theta}{(1 + 3 \cos^2\theta)^2} = \dfrac{5\pi}{8}$

12. $\int_0^{2\pi} \dfrac{d\theta}{(1 + 8 \cos^2\theta)^2}$

13. $\int_0^{2\pi} \dfrac{\cos^2 3\theta \, d\theta}{5 - 4 \cos 2\theta} = \dfrac{3\pi}{8}$

14. $\int_0^{2\pi} \dfrac{\cos^2 3\theta \, d\theta}{5 - 3 \cos 2\theta}$

15. $\int_0^{2\pi} \dfrac{d\theta}{a \cos \theta + b \sin \theta + d} = \dfrac{2\pi}{\sqrt{d^2 - a^2 - b^2}}$,

    where a, b, and d are real and $a^2 + b^2 < d^2$

16. $\int_0^{2\pi} \dfrac{d\theta}{a \cos^2\theta + b \sin^2\theta + d} = \dfrac{2\pi}{\sqrt{(a + d)(b + d)}}$,

    where a, b, and d are real and $a > d$ and $b > d$

17. Compare the complex analysis methods for evaluating trigonometric integrals and the methods learned in calculus.

8.4   Improper Integrals of Rational Functions

An important application of the theory of residues is the evaluation of certain types of improper integrals. Let $f(x)$ be a continuous function of the real variable x on the interval $0 \le x < \infty$. Recall from calculus that the improper integral of f over $[0, \infty)$ is defined by

(1) $\displaystyle \int_0^\infty f(x) \, dx = \lim_{b \to \infty} \int_0^b f(x) \, dx$

provided that the limit exists. If f is defined for all real x, then the integral of f over $(-\infty, \infty)$ is defined by

(2) $\displaystyle \int_{-\infty}^\infty f(x) \, dx = \lim_{a \to -\infty} \int_a^0 f(x) \, dx + \lim_{b \to \infty} \int_0^b f(x) \, dx$

provided that both limits exist. If the integral in equation (2) exists, then its value can be obtained by taking a single limit as follows:

(3) $\displaystyle \int_{-\infty}^\infty f(x) \, dx = \lim_{R \to \infty} \int_{-R}^R f(x) \, dx.$

However, for some functions the limit on the right side of equation (3) exists when definition (2) does not exist.

EXAMPLE 8.11 $\lim_{R\to\infty} \int_{-R}^{R} x\, dx = \lim_{R\to\infty} [R^2/2 - (-R)^2/2] = 0$, but the improper integral of $f(x) = x$ over $(-\infty, \infty)$ does not exist. Therefore equation (3) can be used to extend the notion of the value of an improper integral and motivates us to make the following definition.

Let $f(x)$ be a continuous real valued function for all x. The *Cauchy principal value* (P.V.) of the integral (2) is defined by

$$(4) \quad \text{P.V.} \int_{-\infty}^{\infty} f(x)\, dx = \lim_{R\to\infty} \int_{-R}^{R} f(x)\, dx$$

provided that the limit exists. Therefore Example 8.11 shows that

$$\text{P.V.} \int_{-\infty}^{\infty} x\, dx = 0.$$

EXAMPLE 8.12

$$\int_{-\infty}^{\infty} \frac{dx}{x^2+1} = \lim_{R\to\infty} \int_{-R}^{R} \frac{dx}{x^2+1}$$
$$= \lim_{R\to\infty} [\arctan R - \arctan(-R)]$$
$$= \frac{\pi}{2} - \frac{-\pi}{2} = \pi.$$

If $f(x) = P(x)/Q(x)$, where P and Q are polynomials, then f is called a *rational function*. Techniques in calculus were developed to integrate rational functions. We now show how the residue theorem can be used to obtain the Cauchy principal value of the integral of f over $(-\infty, \infty)$.

Theorem 8.3   *Let $f(z) = P(z)/Q(z)$ where P and Q are polynomials of degree m and n, respectively. If $Q(x) \neq 0$ for all real x and $n \geq m + 2$, then*

$$(5) \quad \text{P.V.} \int_{-\infty}^{\infty} \frac{P(x)}{Q(x)}\, dx = 2\pi i \sum_{j=1}^{k} \text{Res}\left[\frac{P}{Q}, z_j\right]$$

where $z_1, z_2, \ldots, z_{k-1}$, and z_k are the poles of P/Q that lie in the upper half plane. The situation is illustrated in Figure 8.4.

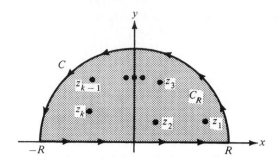

FIGURE 8.4   The poles $z_1, z_2, \ldots, z_{k-1}, z_k$ of P/Q that lie in the upper half plane.

Proof   Since there are a finite number of poles of P/Q that lie in the upper half plane, a real number R can be found such that the poles all lie inside the contour C, which consists of the segment $-R \le x \le R$ of the x axis together with the upper semicircle C_R of radius R shown in Figure 8.4. Property (17) in Section 6.2 can be used to write

$$(6) \qquad \int_{-R}^{R} \frac{P(x)}{Q(x)}\, dx = \int_{C} \frac{P(z)}{Q(z)}\, dz - \int_{C_R} \frac{P(z)}{Q(z)}\, dz.$$

The residue theorem can be used to express equation (6) in the form

$$(7) \qquad \int_{-R}^{R} \frac{P(x)}{Q(x)}\, dx = 2\pi i \sum_{j=1}^{k} \operatorname{Res}\left[\frac{P}{Q}, z_j\right] - \int_{C_R} \frac{P(z)}{Q(z)}\, dz.$$

The result will be established if we can show that the integral of $P(z)/Q(z)$ along C_R on the right side of equation (7) goes to zero as $R \to \infty$. Since we have $n \ge m + 2$, the degree of the polynomial $Q(z)$ is greater than the degree of $zP(z)$. Suppose that

$$P(z) = a_m z^m + a_{m-1} z^{m-1} + \cdots + a_1 z + a_0$$

and

$$Q(z) = b_n z^n + b_{n-1} z^{n-1} + \cdots + b_1 z + b_0.$$

Then

$$P(z) = z^m P_1(z), \quad \text{where}$$
$$P_1(z) = a_m + a_{m-1} z^{-1} + \cdots + a_1 z^{-m+1} + a_0 z^{-m},$$

and

$$Q(z) = z^n Q_1(z), \quad \text{where}$$
$$Q_1(z) = b_n + b_{n-1} z^{-1} + \cdots + b_1 z^{-n+1} + b_0 z^{-n}.$$

Therefore we have

$$(8) \qquad \frac{zP(z)}{Q(z)} = \frac{z^{m+1} P_1(z)}{z^n Q_1(z)}.$$

Since $P_1(z) \to a_m$ and $Q_1(z) \to b_n$ as $|z| \to \infty$ and $n \geq m + 2$, we can use equation (8) to see that

$$\left| \frac{zP(z)}{Q(z)} \right| \to 0 \quad \text{as } |z| \to \infty.$$

Therefore for any $\varepsilon > 0$ we may choose R large enough that

$$\left| \frac{zP(z)}{Q(z)} \right| < \frac{\varepsilon}{\pi}$$

whenever z lies on C_R. Therefore we have

(9) $$\left| \frac{P(z)}{Q(z)} \right| < \frac{\varepsilon}{\pi |z|} = \frac{\varepsilon}{\pi R} \quad \text{whenever } z \text{ lies on } C_R.$$

Using the ML inequality of Section 6.2 and the result of inequality (9), we obtain the estimate

(10) $$\left| \int_{C_R} \frac{P(z)}{Q(z)} \, dz \right| \leq \int_{C_R} \frac{\varepsilon}{\pi R} \, |dz| = \frac{\varepsilon}{\pi R} \pi R = \varepsilon.$$

Since $\varepsilon > 0$ was arbitrary, inequality (10) shows that

(11) $$\lim_{R \to \infty} \int_{C_R} \frac{P(z)}{Q(z)} \, dz = 0.$$

We can use equation (11) in equation (7) and use definition (4) to conclude that

$$\text{P.V.} \int_{-\infty}^{\infty} \frac{P(x)}{Q(x)} \, dx = \lim_{R \to \infty} \int_{-R}^{R} \frac{P(x)}{Q(x)} \, dx = 2\pi i \sum_{j=1}^{k} \text{Res}\left[\frac{P}{Q}, z_j \right],$$

and the theorem is proven.

EXAMPLE 8.13 Show that $\displaystyle\int_{-\infty}^{\infty} \frac{dx}{(x^2 + 1)(x^2 + 4)} = \frac{\pi}{6}$.

Solution The integrand can be written in the form

$$f(z) = \frac{1}{(z + i)(z - i)(z + 2i)(z - 2i)}.$$

We see that f has simple poles at the points $z_1 = i$ and $z_2 = 2i$ in the upper half-plane. Computing the residues, we obtain

$$\text{Res}[f, i] = \frac{-i}{6} \quad \text{and} \quad \text{Res}[f, 2i] = \frac{i}{12}.$$

Using Theorem 8.3, we conclude that

$$\int_{-\infty}^{\infty} \frac{dx}{(x^2 + 1)(x^2 + 4)} = 2\pi i \left(\frac{-i}{6} + \frac{i}{12} \right) = \frac{\pi}{6}.$$

EXAMPLE 8.14   Show that $\displaystyle\int_{-\infty}^{\infty} \frac{dx}{(x^2 + 4)^3} = \frac{3\pi}{256}$.

Solution   The integrand $f(z) = 1/(z^2 + 4)^3$ has a pole of order 3 at the point $z_1 = 2i$. Computing the residue, we find that

$$
\begin{aligned}
\text{Res}[f, 2i] &= \frac{1}{2} \lim_{z \to 2i} \frac{d^2}{dz^2} \frac{1}{(z + 2i)^3} \\
&= \frac{1}{2} \lim_{z \to 2i} \frac{d}{dz} \frac{-3}{(z + 2i)^4} \\
&= \frac{1}{2} \lim_{z \to 2i} \frac{12}{(z + 2i)^5} = \frac{-3i}{512}.
\end{aligned}
$$

Therefore $\displaystyle\int_{-\infty}^{\infty} \frac{dx}{(x^2 + 4)^3} = 2\pi i \frac{-3i}{512} = \frac{3\pi}{256}$.

EXERCISES FOR SECTION 8.4

Use residues to establish the values of the integrals in Exercises 1–15.

1. $\displaystyle\int_{-\infty}^{\infty} \frac{x^2\, dx}{(x^2 + 16)^2} = \frac{\pi}{8}$

2. $\displaystyle\int_{-\infty}^{\infty} \frac{dz}{x^2 + 16}$

3. $\displaystyle\int_{-\infty}^{\infty} \frac{x\, dx}{(x^2 + 9)^2} = 0$

4. $\displaystyle\int_{-\infty}^{\infty} \frac{x + 3}{(x^2 + 9)^2}\, dx$

5. $\displaystyle\int_{-\infty}^{\infty} \frac{2x^2 + 3}{(x^2 + 9)^2}\, dx = \frac{7\pi}{18}$

6. $\displaystyle\int_{-\infty}^{\infty} \frac{dx}{x^4 + 4}$

7. $\displaystyle\int_{-\infty}^{\infty} \frac{x^2\, dx}{x^4 + 4} = \frac{\pi}{2}$

8. $\displaystyle\int_{-\infty}^{\infty} \frac{x^2\, dx}{(x^2 + 4)^3}$

9. $\displaystyle\int_{-\infty}^{\infty} \frac{dx}{(x^2 + 1)^2(x^2 + 4)} = \frac{\pi}{9}$

10. $\displaystyle\int_{-\infty}^{\infty} \frac{x + 2}{(x^2 + 4)(x^2 + 9)}\, dx$

11. $\displaystyle\int_{-\infty}^{\infty} \frac{3x^2 + 2}{(x^2 + 4)(x^2 + 9)}\, dx = \frac{2\pi}{3}$

12. $\displaystyle\int_{-\infty}^{\infty} \frac{dx}{x^6 + 1}$

13. $\displaystyle\int_{-\infty}^{\infty} \frac{x^4\, dx}{x^6 + 1} = \frac{2\pi}{3}$

14. $\displaystyle\int_{-\infty}^{\infty} \frac{dx}{(x^2 + a^2)(x^2 + b^2)} = \frac{\pi}{ab(a + b)}$, where $a > 0$ and $b > 0$

15. $\displaystyle\int_{-\infty}^{\infty} \frac{x^2\, dx}{(x^2 + a^2)^3} = \frac{\pi}{8a^3}$, where $a > 0$

8.5   Improper Integrals Involving Trigonometric Functions

Let P and Q be polynomials of degree m and n, respectively, where $n \geq m + 1$. If $Q(x) \neq 0$ for all real x, then

$$
\text{P.V.} \int_{-\infty}^{\infty} \frac{P(x)}{Q(x)} \cos x\, dx \quad \text{and} \quad \text{P.V.} \int_{-\infty}^{\infty} \frac{P(x)}{Q(x)} \sin x\, dx
$$

are convergent improper integrals. Integrals of this type are sometimes encountered in the study of Fourier transforms and Fourier integrals. We now show how these improper integrals can be evaluated.

It is of particular importance to observe that we will be using identities

(1) $\cos(\alpha x) = \text{Re}[\exp(i\alpha x)]$ and $\sin(\alpha x) = \text{Im}[\exp(i\alpha x)]$

where α is a positive real number. The crucial step in the proof of Theorem 8.4 will not hold if $\cos(\alpha z)$ and $\sin(\alpha z)$ are used instead of $\exp(i\alpha z)$. Lemma 8.3 will give the details.

> **Theorem 8.4**   *Let P and Q be polynomials with real coefficients of degree m and n, respectively, where $n \geq m + 1$, and $Q(x) \neq 0$ for all real x. If $\alpha > 0$ and*
>
> (2) $f(z) = \dfrac{\exp(i\alpha z)P(z)}{Q(z)}$,
>
> *then*
>
> (3) $\text{P.V.} \displaystyle\int_{-\infty}^{\infty} \dfrac{P(x)}{Q(x)} \cos(\alpha x)\, dx = -2\pi \sum_{j=1}^{k} \text{Im}\left(\text{Res}\,[f, z_j] \right)$ *and*
>
> (4) $\text{P.V.} \displaystyle\int_{-\infty}^{\infty} \dfrac{P(x)}{Q(x)} \sin(\alpha x)\, dx = -2\pi \sum_{j=1}^{k} \text{Re}\left(\text{Res}\,[f, z_j] \right)$
>
> *where $z_1, z_2, \ldots, z_{k-1}, z_k$ are the poles of f that lie in the upper half-plane and where $\text{Re}(\text{Res}[f, z_j])$ and $\text{Im}(\text{Res}[f, z_j])$ are the real and imaginary parts of $\text{Res}[f, z_j]$, respectively.*

The proof of the theorem is similar to the proof of Theorem 8.3. Before we turn to the proof, let us first give some examples.

EXAMPLE 8.15   Show that P.V. $\displaystyle\int_{-\infty}^{\infty} \dfrac{x \sin x\, dx}{x^2 + 4} = \dfrac{\pi}{e^2}$.

Solution   The function f in equation (2) is $f(z) = z \exp(iz)/(z^2 + 4)$ and has a simple pole at the point $z_1 = 2i$ in the upper half-plane. Calculating the residue results in

$$\text{Res}[f, 2i] = \lim_{z \to 2i} \frac{z \exp(iz)}{z + 2i} = \frac{2ie^{-2}}{4i} = \frac{1}{2e^2} .$$

Using equation (4), we find that

$$\text{P.V.} \int_{-\infty}^{\infty} \frac{x \sin x\, dx}{x^2 + 4} = 2\pi \,\text{Re}\{\text{Res}[f, 2i]\} = \frac{\pi}{e^2} .$$

EXAMPLE 8.16   Show that $\displaystyle\int_{-\infty}^{\infty} \frac{\cos x \, dx}{x^4 + 4} = \frac{\pi(\cos 1 + \sin 1)}{4e}$.

Solution   The complex function f in equation (2) is $f(z) = \exp(iz)/(z^4 + 4)$ and has simple poles at the points $z_1 = 1 + i$ and $z_2 = -1 + i$ in the upper half-plane. The residues are found with the aid of L'Hôpital's rule.

$$\begin{aligned}
\operatorname{Res}[f, 1 + i] &= \lim_{z \to 1+i} \frac{(z - 1 - i)\exp(iz)}{z^4 + 4} = \frac{0}{0} \\
&= \lim_{z \to 1+i} \frac{[1 + i(z - 1 - i)]\exp(iz)}{4z^3} \\
&= \frac{\exp(-1 + i)}{4(1 + i)^3} = \frac{\sin 1 - \cos 1 - i(\cos 1 + \sin 1)}{16e}.
\end{aligned}$$

Similarly,

$$\operatorname{Res}[f, -1 + i] = \frac{\cos 1 - \sin 1 - i(\cos 1 + \sin 1)}{16e}.$$

Using equation (3), we find that

$$\begin{aligned}
\int_{-\infty}^{\infty} \frac{\cos x \, dx}{x^4 + 4} &= -2\pi[\operatorname{Im}(\operatorname{Res}[f, 1 + i]) + \operatorname{Im}(\operatorname{Res}[f, -1 + i])] \\
&= \frac{\pi(\cos 1 + \sin 1)}{4e}.
\end{aligned}$$

We now turn to the proof of Theorem 8.4, a theorem that depends on the following result.

Lemma 8.3 (Jordan's Lemma)   *Let P and Q be polynomials with real coefficients of degree m and n, respectively, where $n \geq m + 1$. If C_R is the upper semicircle $z = Re^{i\theta}$ for $0 \leq \theta \leq \pi$, then*

(5) $$\lim_{R \to \infty} \int_{C_R} \frac{\exp(iz)P(z)}{Q(z)} \, dz = 0.$$

Proof   Since $n \geq m + 1$, it follows that $|P(z)/Q(z)| \to 0$ as $|z| \to \infty$. Therefore for $\varepsilon > 0$ given there exists an $R_\varepsilon > 0$ such that

(6) $$\left| \frac{P(z)}{Q(z)} \right| < \frac{\varepsilon}{\pi} \quad \text{whenever } |z| \geq R_\varepsilon.$$

Using inequality (22) of Section 6.2 together with inequality (6), we obtain the estimate

(7) $$\left| \int_{C_R} \frac{\exp(iz)P(z)}{Q(z)} \, dz \right| \leq \int_{C_R} \frac{\varepsilon}{\pi} |e^{iz}| |dz|, \quad \text{where } R \geq R_\varepsilon.$$

The parameterization of C_R leads to the equations

(8) $|dz| = R \, d\theta$ and $|e^{iz}| = e^{-y} = e^{-R \sin \theta}$.

Using the trigonometric identity $\sin(\pi - \theta) = \sin \theta$ and equations (8), we can express the integral on the right side of inequality (7) as

(9) $\displaystyle \int_{C_R} \frac{\varepsilon}{\pi} \, |e^{iz}| \, |dz| = \frac{\varepsilon}{\pi} \int_0^\pi e^{-R \sin \theta} R \, d\theta = \frac{2\varepsilon}{\pi} \int_0^{\pi/2} e^{-R \sin \theta} R \, d\theta$.

On the interval $0 \le \theta \le \pi/2$ we can use the inequality

(10) $0 \le \dfrac{2\theta}{\pi} \le \sin \theta$.

We can combine the results of inequalities (7) and (10) and equation (9) to conclude that, for $R \ge R_\varepsilon$,

$$\left| \int_{C_R} \frac{\exp(iz)P(z) \, dz}{Q(z)} \right| \le \frac{2\varepsilon}{\pi} \int_0^{\pi/2} e^{-2R\theta/\pi} R \, d\theta$$

$$= -\varepsilon e^{-2R\theta/\pi} \Big|_0^{\pi/2} < \varepsilon.$$

Since $\varepsilon > 0$ is arbitrary, Lemma 8.3 is proven.

Proof of Theorem 8.4   Let C be the contour that consists of the segment $-R \le x \le R$ of the real axis together with the semicircle C_R of Lemma 8.3. Property (17) of Section 6.2 can be used to write

(11) $\displaystyle \int_{-R}^R \frac{\exp(i\alpha x)P(x) \, dx}{Q(x)} = \int_C \frac{\exp(i\alpha z)P(z) \, dz}{Q(z)} - \int_{C_R} \frac{\exp(i\alpha z)P(z) \, dz}{Q(z)}$.

If R is sufficiently large, then all the poles $z_1, z_2, \ldots, z_k$ of f will lie inside C, and we can use the residue theorem to obtain

(12) $\displaystyle \int_{-R}^R \frac{\exp(i\alpha x)P(x) \, dx}{Q(x)} = 2\pi i \sum_{j=1}^k \mathrm{Res}[f, z_j] - \int_{C_R} \frac{\exp(i\alpha z)P(z) \, dz}{Q(z)}$.

Since α is a positive real number, the change of variables $Z = \alpha z$ shows that Jordan's lemma holds true for the integrand $\exp(i\alpha z)P(z)/Q(z)$. Hence we can let $R \to \infty$ in equation (12) to obtain

(13) $\displaystyle \mathrm{P.V.} \int_{-\infty}^\infty \frac{[\cos(\alpha x) + i \sin(\alpha x)]P(x) \, dx}{Q(x)} = 2\pi i \sum_{j=1}^k \mathrm{Res}[f, z_j]$

$$= -2\pi \sum_{j=1}^k \mathrm{Im}(\mathrm{Res}[f, z_j])$$

$$+ 2\pi i \sum_{j=1}^k \mathrm{Re}(\mathrm{Res}[f, z_j]).$$

Equating the real and imaginary parts of equation (13) results in equations (3) and (4), respectively, and Theorem 8.4 is proven.

EXERCISES FOR SECTION 8.5

Use residues to find the integrals in Exercises 1–12.

1. $\displaystyle\int_{-\infty}^{\infty} \frac{\cos x\, dx}{x^2 + 9} = \frac{\pi}{3e^3}$ and $\displaystyle\int_{-\infty}^{\infty} \frac{\sin x\, dx}{x^2 + 9} = 0$

2. P.V. $\displaystyle\int_{-\infty}^{\infty} \frac{x \cos x\, dx}{x^2 + 9}$ and P.V. $\displaystyle\int_{-\infty}^{\infty} \frac{x \sin x\, dx}{x^2 + 9}$

3. $\displaystyle\int_{-\infty}^{\infty} \frac{x \sin x\, dx}{(x^2 + 4)^2} = \frac{\pi}{4e^2}$

4. $\displaystyle\int_{-\infty}^{\infty} \frac{\cos x\, dx}{(x^2 + 4)^2}$

5. $\displaystyle\int_{-\infty}^{\infty} \frac{\cos x\, dx}{(x^2 + 4)(x^2 + 9)} = \frac{\pi}{5}\left(\frac{1}{2e^2} - \frac{1}{3e^3}\right)$

6. $\displaystyle\int_{-\infty}^{\infty} \frac{\cos x\, dx}{(x^2 + 1)(x^2 + 4)}$

7. $\displaystyle\int_{-\infty}^{\infty} \frac{\cos x\, dx}{x^2 - 2x + 5} = \frac{\pi \cos 1}{2e^2}$

8. $\displaystyle\int_{-\infty}^{\infty} \frac{\cos x\, dx}{x^2 - 4x + 5}$

9. $\displaystyle\int_{-\infty}^{\infty} \frac{x \sin x\, dx}{x^4 + 4} = \frac{\pi \sin 1}{2e}$

10. P.V. $\displaystyle\int_{-\infty}^{\infty} \frac{x^3 \sin x\, dx}{x^4 + 4}$

11. $\displaystyle\int_{-\infty}^{\infty} \frac{\cos 2x\, dx}{x^2 + 2x + 2} = \frac{\pi \cos 2}{e^2}$

12. P.V. $\displaystyle\int_{-\infty}^{\infty} \frac{x^3 \sin 2x\, dx}{x^4 + 4}$

13. Why do we need to use the exponential function when evaluating improper integrals involving the trigonometric functions sine and cosine?

8.6   Indented Contour Integrals

If f is continuous on the interval $b < x \le c$, then the improper integral of f over $(b, c]$ is defined by

$$(1) \quad \int_{b}^{c} f(x)\, dx = \lim_{r \to b^+} \int_{r}^{c} f(x)\, dx$$

provided that the limit exists. Similarly, if f is continuous on the interval $a \le x < b$, then the improper integral of f over $[a, b)$ is defined by

$$(2) \quad \int_{a}^{b} f(x)\, dx = \lim_{R \to b^-} \int_{a}^{R} f(x)\, dx$$

provided that the limit exists. For example,

$$\int_{0}^{9} \frac{dx}{2\sqrt{x}} = \lim_{r \to 0^+} \int_{r}^{9} \frac{dx}{2\sqrt{x}} = \lim_{r \to 0^+} [\sqrt{x}\,|_{r}^{9}] = 3 - \lim_{r \to 0^+} \sqrt{r} = 3.$$

Let f be continuous for all values of x in the interval $[a, c]$, except at the value $x = b$, where $a < b < c$. The *Cauchy principal value* of f over $[a, c]$ is defined by

$$(3) \quad \text{P.V.} \int_{a}^{c} f(x)\, dx = \lim_{r \to 0^+} \left[\int_{a}^{b-r} f(x)\, dx + \int_{b+r}^{c} f(x)\, dx \right]$$

provided that the limit exists.

EXAMPLE 8.17

$$\text{P.V.} \int_{-1}^{8} \frac{dx}{x^{1/3}} = \lim_{r \to 0^+} \left[\int_{-1}^{-r} \frac{dx}{x^{1/3}} + \int_{r}^{8} \frac{dx}{x^{1/3}} \right]$$

$$= \lim_{r \to 0^+} \left[\frac{3}{2} r^{2/3} - \frac{3}{2} + 6 - \frac{3}{2} r^{2/3} \right] = \frac{9}{2}.$$

In this section we extend the results of Sections 8.4 and 8.5 to include the case in which the integrand f has simple poles on the x axis. We now state how residues can be used to find the Cauchy principal value for the integral of f over $(-\infty, \infty)$.

Theorem 8.5   *Let $f(z) = P(z)/Q(z)$ where P and Q are polynomials with real coefficients of degree m and n, respectively, where $n \geq m + 2$. If Q has simple zeros at the points $t_1, t_2, \ldots, t_l$ on the x axis, then*

(4) $$\text{P.V.} \int_{-\infty}^{\infty} \frac{P(x)\,dx}{Q(x)} = 2\pi i \sum_{j=1}^{k} \text{Res}[f, z_j] + \pi i \sum_{j=1}^{l} \text{Res}[f, t_j]$$

where $z_1, z_2, \ldots, z_k$ are the poles of f that lie in the upper half-plane.

Theorem 8.6   *Let P and Q be polynomials of degree m and n, respectively, where $n \geq m + 1$, and let Q have simple zeros at the points $t_1, t_2, \ldots, t_l$ on the x axis. If α is a positive real number and if*

(5) $$f(z) = \frac{\exp(i\alpha z)P(z)}{Q(z)},$$

then

(6) $$\text{P.V.} \int_{-\infty}^{\infty} \frac{P(x)}{Q(x)} \cos \alpha x \, dx = -2\pi \sum_{j=1}^{k} \text{Im}(\text{Res}[f, z_j]) - \pi \sum_{j=1}^{l} \text{Im}(\text{Res}[f, t_j])$$

and

(7) $$\text{P.V.} \int_{-\infty}^{\infty} \frac{P(x)}{Q(x)} \sin \alpha x \, dx = 2\pi \sum_{j=1}^{k} \text{Re}(\text{Res}[f, z_j]) + \pi \sum_{j=1}^{l} \text{Re}(\text{Res}[f, t_j]),$$

where $z_1, z_2, \ldots, z_k$ are the poles of f that lie in the upper half-plane.

Before we prove Theorems 8.5 and 8.6, let us make some observations and look at some examples. First, the formulas in equations (4), (6), and (7) give the Cauchy principal value in the integral. This answer is special because of the manner in which the limit in equation (3) is taken. Second, the formulas are similar to those in Sections 8.4 and 8.5, except that here we add one-half of the value of each residue at the points $t_1, t_2, \ldots, t_l$ on the x axis.

EXAMPLE 8.18 Show that P.V. $\displaystyle\int_{-\infty}^{\infty} \frac{x\,dx}{x^3 - 8} = \frac{\pi\sqrt{3}}{6}$.

Solution   The complex integrand

$$f(z) = \frac{z}{z^3 - 8} = \frac{z}{(z - 2)(z + 1 + i\sqrt{3})\,(z + 1 - i\sqrt{3})}$$

has simple poles at the points $t_1 = 2$ on the x axis and $z_1 = -1 + i\sqrt{3}$ in the upper half-plane. Now equation (4) gives

$$\text{P.V.} \int_{-\infty}^{\infty} \frac{x\,dx}{x^3 - 8} = 2\pi i \ \text{Res}[f, z_1] + \pi i \ \text{Res}[f, t_1]$$

$$= 2\pi i \ \frac{-1 - i\sqrt{3}}{12} + \pi i \ \frac{1}{6} = \frac{\pi\sqrt{3}}{6} \ .$$

EXAMPLE 8.18* Show that $\displaystyle\int_{-\infty}^{\infty} \frac{t\,dt}{t^3 - 8} = \frac{\pi\sqrt{3}}{6}$.

Solution   Using a Computer Algebra System
Mathematica and MAPLE give the following indefinite integral:

$$g(t) = \frac{\arctan\left(\dfrac{1 + t}{\sqrt{3}}\right)}{2\sqrt{3}} + \frac{\text{Log}(t - 2)}{6} + \frac{\text{Log}(t^2 + 2t + 4)}{12} \ .$$

However, for real numbers, the second term should be rewritten as $\dfrac{\text{Log}[(t - 2)^2]}{12}$

and we can use the following equivalent formula:

$$g(t) = \frac{\arctan\left(\dfrac{1 + t}{\sqrt{3}}\right)}{2\sqrt{3}} + \frac{\text{Log}[(t - 2)^2]}{12} + \frac{\text{Log}(t^2 + 2t + 4)}{12} \ .$$

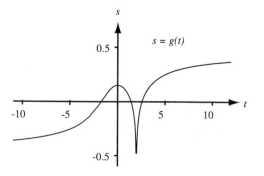

FIGURE 8.5 Graph of $s = g(t) = \displaystyle\int \frac{t\,dt}{t^3 - 8}$.

This choice produces a graph that is continuous in the extended real numbers, i.e., $\lim_{t \to 2} g(t) = -\infty$, as shown in Figure 8.5. The limits when t approaches infinity are

$$\lim_{t \to -\infty} g(t) = \frac{\pi\sqrt{3}}{12} \quad \text{and} \quad \lim_{t \to -\infty} g(t) = \frac{-\pi\sqrt{3}}{12},$$

and the Cauchy principal limit at $t = 2$, as $r \to 0$ can be shown to be

$$\lim_{r \to 0^+} [g(2 + r) - g(2 - r)] = 0.$$

Therefore, the Cauchy principal value of the improper integral is

$$\text{P.V.} \int_{-\infty}^{\infty} \frac{t \, dt}{t^3 - 8} = \lim_{r \to 0^+} \left[\int_{-\infty}^{2-r} \frac{t \, dt}{t^3 - 8} + \int_{2+r}^{\infty} \frac{t \, dt}{t^3 - 8} \right]$$

$$= \lim_{t \to \infty} g(t) - \lim_{r \to 0^+} [g(2 + r) - g(2 - r)] - \lim_{t \to -\infty} g(t)$$

$$= \frac{\pi\sqrt{3}}{12} - 0 + \frac{\pi\sqrt{3}}{12} = \frac{\pi\sqrt{3}}{6}.$$

EXAMPLE 8.19   Show that P.V. $\int_{-\infty}^{\infty} \frac{\sin x \, dx}{(x - 1)(x^2 + 4)} = \frac{\pi}{5} \left(\cos 1 - \frac{1}{e^2} \right).$

Solution   The complex integrand $f(z) = \exp(iz)/[(z - 1)(z^2 + 4)]$ has simple poles at the points $t_1 = 1$ on the x axis and $z_1 = 2i$ in the upper half-plane. Now equation (7) gives

$$\text{P.V.} \int_{-\infty}^{\infty} \frac{\sin x \, dx}{(x - 1)(x^2 + 4)} = 2\pi \, \text{Re}(\text{Res}[f, z_1]) + \pi \, \text{Re}(\text{Res}[f, t_1])$$

$$= 2\pi \, \text{Re}\left(\frac{-2 + i}{20e^2} \right) + \pi \, \text{Re}\left(\frac{\cos 1 + i \sin 1}{5} \right)$$

$$= \frac{\pi}{5} \left(\cos 1 - \frac{1}{e^2} \right).$$

The proofs of Theorems 8.5 and 8.6 depend on the following result.

Lemma 8.4   *Let f have a simple pole at the point t_0 on the x axis. If the contour is $C: z = t_0 + re^{i\theta}$ for $0 \le \theta \le \pi$, then*

(8) $\lim_{r \to 0} \int_C f(z) \, dz = i\pi \, \text{Res}[f, t_0].$

Proof   The Laurent series for f at $z = t_0$ has the form

(9) $f(z) = \dfrac{\text{Res}[f, t_0]}{z - t_0} + g(z),$

where g is analytic at $z = t_0$. Using the parameterization of C and equation (9), we can write

(10) $$\int_C f(z)\, dz = \text{Res}[f, t_0] \int_0^\pi \frac{ire^{i\theta}\, d\theta}{re^{i\theta}} + ir \int_0^\pi g(t_0 + re^{i\theta})e^{i\theta}\, d\theta$$

$$= i\pi\, \text{Res}[f, t_0] + ir \int_0^\pi g(t_0 + re^{i\theta})e^{i\theta}\, d\theta.$$

Since g is continuous at t_0, there is an $M > 0$ so that $\left| g(t_0 + re^{i\theta}) \right| \le M$. Hence

(11) $$\left| \lim_{r \to 0} ir \int_0^\pi g(t_0 + re^{i\theta})e^{i\theta}\, d\theta \right| \le \lim_{r \to 0} r \int_0^\pi M\, d\theta = \lim_{r \to 0} r\pi M = 0.$$

When inequality (11) is used in equation (10), the resulting limit is given by equation (8), and Lemma 8.4 is proven.

Proof of Theorems 8.5 and 8.6   Since f has only a finite number of poles, we can choose r small enough so that the semicircles

$$C_j\colon z = t_j + re^{i\theta} \quad \text{for } 0 \le \theta \le \pi \text{ and } j = 1, 2, \ldots, l$$

are disjoint and the poles $z_1, z_2, \ldots, z_k$ of f in the upper half-plane lie above them as shown in Figure 8.6.

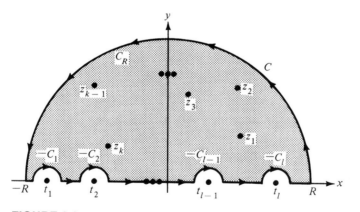

FIGURE 8.6   The poles $t_1, t_2, \ldots, t_l$ of f that lie on the x axis and the poles z_1, $z_2, \ldots, z_k$ that lie above the semicircles $C_1, C_2, \ldots, C_l$.

Let R be chosen large enough so that the poles of f in the upper half-plane lie under the semicircle $C_R\colon z = Re^{i\theta}$ for $0 \le \theta \le \pi$ and the poles of f on the x axis lie in the interval $-R \le x \le R$. Let C be the simple closed positively oriented contour that consists of C_R and $-C_1, -C_2, \ldots, -C_l$ and the segments of the real axis that lie between the semicircles as shown in Figure 8.6. The residue theorem can be used to obtain

(12) $\int_C f(z)\,dz = 2\pi i \sum_{j=1}^{k} \text{Res}[f, z_j].$

Equation (12) can be written as

(13) $\int_{I_R} f(x)\,dx = 2\pi i \sum_{j=1}^{k} \text{Res}[f, z_j] + \sum_{j=1}^{l} \int_{C_j} f(z)\,dz - \int_{C_R} f(z)\,dz$

where I_R is the portion of the interval $-R \le x \le R$ that lies outside the intervals $(t_j - r, t_j + r)$ for $j = 1, 2, \ldots, l$. The proofs of Theorems 8.3 and 8.4 show that

(14) $\lim_{R \to \infty} \int_{C_R} f(z)\,dz = 0.$

If we let $R \to \infty$ and $r \to 0$ in equation (13) and use the result of equation (14) and Lemma 8.4, then we obtain

(15)   P.V. $\int_{-\infty}^{\infty} f(x)\,dx = 2\pi i \sum_{j=1}^{k} \text{Res}[f, z_j] + \pi i \sum_{j=1}^{l} \text{Res}[f, t_j].$

If f is given in Theorem 8.5, the equation (15) becomes equation (4). If f is given in Theorem 8.6, then equating the real and imaginary parts of equation (15) results in equations (6) and (7), respectively, and the theorems are established.

EXERCISES FOR SECTION 8.6

Use residues to compute or verify the integrals in Exercises 1–15.

1. P.V. $\int_{-\infty}^{\infty} \dfrac{dx}{x(x-1)(x-2)} = 0$

2. P.V. $\int_{-\infty}^{\infty} \dfrac{dx}{x^3 + x}$

3. P.V. $\int_{-\infty}^{\infty} \dfrac{x\,dx}{x^3 + 1} = \dfrac{\pi}{\sqrt{3}}$

4. P.V. $\int_{-\infty}^{\infty} \dfrac{dx}{x^3 + 1}$

5. P.V. $\int_{-\infty}^{\infty} \dfrac{x^2\,dx}{x^4 - 1} = \dfrac{\pi}{2}$

6. P.V. $\int_{-\infty}^{\infty} \dfrac{x^4\,dx}{x^6 - 1}$

7. P.V. $\int_{-\infty}^{\infty} \dfrac{\sin x\,dx}{x} = \pi$

8. P.V. $\int_{-\infty}^{\infty} \dfrac{\cos x\,dx}{x^2 - x}$

9. P.V. $\int_{-\infty}^{\infty} \dfrac{\sin x\,dx}{x(\pi^2 - x^2)} = \dfrac{2}{\pi}$

10. P.V. $\int_{-\infty}^{\infty} \dfrac{\cos x\,dx}{\pi^2 - 4x^2}$

11. P.V. $\int_{-\infty}^{\infty} \dfrac{\sin x\,dx}{x(x^2 + 1)} = \pi\left(1 - \dfrac{1}{e}\right)$

12. P.V. $\int_{-\infty}^{\infty} \dfrac{x \cos x\,dx}{x^2 + 3x + 2}$

13. P.V. $\int_{-\infty}^{\infty} \dfrac{\sin x\,dx}{x(1 - x^2)} = \pi(1 - \cos 1)$

14. P.V. $\int_{-\infty}^{\infty} \dfrac{\cos x\,dx}{a^2 - x^2} = \dfrac{\pi \sin a}{a}$

15. P.V. $\int_{-\infty}^{\infty} \dfrac{\sin^2 x\,dx}{x^2} = \pi.$ *Hint*: Use the trigonometric identity $\sin^2 x = \dfrac{1}{2} - \dfrac{1}{2}\cos 2x.$

8.7   Integrands with Branch Points

We now show how to evaluate certain improper real integrals involving the integrand $x^\alpha P(x)/Q(x)$. Since the complex function z^α is multivalued, we must first specify the branch that we will be using.

Let α be a real number with $0 < \alpha < 1$. Then in this section we will use the branch of z^α defined as follows:

(1) $z^\alpha = e^{\alpha(\ln r + i\theta)} = r^\alpha(\cos \alpha\theta + i \sin \alpha\theta)$,   where $0 \le \theta < 2\pi$.

Using definition (1), we see that z^α is analytic in the domain $r > 0, 0 < \theta < 2\pi$.

> **Theorem 8.7**   *Let P and Q be polynomials of degree m and n, respectively, where $n \ge m + 2$. If $Q(x) \ne 0$ for $x > 0$ and Q has a zero of order at most 1 at the origin and*
>
> (2) $f(z) = \dfrac{z^\alpha P(z)}{Q(z)}$,   *where $0 < \alpha < 1$,*
>
> *then*
>
> (3)   P.V. $\displaystyle\int_0^\infty \dfrac{x^\alpha P(x)\, dx}{Q(x)} = \dfrac{2\pi i}{1 - e^{i\alpha 2\pi}} \sum_{j=1}^{k} \text{Res}[\,f, z_j\,]$
>
> *where $z_1, z_2, \ldots, z_k$ are the nonzero poles of P/Q.*

Proof   Let C denote the simple, closed, positively oriented contour that consists of the portions of the circles $|z| = r$ and $|z| = R$ and the horizontal segments joining them as shown in Figure 8.7. A small value of r and a large value of R can be selected so that the nonzero poles $z_1, z_2, \ldots, z_k$ of P/Q lie inside C.

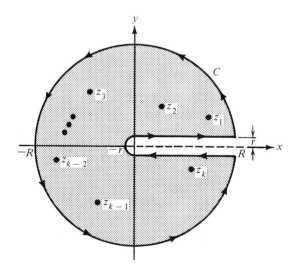

FIGURE 8.7   The contour C that encloses all the nonzero poles $z_1, z_2, \ldots, z_k$ of P/Q.

Using the residue theorem, we can write

$$(4) \quad \int_C f(z) \, dz = 2\pi i \sum_{j=1}^{k} \text{Res}[f, z_j].$$

If we let $r \to 0$ in equation (4) and use property (17) of Section 6.2 to express the limiting value of the integral on the lower segment, we find that equation (4) becomes

$$\int_0^R \frac{x^a P(x) \, dx}{Q(x)} - \int_0^R \frac{x^a e^{ia2\pi} P(x) \, dx}{Q(x)} = 2\pi i \sum_{j=1}^{k} \text{Res}[f, z_j] - \int_{C_R} f(z) \, dz,$$

which can be written as

$$(5) \quad \int_0^R \frac{x^a P(x) \, dx}{Q(x)} = \frac{2\pi i}{1 - e^{ia2\pi}} \sum_{j=1}^{k} \text{Res}[f, z_j] - \frac{1}{1 - e^{ia2\pi}} \int_{C_R} f(z) \, dz.$$

Letting $R \to \infty$ in equation (5) results in equation (3), and Theorem 8.7 is established.

EXAMPLE 8.20 Show that P.V. $\displaystyle\int_0^\infty \frac{x^a \, dx}{x(x + 1)} = \frac{\pi}{\sin a\pi}$, where $0 < a < 1$.

Solution The complex function $f(z) = z^a/[z(z + 1)]$ has a nonzero pole at the point $z_1 = -1$. Using equation (3) we find that

$$\int_0^\infty \frac{x^a \, dx}{x(x + 1)} = \frac{2\pi i}{1 - e^{ia2\pi}} \text{Res}[f, -1] = \frac{2\pi i}{1 - e^{ia2\pi}} \left(\frac{e^{ia\pi}}{-1} \right)$$

$$= \frac{\pi}{\dfrac{e^{ia\pi} - e^{-ia\pi}}{2i}} = \frac{\pi}{\sin a\pi}.$$

The preceding ideas can be applied to other multivalued functions.

EXAMPLE 8.21 Show that P.V. $\displaystyle\int_0^\infty \frac{\ln x \, dx}{x^2 + a^2} = \frac{\pi \ln a}{2a}$, where $a > 0$.

Solution Here we use the complex function $f(z) = \text{Log } z/(z^2 + a^2)$. The path C of integration will consist of the segments $[-R, -r]$ and $[r, R]$ of the x axis together with the upper semicircles C_r: $z = re^{i\theta}$ and C_R: $z = Re^{i\theta}$ for $0 \le \theta \le \pi$ as shown in Figure 8.8.

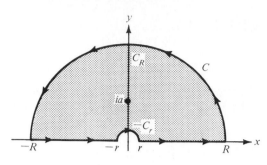

FIGURE 8.8   The contour C for the integrand $f(z) = (\text{Log } z)/(z^2 + a^2)$.

The residue theorem can be used to write

(6) $\displaystyle \int_C f(z)\, dz = 2\pi i \, \text{Res}[f, ai] = \frac{\pi \ln a}{a} + i\frac{\pi^2}{2a}.$

The inequality

$$\left| \int_0^\pi \frac{\ln R + i\theta}{R^2 e^{i2\theta} + a^2} \, i\, Re^{i\theta}\, d\theta \right| \le \frac{R(\ln R + \pi)\pi}{R^2 - a^2}$$

and L'Hôpital's rule can be used to show that

(7) $\displaystyle \lim_{R \to \infty} \int_{C_R} f(z)\, dz = 0.$

A similar computation will show that

(8) $\displaystyle \lim_{r \to 0} \int_{C_r} f(z)\, dz = 0.$

We can use the results of equations (7) and (8) in equation (6) to obtain

(9)   P.V. $\displaystyle \left(\int_{-\infty}^0 \frac{\ln|x| + i\pi}{x^2 + a^2} \, dx + \int_0^\infty \frac{\ln x\, dx}{x^2 + a^2} \right) = \frac{\pi \ln a}{a} + i\frac{\pi^2}{2a}.$

Equating the real parts in equation (9), we obtain

P.V. $\displaystyle \int_0^\infty \frac{2 \ln x\, dx}{x^2 + a^2} = \frac{\pi \ln a}{a},$

and the result is established.

EXERCISES FOR SECTION 8.7

Use residues to compute or verify the integrals in Exercises 1–11.

1. P.V. $\displaystyle\int_0^\infty \frac{dx}{x^{2/3}(1 + x)} = \frac{2\pi}{\sqrt{3}}$

2. P.V. $\displaystyle\int_0^\infty \frac{dx}{x^{1/2}(1 + x)}$

3. P.V. $\displaystyle\int_0^\infty \frac{x^{1/2}\,dx}{(1 + x)^2} = \frac{\pi}{2}$

4. P.V. $\displaystyle\int_0^\infty \frac{x^{1/2}\,dx}{1 + x^2}$

5. P.V. $\displaystyle\int_0^\infty \frac{\ln(x^2 + 1)\,dx}{x^2 + 1} = \pi \ln 2$. Use $f(z) = \dfrac{\text{Log}(z + i)}{z^2 + 1}$.

6. P.V. $\displaystyle\int_0^\infty \frac{\ln x\,dx}{(1 + x^2)^2}$

7. P.V. $\displaystyle\int_0^\infty \frac{\ln(1 + x)}{x^{1+a}} = \frac{\pi}{a \sin \pi a}$, where $0 < a < 1$

8. P.V. $\displaystyle\int_0^\infty \frac{\ln x\,dx}{(x + a)^2}$, where $a > 0$

9. P.V. $\displaystyle\int_{-\infty}^\infty \frac{\sin x}{x}\,dx = \pi$

Hint: Use the integrand $f(z) = \exp(iz)/z$ and the contour C in Figure 8.8, and let $r \to 0$ and $R \to \infty$.

10. P.V. $\displaystyle\int_{-\infty}^\infty \frac{\sin^2 x}{x^2}\,dx = \pi$

Hint: Use the integrand $f(z) = [1 - \exp(i2z)]/z^2$ and the contour C in Figure 8.8, and let $r \to 0$ and $R \to \infty$.

11. The Fresnel integrals

$$\text{P.V.} \int_0^\infty \cos(x^2)\,dx = \text{P.V.} \int_0^\infty \sin(x^2)\,dx = \frac{\sqrt{\pi}}{2\sqrt{2}}$$

are important in the study of optics. Use the integrand $f(z) = \exp(-z^2)$ and the contour C shown in Figure 8.9, and let $R \to \infty$; then establish these integrals. Also use the fact from calculus that P.V. $\int_0^\infty e^{-x^2}\,dx = \sqrt{\pi}/2$.

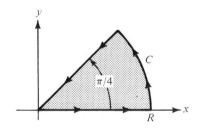

FIGURE 8.9   Accompanies Exercise 11.

8.8   The Argument Principle and Rouché's Theorem

We will now derive two results based on Cauchy's residue theorem. They have important practical applications and pertain only to functions all of whose isolated singularities are poles.

Definition 8.1   *A function f(z) is said to be meromorphic in a domain D provided that the only singularities of f(z) are isolated poles (and removable singularities).*

Observe that analytic functions are a special case of meromorphic functions. Rational functions $f(z) = P(z)/Q(z)$, where $P(z)$ and $Q(z)$ are polynomials, are meromorphic in the entire complex plane. A meromorphic function does not have essential singularities!

Suppose that $f(z)$ is analytic at each point on a simple closed contour C and $f(z)$ is meromorphic in the domain that is the interior of C. An extension of theorem 7.14 can be made that shows that $f(z)$ has at most finitely many zeros that lie inside C. Since the function $g(z) = 1/f(z)$ is also meromorphic, it can have only finitely many zeros inside C. Therefore $f(z)$ can have at most a finite number of poles that lie inside C.

An application of the residue theorem that is useful in determining the number of zeros and poles of a function is called the argument principle.

Theorem 8.8 (Argument Principle)   *Let f(z) be meromorphic in the simply connected domain D. Let C be a simple closed positively oriented contour in D along which $f(z) \neq 0$ and $f(z) \neq \infty$. Then*

(1) $$\frac{1}{2\pi i} \int_C \frac{f'(z)}{f(z)} \, dz = N - P$$

where N is the number of zeros of f(z) that lie inside C and P is the number of poles that lie inside C.

Proof   Let $a_1, a_2, \ldots, a_N$ be the zeros of $f(z)$ inside C counted according to multiplicity and let $b_1, b_2, \ldots, b_P$ be the poles of $f(z)$ inside C counted according to multiplicity. Then $f(z)$ has the representation

(2) $$f(z) = \frac{(z - a_1)(z - a_2) \cdots (z - a_N)}{(z - b_1)(z - b_2) \cdots (z - b_P)} g(z),$$

where $g(z)$ is analytic and nonzero on C and inside C. An elementary calculation shows that

(3) $$\frac{f'(z)}{f(z)} = \frac{1}{(z - a_1)} + \frac{1}{(z - a_2)} + \cdots + \frac{1}{(z - a_N)}$$
$$- \frac{1}{(z - b_1)} - \frac{1}{(z - b_2)} - \cdots - \frac{1}{(z - b_P)} + \frac{g'(z)}{g(z)}.$$

According to Example 6.14, we have

$$\int_C \frac{dz}{(z - a_j)} = 2\pi i \quad \text{for } j = 1, 2, \ldots, N$$

and

$$\int_C \frac{dz}{(z - b_k)} = 2\pi i \quad \text{for } k = 1, 2, \ldots, P.$$

Since $g'(z)/g(z)$ is analytic inside and on C, it follows from the Cauchy-Goursat theorem that

$$\int_C \frac{g'(z)dz}{g(z)} = 0.$$

These facts can be used to integrate both sides of equation (3) over C. The result is equation (1), and the theorem is proven.

Corollary 8.1   *Suppose that $f(z)$ is analytic in the simply connected domain D. Let C be a simple closed positively oriented contour in D along which $f(z) \neq 0$. Then*

(4) $$\frac{1}{2\pi i} \int_C \frac{f'(z)}{f(z)} dz = N$$

where N is the number of zeros of $f(z)$ that lie inside C.

Theorem 8.9 (Roché's Theorem)   *Let $f(z)$ and $g(z)$ be analytic functions defined in the simply connected domain D. Let C be a simply closed contour in D. If the strict inequality*

(5) $$|f(z) - g(z)| < |f(z)| \quad \text{holds for all } z \text{ on } C,$$

then $f(z)$ and $g(z)$ have the same number of zeros inside C (counting multiplicity).

Proof   The condition $|f(z) - g(z)| < |f(z)|$ precludes the possibility of $f(z)$ or $g(z)$ having zeros on the contour C. Therefore division by $f(z)$ is permitted, and we obtain

(6) $$\left| \frac{g(z)}{f(z)} - 1 \right| < 1 \quad \text{for all } z \text{ on } C.$$

Let $F(z) = g(z)/f(z)$. Then $F(C)$, the image of the curve C under the mapping $w = F(z)$, is contained in the disk $|w - 1| < 1$ in the w plane. Therefore $F(C)$ is a closed curve that does not wind around $w = 0$. Hence $1/w$ is analytic on the curve $f(C)$, and we obtain

(7) $$\int_{F(C)} \frac{dw}{w} = 0.$$

Using the change of variable $w = f(z)$ and $dw = F'(z)dz$, we see that the integral in equation (7) can be expressed as

$$(8) \quad \int_C \frac{F'(z)}{F(z)} \, dz = 0.$$

Since $F'(z) = [g'(z)f(z) - f'(z)g(z)]/[f(z)]^2$, it follows that

$$(9) \quad \frac{F'(z)}{F(z)} = \frac{g'(z)}{g(z)} - \frac{f'(z)}{f(z)}.$$

Hence equations (8) and (9) can be used to obtain

$$(10) \quad \frac{1}{2\pi i} \int_C \frac{f'(z)}{f(z)} \, dz = \frac{1}{2\pi i} \int_C \frac{g'(z)}{g(z)} \, dz.$$

Corollary 8.1 and equation (10) imply that the number of zeros of $f(z)$ inside C equals the number of zeros of $g(z)$ inside C, and the theorem is proven.

One can use Rouché's theorem to gain information about the location of the zeros of an analytic function.

EXAMPLE 8.22   Show that all four zeros of the polynomial

$$g(z) = z^4 - 7z - 1$$

lie in the disk $|z| < 2$.

Solution   Let $f(z) = z^4$, then $f(z) - g(z) = 7z + 1$. At points on the circle $|z| = 2$ we have the relation

$$|f(z) - g(z)| = |7z + 1| \le |7z| + 1 = 7(2) + 1 = 15 < 16 = |f(z)|.$$

The function $f(z)$ has a zero of order 4 at the origin, and the hypothesis of Rouché's theorem holds true for the circle $|z| = 2$. Therefore $g(z)$ has four zeros inside $|z| = 2$.

EXAMPLE 8.23   Show that the polynomial $g(z) = z^4 - 7z - 1$ has one zero in the disk $|z| < 1$.

Solution   Let $f(z) = -7z - 1$, then $f(z) - g(z) = -z^4$. At points on the circle $|z| = 1$ we have the relation

$$|f(z) - g(z)| = |-z^4| = 1 < 6 = |7 - 1| = ||7z| - |-1||$$
$$\le |7z - 1| = |f(z)|.$$

The function $f(z)$ has one zero at $z = -1/7$ in the disk $|z| < 1$, and the hypothesis of Rouché's theorem holds true on the circle $|z| = 1$. Therefore $g(z)$ has one zero inside $|z| = 1$.

Certain feedback control systems in engineering must be stable. A test for stability involves the function $G(z) = 1 + F(z)$, where $F(z)$ is a rational function. If $G(z)$ does not have any zeros for $\text{Re}(z) \geq 0$, then the system is stable. The number of zeros of $G(z)$ can be determined by writing $F(z) = P(z)/Q(z)$, where $P(z)$ and $Q(z)$ are polynomials with no common zero. Then $G(z) = [Q(z) + P(z)]/Q(z)$. We can check for zeros of $Q(z) + P(z)$ using Theorem 8.8. A value R is selected so that $G(z) \neq 0$ for $|z| > R$. Contour integration is then performed along the contour consisting of the right half of the circle $|z| = R$ and the line segment between iR and $-iR$. The method is known as the Nyquist stability criterion.

The Winding Number

Suppose that $C: z(t) = x(t) + iy(t)$ for $a \leq t \leq b$ is a simple closed contour. Let $a = t_0 < t_1 < \cdots < t_n = b$ be a partition of the interval and let $z_k = z(t_k)$ (for $k = 0, 1, \ldots, n$) denote points on C where $z_0 = z_n$. If z^* lies inside C, then $z(t)$ winds around z^* once as t goes from a to b (see Figure 8.10).

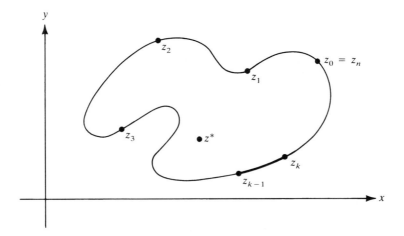

FIGURE 8.10   The points z_k on the contour C that winds around z^*.

Now suppose that $f(z)$ is analytic at each point on C and meromorphic inside C. Then $f(C)$ is a closed curve in the w plane that passes through $w_k = f(z_k)$ (for $k = 0, 1, \ldots, n$), where $w_0 = w_n$. The subintervals $[t_{k-1}, t_k]$ can be chosen small enough so that a continuous branch $\log w = \ln|w| + i \arg w = \ln \rho + i\phi$ can be defined on the portion of $f(C)$ between w_{k-1} and w_k (see Figure 8.11). Then

(11) $\log f(z_k) - \log f(z_{k-1}) = \ln \rho_k - \ln \rho_{k-1} + i\Delta\phi_k,$

where $\Delta\phi_k = \phi_k - \phi_{k-1}$ measures the amount that the portion of the curve $f(C)$ between w_k and w_{k-1} winds around the origin $w = 0$.

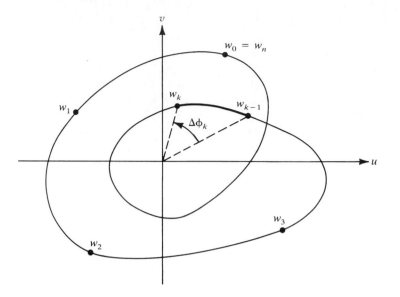

FIGURE 8.11   The points w_k on the coutour $f(C)$ that winds around 0.

Formula (1) will now be shown to be the number of times that $f(C)$ winds around $w = 0$. The parameterization given above together with the appropriate branches of log w are used to write

$$\int_C \frac{f'(z)}{f(z)}\, dz = \sum_{k=1}^{n} \int_{t_{k-1}}^{t_k} \frac{f'(z(t))}{f(z(t))}\, z'(t)\, dt$$

$$= \sum_{k=1}^{n} [\log w_k - \log w_{k-1}],$$

which in turn can be written as

$$(12) \quad \int_C \frac{f'(z)}{f(z)}\, dz = \sum_{k=1}^{n} [\ln \rho_k - \ln \rho_{k-1}] + i \sum_{k=1}^{n} \Delta\phi_k.$$

By using the fact that $\rho_0 = \rho_n$ the first summation in equation (12) vanishes. The summation of the quantities $\Delta\phi_k$ is the total amount that $f(C)$ winds around $w = 0$ in radians. When the quantities in equation (12) are divided by $2\pi i$, we are left with an integer that is the number of times $f(C)$ winds around $w = 0$. For example, the image of the circle $C: |z| = 2$ under the mapping $w = f(z) = z^2 + z$ is the curve $x = 4 \cos 2t + 2 \cos t$, $y = 4 \sin 2t + 2 \sin t$ for $0 < t \le 2\pi$ that is shown in Figure 8.12. Notice that the image curve $f(C)$ winds twice around the origin $w = 0$.

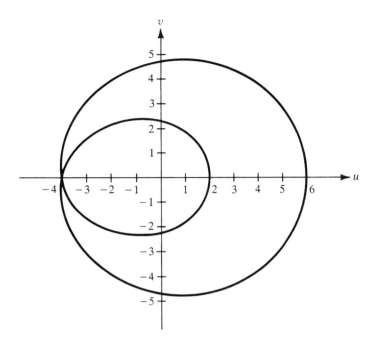

FIGURE 8.12   The image curve $f(C)$ of the circle C: $|z| = 2$ under the mapping $w = f(z) = z^2 + z$.

EXERCISES FOR SECTION 8.8

For Exercises 1–5, use Rouché's theorem to show that the roots lie in the indicated region.

1. Let $P(z) = z^5 + 4z - 15$.
   (a) Show that there are no roots in $|z| < 1$. *Hint*: Use $f(z) = 15$.
   (b) Show that there are five roots in $|z| < 2$. *Hint*: $f(z) = z^5$.
   Remark: A factorization of the polynomial using numerical approximations for the co-efficients is

   $$(z - 1.546)(z^2 - 1.340z + 2.857)(z^2 + 2.885z + 3.397).$$

2. Let $P(z) = z^3 + 9z + 27$.
   (a) Show that there are no roots in $|z| < 2$. *Hint*: Use $f(z) = 27$.
   (b) Show that there are three roots in $|z| < 4$. *Hint*: $f(z) = z^3$.
   Remark: A factorization of the polynomial using numerical approximations for the coefficients is

   $$(z + 2.047)(z^2 - 2.047z + 13.19).$$

3. Let $P(z) = z^5 + 6z^2 + 2z + 1$.
   (a) Show that there are two roots in $|z| < 1$. *Hint*: Use $f(z) = 6z^2$.
   (b) Show that there are five roots in $|z| < 2$.

4. Let $P(z) = z^6 - 5z^4 + 10$.
   (a) Show that there are no roots in $|z| < 1$.
   (b) Show that there are four roots in $|z| < 2$. *Hint*: Use $f(z) = 5z^4$.
   (c) Show that there are six roots in $|z| < 3$.

5. Let $P(z) = 3z^3 - 2iz^2 + iz - 7$.
   (a) Show that there are no roots in $|z| < 1$.
   (b) Show that there are three roots in $|z| < 2$.

6. Use Rouché's theorem to prove the fundamental theorem of algebra. *Hint*: Let $f(z) = -a_n z^n$ and $g(z) = a_0 + a_1 z + \cdots + a_{n-1} z^{n-1}$. Then show that for points z on the circle $|z| = R$ we have

$$\left| \frac{g(z)}{f(z)} \right| < \frac{|a_0| + |a_1| + \cdots + |a_{n-1}|}{|a_n| R},$$

   and see what happens when R is made large.

7. Use Rouché's theorem to prove the following. If $h(z)$ is analytic and nonzero and $|h(z)| < 1$ for $|z| < 1$, then $h(z) - z^n$ has n roots inside the unit circle $|z| = 1$.

8. Suppose that $f(z)$ is analytic inside and on the simple closed contour C. If $f(z)$ is a one-to-one function at points z on C, then prove that $f(z)$ is one-to-one inside C. *Hint*: Consider the image of C.

9. Look up the articles on Rouché's theorem and discuss what you found. Resources include bibliographical items 68 and 172.

10. Write a report on the winding number. Include ideas and examples not mentioned in the text. Resources include bibliographical items 6, 51, 88, 141, and 166.

9

Conformal Mapping

9.1 Basic Properties of Conformal Mappings

Let f be an analytic function in the domain D, and let z_0 be a point in D. If $f'(z_0) \neq 0$, then we can express f in the form

(1) $\quad f(z) = f(z_0) + f'(z_0)(z - z_0) + \eta(z)(z - z_0), \quad$ where $\eta(z) \to 0$ as $z \to z_0$.

If z is near z_0, then the transformation $w = f(z)$ has the *linear approximation*

(2) $\quad S(z) = A + B(z - z_0), \quad$ where $A = f(z_0)$ and $B = f'(z_0)$.

Since $\eta(z) \to 0$ when $z \to z_0$, it is reasonable that for points near z_0 the transformation $w = f(z)$ has an effect much like the linear mapping $w = S(z)$. The effect of the linear mapping S is a rotation of the plane through the angle $\alpha = \arg[f'(z_0)]$, followed by a magnification by the factor $|f'(z_0)|$, followed by a rigid translation by the vector $A - Bz_0$. Consequently, the mapping $w = S(z)$ preserves angles at the point z_0. We now show that the mapping $w = f(z)$ also preserves angles at z_0.

Let $C: z(t) = x(t) + iy(t)$, $-1 \leq t \leq 1$ denote a smooth curve that passes through the point $z(0) = z_0$. A vector $\mathbf{T}$ tangent to C at the point z_0 is given by

(3) $\quad \mathbf{T} = z'(0)$,

where the complex number $z'(0)$ has assumed its vector interpretation.

The angle of inclination of $\mathbf{T}$ with respect to the positive x axis is

(4) $\quad \beta = \arg z'(0)$.

The image of C under the mapping $w = f(z)$ is the curve K given by the formula $K: w(t) = u(x(t), y(t)) + iv(x(t), y(t))$. The chain rule can be used to show that a vector $\mathbf{T}^*$ tangent to K at the point $w_0 = f(z_0)$ is given by

(5) $\quad \mathbf{T}^* = w'(0) = f'(z_0)z'(0)$.

The angle of inclination of $\mathbf{T}^*$ with respect to the positive u axis is

(6) $\gamma = \arg[f'(z_0)] + \arg[z'(0)] = \alpha + \beta,$ where $\alpha = \arg[f'(z_0)]$.

Therefore the effect of the transformation $w = f(z)$ is to rotate the angle of inclination of the tangent vector $\mathbf{T}$ at z_0 through the angle $\alpha = \arg[f'(z_0)]$ to obtain the angle of inclination of the tangent vector $\mathbf{T}^*$ at w_0. The situation is illustrated in Figure 9.1.

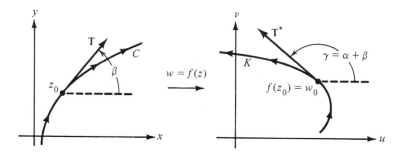

FIGURE 9.1   The tangents at the points z_0 and w_0, where f is an analytic function and $f'(z_0) \neq 0$.

A mapping $w = f(z)$ is said to be angle preserving, or *conformal at* z_0, if it preserves angles between oriented curves in magnitude as well as in orientation. The following result shows where a mapping by an analytic function is conformal.

Theorem 9.1   *Let f be an analytic function in the domain D, and let z_0 be a point in D. If $f'(z_0) \neq 0$, then f is conformal at z_0.*

Proof   Let C_1 and C_2 be two smooth curves passing through z_0 with tangents given by $\mathbf{T}_1$ and $\mathbf{T}_2$, respectively. Let β_1 and β_2 denote the angles of inclination of $\mathbf{T}_1$ and $\mathbf{T}_2$, respectively. The image curves K_1 and K_2 that pass through the point $w_0 = f(z_0)$ will have tangents denoted by $\mathbf{T}_1^*$ and $\mathbf{T}_2^*$, respectively. Using equation (6), we see that the angles of inclination γ_1 and γ_2 of $\mathbf{T}_1^*$ and $\mathbf{T}_2^*$ are related to β_1 and β_2 by the equations

(7) $\gamma_1 = \alpha + \beta_1$ and $\gamma_2 = \alpha + \beta_2,$

where $\alpha = \arg f'(z_0)$. Hence from equations (7) we conclude that

(8) $\gamma_2 - \gamma_1 = \beta_2 - \beta_1.$

That is, the angle $\gamma_2 - \gamma_1$ from K_1 to K_2 is the same in magnitude and orientation as the angle $\beta_2 - \beta_1$ from C_1 to C_2. Therefore the mapping $w = f(z)$ is conformal at z_0. The situation is shown in Figure 9.2.

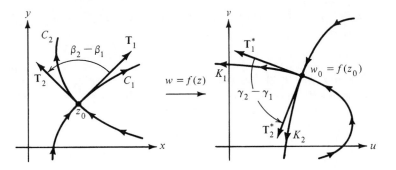

FIGURE 9.2   The analytic mapping $w = f(z)$ is conformal at the point z_0, where $f'(z_0) \neq 0$.

EXAMPLE 9.1   Show that the mapping $w = f(z) = \cos z$ is conformal at the points $z_1 = i$, $z_2 = 1$, and $z_3 = \pi + i$, and determine the angle of rotation given by $\alpha = \arg f'(z)$ at the given points.

Solution   Since $f'(z) = -\sin z$, we conclude that the mapping $w = \cos z$ is conformal at all points except $z = n\pi$, where n is an integer. Calculation reveals that

$$f'(i) = -i \sinh 1, \quad f'(1) = -\sin 1, \quad \text{and} \quad f'(\pi + i) = i \sinh 1.$$

Therefore the angle of rotation is given by

$$\alpha_1 = \arg f'(i) = \frac{-\pi}{2}, \quad \alpha_2 = \arg f'(1) = \pi, \quad \text{and}$$

$$\alpha_3 = \arg f'(\pi + i) = \frac{\pi}{2}, \text{ respectively.}$$

Let f be a nonconstant analytic function. If $f'(z_0) = 0$, then z_0 is called a *critical point* of f, and the mapping $w = f(z)$ is not conformal at z_0. The next result shows what happens at a critical point.

Theorem 9.2   *Let f be analytic at z_0. If $f'(z_0) = 0, \ldots, f^{(k-1)}(z_0) = 0$, and $f^{(k)}(z_0) \neq 0$, then the mapping $w = f(z)$ magnifies angles at the vertex z_0 by the factor k.*

Proof   Since f is analytic at z_0, it has the representation

(9) $f(z) = f(z_0) + a_k(z - z_0)^k + a_{k+1}(z - z_0)^{k+1} + \cdots.$

From (9) we conclude that

(10) $f(z) - f(z_0) = (z - z_0)^k g(z),$

where g is analytic at z_0 and $g(z_0) \neq 0$. Consequently, if $w = f(z)$ and $w_0 = f(z_0)$, then using equation (10), we obtain

(11) $\arg(w - w_0) = \arg[f(z) - f(z_0)] = k \arg(z - z_0) + \arg[g(z)]$.

Let C be a smooth curve that passes through z_0. If $z \to z_0$, along C, then $w \to w_0$ along the image curve K, and the angle of inclination of the tangents **T** to C and **T*** to K are given, respectively, by the following limits:

(12) $\beta = \lim_{z \to z_0} \arg(z - z_0)$ and $\gamma = \lim_{w \to w_0} \arg(w - w_0)$.

From equations (11) and (12) it follows that

(13) $\gamma = \lim_{z \to z_0} (k \arg(z - z_0) + \arg[g(z)]) = k\beta + \delta$,

where $\delta = \arg[g(z_0)] = \arg a_k$.

Let C_1 and C_2 be two smooth curves that pass through z_0, and let K_1 and K_2 be their images. Then from equation (13) it follows that

(14) $\Delta\gamma = \gamma_2 - \gamma_1 = k(\beta_2 - \beta_1) = k\Delta\beta$.

That is, the angle $\Delta\gamma$ from K_1 to K_2 is k times as large as the angle $\Delta\beta$ from C_1 to C_2. Therefore angles at the vertex z_0 are magnified by the factor k. The situation is shown in Figure 9.3.

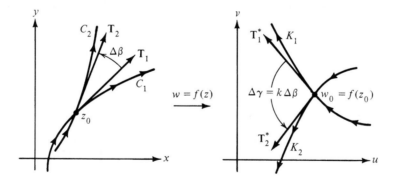

FIGURE 9.3   The analytic mapping $w = f(z)$ at a point z_0, where $f'(z_0) = 0, \ldots, f^{(k-1)}(z_0) = 0$ and $f^{(k)}(z_0) \neq 0$.

EXAMPLE 9.2   The mapping $w = f(z) = z^2$ maps the square $S = \{x + iy: 0 < x < 1, 0 < y < 1\}$ onto the region in the upper half plane $\text{Im}(w) > 0$, which lies under the parabolas

$$u = 1 - \tfrac{1}{4}v^2 \quad \text{and} \quad u = -1 + \tfrac{1}{4}v^2,$$

as shown in Figure 9.4. The derivative is $f'(z) = 2z$, and we conclude that the mapping $w = z^2$ is conformal for all $z \neq 0$. It is worthwhile to observe that the right angles at the vertices $z_1 = 1$, $z_2 = 1 + i$, and $z_3 = i$ are mapped onto right angles at the vertices $w_1 = 1$, $w_2 = 2i$, and $w_3 = -1$, respectively. At the point $z_0 = 0$ we have $f'(0) = 0$ and $f''(0) \neq 0$. Hence angles at the vertex $z_0 = 0$ are magnified by the factor $k = 2$. In particular, we see that the right angle at $z_0 = 0$ is mapped onto the straight angle at $w_0 = 0$.

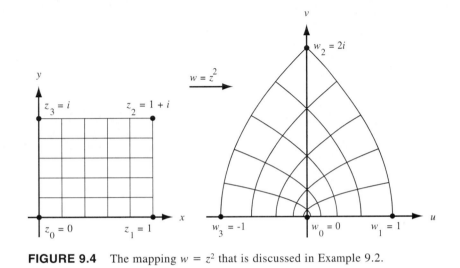

FIGURE 9.4   The mapping $w = z^2$ that is discussed in Example 9.2.

Another property of a conformal mapping $w = f(z)$ is obtained by considering the modulus of $f'(z_0)$. If z_1 is near z_0, we can use equation (1) and neglect the term $\eta(z_1)(z_1 - z_0)$. We then have the approximation

(15) $w_1 - w_0 = f(z_1) - f(z_0) \approx f'(z_0)(z_1 - z_0)$.

Using equation (15), we see that the distance $|w_1 - w_0|$ between the images of the points z_1 and z_0 is given approximately by $|f'(z_0)| \, |z_1 - z_0|$. Therefore we say that the transformation $w = f(z)$ changes small distances near z_0 by the *scale factor* $|f'(z_0)|$. For example, the scale factor of the transformation $w = f(z) = z^2$ near the point $z_0 = 1 + i$ is $|f'(1 + i)| = |2(1 + i)| = 2\sqrt{2}$.

It is also necessary to say a few things about the inverse transformation $z = g(w)$ of a conformal mapping $w = f(z)$ near a point z_0, where $f'(z_0) \neq 0$. A complete justification of the following relies on theorems studied in advanced calculus.* Let the mapping $w = f(z)$ be expressed in the coordinate form

(16) $u = u(x, y)$ and $v = v(x, y)$.

---

*See, for instance, R. Creighton Buck, *Advanced Calculus*, 3rd ed. (New York, McGraw-Hill Book Company), pp. 358–361, 1978.

The mapping in equations (16) represents a transformation from the xy plane into the uv plane, and the *Jacobian determinant* $J(x, y)$ is defined by

$$(17) \quad J(x, y) = \begin{vmatrix} u_x(x, y) & u_y(x, y) \\ v_x(x, y) & v_y(x, y) \end{vmatrix}.$$

It is known that the transformation in equations (16) has a local inverse provided that $J(x, y) \neq 0$. Expanding equation (17) and using the Cauchy-Riemann equations, we obtain

$$(18) \quad J(x_0, y_0) = u_x(x_0, y_0)v_y(x_0, y_0) - v_x(x_0, y_0)u_y(x_0, y_0)$$
$$= u_x^2(x_0, y_0) + v_x^2(x_0, y_0) = |f'(z_0)|^2 \neq 0.$$

Consequently, equations (17) and (18) imply that a local inverse $z = g(w)$ exists in a neighborhood of the point w_0. The derivative of g at w_0 is given by the familiar computation

$$(19) \quad g'(w_0) = \lim_{w \to w_0} \frac{g(w) - g(w_0)}{w - w_0}$$

$$= \lim_{z \to z_0} \frac{z - z_0}{f(z) - f(z_0)} = \frac{1}{f'(z_0)} = \frac{1}{f'(g(w_0))}.$$

EXERCISES FOR SECTION 9.1

1. State where the following mappings are conformal.
   (a) $w = \exp z$
   (b) $w = \sin z$
   (c) $w = z^2 + 2z$
   (d) $w = \exp(z^2 + 1)$
   (e) $w = \dfrac{1}{z}$
   (f) $w = \dfrac{z + 1}{z - 1}$

For Exercises 2–5, find the angle of rotation $\alpha = \arg f'(z)$ and the scale factor $|f'(z)|$ of the mapping $w = f(z)$ at the indicated points.

2. $w = 1/z$ at the points 1, $1 + i$, and i
3. $w = \ln r + i\theta$, where $-\pi/2 < \theta < 3\pi/2$ at the points 1, $1 + i$, i, and -1
4. $w = r^{1/2}\cos(\theta/2) + ir^{1/2}\sin(\theta/2)$, where $-\pi < \theta < \pi$, at the points i, 1, $-i$, and $3 + 4i$
5. $w = \sin z$ at the points $\pi/2 + i$, 0, and $-\pi/2 + i$
6. Consider the mapping $w = z^2$. If $a \neq 0$ and $b \neq 0$, show that the lines $x = a$ and $y = b$ are mapped onto orthogonal parabolas.
7. Consider the mapping $w = z^{1/2}$, where $z^{1/2}$ denotes the principal branch of the square root function. If $a > 0$ and $b > 0$, show that the lines $x = a$ and $y = b$ are mapped onto orthogonal curves.
8. Consider the mapping $w = \exp z$. Show that the lines $x = a$ and $y = b$ are mapped onto orthogonal curves.
9. Consider the mapping $w = \sin z$. Show that the line segment $-\pi/2 < x < \pi/2$, $y = 0$, and the vertical line $x = a$, where $|a| < \pi/2$ are mapped onto orthogonal curves.
10. Consider the mapping $w = \text{Log } z$, where $\text{Log } z$ denotes the principal branch of the logarithm function. Show that the positive x axis and the vertical line $x = 1$ are mapped onto orthogonal curves.

11. Let f be analytic at z_0 and $f'(z_0) \neq 0$. Show that the function $g(z) = \overline{f(z)}$ preserves the magnitude, but reverses the sense, of angles at z_0.
12. If $w = f(z)$ is a mapping, where $f(z)$ is not analytic, then what behavior would one expect regarding the angles between curves.
13. Write a report on conformal mapping. Your research could be theoretical and develop ideas not found in the text or practical and involve applications and/or computers. Resources include bibliographical items 33, 34, 35, 37, 41, 47, 48, 75, 92, 93, 96, 130, 136, 146, 154, 159, 164, 176, 180, and 182.

9.2   Bilinear Transformations

Another important class of elementary mappings was studied by Augustus Ferdinand Möbius (1790–1868). These mappings are conveniently expressed as the quotient of two linear expressions and are commonly known as linear fractional or bilinear transformations. They arise naturally in mapping problems involving the function arctan z. In this section we will show how they are used to map a disk one-to-one and onto a half plane.

Let a, b, c, and d denote four complex constants with the restriction that $ad \neq bc$. Then the function

$$(1) \quad w = S(z) = \frac{az + b}{cz + d}$$

is called a *bilinear transformation* or *Möbius transformation* or *linear fractional transformation*. If the expression for S in equation (1) is multiplied through by the quantity $cz + d$, then the resulting expression has the bilinear form $cwz - az + dw - b = 0$. We can collect terms involving z and write $z(cw - a) = -dw + b$. For values of $w \neq a/c$ the inverse transformation is given by

$$(2) \quad z = S^{-1}(w) = \frac{-dw + b}{cw - a}.$$

We can extend S and S^{-1} to mappings in the extended complex plane. The value $S(\infty)$ should be chosen to equal the limit of $S(z)$ as $z \to \infty$. Therefore we define

$$(3) \quad S(\infty) = \lim_{z \to \infty} S(z) = \lim_{z \to \infty} \frac{a + (b/z)}{c + (d/z)} = \frac{a}{c},$$

and the inverse is $S^{-1}(a/c) = \infty$. Similarly, the value $S^{-1}(\infty)$ is obtained by

$$(4) \quad S^{-1}(\infty) = \lim_{w \to \infty} S^{-1}(w) = \lim_{w \to \infty} \frac{-d + (b/w)}{c - (a/w)} = \frac{-d}{c},$$

and the inverse is $S(-d/c) = \infty$. With these extensions we conclude that the transformation $w = S(z)$ is a one-to-one mapping of the extended complex z plane onto the extended complex w plane.

We now show that a bilinear transformation carries the class of circles and lines onto itself. Let S be an arbitrary bilinear transformation given by equation (1).

If $c = 0$, then S reduces to a linear transformation, which carries lines onto lines and circles onto circles. If $c \neq 0$, then we can write S in the form

$$(5) \quad S(z) = \frac{a(cz + d) + bc - ad}{c(cz + d)} = \frac{a}{c} + \frac{bc - ad}{c} \frac{1}{cz + d}.$$

The condition $ad \neq bc$ precludes the possibility that S reduces to a constant. It is easy to see from equation (5) that S can be considered as a composition of functions. It is a linear mapping $\xi = cz + d$, followed by the reciprocal transformation $Z = 1/\xi$, followed by $w = (a/c) + [(bc - ad)/c]Z$. It was shown in Chapter 2 that each function in the composition maps the class of circles and lines onto itself, it follows that the bilinear transformation S has this property. A half plane can be considered a family of parallel lines and a disk as a family of circles. Therefore it is reasonable to conclude that a bilinear transformation maps the class of half planes and disks onto itself. Example 9.3 illustrates this idea.

EXAMPLE 9.3 Show that $w = S(z) = i(1 - z)/(1 + z)$ maps the unit disk $|z| < 1$ one-to-one and onto the upper half plane $\text{Im}(w) > 0$.

Solution Let us first consider the unit circle C: $|z| = 1$, which forms the boundary of the disk, and find its image in the w plane. If we write $S(z) = (-iz + i)/(z + 1)$, then we see that $a = -i$, $b = i$, $c = 1$, and $d = 1$. Using equation (2), we find that the inverse is given by

$$(6) \quad z = S^{-1}(w) = \frac{-dw + b}{cw - a} = \frac{-w + i}{w + i}.$$

If $|z| = 1$, then equation (6) implies that the images of points on the unit circle satisfy the equation

$$(7) \quad |w + i| = |-w + i|.$$

Squaring both sides of equation (7), we obtain $u^2 + (1 + v)^2 = u^2 + (1 - v)^2$, which can be simplified to yield $v = 0$, which is the equation of the u axis in the w plane.

The circle C divides the z plane into two portions, and its image is the u axis, which divides the w plane into two portions. Since the image of the point $z = 0$ is $w = S(0) = i$, we expect that the interior of the circle C is mapped onto the portion of the w plane that lies above the u axis. To show that this is true, we let $|z| < 1$. Then equation (6) implies that the image values must satisfy the inequality $|-w + i| < |w + i|$, which can be written as

$$(8) \quad d_1 = |w - i| < |w - (-i)| = d_2.$$

If we interpret d_1 as the distance from w to i and d_2 as the distance from w to $-i$, then a geometric argument shows that the image point w must lie in the upper half plane $\text{Im}(w) > 0$, as shown in Figure 9.5. Since S is one-to-one and onto in the extended complex plane, it follows that S maps the disk onto the half plane.

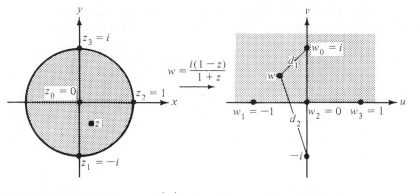

FIGURE 9.5 The image of $|z| < 1$ under $w = i(1 - z)/(1 + z)$.

The general formula (1) of a bilinear transformation appears to involve four independent coefficients a, b, c, d. But since $S(z) \neq K$, either $a \neq 0$ or $c \neq 0$, the transformation can be expressed with three unknown coefficients and can be written either

$$S(z) = \frac{z + b/a}{cz/a + d/a} \quad \text{or} \quad S(z) = \frac{az/c + b/c}{z + d/c},$$

respectively. This permits us to uniquely determine a bilinear transformation if three distinct image values $S(z_1) = w_1$, $S(z_2) = w_2$, and $S(z_3) = w_3$ are specified. To determine such a mapping, it is convenient to use an implicit formula involving z and w.

Theorem 9.3 (The Implicit Formula) *There exists a unique bilinear transformation that maps three distinct points z_1, z_2, and z_3 onto three distinct points w_1, w_2, and w_3, respectively. An implicit formula for the mapping is given by the equation*

(9)
$$\frac{z - z_1}{z - z_3} \frac{z_2 - z_3}{z_2 - z_1} = \frac{w - w_1}{w - w_3} \frac{w_2 - w_3}{w_2 - w_1}.$$

Proof Equation (9) can be algebraically manipulated, and we can solve for w in terms of z. The result will be an expression for w that has the form of equation (1), where the coefficients a, b, c, and d involve the values z_1, z_2, z_3, w_1, w_2, and w_3. The details are left as an exercise.

If we set $z = z_1$ and $w = w_1$ in equation (9), then both sides of the equation are zero. This shows that w_1 is the image of z_1. If we set $z = z_2$ and $w = w_2$ in equation (9), then both sides of the equation take on the value 1. Hence w_2 is the image of z_2. Taking reciprocals, we can write equation (9) in the form

(10)
$$\frac{z - z_3}{z - z_1} \frac{z_2 - z_1}{z_2 - z_3} = \frac{w - w_3}{w - w_1} \frac{w_2 - w_1}{w_2 - w_3}.$$

If we set $z = z_3$ and $w = w_3$ in equation (10), then both sides of the equation are zero. Therefore, w_3 is the image of z_3, and we have shown that the transformation has the required mapping properties.

EXAMPLE 9.4 Construct the bilinear transformation $w = S(z)$ that maps the points $z_1 = -i$, $z_2 = 1$, $z_3 = i$ onto the points $w_1 = -1$, $w_2 = 0$, $w_3 = 1$, respectively.

Solution We can use the implicit formula (10) and write

$$(11) \qquad \frac{z - i}{z + i} \frac{1 + i}{1 - i} = \frac{w - 1}{w + 1} \frac{0 + 1}{0 - 1} = \frac{-w + 1}{w + 1}.$$

Working with the left and right sides of equation (11), we obtain

$$(12) \qquad \begin{aligned} (1 + i)zw &+ (1 - i)w + (1 + i)z + (1 - i) \\ &= (-1 + i)zw + (-1 - i)w + (1 - i)z + (1 + i). \end{aligned}$$

Collecting terms involving w and zw on the left results in

$$(13) \qquad 2w + 2zw = 2i - 2iz.$$

After the 2's are cancelled in equation (13), we obtain $w(1 + z) = i(1 - z)$. Therefore the desired bilinear transformation is

$$w = S(z) = \frac{i(1 - z)}{1 + z}.$$

EXAMPLE 9.5 Find the bilinear transformation $w = S(z)$ that maps the points $z_1 = -2$, $z_2 = -1 - i$, and $z_3 = 0$ onto $w_1 = -1$, $w_2 = 0$, and $w_3 = 1$, respectively.

Solution We can use the implicit formula (9) and write

$$(14) \qquad \frac{z - (-2)}{z - 0} \frac{-1 - i - 0}{-1 - i - (-2)} = \frac{w - (-1)}{w - 1} \frac{0 - 1}{0 - (-1)}.$$

From the fact that $(-1 - i)/(1 - i) = 1/i$, equation (14) can be written as

$$(15) \qquad \frac{z + 2}{iz} = \frac{1 + w}{1 - w}.$$

Equation (15) is equivalent to $z + 2 - zw - 2w = iz + izw$, which can be solved for w in terms of z, giving the solution

$$w = S(z) = \frac{(1 - i)z + 2}{(1 + i)z + 2}.$$

Let D be a region in the z plane that is bounded by either a circle or a straight line C. Let z_1, z_2, and z_3 be three distinct points that lie on C with the property that an observer moving along C from z_1 to z_3 through z_2 finds the region D on the left. In the case that C is a circle and D is the interior of C we say that C is positively oriented. Conversely, the ordered triple z_1, z_2, z_3 uniquely determines a region that lies on the left of C.

Let G be a region in the w plane that is bounded by either a circle or a straight line K. Let w_1, w_2, and w_3 be three distinct points that lie on K such that an observer moving along K from w_1 to w_3 through w_2 finds the region G on the left. Since a bilinear transformation is a conformal mapping that maps the class of circles and straight lines onto itself, we can use the implicit formula (9) to construct a bilinear transformation $w = S(z)$ that is a one-to-one mapping of D onto G.

EXAMPLE 9.6   Show that the mapping

$$w = S(z) = \frac{(1 - i)z + 2}{(1 + i)z + 2}$$

maps the disk D: $\left|z + 1\right| < 1$ onto the upper half plane $\text{Im}(w) > 0$.

Solution   For convenience we choose the ordered triple $z_1 = -2$, $z_2 = -1 - i$, $z_3 = 0$, which will give the circle C: $\left|z + 1\right| = 1$ a positive orientation and the disk D a "left orientation." We saw in Example 9.5 that the corresponding image points are

$$w_1 = S(z_1) = -1, \quad w_2 = S(z_2) = 0, \quad \text{and} \quad w_3 = S(z_3) = 1.$$

Since the ordered triple of points w_1, w_2, and w_3 lie on the u axis, it follows that the image of the circle C is the u axis. The points w_1, w_2, and w_3 give the upper half plane G: $\text{Im}(w) > 0$ a "left orientation." Therefore $w = S(z)$ maps the disk D onto the upper half plane G. To check our work, we choose a point z_0 that lies in D and find the half plane where its image w_0 lies. The choice $z_0 = -1$ yields $w_0 = S(-1) = i$. Hence the upper half plane is the correct image. The situation is illustrated in Figure 9.6.

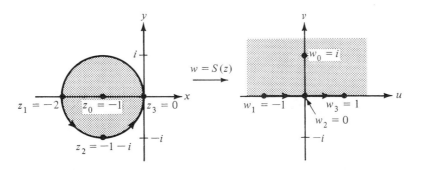

FIGURE 9.6   The bilinear mapping $w = S(z) = [(1 - i)z + 2]/[(1 + i)z + 2]$.

In equation (9) the point at infinity can be introduced as one of the prescribed points in either the z plane or the w plane. For example, if $w_3 = \infty$, then we are permitted to write

$$(16) \qquad \frac{w_2 - w_3}{w - w_3} = \frac{w_2 - \infty}{w - \infty} = 1,$$

and substitution of equation (16) into equation (9) yields

$$(17) \qquad \frac{z - z_1}{z - z_3} \frac{z_2 - z_3}{z_2 - z_1} = \frac{w - w_1}{w_2 - w_1}, \qquad \text{where } w_3 = \infty.$$

Equation (17) is sometimes used to map the crescent-shaped region that lies between tangent circles onto an infinite strip.

EXAMPLE 9.7   Find a bilinear transformation that maps the crescent-shaped region that lies inside the disk $|z - 2| < 2$ and outside the circle $|z - 1| = 1$ onto a horizontal strip.

Solution   For convenience we choose $z_1 = 4$, $z_2 = 2 + 2i$, and $z_3 = 0$ and the image values $w_1 = 0$, $w_2 = 1$, and $w_3 = \infty$, respectively. The ordered triple z_1, z_2, and z_3 gives the circle $|z - 2| = 2$ a positive orientation and the disk $|z - 2| < 2$ has a "left orientation." The image points w_1, w_2, and w_3 all lie on the extended u axis, and they determine a left orientation for the upper half plane $\text{Im}(w) > 0$. Therefore we can use the implicit formula (17) to write

$$(18) \qquad \frac{z - 4}{z - 0} \frac{2 + 2i - 0}{2 + 2i - 4} = \frac{w - 0}{1 - 0},$$

which determines a mapping of the disk $|z - 2| < 2$ onto the upper half plane $\text{Im}(w) > 0$. We can simplify equation (18) to obtain the desired solution

$$w = S(z) = \frac{-iz + 4i}{z}.$$

A straightforward calculation shows that the points $z_4 = 1 - i$, $z_5 = 2$, and $z_6 = 1 + i$ are mapped respectively onto the points

$$w_4 = S(1 - i) = -2 + i, \quad w_5 = S(2) = i, \quad \text{and} \quad w_6 = S(1 + i) = 2 + i$$

The points w_4, w_5, and w_6 lie on the horizontal line $\text{Im}(w) = 1$ in the upper half plane. Therefore the crescent-shaped region is mapped onto the horizontal strip $0 < \text{Im}(w) < 1$ as shown in Figure 9.7.

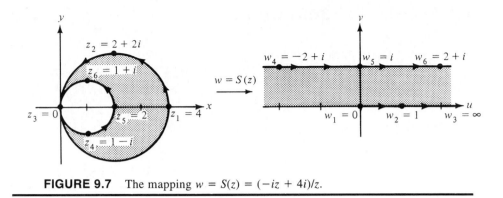

FIGURE 9.7 The mapping $w = S(z) = (-iz + 4i)/z$.

Lines of Flux

In the study of electronics, images of certain lines represent lines of electric flux, which are the trajectory of an electron that is placed in an electric field. Consider the bilinear transformation

$$w = S(z) = \frac{z}{z - a} \quad \text{and} \quad z = S^{-1}(w) = \frac{aw}{w - 1}.$$

The half rays $\{\arg(w) = c\}$, where c is a constant, that meet at the origin $w = 0$, represent the lines of electric flux produced by a source located at $w = 0$ (and a sink at $w = \infty$). The preimage of this family of lines is a family of circles that pass through the points $z = 0$ and $z = a$. We visualize these circles as the lines of electric flux from one point charge to another. The limiting case as $a \to 0$ is called a dipole and is discussed in Exercise 6 in Section 10.11. The graphs for the cases $a = 1$, $a = 0.5$, and $a = 0.1$ are shown in Figure 9.8

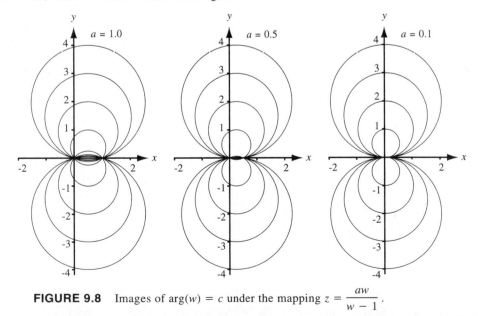

FIGURE 9.8 Images of $\arg(w) = c$ under the mapping $z = \dfrac{aw}{w - 1}$.

EXERCISES FOR SECTION 9.2

1. Let $w = S(z) = [(1 - i)z + 2]/[(1 + i)z + 2]$. Find $S^{-1}(w)$.
2. Let $w = S(z) = (i + z)/(i - z)$. Find $S^{-1}(w)$.
3. Find the image of the right half plane $\text{Re}(z) > 0$ under $w = i(1 - z)/(1 + z)$.
4. Show that the bilinear transformation $w = i(1 - z)/(1 + z)$ maps the portion of the disk $|z| < 1$ that lies in the upper half plane $\text{Im}(z) > 0$ onto the first quadrant $u > 0$, $v > 0$.
5. Find the image of the upper half plane $\text{Im}(z) > 0$ under the transformation

$$w = \frac{(1 - i)z + 2}{(1 + i)z + 2}.$$

6. Find the bilinear transformation $w = S(z)$ that maps the points $z_1 = 0$, $z_2 = i$, and $z_3 = -i$ onto $w_1 = -1$, $w_2 = 1$, and $w_3 = 0$, respectively.
7. Find the bilinear transformation $w = S(z)$ that maps the points $z_1 = -i$, $z_2 = 0$, and $z_3 = i$ onto $w_1 = -1$, $w_2 = i$, and $w_3 = 1$, respectively.
8. Find the bilinear transformation $w = S(z)$ that maps the points $z_1 = 0$, $z_2 = 1$, and $z_3 = 2$ onto $w_1 = 0$, $w_2 = 1$, and $w_3 = \infty$, respectively.
9. Find the bilinear transformation $w = S(z)$ that maps the points $z_1 = 1$, $z_2 = i$, and $z_3 = -1$ onto $w_1 = 0$, $w_2 = 1$, and $w_3 = \infty$, respectively.
10. Show that the transformation $w = (i + z)/(i - z)$ maps the unit disk $|z| < 1$ onto the right half plane $\text{Re}(w) > 0$.
11. Find the image of the lower half plane $\text{Im}(z) < 0$ under $w = (i + z)/(i - z)$.
12. Let $S_1(z) = (z - 2)/(z + 1)$ and $S_2(z) = z/(z + 3)$. Find $S_1(S_2(z))$ and $S_2(S_1(z))$.
13. Find the image of the quadrant $x > 0$, $y > 0$ under $w = (z - 1)/(z + 1)$.
14. Find the image of the horizontal strip $0 < y < 2$ under $w = z/(z - i)$.
15. Show that equation (9) can be written in the form of equation (1).
16. Show that the bilinear transformation $w = S(z) = (az + b)/(cz + d)$ is conformal at all points $z \neq -d/c$.
17. A *fixed point* of a mapping $w = f(z)$ is a point z_0 such that $f(z_0) = z_0$. Show that a bilinear transformation can have at most two fixed points.
18. (a) Find the fixed points of $w = (z - 1)/(z + 1)$.
    (b) Find the fixed points of $w = (4z + 3)/(2z - 1)$.
19. Write a report on bilinear transformations. Include some ideas not presented in the text. Resources include bibliographical items 12, 23, 24, 30, 36, and 43.

9.3   Mappings Involving Elementary Functions

In Section 5.1 we saw that the function $w = f(z) = \exp z$ is a one-to-one mapping of the fundamental period strip $-\pi < y \leq \pi$ in the z plane onto the w plane with the point $w = 0$ deleted. Since $f'(z) \neq 0$, the mapping $w = \exp z$ is a conformal mapping at each point z in the complex plane. The family of horizontal lines $y = c$

and $-\pi < c \le \pi$, and the segments $x = a$ and $-\pi < y \le \pi$ form an orthogonal grid in the fundamental period strip. Their images under the mapping $w = \exp z$ are the rays $\rho > 0$ and $\phi = c$ and the circles $|w| = e^a$, respectively. These images form an orthogonal curvilinear grid in the w plane, as shown in Figure 9.9. If $-\pi < c < d \le \pi$, then the rectangle $R = \{x + iy: a < x < b, c < y < d\}$ is mapped one-to-one and onto the region $G = \{\rho e^{i\phi}: e^a < \rho < e^b, c < \phi < d\}$. The inverse mapping is the principal branch of the logarithm $z = \text{Log } w$.

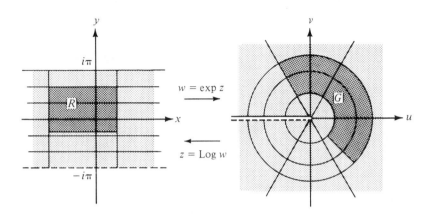

FIGURE 9.9   The conformal mapping $w = \exp z$.

In this section we will show how compositions of conformal transformations are used to construct mappings with specified characteristics.

EXAMPLE 9.8   The transformation $w = f(z) = (e^z - i)/(e^z + i)$ is a one-to-one conformal mapping of the horizontal strip $0 < y < \pi$ onto the disk $|w| < 1$. Furthermore, the x axis is mapped onto the lower semicircle bounding the disk, and the line $y = \pi$ is mapped onto the upper semicircle.

Solution   The function $w = f(z)$ can be considered as a composition of the exponential mapping $Z = \exp z$ followed by the bilinear transformation $w = (Z - i)/(Z + i)$. The image of the horizontal strip $0 < y < \pi$ under the mapping $Z = \exp z$ is the upper half plane $\text{Im}(Z) > 0$; the x axis is mapped onto the positive X axis; and the line $y = \pi$ is mapped onto the negative X axis. The bilinear transformation $w = (Z - i)/(Z + i)$ then maps the upper half plane $\text{Im}(Z) > 0$ onto the disk $|w| < 1$; the positive X axis is mapped onto the lower semicircle; and the negative X axis onto the upper semicircle. Figure 9.10 illustrates the composite mapping.

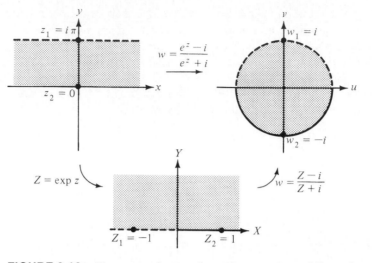

FIGURE 9.10   The composite transformation $w = (e^z - i)/(e^z + i)$.

EXAMPLE 9.9   The transformation $w = f(z) = \text{Log}[(1 + z)/(1 - z)]$ is a one-to-one conformal mapping of the unit disk $|z| < 1$ onto the horizontal strip $|v| < \pi/2$. Furthermore, the upper semicircle of the disk is mapped onto the line $v = \pi/2$ and the lower semicircle onto $v = -\pi/2$.

Solution   The function $w = f(z)$ is the composition of the bilinear transformation $Z = (1 + z)/(1 - z)$ followed by the logarithmic mapping $w = \text{Log } Z$. The image of the disk $|z| < 1$ under the bilinear mapping $Z = (1 + z)/(1 - z)$ is the right half plane $\text{Re}(Z) > 0$; the upper semicircle is mapped onto the positive Y axis; and the lower semicircle is mapped onto the negative Y axis. The logarithmic function $w = \text{Log } Z$ then maps the right half plane onto the horizontal strip; the image of the positive Y axis is the line $v = \pi/2$; and the image of the negative Y axis is the line $v = -\pi/2$. Figure 9.11 shows the composite mapping.

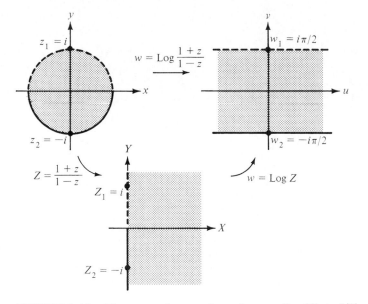

FIGURE 9.11   The composite transformation $w = \text{Log}[(1 + z)/(1 - z)]$.

EXAMPLE 9.10   The transformation $w = f(z) = (1 + z)^2/(1 - z)^2$ is a one-to-one conformal mapping of the portion of the disk $|z| < 1$ that lies in the upper half plane $\text{Im}(z) > 0$ onto the upper half plane $\text{Im}(w) > 0$. Furthermore, the image of the semicircular portion of the boundary is mapped onto the negative u axis, and the segment $-1 < x < 1$, $y = 0$ is mapped onto the positive u axis.

Solution   The function $w = f(z)$ is the composition of the bilinear transformation $Z = (1 + z)/(1 - z)$ followed by the mapping $w = Z^2$. The image of the half-disk under the bilinear mapping $Z = (1 + z)/(1 - z)$ is the first quadrant $X > 0$, $Y > 0$; the image of the segment $y = 0$, $-1 < x < 1$, is the positive X axis; and the image of the semicircle is the positive Y axis. The mapping $w = Z^2$ then maps the first quadrant in the Z plane onto the upper half plane $\text{Im}(w) > 0$, as shown in Figure 9.12.

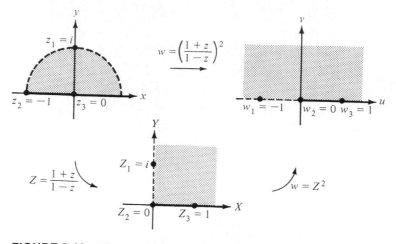

FIGURE 9.12   The composite transformation $w = [(1 + z)/(1 - z)]^2$.

EXAMPLE 9.11   Consider the function $w = f(z) = (z^2 - 1)^{1/2}$, which is the composition of the functions $Z = z^2 - 1$ and $w = Z^{1/2}$, where the branch of the square root is $Z^{1/2} = R^{1/2}[\cos(\theta/2) + i\sin(\theta/2)]$, ($\theta = \arg Z$ and $0 \le \theta < 2\pi$). Then the transformation $w = f(z)$ maps the upper half plane $\text{Im}(z) > 0$ one-to-one and onto the upper half plane $\text{Im}(w) > 0$ slit along the segment $u = 0$, $0 < v \le 1$.

Solution   The function $Z = z^2 - 1$ maps the upper half plane $\text{Im}(z) > 0$ one-to-one and onto the Z-plane slit along the ray $Y = 0$, $X \ge -1$. Then the function $w = Z^{1/2}$ maps the slit plane onto the slit half plane, as shown in Figure 9.13.

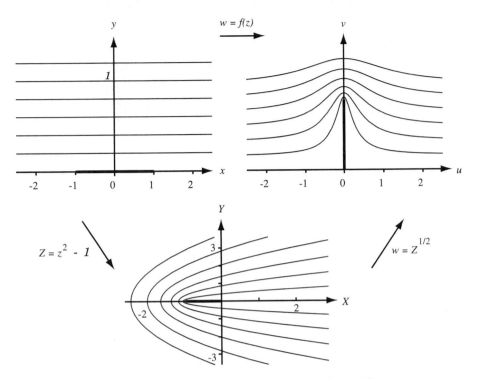

FIGURE 9.13   The composite transformation $w = f(z) = (z^2 - 1)^{1/2}$, and the intermediate steps $Z = z^2 - 1$ and $w = Z^{1/2}$.

Remark   The images of the horizontal lines $y = b$ are curves in the w plane that bend around the segment from 0 to i. The curves represent the streamlines of a fluid flowing across the w plane. We will study fluid flows in more detail in Section 10.7.

The Mapping $w = (z^2 - 1)^{1/2}$

The double-valued function $f(z) = (z^2 - 1)^{1/2}$ has a branch that is continuous for values of z distant from the origin. This feature is motivated by our desire for the approximation $(z^2 - 1)^{1/2} \approx z$ to hold for values of z distant from the origin. Let us express $(z^2 - 1)^{1/2}$ in the following form:

$$(1) \quad w = f_1(z) = (z - 1)^{1/2}(z + 1)^{1/2},$$

where the principal branch of the square root function is used in both factors. We claim that the mapping $w = f_1(z)$ is a one-to-one conformal mapping from the domain set D_1 consisting of the z plane slit along the segment $-1 \le x \le 1, y = 0$, onto the range set H_1 consisting of the w plane slit along the segment $u = 0, -1 \le v \le 1$.

To verify this we investigate the two formulas on the right side of equation (1), and express them in the form

(2) $(z - 1)^{1/2} = \sqrt{r_1}e^{i\theta_1/2}$,   where $r_1 = |z - 1|$, $\theta_1 = \text{Arg}(z - 1)$,   and

(3) $(z + 1)^{1/2} = \sqrt{r_2}e^{i\theta_2/2}$,   where $r_2 = |z + 1|$, $\theta_2 = \text{Arg}(z + 1)$.

The discontinuities of $\text{Arg}(z - 1)$ and $\text{Arg}(z + 1)$ are points on the real axis such that $x \leq 1$ and $x \leq -1$, respectively. We now show that $f_1(z)$ is continuous on the ray $x < -1$, $y = 0$.

Let $z_0 = x_0 + iy_0$ denote a point on the ray $x < -1$, $y = 0$, then we obtain the following limits as z approaches z_0 from the upper half plane and the lower half plane, respectively:

(4) $$\lim_{\substack{z \to z_0 \\ \text{Im}(z) > 0}} f_1(z) = \lim_{\substack{r_1 \to |x_0 - 1| \\ \theta_1 \to \pi}} \sqrt{r_1}e^{i\theta_1/2} \lim_{\substack{r_2 \to |x_0 + 1| \\ \theta_2 \to \pi}} \sqrt{r_2}e^{i\theta_2/2}$$

$$= \sqrt{|x_0 - 1|}\,(i)\,\sqrt{|x_0 + 1|}\,(i)$$

$$= -\sqrt{|x_0^2 - 1|},\quad \text{and}$$

(5) $$\lim_{\substack{z \to z_0 \\ \text{Im}(z) < 0}} f_1(z) = \lim_{\substack{r_1 \to |x_0 - 1| \\ \theta_1 \to -\pi}} \sqrt{r_1}e^{i\theta_1/2} \lim_{\substack{r_2 \to |x_0 + 1| \\ \theta_2 \to -\pi}} \sqrt{r_2}e^{i\theta_2/2}$$

$$= \sqrt{|x_0 - 1|}\,(-i)\,\sqrt{|x_0 + 1|}\,(-i)$$

$$= -\sqrt{|x_0^2 - 1|}.$$

Since both limits agree with the value of $f_1(z_0)$ it follows that $f_1(z)$ is continuous along the ray $x < -1$, $y = 0$.

The inverse mapping is easily found and can be expressed in a similar form:

(6) $z = g_1(w) = (w^2 + 1)^{1/2} = (w + i)^{1/2}(w - i)^{1/2}$,

where the branches of the square root function are given by

(7) $(w + i)^{1/2} = \sqrt{\rho_1}e^{i\phi_1/2}$,   where $\rho_1 = |w + i|$, $\phi_1 = \arg(w + i)$,

$$\text{and } \frac{-\pi}{2} < \arg(w + i) < \frac{-3\pi}{2},\quad \text{and}$$

(8) $(w - i)^{1/2} = \sqrt{\rho_2}e^{i\phi_2/2}$,   where $\rho_2 = |w - i|$, $\phi_2 = \arg(w - i)$,

$$\text{and } \frac{-\pi}{2} < \arg(w + i) < \frac{-3\pi}{2}.$$

A similar argument will show that $g_1(w)$ is continuous for all w except those points that lie on the segment $u = 0$, $-1 \leq v \leq 1$. It is straightforward to verify that

(9) $g_1(f_1(z)) = z$ and $f_1(g_1(w)) = w$,

hold for z in D_1 and w in H_1, respectively, Therefore we conclude that $w = f_1(z)$ is a one-to-one mapping from D_1 onto H_1. It is tedious to verify that $f_1(z)$ is also analytic on the ray $x < -1$, $y = 0$. We leave this verification as a challenging exercise.

The Riemann Surface for $w = (z^2 - 1)^{1/2}$

Using the other branch of the square root, we find that $w = f_2(z) = -f_1(z)$, is a one-to-one conformal mapping from the domain set D_2 consisting of the z plane slit along the segment $-1 \le x \le 1$, $y = 0$, onto the range set H_2 consisting of the w plane slit along the segment $u = 0$, $-1 \le v \le 1$. The sets D_1 and H_2 for $f_1(z)$ and D_2 and H_2 for $f_2(z)$ are shown in Figure 9.14.

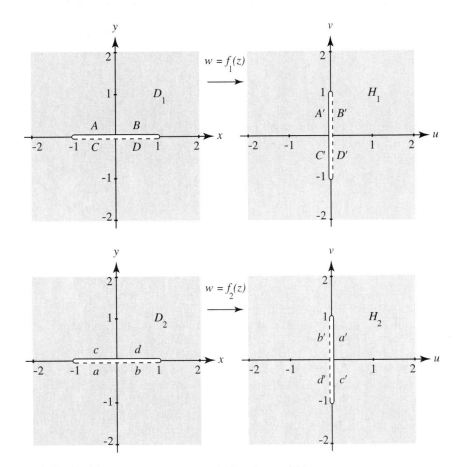

FIGURE 9.14 The mappings $w = f_1(z)$ and $w = f_2(z)$.

The Riemann surface for $w = (z^2 - 1)^{1/2}$ is obtained by gluing the edges of D_1 and D_2 together and H_1 and H_2 together in the following manner. In the domain set, glue edges A to a, B to b, C to c, and D to d. In the image set, glue edges A' to a', B' to b', C' to c', and D' to d'. The result is a Riemann domain surface and Riemann image surface for the mapping, see Figures 9.15(a) and 9.15(b), respectively.

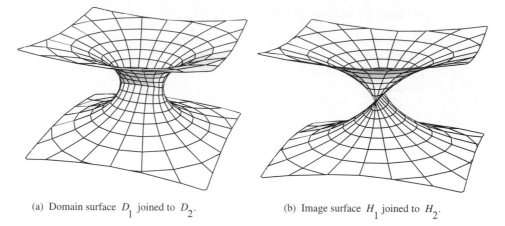

(a) Domain surface D_1 joined to D_2.   (b) Image surface H_1 joined to H_2.

FIGURE 9.15   The Riemann surfaces for the mapping $w = (z^2 - 1)^{1/2}$.

EXERCISES FOR SECTION 9.3

1. Find the image of the semi-infinite strip $0 < x < \pi/2$, $y > 0$, under the transformation $w = \exp iz$.
2. Find the image of the rectangle $0 < x < \ln 2$, $0 < y < \pi/2$, under the transformation $w = \exp z$.
3. Find the image of the first quadrant $x > 0$, $y > 0$, under $w = (2/\pi) \operatorname{Log} z$.
4. Find the image of the annulus $1 < |z| < e$ under $w = \operatorname{Log} z$.
5. Show that the multivalued function $w = \log z$ maps the annulus $1 < |z| < e$ onto the vertical strip $0 < \operatorname{Re}(w) < 1$.
6. Show that $w = (2 - z^2)/z^2$ maps the portion of the right half plane $\operatorname{Re}(z) > 0$ that lies to the right of the hyperbola $x^2 - y^2 = 1$ onto the unit disk $|w| < 1$.
7. Show that the function $w = (e^z - i)/(e^z + i)$ maps the horizontal strip $-\pi < \operatorname{Im}(z) < 0$ onto the region $1 < |w|$.
8. Show that $w = (e^z - 1)/(e^z + 1)$ maps the horizontal strip $|y| < \pi/2$ onto the unit disk $|w| < 1$.
9. Find the image of the upper half plane $\operatorname{Im}(z) > 0$ under $w = \operatorname{Log}[(1 + z)/(1 - z)]$.
10. Find the image of the portion of the upper half plane $\operatorname{Im}(z) > 0$ that lies outside the circle $|z| = 1$ under the transformation $w = \operatorname{Log}[(1 + z)/(1 - z)]$.
11. Show that the function $w = (1 + z)^2/(1 - z)^2$ maps the portion of the disk $|z| < 1$ that lies in the first quadrant onto the portion of the upper half plane $\operatorname{Im}(w) > 0$ that lies outside the unit disk.

12. Find the image of the upper half plane $\text{Im}(z) > 0$ under $w = \text{Log}(1 - z^2)$.
13. Find the branch of $w = (z^2 + 1)^{1/2}$ that maps the right half plane $\text{Re}(z) > 0$ onto the right half plane $\text{Re}(w) > 0$ slit along the segment $0 < u \leq 1$, $v = 0$.
14. Show that the transformation $w = (z^2 - 1)/(z^2 + 1)$ maps the portion of the first quadrant $x > 0$, $y > 0$, that lies outside the circle $|z| = 1$ onto the first quadrant $u > 0$, $v > 0$.
15. Find the image of the sector $r > 0$, $0 < \theta < \pi/4$, under $w = (i - z^4)/(i + z^4)$.
16. Write a report on Riemann surfaces. Resources include bibliographical items 99, 128, and 129.
17. Show that the function $f_1(z)$ in equation (1) is analytic on the ray $x < -1$, $y = 0$.

9.4   Mapping by Trigonometric Functions

The trigonometric functions can be expressed with compositions that involve the exponential function followed by a bilinear function. We will be able to find images of certain regions by following the shapes of successive images in the composite mapping.

EXAMPLE 9.12   The transformation $w = \tan z$ is a one-to-one conformal mapping of the vertical strip $|x| < \pi/4$ onto the unit disk $|w| < 1$.

Solution   Using identities (11) and (12) in Section 5.4, we write

$$(1) \quad w = \tan z = \frac{1}{i} \frac{e^{iz} - e^{-iz}}{e^{iz} + e^{-iz}} = \frac{-ie^{i2z} + i}{e^{i2z} + 1}.$$

From equation (1) it is easy to see that the mapping $w = \tan z$ can be considered as the composition

$$(2) \quad w = \frac{-iZ + i}{Z + 1} \quad \text{and} \quad Z = e^{i2z}.$$

The function $Z = \exp(i2z)$ maps the vertical strip $|x| < \pi/4$ one-to-one and onto the right half plane $\text{Re}(Z) > 0$. Then the bilinear transformation $w = (-iZ + i)/(Z + 1)$ maps the half plane one-to-one and onto the disk as shown in Figure 9.16.

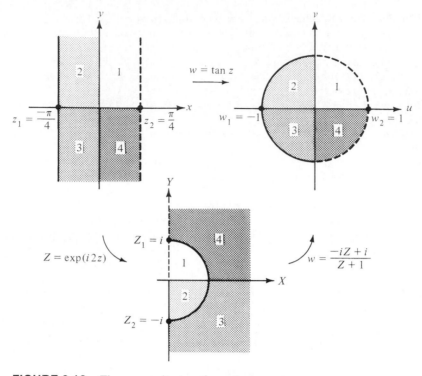

FIGURE 9.16 The composite transformation $w = \tan z$.

EXAMPLE 9.13 The transformation $w = f(z) = \sin z$ is a one-to-one conformal mapping of the vertical strip $|x| < \pi/2$ onto the w plane slit along the rays $u \le -1$, $v = 0$ and $u \ge 1$, $v = 0$.

Solution Since $f'(z) = \cos z \ne 0$ for values of z satisfying the inequality $-\pi/2 < \text{Re}(z) < \pi/2$, it follows that $w = \sin z$ is a conformal mapping. Using equation (14) in Section 5.2, we write

$$(3) \quad u + iv = \sin z = \sin x \cosh y + i \cos x \sinh y.$$

If $|a| < \pi/2$, then the image of the vertical line $x = a$ is the curve in the w plane given by the parametric equations

$$(4) \quad u = \sin a \cosh y \quad \text{and} \quad v = \cos a \sinh y$$

for $-\infty < y < \infty$. We can rewrite equations (4) in the form

$$(5) \quad \cosh y = \frac{u}{\sin a} \quad \text{and} \quad \sinh y = \frac{v}{\cos a}.$$

We can eliminate y in equations (5) by squaring and using the hyperbolic identity $\cosh^2 y - \sinh^2 y = 1$, and the result is the single equation

(6) $$\frac{u^2}{\sin^2 a} - \frac{v^2}{\cos^2 a} = 1.$$

The curve given by equation (6) is identified as a hyperbola in the (u, v) plane that has foci at the points $(\pm 1, 0)$. Therefore the vertical line $x = a$ is mapped in a one-to-one manner onto the branch of the hyperbola given by equation (6) that passes through the point $(\sin a, 0)$. If $0 < a < \pi/2$, then it is the right branch; and if $-\pi/2 < a < 0$, it is the left branch. The image of the y axis, which is the line $x = 0$, is the v axis. The images of several vertical lines are shown in Figure 9.17.

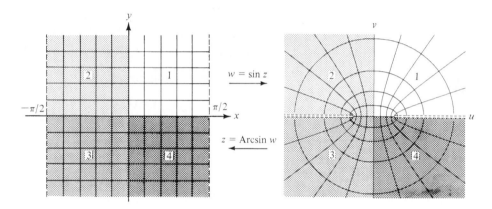

FIGURE 9.17   The transformation $w = \sin z$.

The image of the horizontal segment $-\pi/2 < x < \pi/2$, $y = b$ is the curve in the w plane given by the parametric equations

(7) $u = \sin x \cosh b$ and $v = \cos x \sinh b$

for $-\pi/2 < x < \pi/2$. We can rewrite equations (7) in the form

(8) $\sin x = \dfrac{u}{\cosh b}$ and $\cos x = \dfrac{v}{\sinh b}.$

We can eliminate x in equations (8) by squaring and using the trigonometric identity $\sin^2 x + \cos^2 x = 1$, and the result is the single equation

(9) $$\frac{u^2}{\cosh^2 b} + \frac{v^2}{\sinh^2 b} = 1.$$

The curve given by equation (9) is identified as an ellipse in the (u, v) plane that passes through the points $(\pm \cosh b, 0)$ and $(0, \pm \sinh b)$ and has foci at the points $(\pm 1, 0)$. Therefore if $b > 0$, then $v = \cos x \sinh b > 0$, and the image of the horizontal segment is the portion of the ellipse given by equation (9) that lies in the upper half plane $\text{Im}(w) > 0$. If $b < 0$, then it is the portion that lies in the lower half plane. The images of several segments are shown in Figure 9.17.

We now develop explicit formulas for the real and imaginary parts of the principal value of the arcsine function $w = f(z) = \text{Arcsin } z$. This mapping will be used to solve certain problems involving steady temperatures and ideal fluid flow in Section 10.7. The mapping is found by solving the equation

(10) $x + iy = \sin w = \sin u \cosh v + i \cos u \sinh v$

for u and v expressed as functions of x and y. To solve for u, we first equate the real and imaginary parts of equation (10) and obtain the system of equations

(11) $\cosh v = \dfrac{x}{\sin u}$ and $\sinh v = \dfrac{y}{\cos u}$.

Then we eliminate v in equations (11) and obtain the single equation

(12) $\dfrac{x^2}{\sin^2 u} - \dfrac{y^2}{\cos^2 u} = 1$.

If we treat u as a constant, then equation (12) represents a hyperbola in the (x, y) plane, the foci occur at the points $(\pm 1, 0)$, and the traverse axis is given by $2 \sin u$. Therefore a point (x, y) on the hyperbola must satisfy the equation

(13) $2 \sin u = \sqrt{(x + 1)^2 + y^2} - \sqrt{(x - 1)^2 + y^2}$.

The quantity on the right side of equation (13) represents the difference of the distances from (x, y) to $(-1, 0)$ and from (x, y) to $(1, 0)$.

Solving equation (13) for u yields the real part

(14) $u(x, y) = \arcsin \left[\dfrac{\sqrt{(x + 1)^2 + y^2} - \sqrt{(x - 1)^2 + y^2}}{2} \right]$.

The principal branch of the real function $\arcsin t$ is used in equation (14), where the range values satisfy the inequality $-\pi/2 < \arcsin t < \pi/2$.

Similarly, we can start with equation (10) and obtain the system of equations

(15) $\sin u = \dfrac{x}{\cosh v}$ and $\cos u = \dfrac{y}{\sinh v}$.

Then we eliminate u in equations (15) and obtain the single equation

(16) $\dfrac{x^2}{\cosh^2 v} + \dfrac{y^2}{\sinh^2 v} = 1$.

If we treat v as a constant, then equation (16) represents an ellipse in the (x, y) plane, the foci occur at the points $(\pm 1, 0)$, and the major axis has length $2 \cosh v$. Therefore a point (x, y) on this ellipse must satisfy the equation

(17) $2 \cosh v = \sqrt{(x + 1)^2 + y^2} + \sqrt{(x - 1)^2 + y^2}$.

The quantity on the right side of equation (17) represents the sum of the distances from (x, y) to $(-1, 0)$ and from (x, y) to $(1, 0)$.

The function $z = \sin w$ maps points in the upper-half (lower-half) of the vertical strip $-\pi/2 < u < \pi/2$ onto the upper half plane (lower half plane), respectively. Hence we can solve equation (17) to obtain v as a function of x and y:

$$(18) \quad v(x, y) = (\text{sign } y)\text{arccosh}\left[\frac{\sqrt{(x + 1)^2 + y^2} + \sqrt{(x - 1)^2 + y^2}}{2}\right],$$

where sign $y = +1$ if $y \geq 0$ and sign $y = -1$ if $y < 0$. The real function given by arccosh $t = \ln(t + \sqrt{t^2 - 1})$ with $t \geq 1$ is used in equation (18).

Therefore the mapping $w = \text{Arcsin } z$ is a one-to-one conformal mapping of the z plane cut along the rays $x \leq -1$, $y = 0$ and $x \geq 1$, $y = 0$ onto the vertical strip $-\pi/2 < u < \pi/2$ in the w plane. The Arcsine transformation is indicated in Figure 9.17. The formulas in equations (14) and (18) are also convenient for evaluating Arcsin z as shown in Example 9.14.

Therefore, the mapping $w = \text{Arcsin } z$ is one-to-one conformal mapping of the z plane cut along the rays $x \leq -1$, $y = 0$ and $x \geq 1$, $y = 0$ onto the vertical strip $\dfrac{-\pi}{2} \leq u \leq \dfrac{\pi}{2}$ in the w plane, and this can be construed from Figure 9.17 if we interchange the roles of the z and w planes. The image of the square $0 \leq x \leq 4$, $0 \leq y \leq 4$ under $w = \text{Arcsin } z$ is shown in Figure 9.18 and was obtained by plotting two families of curves $\{(u(c, t), v(c, t)): 0 \leq t \leq 4\}$ and $\{(u(t, c), v(t, c)): 0 \leq t \leq 4\}$, where $c = \dfrac{k}{5}$, $k = 0, 1, \ldots, 20$. Formulas (14) and (18) are also convenient for evaluating Arcsin z, as shown in Example 9.14.

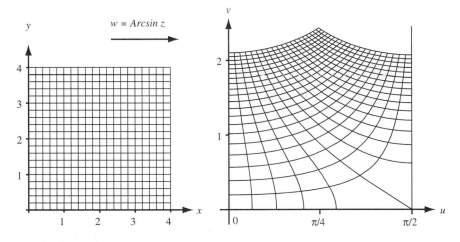

FIGURE 9.18   The mapping $w = \text{Arcsin } z$.

EXAMPLE 9.14   Find the principal value Arcsin(1 + *i*).

Solution   Using formulas (14) and (18), we find that

$$u(1, 1) = \arcsin \frac{\sqrt{5} - 1}{2} \approx 0.666239432 \quad \text{and}$$

$$v(1, 1) = \text{arccosh} \frac{\sqrt{5} + 1}{2} \approx 1.061275062.$$

Therefore we obtain

$$\arcsin(1 + i) \approx 0.666239432 + i\, 1.061275062.$$

Is there any reason to assume that there exists a conformal mapping for some specified domain *D* onto another domain *G*? The theorem concerning the existence of conformal mappings is attributed to Riemann and can be found in Lars V. Ahlfors, *Complex Analysis* (New York: McGraw-Hill Book Co.) Chapter 6, 1966.

Theorem 9.4 (Riemann Mapping Theorem)   *If D is any simply connected domain in the plane (other than the entire plane itself), then there exists a one-to-one conformal mapping w = f(z) that maps D onto the unit disk* $|w| < 1$.

EXERCISES FOR SECTION 9.4

1. Find the image of the semi-infinite strip $-\pi/4 < x < 0$, $y > 0$ under the mapping $w = \tan z$.
2. Find the image of the vertical strip $0 < \text{Re}(z) < \pi/2$ under the mapping $w = \tan z$.
3. Find the image of the vertical line $x = \pi/4$ under the transformation $w = \sin z$.
4. Find the image of the horizontal line $y = 1$ under the transformation $w = \sin z$.
5. Find the image of the rectangle $R = \{x + iy: 0 < x < \pi/4, 0 < y < 1\}$ under the transformation $w = \sin z$.
6. Find the image of the semi-infinite strip $-\pi/2 < x < 0$, $y > 0$ under the mapping $w = \sin z$.
7. (a) Find $\lim\limits_{y \to +\infty} \text{Arg}(\sin[(\pi/6) + iy])$.
   (b) Find $\lim\limits_{y \to +\infty} \text{Arg}(\sin[(-2\pi/3) + iy])$.
8. Use formulas (14) and (18) to find the following:
   (a) Arcsin(2 + 2*i*)   (b) Arcsin(−2 + *i*)
   (c) Arcsin(1 − 3*i*)   (d) Arcsin(−4 − *i*)
9. Show that the function $w = \sin z$ maps the rectangle $R = \{x + iy: -\pi/2 < x < \pi/2, 0 < y < b\}$ one-to-one and onto the portion of the upper half plane $\text{Im}(w) > 0$ that lies inside the ellipse

$$\frac{u^2}{\cosh^2 b} + \frac{v^2}{\sinh^2 b} = 1.$$

10. Find the image of the vertical strip $-\pi/2 < x < 0$ under the mapping $w = \cos z$.
11. Find the image of the horizontal strip $0 < \text{Im}(z) < \pi/2$ under the mapping $w = \sinh z$.
12. Find the image of the right half plane $\text{Re}(z) > 0$ under the mapping

$$w = \arctan z = \frac{i}{2} \text{Log} \frac{i+z}{i-z}.$$

13. Find the image of the first quadrant $x > 0$, $y > 0$ under $w = \text{Arcsin } z$.
14. Find the image of the first quadrant $x > 0$, $y > 0$ under the mapping $w = \text{Arcsin } z^2$.
15. Show that the transformation $w = \sin^2 z$ is a one-to-one conformal mapping of the semi-infinite strip $0 < x < \pi/2$, $y > 0$ onto the upper half plane $\text{Im}(w) > 0$.
16. Find the image of the semi-infinite strip $|x| < \pi/2$, $y > 0$ under the mapping $w = \text{Log}(\sin z)$.
17. Write a report on Riemann mapping theorem. Resources include bibliographical items 4, 88, 106, 129, and 179.
18. Write a report on the topic of analytic continuation. Be sure to discuss the chain of power series and disks of convergence. Resources include bibliographical items 4, 19, 46, 51, 52, 93, 106, 128, 129, 141, 145, and 166.

10

Applications of Harmonic Functions

10.1 Preliminaries

In most applications involving harmonic functions it is required to find a harmonic function that takes on prescribed values along certain contours. We will assume that the reader is familiar with the material in Sections 2.5, 3.3, 5.1, and 5.2.

EXAMPLE 10.1 Find the function $u(x, y)$ that is harmonic in the vertical strip $a \le \text{Re}(z) \le b$ and takes on the boundary values

(1) $u(a, y) = U_1$ and $u(b, y) = U_2$

along the vertical lines $x = a$ and $x = b$, respectively.

   Solution   Intuition suggests that we should seek a solution that takes on constant values along the vertical lines $x = x_0$ and that $u(x, y)$ should be a function of x alone; that is,

(2) $u(x, y) = P(x)$ for $a \le x \le b$ and for all y.

Laplace's equation, $u_{xx}(x, y) + u_{yy}(x, y) = 0$, implies that $P''(x) = 0$, so $P(x) = mx + c$, where m and c are constants. The boundary conditions $u(a, y) = P(a) = U_1$ and $u(b, y) = P(b) = U_2$ lead to the solution

(3) $u(x, y) = U_1 + \dfrac{U_2 - U_1}{b - a} (x - a).$

The level curves $u(x, y) =$ constant are vertical lines as indicated in Figure 10.1.

EXAMPLE 10.2 Find the function $\Psi(x, y)$ that is harmonic in the sector $0 < \text{Arg } z < \alpha$, where $\alpha \le \pi$, and takes on the boundary values

(4) $\Psi(x, 0) = C_1$ for $x > 0$ and
        $\Psi(x, y) = C_2$ at points on the ray $r > 0$, $\theta = \alpha$.

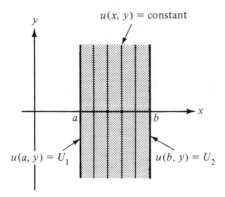

FIGURE 10.1 The harmonic function
$u(x, y) = U_1 + (U_2 - U_1)(x - a)/(b - a)$.

Solution If we recall that the function Arg z is harmonic and takes on constant values along rays emanating from the origin, then we see that a solution has the form

(5) $\quad \Psi(x, y) = a + b \text{ Arg } z$,

where a and b are constants. Boundary conditions (4) lead to

(6) $\quad \Psi(x, y) = C_1 + \dfrac{C_2 - C_1}{\alpha} \text{ Arg } z$.

The situation is shown in Figure 10.2.

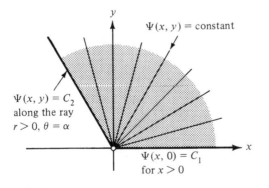

FIGURE 10.2 The harmonic function
$\Psi(x, y) = C_1 + (C_2 - C_1)(1/\alpha) \text{ Arg } z$.

EXAMPLE 10.3 Find the function $\Phi(x, y)$ that is harmonic in the annulus $1 < |z| < R$ and takes on the boundary values

(7) $\quad \Phi(x, y) = K_1$, when $|z| = 1$, and
$\quad \Phi(x, y) = K_2$, when $|z| = R$.

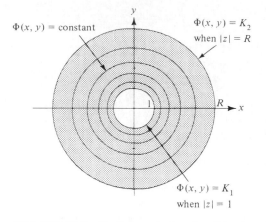

FIGURE 10.3   The harmonic function $\Phi(x, y) = K_1 + \ln |z| (K_2 - K_1)/\ln R$.

Solution   This is a companion problem to Example 10.2. Here we use the fact that $\ln |z|$ is a harmonic function for all $z \neq 0$. Let us announce that the solution is

(8) $$\Phi(x, y) = K_1 + \frac{K_2 - K_1}{\ln R} \ln |z|$$

and that the level curves $\Phi(x, y) = $ constant are concentric circles, as illustrated in Figure 10.3.

10.2   Invariance of Laplace's Equation and the Dirichlet Problem

Theorem 10.1   *Let $\Phi(u, v)$ be harmonic in a domain G in the w plane. Then Φ satisfies Laplace's equation*

(1) $\Phi_{uu}(u, v) + \Phi_{vv}(u, v) = 0$

at each point $w = u + iv$ in G. If

(2) $w = f(z) = u(x, y) + iv(x, y)$

is a conformal mapping from a domain D in the z plane onto G, then the composition

(3) $\phi(x, y) = \Phi(u(x, y), v(x, y))$

is harmonic in D, and ϕ satisfies Laplace's equation

(4) $\phi_{xx}(x, y) + \phi_{yy}(x, y) = 0$

at each point $z = x + iy$ in D.

Proof   Equations (1) and (4) are facts about the harmonic functions Φ and ϕ that were studied in Section 3.3. A direct proof that the function ϕ in equation (3) is harmonic would involve a tedious calculation of the partial derivatives ϕ_{xx} and ϕ_{yy}. An easier proof uses a complex variable technique. Let us assume that there is a harmonic conjugate $\Psi(u, v)$ so that the function

$$(5)\quad g(w) = \Phi(u, v) + i\,\Psi(u, v)$$

is analytic in a neighborhood of the point $w_0 = f(z_0)$. Then the composition $h(z) = g(f(z))$ is analytic in a neighborhood of z_0 and can be written as

$$(6)\quad h(z) = \Phi(u(x, y), v(x, y)) + i\,\Psi(u(x, y), v(x, y)).$$

If we use Theorem 3.5, it follows that $\Phi(u(x, y), v(x, y))$ is harmonic in a neighborhood of z_0, and Theorem 10.1 is established.

EXAMPLE 10.4   Show that $\phi(x, y) = \arctan[2x/(x^2 + y^2 - 1)]$ is harmonic in the disk $|z| < 1$, where $-\pi/2 < \arctan t < \pi/2$.

Solution   The results of Exercise 10 of Section 9.2 show that the function

$$(7)\quad f(z) = \frac{i + z}{i - z} = \frac{1 - x^2 - y^2}{x^2 + (y - 1)^2} - \frac{i2x}{x^2 + (y - 1)^2}$$

is a conformal mapping of the disk $|z| < 1$ onto the right half plane $\mathrm{Re}(w) > 0$. The results from Exercise 12 in Section 5.2 show that the function

$$(8)\quad \Phi(u, v) = \arctan\frac{v}{u} = \mathrm{Arg}(u + iv)$$

is harmonic in the right half plane $\mathrm{Re}(w) > 0$. We can use equation (7) to write

$$(9)\quad u(x, y) = \frac{1 - x^2 - y^2}{x^2 + (y - 1)^2} \quad \text{and} \quad v(x, y) = \frac{-2x}{x^2 + (y - 1)^2}.$$

Substituting equation (9) into equation (8) and using equation (3), we see that $\phi(x, y) = \arctan(v(x, y)/u(x, y)) = \arctan(2x/(x^2 + y^2 - 1))$ is harmonic for $|z| < 1$.

Let D be a domain whose boundary is made up of piecewise smooth contours joined end to end. The *Dirichlet problem* is to find a function ϕ that is harmonic in D such that ϕ takes on prescribed values at points on the boundary. Let us first study this problem in the upper half plane.

EXAMPLE 10.5   Show that the function

$$(10)\quad \Phi(u, v) = \frac{1}{\pi}\,\mathrm{Arctan}\,\frac{v}{u - u_0} = \frac{1}{\pi}\,\mathrm{Arg}(w - u_0)$$

is harmonic in the upper half plane $\text{Im}(w) > 0$ and takes on the boundary values

(11) $\Phi(u, 0) = 0$ for $u > u_0$ and
  $\Phi(u, 0) = 1$ for $u < u_0$.

Solution The function

(12) $g(w) = \dfrac{1}{\pi} \text{Log}(w - u_0) = \dfrac{1}{\pi} \ln|w - u_0| + \dfrac{i}{\pi} \text{Arg}(w - u_0)$

is analytic in the upper half plane $\text{Im}(w) > 0$, and its imaginary part is the harmonic function $(1/\pi) \text{Arg}(w - u_0)$.

Remark Let t be a real number. We shall use the convention $\text{Arctan}(\pm\infty) = \pi/2$ so that the function $\text{Arctan } t$ denotes the branch of the inverse tangent that lies in the range $0 < \text{Arctan } t < \pi$. This will permit us to write the solution in equation (10) as $\Phi(u, v) = (1/\pi) \text{Arctan}(v/(u - u_0))$.

Theorem 10.2 (*N*-Value Dirichlet Problem for the Upper Half Plane) *Let $u_1 < u_2 < \cdots < u_{N-1}$ denote $N - 1$ real constants. The function*

(13) $\Phi(u, v) = a_{N-1} + \dfrac{1}{\pi} \displaystyle\sum_{k=1}^{N-1} (a_{k-1} - a_k) \text{Arg}(w - u_k)$

  $= a_{N-1} + \dfrac{1}{\pi} \displaystyle\sum_{k=1}^{N-1} (a_{k-1} - a_k) \text{Arctan} \dfrac{v}{u - u_k}$

is harmonic in the upper half plane $\text{Im}(w) > 0$ and takes on the boundary values

(14) $\Phi(u, 0) = a_0$ *for $u < u_1$,*
  $\Phi(u, 0) = a_k$ *for $u_k < u < u_{k+1}$ for $k = 1, 2, \ldots, N - 2$,*
  $\Phi(u, 0) = a_{N-1}$ *for $u > u_{N-1}$.*

The situation is illustrated in Figure 10.4.

Proof Since each term in the sum in equation (13) is harmonic, it follows that Φ is harmonic for $\text{Im}(w) > 0$. To show that Φ has the prescribed boundary conditions, we fix j and let $u_j < u < u_{j+1}$. Using Example 10.5, we see that

(15) $\dfrac{1}{\pi} \text{Arg}(u - u_k) = 0$ if $k \le j$ and $\dfrac{1}{\pi} \text{Arg}(u - u_k) = 1$ if $k > j$.

Using equations (15) in equation (13) results in

(16) $\Phi(u, 0) = a_{N-1} + \displaystyle\sum_{k=1}^{j} (a_{k-1} - a_k)(0) + \sum_{k=j+1}^{N-1} (a_{k-1} - a_k)(1)$

  $= a_{N-1} + (a_{N-2} - a_{N-1}) + \cdots + (a_{j+1} - a_{j+2}) + (a_j - a_{j+1})$

  $= a_j$ for $u_j < u < u_{j+1}$.

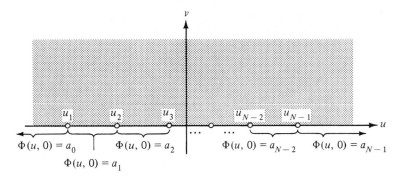

FIGURE 10.4   The boundary conditions for the harmonic function $\Phi(u, v)$ in the statement of Theorem 10.2.

The reader can verify that the boundary conditions are correct for $u < u_1$ and $u > u_{N-1}$, and the result will be established.

EXAMPLE 10.6   Find the function $\phi(x, y)$ that is harmonic in the upper half plane $\text{Re}(z) > 0$, which takes on the boundary values indicated in Figure 10.5.

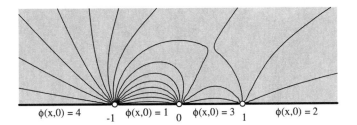

FIGURE 10.5   The boundary values for the Dirichlet problem in Example 10.6.

Solution   This is a four-value Dirichlet problem in the upper half plane $\text{Im}(z) > 0$. For the z plane the solution in equation (13) becomes

$$(17) \quad \phi(x, y) = a_3 + \frac{1}{\pi} \sum_{k=1}^{3} (a_{k-1} - a_k) \, \text{Arg}(z - x_k).$$

Here we have $a_0 = 4$, $a_1 = 1$, $a_2 = 3$, $a_3 = 2$ and $x_1 = -1$, $x_2 = 0$, $x_3 = 1$, which can be substituted into equation (17) to obtain

$$\phi(x, y) = 2 + \frac{4 - 1}{\pi} \, \text{Arg}(z + 1) + \frac{1 - 3}{\pi} \, \text{Arg}(z - 0) + \frac{3 - 2}{\pi} \, \text{Arg}(z - 1)$$

$$= 2 + \frac{3}{\pi} \, \text{Arctan} \, \frac{y}{x + 1} - \frac{2}{\pi} \, \text{Arctan} \, \frac{y}{x} + \frac{1}{\pi} \, \text{Arctan} \, \frac{y}{x - 1}.$$

EXAMPLE 10.7   Find the function $\phi(x, y)$ that is harmonic in the upper half plane $\text{Re}(z) > 0$, which takes on the boundary values

$$\phi(x, 0) = 1 \quad \text{for } |x| < 1,$$
$$\phi(x, 0) = 0 \quad \text{for } |x| > 1.$$

Solution   This is a three-value Dirichlet problem with $a_0 = 0$, $a_1 = 1$, $a_2 = 0$ and $x_1 = -1$ and $x_2 = 1$. Applying formula (13) the solution is

$$\phi(x, y) = 0 + \frac{0 - 1}{\pi} \text{Arg}(z + 1) + \frac{1 - 0}{\pi} \text{Arg}(z - 1)$$

$$= -\frac{1}{\pi} \text{Arctan} \frac{y}{x + 1} + \frac{1}{\pi} \text{Arctan} \frac{y}{x - 1},$$

and a three-dimensional graph $u = \phi(x, y)$ is shown in Figure 10.6.

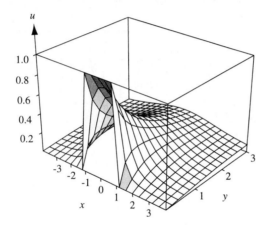

FIGURE 10.6   The graph of $u = \phi(x, y)$ with the boundary values $\phi(x, 0) = 1$ for $|x| < 1$ and $\phi(x, 0) = 0$ for $|x| > 1$.

We now state the *N-value Dirichlet problem* for a *simply connected domain*. Let D be a simply connected domain bounded by the simple closed contour C, and let $z_1, z_2, \ldots, z_N$ denote N points that lie along C in this specified order as C is traversed in the positive (counterclockwise) sense (see Figure 10.7). Let C_k denote the portion of C that lies strictly between z_k and z_{k+1} (for $k = 1, 2, \ldots, N - 1$), and let C_N denote the portion that lies strictly between z_N and z_1. Let $a_1, a_2, \ldots, a_N$ be real constants. We want to find a function $\phi(x, y)$ that is harmonic in D and continuous on $D \cup C_1 \cup C_2 \cup \cdots \cup C_N$ that takes on the boundary values:

(18) $\phi(x, y) = a_1 \quad \text{for } z = x + iy \text{ on } C_1,$
$\phi(x, y) = a_2 \quad \text{for } z = x + iy \text{ on } C_2,$

$$\vdots$$

$\phi(x, y) = a_N \quad \text{for } z = x + iy \text{ on } C_N.$

The situation is illustrated in Figure 10.7.

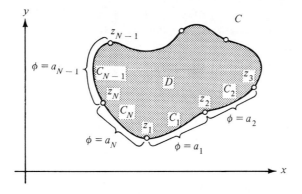

FIGURE 10.7   The boundary values for $\phi(x, y)$ for the Dirichlet problem in the simply connected domain D.

One method for finding ϕ is to find a conformal mapping

(19) $w = f(z) = u(x, y) + iv(x, y)$

of D onto the upper half plane Im $(w) > 0$, such that the N points $z_1, z_2, \ldots, z_N$ are mapped onto the points $u_k = f(z_k)$ for $k = 1, 2, \ldots, N - 1$ and z_N is mapped onto $u_N = +\infty$ along the u axis in the w plane.

Using Theorem 10.1, we see that the mapping in equation (19) gives rise to a new N-value Dirichlet problem in the upper half plane Im$(w) > 0$ for which the solution is given by Theorem 10.2. If we set $a_0 = a_N$, then the solution to the Dirichlet problem in D with boundary values (18) is

$$(20) \quad \phi(x, y) = a_{N-1} + \frac{1}{\pi} \sum_{k=1}^{N-1} (a_{k-1} - a_k) \operatorname{Arg}[f(z) - u_k]$$

$$= a_{N-1} + \frac{1}{\pi} \sum_{k=1}^{N-1} (a_{k-1} - a_k) \operatorname{Arctan} \frac{v(x, y)}{u(x, y) - u_k}.$$

This method relies on our ability to construct a conformal mapping from D onto the upper half plane Im$(w) > 0$. Theorem 9.4 guarantees the existence of such a conformal mapping.

EXAMPLE 10.8   Find a function $\phi(x, y)$ that is harmonic in the unit disk $|z| < 1$ and takes on the boundary values

(21) $\phi(x, y) = 0$ for $x + iy = e^{i\theta}, 0 < \theta < \pi$,
$\phi(x, y) = 1$ for $x + iy = e^{i\theta}, \pi < \theta < 2\pi$.

Solution   Example 9.3 showed that the function

$$(22) \quad u + iv = \frac{i(1 - z)}{1 + z} = \frac{2y}{(x + 1)^2 + y^2} + i\frac{1 - x^2 - y^2}{(x + 1)^2 + y^2}$$

is a one-to-one conformal mapping of the unit disk $|z| < 1$ onto the upper half plane $\mathrm{Im}(w) > 0$. Using equation (22), we see that the points $z = x + iy$ that lie on the upper semicircle $y > 0$, $1 - x^2 - y^2 = 0$ are mapped onto the positive u axis. Similarly, the lower semicircle is mapped onto the negative u axis as shown in Figure 10.8.

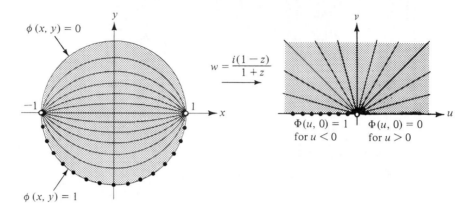

FIGURE 10.8   The Dirichlet problems for $|z| < 1$ and $\mathrm{Im}(w) > 0$ in the solution of Example 10.8.

The mapping (22) gives rise to a new Dirichlet problem of finding a harmonic function $\Phi(u, v)$ that has the boundary values

(23) $\Phi(u, 0) = 0$ for $u > 0$ and $\Phi(u, 0) = 1$ for $u < 0$,

as shown in Figure 10.8. Using the result of Example 10.5 and the function u and v in the mapping (22), we find that the solution to equation (21) is

$$\phi(x, y) = \frac{1}{\pi} \, \mathrm{Arctan} \, \frac{v(x, y)}{u(x, y)} = \frac{1}{\pi} \, \mathrm{Arctan} \, \frac{1 - x^2 - y^2}{2y}.$$

EXAMPLE 10.9   Find a function $\phi(x, y)$ that is harmonic in the upper half-disk H: $y > 0$, $|z| < 1$ and takes on the boundary values

(24) $\phi(x, y) = 0$ for $x + iy = e^{i\theta}$, $0 < \theta < \pi$,
$\phi(x, 0) = 1$ for $-1 < x < 1$.

Solution   By using the result of Exercise 4 in Section 9.2 the function in (22) is seen to map the upper half-disk H onto the first quadrant Q: $u > 0$, $v > 0$. Using the conformal mapping (22), we see that points $z = x + iy$ that lie on the segment $y = 0$, $-1 < x < 1$ are mapped onto the positive v axis.

Mapping (22) gives rise to a new Dirichlet problem of finding a harmonic function $\Phi(u, v)$ in Q that has the boundary values

(25) $\Phi(u, 0) = 0$ for $u > 0$ and $\Phi(0, v) = 1$ for $v > 0$,

as shown in Figure 10.9. In this case the method in Example 10.2 can be used to see that $\Phi(u, v)$ is given by

(26) $\Phi(u, v) = 0 + \dfrac{1 - 0}{\pi/2} \text{Arg } w = \dfrac{2}{\pi} \text{Arg } w = \dfrac{2}{\pi} \text{Arctan } \dfrac{v}{u}.$

Using the functions u and v in mapping (22) in equation (26), we find that the solution of the Dirichlet problem in H is

$$\phi(x, y) = \dfrac{2}{\pi} \text{Arctan } \dfrac{v(x, y)}{u(x, y)} = \dfrac{2}{\pi} \text{Arctan } \dfrac{1 - x^2 - y^2}{2y}.$$

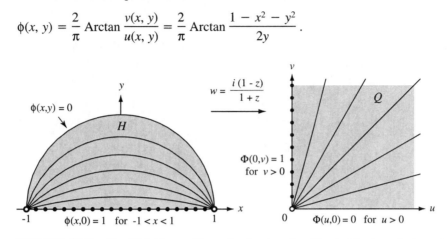

FIGURE 10.9   The Dirichlet problems for the domains H and Q in the solution of Example 10.9.

A three-dimensional graph $u = \phi(x, y)$, in cylindrical coordinates is shown in Figure 10.10.

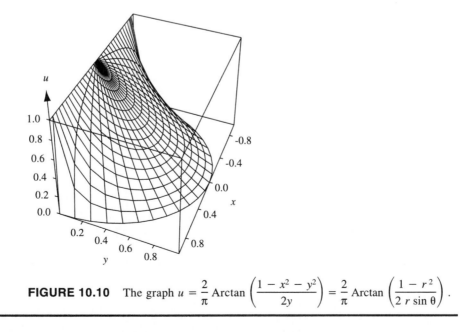

FIGURE 10.10   The graph $u = \dfrac{2}{\pi} \text{Arctan}\left(\dfrac{1 - x^2 - y^2}{2y}\right) = \dfrac{2}{\pi} \text{Arctan}\left(\dfrac{1 - r^2}{2\,r \sin \theta}\right).$

EXAMPLE 10.10 Find a function $\phi(x, y)$ that is harmonic in the quarter disk $G: x > 0, y > 0, |z| < 1$ and takes on the boundary values

(27) $\quad \phi(x, y) = 0 \quad$ for $z = e^{i\theta}, 0 < \theta < \pi/2,$
$\quad\quad\quad \phi(x, 0) = 1 \quad$ for $0 \leq x < 1,$
$\quad\quad\quad \phi(0, y) = 1 \quad$ for $0 \leq y < 1.$

Solution The function

(28) $\quad u + iv = z^2 = x^2 - y^2 + i2xy$

maps the quarter disk onto the upper half-disk $H: v > 0, |w| < 1$. The new Dirichlet problem in H is shown in Figure 10.11. From the result of Example 10.9 the solution $\Phi(u, v)$ in H is

(29) $\quad \Phi(u, v) = \dfrac{2}{\pi} \text{Arctan} \dfrac{1 - u^2 - v^2}{2v}.$

Using equation (28), one can show that $u^2 + v^2 = (x^2 + y^2)^2$ and $2v = 4xy$, which can be used in equation (29) to show that the solution ϕ in G is

$$\phi(x, y) = \dfrac{2}{\pi} \text{Arctan} \dfrac{1 - (x^2 + y^2)^2}{4xy}.$$

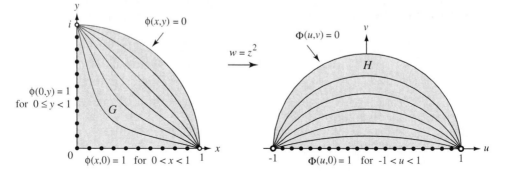

FIGURE 10.11 The Dirichlet problems for the domains G and H in the solution of Example 10.10.

A three-dimensional graph $u = \phi(x, y)$ in cylindrical coordinates is shown in Figure 10.12.

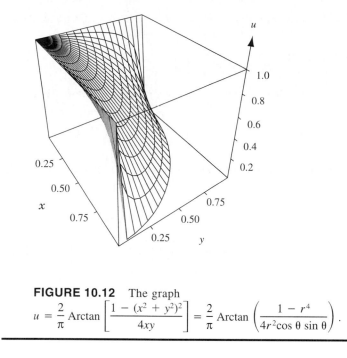

FIGURE 10.12   The graph
$$u = \frac{2}{\pi} \text{Arctan} \left[\frac{1 - (x^2 + y^2)^2}{4xy} \right] = \frac{2}{\pi} \text{Arctan} \left(\frac{1 - r^4}{4r^2\cos\theta \sin\theta} \right).$$

EXERCISES FOR SECTION 10.2

For all of the following exercises, find a solution $\phi(x, y)$ of the Dirichlet problem in the domain indicated that takes on the prescribed boundary values.

1. Find the function $\phi(x, y)$ that is harmonic in the horizontal strip $1 < \text{Im}(z) < 2$ and has the boundary values

    $\phi(x, 1) = 6$ for all x, $\phi(x, 2) = -3$ for all x.

2. Find the function $\phi(x, y)$ that is harmonic in the sector $0 < \text{Arg } z < \pi/3$ and has the boundary values

    $\phi(x, y) = 2$ for Arg $z = \pi/3$, $\phi(x, 0) = 1$ for $x > 0$.

3. Find the function $\phi(x, y)$ that is harmonic in the annulus $1 < |z| < 2$ and has the boundary values

    $\phi(x, y) = 5$ when $|z| = 1$, $\phi(x, y) = 8$ when $|z| = 2$.

4. Find the function $\phi(x, y)$ that is harmonic in the upper half plane $\text{Im}(z) > 0$ and has the boundary values

    $\phi(x, 0) = 1$ for $-1 < x < 1$, $\phi(x, 0) = 0$ for $|x| > 1$.

5. Find the function $\phi(x, y)$ that is harmonic in the upper half plane $\text{Im}(z) > 0$ and has the boundary values

   $\phi(x, 0) = 3$ for $x < -3$, $\phi(x, 0) = 7$ for $-3 < x < -1$,
   $\phi(x, 0) = 1$ for $-1 < x < 2$, $\phi(x, 0) = 4$ for $x > 2$.

6. Find the function $\phi(x, y)$ that is harmonic in the first quadrant $x > 0$, $y > 0$ and has the boundary values

   $\phi(0, y) = 0$ for $y > 1$, $\phi(0, y) = 1$ for $0 < y < 1$,
   $\phi(x, 0) = 1$ for $0 \le x < 1$, $\phi(x, 0) = 0$ for $x > 1$.

7. Find the function $\phi(x, y)$ that is harmonic in the unit disk $|z| < 1$ and has the boundary values

   $\phi(x, y) = 0$ for $z = e^{i\theta}, 0 < \theta < \pi$,
   $\phi(x, y) = 5$ for $z = e^{i\theta}, \pi < \theta < 2\pi$.

8. Find the function $\phi(x, y)$ that is harmonic in the unit disk $|z| < 1$ and has the boundary values

   $\phi(x, y) = 8$ for $z = e^{i\theta}, 0 < \theta < \pi$,
   $\phi(x, y) = 4$ for $z = e^{i\theta}, \pi < \theta < 2\pi$.

9. Find the function $\phi(x, y)$ that is harmonic in the upper half-disk $y > 0$, $|z| < 1$ and has the boundary values

   $\phi(x, y) = 5$ for $z = e^{i\theta}, 0 < \theta < \pi$,
   $\phi(x, 0) = -5$ for $-1 < x < 1$.

10. Find the function $\phi(x, y)$ that is harmonic in the portion of the upper half plane $\text{Im}(z) > 0$ that lies outside the circle $|z| = 1$ and has the boundary values

    $\phi(x, y) = 1$ for $z = e^{i\theta}, 0 < \theta < \pi$,
    $\phi(x, 0) = 0$ for $|x| > 1$.

    Hint: Use the mapping $w = -1/z$ and the result of Example 10.9.

11. Find the function $\phi(x, y)$ that is harmonic in the quarter disk $x > 0$, $y > 0$, $|z| < 1$ and has the boundary values

    $\phi(x, y) = 3$ for $z = e^{i\theta}, 0 < \theta < \pi/2$,
    $\phi(x, 0) = -3$ for $0 \le x < 1$,
    $\phi(0, y) = -3$ for $0 < y < 1$.

12. Find the function $\phi(x, y)$ that is harmonic in the unit disk $|z| < 1$ and has the boundary values

    $\phi(x, y) = 1$ for $z = e^{i\theta}, -\pi/2 < \theta < \pi/2$,
    $\phi(x, y) = 0$ for $z = e^{i\theta}, \pi/2 < \theta < 3\pi/2$.

13. Look up the article on the Poisson integral formula and discuss what you found. Use bibliographical item 115.

14. Write a report on how computer graphics are used for graphing harmonic functions, complex functions, and conformal mappings. Resources include bibliographical items 33, 34, 109, and 146.

10.3   Poisson's Integral Formula for the Upper Half Plane

The Dirichlet problem for the upper half plane $\text{Im}(z) > 0$ is to find a function $\phi(x, y)$ that is harmonic in the upper half plane and has the boundary values $\phi(x, 0) = U(x)$, where $U(x)$ is a real-valued function of the real variable x.

Theorem 10.3 (Poisson's Integral Formula)   *Let $U(t)$ be a real-valued function that is piecewise continuous and bounded for all real t. The function*

(1) $$\phi(x, y) = \frac{y}{\pi} \int_{-\infty}^{\infty} \frac{U(t)\, dt}{(x - t)^2 + y^2}$$

is harmonic in the upper half plane $\text{Im}(z) > 0$ and has the boundary values

(2) $\phi(x, 0) = U(x)$ *wherever U is continuous.*

Proof   The integral formula (1) is easy to motivate from the results of Theorem 10.2 regarding the Dirichlet problem. Let $t_1 < t_2 < \cdots < t_N$ denote N points that lie along the x axis. Let $t_0^* < t_1^* < \cdots < t_N^*$ be $N + 1$ points that are chosen so that $t_{k-1}^* < t_k < t_k^*$ (for $k = 1, 2, \ldots, N$) and $U(t)$ is continuous at each value t_k^*. Then according to Theorem 10.2, the function

(3) $$\Phi(x, y) = U(t_N^*) + \frac{1}{\pi} \sum_{k=1}^{N} [U(t_{k-1}^*) - U(t_k^*)]\, \text{Arg}(z - t_k)$$

is harmonic in the upper half plane and takes on the boundary values

(4) $\Phi(x, 0) = U(t_0^*)$ for $x < t_1$,

$\quad\quad \Phi(x, 0) = U(t_k^*)$ for $t_k < x < t_{k+1}$,

$\quad\quad \Phi(x, 0) = U(t_N^*)$ for $x > t_N$,

as shown in Figure 10.13.

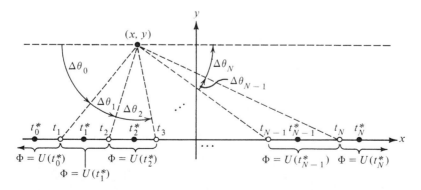

FIGURE 10.13   The boundary values for Φ in the proof of Theorem 10.3.

We can use properties of the argument of a complex number in Section 1.4 to write equation (3) in the form

$$(5) \quad \Phi(x, y) = \frac{1}{\pi} U(t_0^*) \, \text{Arg}(z - t_1) + \frac{1}{\pi} \sum_{k=1}^{N-1} U(t_k^*) \, \text{Arg}\left(\frac{z - t_{k+1}}{z - t_k}\right)$$

$$+ \frac{1}{\pi} U(t_N^*)[\pi - \text{Arg}(z - t_N)].$$

The value $\Phi(x, y)$ in equation (5) is given by the weighted mean

$$(6) \quad \Phi(x, y) = \frac{1}{\pi} \sum_{k=0}^{N} U(t_k^*) \, \Delta\theta_k,$$

where the angles $\Delta\theta_k$ $(k = 0, 1, \ldots, N)$ sum up to π and are shown in Figure 10.13.

Using the substitutions

$$(7) \quad \theta = \text{Arg}(z - t) = \text{Arctan}\left(\frac{y}{x - t}\right) \quad \text{and} \quad d\theta = \frac{y \, dt}{(x - t)^2 + y^2},$$

we can write equation (6) as

$$(8) \quad \Phi(x, y) = \frac{y}{\pi} \sum_{k=0}^{N} \frac{U(t_k^*) \, \Delta t_k}{(x - t_k^*)^2 + y^2}.$$

The limit of the Riemann sum, equation (8), becomes an improper integral

$$\phi(x, y) = \frac{y}{\pi} \int_{-\infty}^{\infty} \frac{U(t) \, dt}{(x - t)^2 + y^2}$$

and the result is established.

EXAMPLE 10.11 Find the function $\phi(x, y)$ that is harmonic in the upper half plane $\text{Im}(z) > 0$ and has the boundary values

$$(9) \quad \phi(x, 0) = 1 \quad \text{for } -1 < x < 1, \quad \phi(x, 0) = 0 \quad \text{for } |x| > 1.$$

Solution Using formula (1), we obtain

$$(10) \quad \phi(x, y) = \frac{y}{\pi} \int_{-1}^{1} \frac{dt}{(x - t)^2 + y^2} = \frac{1}{\pi} \int_{-1}^{1} \frac{y \, dt}{(x - t)^2 + y^2}.$$

Using the antiderivative in equation (7), we can write the solution in equation (10) as

$$\phi(x, y) = \frac{1}{\pi} \text{Arctan}\left(\frac{y}{x - t}\right) \Bigg|_{t=-1}^{t=1}$$

$$= \frac{1}{\pi} \text{Arctan}\left(\frac{y}{x - 1}\right) - \frac{1}{\pi} \text{Arctan}\left(\frac{y}{x + 1}\right).$$

EXAMPLE 10.12   Find the function $\phi(x, y)$ that is harmonic in the upper half plane $\text{Im}(z) > 0$ and has the boundary values

(11) $\phi(x, 0) = x$ for $-1 < x < 1$, $\phi(x, 0) = 0$ for $|x| > 1$.

Solution   Using formula (1), we obtain

(12) $$\phi(x, y) = \frac{y}{\pi} \int_{-1}^{1} \frac{t \, dt}{(x - t)^2 + y^2}$$
$$= \frac{y}{\pi} \int_{-1}^{1} \frac{(x - t)(-1) \, dt}{(x - t)^2 + y^2} + \frac{x}{\pi} \int_{-1}^{1} \frac{y \, dt}{(x - t)^2 + y^2}.$$

Using techniques of calculus and equations (7), we find that the solution in equation (12) is

$$\phi(x, y) = \frac{y}{2\pi} \ln \frac{(x - 1)^2 + y^2}{(x + 1)^2 + y^2} + \frac{x}{\pi} \text{Arctan} \frac{y}{x - 1} - \frac{x}{\pi} \text{Arctan} \frac{y}{x + 1}.$$

The function $\phi(x, y)$ is continuous in the upper half plane and on the boundary $\phi(x, 0)$ has discontinuities at $x = \pm 1$ on the real axis. The graph in Figure 10.14 shows this phenomenon.

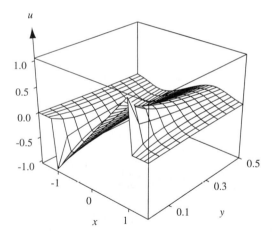

FIGURE 10.14   The graph of $u = \phi(x, y)$ with the boundary values $\phi(x, 0) = x$ for $|x| < 1$ and $\phi(x, 0) = 0$ for $|x| > 1$.

EXAMPLE 10.13   Find $\phi(x, y)$, harmonic in the upper half plane $\text{Im}(z) > 0$, that has the boundary values $\phi(x, 0) = x$ for $|x| < 1$, $\phi(x, 0) = -1$ for $x < -1$, and $\phi(x, 0) = 1$ for $x > 1$.

Solution   Using techniques from Section 10.2, the function

$$v(x, y) = 1 - \frac{1}{\pi} \text{Arctan} \frac{y}{x + 1} - \frac{1}{\pi} \text{Arctan} \frac{y}{x - 1}$$

is harmonic in the upper half plane and has the boundary values $v(x, 0) = 0$ for $|x| < 1$, $v(x, 0) = -1$ for $x < -1$, and $v(x, 0) = 1$ for $x > 1$. This function can be added to the one in Example 10.12 to obtain the desired result, which is

$$\phi(x, y) = 1 + \frac{y}{2\pi} \ln\left[\frac{(x - 1)^2 + y^2}{(x + 1)^2 + y^2}\right] + \frac{x - 1}{\pi} \text{Arctan} \frac{y}{x - 1}$$
$$- \frac{x + 1}{\pi} \text{Arctan} \frac{y}{x + 1}.$$

Figure 10.15 shows the graph of $\phi(x, y)$.

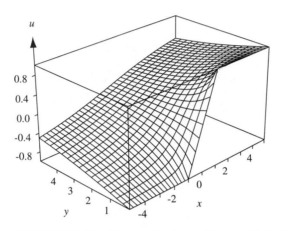

FIGURE 10.15   The graph of $u = \phi(x, y)$ with the boundary values $\phi(x, 0) = x$ for $|x| < 1$, $\phi(x, 0) = -1$ for $x < -1$ and $\phi(x, 0) = 1$ for $x > 1$.

EXERCISES FOR SECTION 10.3

1. Use Poisson's integral formula to find the harmonic function $\phi(x, y)$ in the upper half plane that takes on the boundary values

   $\phi(t, 0) = U(t) = 0$ for $t < 0$,
   $\phi(t, 0) = U(t) = t$ for $0 < t < 1$,
   $\phi(t, 0) = U(t) = 0$ for $1 < t$.

2. Use Poisson's integral formula to find the harmonic function $\phi(x, y)$ in the upper half plane that takes on the boundary values

   $\phi(t, 0) = U(t) = 0$ for $t < 0$,
   $\phi(t, 0) = U(t) = t$ for $0 < t < 1$,
   $\phi(t, 0) = U(t) = 1$ for $1 < t$.

3. Use Poisson's integral formula for the upper half plane to conclude that

   $$\phi(x, y) = e^{-y}\cos x = \frac{y}{\pi} \int_{-\infty}^{\infty} \frac{\cos t \, dt}{(x - t)^2 + y^2}.$$

4. Use Poisson's integral formula for the upper half plane to conclude that

$$\phi(x, y) = e^{-y}\sin x = \frac{y}{\pi} \int_{-\infty}^{\infty} \frac{\sin t \, dt}{(x - t)^2 + y^2} \, .$$

5. Show that the function $\phi(x, y)$ given by Poisson's integral formula is harmonic by applying Leibniz's rule, which permits us to write

$$\left(\frac{\partial^2}{\partial x^2} + \frac{\partial^2}{\partial y^2}\right)\phi(x, y) = \frac{1}{\pi} \int_{-\infty}^{\infty} U(t)\left[\left(\frac{\partial^2}{\partial x^2} + \frac{\partial^2}{\partial y^2}\right)\frac{y}{(x - t)^2 + y^2}\right] dt.$$

6. Let $U(t)$ be a real-valued function that satisfies the conditions for Poisson's integral formula for the upper half plane. If $U(t)$ is an even function, that is, $U(-t) = U(t)$, then show that the harmonic function $\phi(x, y)$ has the property $\phi(-x, y) = \phi(x, y)$.

7. Let $U(t)$ be a real-valued function that satisfies the conditions for Poisson's integral formula for the upper half plane. If $U(t)$ is an odd function, that is, $U(-t) = -U(t)$, then show that the harmonic function $\phi(x, y)$ has the property $\phi(-x, y) = -\phi(x, y)$.

8. Write a report on the Dirichlet problem and include some applications. Resources include bibliographical items 70, 71, 76, 77, 85, 98, 135, and 138.

10.4   Two-Dimensional Mathematical Models

We now turn our attention to problems involving steady state heat flow, electrostatics, and ideal fluid flow that can be solved by conformal mapping techniques. The method uses conformal mapping to transform a region in which the problem is posed to one in which the solution is easy to obtain. Since our solutions will involve only two independent variables, x and y, we first mention a basic assumption needed for the validity of the model.

The physical problems we just mentioned are real-world applications and involve solutions in three-dimensional Cartesian space. Such problems generally would involve the Laplacian in three variables and the divergence and curl of three-dimensional vector functions. Since complex analysis involves only x and y, we consider the special case in which the solution does not vary with the coordinate along the axis perpendicular to the xy plane. For steady state heat flow and electrostatics this assumption will mean that the temperature T, or the potential V, varies only with x and y. For the flow of ideal fluids this means that the fluid motion is the same in any plane that is parallel to the z plane. Curves drawn in the z plane are to be interpreted as cross sections that correspond to infinite cylinders perpendicular to the z plane. Since an infinite cylinder is the limiting case of a "long" physical cylinder, the mathematical model that we present is valid provided that the three-dimensional problem involves a physical cylinder long enough that the effects at the ends can be reasonably neglected.

In Sections 10.1 and 10.2 we learned how to obtain solutions $\phi(x, y)$ for harmonic functions. For applications it is important to consider the family of level curves

(1) $\{\phi(x, y) = K_1 \colon K_1 \text{ is a real constant}\}$

and the conjugate harmonic function $\psi(x, y)$ and its family of level curves

(2) $\{\psi(x, y) = K_2: K_2 \text{ is a real constant}\}$.

It is convenient to introduce the terminology *complex potential* for the analytic function

(3) $F(z) = \phi(x, y) + i\psi(x, y)$.

The following result regarding the orthogonality of the above mentioned families of level curves will be used in developing ideas concerning the physical applications.

> **Theorem 10.4 (Orthogonal Families of Level Curves)**   *Let $\phi(x, y)$ be harmonic in a domain D. Let $\psi(x, y)$ be the harmonic conjugate, and let $F(z) = \phi(x, y) + i\psi(x, y)$ be the complex potential. Then the two families of level curves given in* (1) *and* (2), *respectively, are orthogonal in the sense that if (a, b) is a point common to the two curves $\phi(x, y) = K_1$ and $\psi(x, y) = K_2$, and if $F'(a + ib) \neq 0$, then these two curves intersect orthogonally.*

Proof   Since $\phi(x, y) = K_1$ is an implicit equation of a plane curve, the gradient vector grad ϕ, evaluated at (a, b), is perpendicular to the curve at (a, b). This vector is given by

(4) $\mathbf{N}_1 = \phi_x(a, b) + i\phi_y(a, b)$.

Similarly, the vector $\mathbf{N}_2$ defined by

(5) $\mathbf{N}_2 = \psi_x(a, b) + i\psi_y(a, b)$

is orthogonal to the curve $\psi(x, y) = K_2$ at (a, b). Using the Cauchy-Riemann equations, $\phi_x = \psi_y$ and $\phi_y = -\psi_x$, we have

(6) $\mathbf{N}_1 \cdot \mathbf{N}_2 = \phi_x(a, b)[\psi_x(a, b)] + \phi_y(a, b)[\psi_y(a, b)]$
$= \phi_x(a, b)[-\phi_y(a, b)] + \phi_y(a, b)[\phi_x(a, b)] = 0$.

In addition, since $F'(a + ib) \neq 0$, we have

(7) $\phi_x(a, b) + i\psi_x(a, b) \neq 0$.

The Cauchy-Riemann equations and inequality (7) imply that both $\mathbf{N}_1$ and $\mathbf{N}_2$ are nonzero. Therefore equation (6) implies that $\mathbf{N}_1$ is perpendicular to $\mathbf{N}_2$, and hence the curves are orthogonal.

The complex potential $F(z) = \phi(x, y) + i\psi(x, y)$ has many physical interpretations. Suppose, for example, that we have solved a problem in steady state temperatures; then a similar problem with the same boundary conditions in electrostatics is obtained by interpreting the isothermals as equipotential curves and the heat flow lines as flux lines. This implies that heat flow and electrostatics correspond directly.

Or suppose we have solved a fluid flow problem; then an analogous problem in heat flow is obtained by interpreting the equipotentials as isothermals and stream-lines as heat flow lines. Various interpretations of the families of level curves given in expressions (1) and (2) and correspondences between families are summarized in Table 10.1.

Table 10.1   Interpretations for Level Curves

Physical Phenomenon	$\phi(x, y) =$ **constant**	$\psi(x, y) =$ **constant**
Heat flow	Isothermals	Heat flow lines
Electrostatics	Equipotential curves	Flux lines
Fluid flow	Equipotentials	Streamlines
Gravitational field	Gravitational potential	Lines of force
Magnetism	Potential	Lines of force
Diffusion	Concentration	Lines of flow
Elasticity	Strain function	Stress lines
Current flow	Potential	Lines of flow

10.5   Steady State Temperatures

In the theory of heat conduction the assumption is made that heat flows in the direction of decreasing temperature. We also assume that the time rate at which heat flows across a surface area is proportional to the component of the temperature gradient in the direction perpendicular to the surface area. If the temperature $T(x, y)$ does not depend on time, then the heat flow at the point (x, y) is given by the vector

$$(1) \quad \mathbf{V}(x, y) = -K \operatorname{grad} T(x, y) = -K[T_x(x, y) + iT_y(x, y)],$$

where K is the thermal conductivity of the medium and is assumed to be constant. If Δz denotes a straight line segment of length Δs, then the amount of heat flowing across the segment per unit of time is

$$(2) \quad \mathbf{V} \cdot \mathbf{N} \, \Delta s,$$

where $\mathbf{N}$ is a unit vector perpendicular to the segment.

If we assume that no thermal energy is created or destroyed within the region, then the net amount of heat flowing through any small rectangle with sides of length Δx and Δy is identically zero (see Figure 10.16(a)). This leads to the conclusion that $T(x, y)$ is a harmonic function. The following heuristic argument is often used to suggest that $T(x, y)$ satisfies Laplace's equation. Using expression (2), we find that the amount of heat flowing out of the right edge of the rectangle in Figure 10.16(a) is approximately

$$(3) \quad \mathbf{V} \cdot \mathbf{N}_1 \, \Delta s_1 = -K[T_x(x + \Delta x, y) + iT_y(x + \Delta x, y)] \cdot (1 + 0i) \, \Delta y$$
$$= -KT_x(x + \Delta x, y) \, \Delta y,$$

and the amount of heat flowing out of the left edge is

(4) $\mathbf{V} \cdot \mathbf{N}_2 \, \Delta s_2 = -K[T_x(x, y) + iT_y(x, y)] \cdot (-1 + 0i) \, \Delta y = KT_x(x, y) \, \Delta y.$

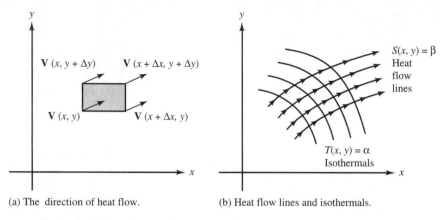

(a) The direction of heat flow.   (b) Heat flow lines and isothermals.

FIGURE 10.16   Steady state temperatures.

If we add the contributions in equations (3) and (4), the result is

(5) $-K\left[\dfrac{T_x(x + \Delta x, y) - T_x(x, y)}{\Delta x}\right] \Delta x \, \Delta y \approx -KT_{xx}(x, y) \, \Delta x \, \Delta y.$

In a similar fashion it is found that the contribution for the amount of heat flowing out of the top and bottom edges is

(6) $-K\left[\dfrac{T_y(x, y + \Delta y) - T_y(x, y)}{\Delta y}\right] \Delta x \, \Delta y \approx -KT_{yy}(x, y) \, \Delta x \, \Delta y.$

Adding the quantities in equations (5) and (6), we find that the net heat flowing out of the rectangle is approximated by the equation

(7) $-K[T_{xx}(x, y) + T_{yy}(x, y)] \, \Delta x \, \Delta y = 0,$

which implies that $T(x, y)$ satisfies Laplace's equation and is a harmonic function.

If the domain in which $T(x, y)$ is defined is simply connected, then a conjugate harmonic function $S(x, y)$ exists, and

(8) $F(z) = T(x, y) + iS(x, y)$

is an analytic function. The curves $T(x, y) = K_1$ are called *isothermals* and are lines connecting points of the same temperature. The curves $S(x, y) = K_2$ are called the *heat flow lines*, and one can visualize the heat flowing along these curves from points of higher temperature to points of lower temperature. The situation is illustrated in Figure 10.16(b).

Boundary value problems for steady state temperatures are realizations of the Dirichlet problem where the value of the harmonic function $T(x, y)$ is interpreted as the temperature at the point (x, y).

EXAMPLE 10.14   Suppose that two parallel planes are perpendicular to the
z plane and pass through the horizontal lines $y = a$ and $y = b$ and that the temperature
is held constant at the values $T(x, a) = T_1$ and $T(x, b) = T_2$, respectively, on these
planes. Then T is given by

$$(9) \quad T(x, y) = T_1 + \frac{T_2 - T_1}{b - a}(y - a).$$

Solution   It is reasonable to make the assumption that the temperature at
all points on the plane passing through the line $y = y_0$ is constant. Hence $T(x, y) =$
$t(y)$, where $t(y)$ is a function of y alone. Laplace's equation implies that $t''(y) = 0$,
and an argument similar to that in Example 10.1 will show that the solution $T(x, y)$
has the form given in equation (9).

The isothermals $T(x, y) = \alpha$ are easily seen to be horizontal lines. The conjugate harmonic function is

$$S(x, y) = \frac{T_1 - T_2}{b - a} x,$$

and the heat flow lines $S(x, y) = \beta$ are vertical segments between the horizontal
lines. If $T_1 > T_2$, then the heat flows along these segments from the plane through
$y = a$ to the plane through $y = b$ as illustrated in Figure 10.17.

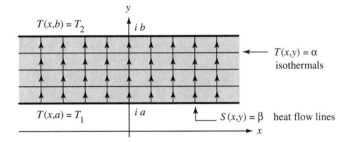

FIGURE 10.17   The temperature between parallel planes where $T_1 > T_2$.

EXAMPLE 10.15   Find the temperature $T(x, y)$ at each point in the upper half
plane $\text{Im}(z) > 0$ if the temperature along the x axis satisfies

$$(10) \quad T(x, 0) = T_1 \quad \text{for } x > 0 \quad \text{and} \quad T(x, 0) = T_2 \quad \text{for } x < 0.$$

Solution   Since $T(x, y)$ is a harmonic function, this is an example of a
Dirichlet problem. From Example 10.2 it follows that the solution is

$$(11) \quad T(x, y) = T_1 + \frac{T_2 - T_1}{\pi} \text{Arg } z.$$

The isothermals $T(x, y) = \alpha$ are rays emanating from the origin. The conjugate harmonic function is $S(x, y) = (1/\pi)(T_1 - T_2) \ln|z|$, and the heat flow lines $S(x, y) = \beta$ are semicircles centered at the origin. If $T_1 > T_2$, then the heat flows counterclockwise along the semicircles as shown in Figure 10.18.

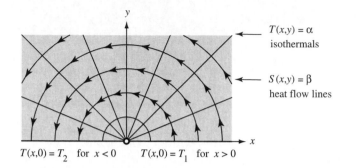

$$T(x,0) = T_2 \quad \text{for } x < 0 \qquad T(x,0) = T_1 \quad \text{for } x > 0$$

FIGURE 10.18 The temperature $T(x, y)$ in the upper half plane where $T_1 > T_2$.

EXAMPLE 10.16 Find the temperature $T(x, y)$ at each point in the upper half-disk H: $\mathrm{Im}(z) > 0$, $|z| < 1$ if the temperature at points on the boundary satisfies

(12) $\quad T(x, y) = 100 \quad$ for $z = e^{i\theta}, 0 < \theta < \pi$,

$\qquad T(x, 0) = 50 \quad$ for $-1 < x < 1$.

Solution As discussed in Example 10.9, the function

(13) $\quad u + iv = \dfrac{i(1 - z)}{1 + z} = \dfrac{2y}{(x + 1)^2 + y^2} + i\dfrac{1 - x^2 - y^2}{(x + 1)^2 + y^2}$

is a one-to-one conformal mapping of the half-disk H onto the first quadrant Q: $u > 0$, $v > 0$. The conformal map (13) gives rise to a new problem of finding the temperature $T^*(u, v)$ that satisfies the boundary conditions

(14) $\quad T^*(u, 0) = 100 \quad$ for $u > 0 \quad$ and $\quad T^*(0, v) = 50 \quad$ for $v > 0$.

If we use Example 10.2, the harmonic function $T^*(u, v)$ is given by

(15) $\quad T^*(u, v) = 100 + \dfrac{50 - 100}{\pi/2} \, \mathrm{Arg}\, w = 100 - \dfrac{100}{\pi} \, \mathrm{Arctan}\, \dfrac{v}{u}$.

Substituting the expressions for u and v in mapping (13) into equation (15) yields the desired solution

$$T(x, y) = 100 - \frac{100}{\pi} \text{Arctan} \frac{1 - x^2 - y^2}{2y}.$$

The isothermals $T(x, y)$ = constant are circles that pass through the points ± 1 as shown in Figure 10.19.

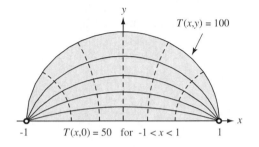

FIGURE 10.19   The temperature $T(x, y)$ in a half-disk.

We now turn our attention to the problem of finding the steady state temperature function $T(x, y)$ inside the simply connected domain D whose boundary consists of three adjacent curves C_1, C_2, and C_3, where $T(x, y) = T_1$ along C_1, $T(x, y) = T_2$ along C_2, and the region is insulated along C_3. Zero heat flowing across C_3 implies that

(16) $\mathbf{V}(x, y) \cdot \mathbf{N}(x, y) = -K\mathbf{N}(x, y) \cdot \text{grad } T(x, y) = 0,$

where $\mathbf{N}(x, y)$ is perpendicular to C_3. This means that the direction of heat flow must be parallel to this portion of the boundary. In other words, C_3 must be part of a heat flow line $S(x, y)$ = constant and the isothermals $T(x, y)$ = constant intersect C_3 orthogonally.

This problem can be solved if we can find a conformal mapping

(17) $w = f(z) = u(x, y) + iv(x, y)$

from D onto the semi-infinite strip G: $0 < u < 1$, $v > 0$ so that the image of the curve C_1 is the ray $u = 0$, $v > 0$; the image of the curve C_2 is the ray given by $u = 1$, $v > 0$; and the thermally insulated curve C_3 is mapped onto the thermally insulated segment $0 < u < 1$ of the u axis, as shown in Figure 10.20.

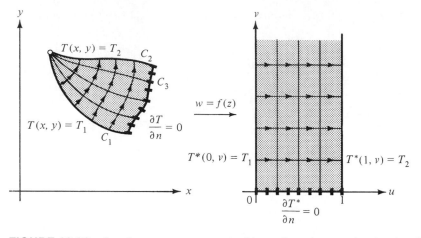

FIGURE 10.20   Steady state temperatures with one boundary portion insulated.

The new problem in G is to find the steady state temperature function $T^*(u, v)$ so that along the rays we have the boundary values

(18) $T^*(0, v) = T_1$ for $v > 0$ and $T^*(1, v) = T_2$ for $v > 0$.

The condition that a segment of the boundary is insulated can be expressed mathematically by saying that the normal derivative of $T^*(u, v)$ is zero. That is,

(19) $\dfrac{\partial T^*}{\partial n} = T_v^*(u, 0) = 0$

where n is a coordinate measured perpendicular to the segment.

It is easy to verify that the function

(20) $T^*(u, v) = T_1 + (T_2 - T_1)u$

satisfies the conditions (19) and (20) for the region G. Therefore using (17), we find that the solution in D is

(21) $T(x, y) = T_1 + (T_2 - T_1)u(x, y)$.

The isothermals $T(x, y) = $ constant, and their images under $w = f(z)$ are illustrated in Figure 10.20.

EXAMPLE 10.17   Find the steady state temperature $T(x, y)$ for the domain D consisting of the upper half plane $\text{Im}(z) > 0$ where $T(x, y)$ has the boundary conditions

(22) $T(x, 0) = 1$ for $x > 1$ and $T(x, 0) = -1$ for $x < -1$ and

$\dfrac{\partial T}{\partial n} = T_y(x, 0) = 0$ for $-1 < x < 1$.

Solution   The mapping $w =$ Arcsin z conformally maps D onto the semi-infinite strip $v > 0$, $-\pi/2 < u < \pi/2$, where the new problem is to find the steady state temperature $T^*(u, v)$ that has the boundary conditions

(23) $T^*\left(\dfrac{\pi}{2}, v\right) = 1$ for $v > 0$ and $T^*\left(\dfrac{-\pi}{2}, v\right) = -1$ for $v > 0$

and $\dfrac{\partial T^*}{\partial n} = T_v^*(u, 0) = 0$ for $\dfrac{-\pi}{2} < u < \dfrac{\pi}{2}$.

By using the result of Example 10.1 it is easy to obtain the solution

(24) $T^*(u, v) = \dfrac{2}{\pi} u.$

Therefore the solution in D is

(25) $T(x, y) = \dfrac{2}{\pi} \operatorname{Re}(\text{Arcsin } z).$

If an explicit solution is required, then we can use formula (14) in Section 9.4 to obtain

(26) $T(x, y) = \dfrac{2}{\pi} \arcsin\left[\dfrac{\sqrt{(x + 1)^2 + y^2} - \sqrt{(x - 1)^2 + y^2}}{2}\right],$

where the real function arcsin t has range values satisfying $-\pi/2 < \arcsin t < \pi/2$, see Figure 10.21.

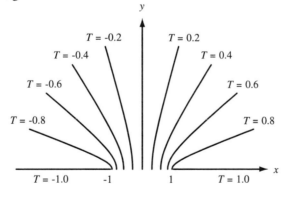

FIGURE 10.21   The temperature $T(x, y)$ with $T_y(x, 0) = 0$ for $-1 < x < 1$, and boundary values $T(x, 0) = -1$ for $x < -1$ and $T(x, 0) = 1$ for $x > 1$.

EXERCISES FOR SECTION 10.5

1. Show that $H(x, y, z) = 1/\sqrt{x^2 + y^2 + z^2}$ satisfies Laplace's equation $H_{xx} + H_{yy} + H_{zz} = 0$ in three-dimensional Cartesian space but that $h(x, y) = 1/\sqrt{x^2 + y^2}$ does not satisfy equation $h_{xx} + h_{yy} = 0$ in two-dimensional Cartesian space.

2. Find the temperature function $T(x, y)$ in the infinite strip bounded by the lines $y = -x$ and $y = 1 - x$ that satisfies the boundary values in Figure 10.22.

$$T(x, -x) = 25 \quad \text{for all } x,$$
$$T(x, 1 - x) = 75 \quad \text{for all } x.$$

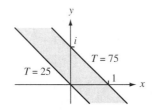

FIGURE 10.22   Accompanies Exercise 2.

3. Find the temperature function $T(x, y)$ in the first quadrant $x > 0$, $y > 0$ that satisfies the boundary values in Figure 10.23. *Hint*: Use $w = z^2$.

$$T(x, 0) = 10 \quad \text{for } x > 1,$$
$$T(x, 0) = 20 \quad \text{for } 0 < x < 1,$$
$$T(0, y) = 20 \quad \text{for } 0 \le y < 1,$$
$$T(0, y) = 10 \quad \text{for } y > 1.$$

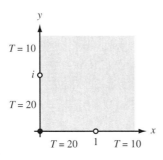

FIGURE 10.23   Accompanies Exercise 3.

4. Find the temperature function $T(x, y)$ inside the unit disk $|z| < 1$ that satisfies the boundary values in Figure 10.24. *Hint*: Use $w = i(1 - z)/(1 + z)$.

$$T(x, y) = 20 \quad \text{for } z = e^{i\theta}, 0 < \theta < \frac{\pi}{2},$$

$$T(x, y) = 60 \quad \text{for } z = e^{i\theta}, \frac{\pi}{2} < \theta < 2\pi.$$

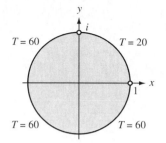

FIGURE 10.24   Accompanies Exercise 4.

5. Find the temperature function $T(x, y)$ in the semi-infinite strip $-\pi/2 < x < \pi/2, y > 0$ that satisfies the boundary values in Figure 10.25. *Hint*: Use $w = \sin z$.

$$T\left(\frac{\pi}{2}, y\right) = 100 \quad \text{for } y > 0,$$

$$T(x, y) = 0 \quad \text{for } \frac{-\pi}{2} < x < \frac{\pi}{2},$$

$$T\left(\frac{-\pi}{2}, y\right) = 100 \quad \text{for } y > 0.$$

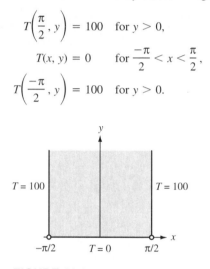

FIGURE 10.25   Accompanies Exercise 5.

6. Find the temperature function $T(x, y)$ in the domain $r > 1, 0 < \theta < \pi$ that satisfies the boundary values in Figure 10.26. *Hint*: $w = i(1 - z)/(1 + z)$.

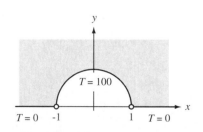

FIGURE 10.26   Accompanies Exercise 6.

$T(x, 0) = 0 \quad \text{for } x > 1,$
$T(x, 0) = 0 \quad \text{for } x < -1,$
$T(x, y) = 100 \quad \text{if } z = e^{i\theta}, 0 < \theta < \pi.$

7. Find the temperature function $T(x, y)$ in the domain $1 < r < 2, 0 < \theta < \pi/2$ that satisfies the boundary conditions in Figure 10.27. *Hint*: Use $w = \text{Log } z$.

$$T(x, y) = 0 \quad \text{for } r = e^{i\theta}, 0 < \theta < \frac{\pi}{2},$$

$$T(x, y) = 50 \quad \text{for } r = 2e^{i\theta}, 0 < \theta < \frac{\pi}{2},$$

$$\frac{\partial T}{\partial n} = T_y(x, 0) = 0 \quad \text{for } 1 < x < 2,$$

$$\frac{\partial T}{\partial n} = T_x(0, y) = 0 \quad \text{for } 1 < y < 2.$$

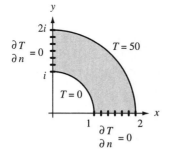

FIGURE 10.27   Accompanies Exercise 7.

8. Find the temperature function $T(x, y)$ in the domain $0 < r < 1, 0 < \text{Arg } z < \alpha$ that satisfies the boundary conditions in Figure 10.28. *Hint*: Use $w = \text{Log } z$.

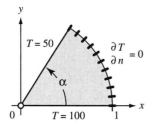

FIGURE 10.28   Accompanies Exercise 8.

$$T(x, 0) = 100 \quad \text{for } 0 < x < 1,$$
$$T(x, y) = 50 \quad \text{for } z = re^{i\alpha}, 0 < r < 1,$$
$$\frac{\partial T}{\partial n} = 0 \quad \text{for } z = e^{i\theta}, 0 < \theta < \alpha.$$

9. Find the temperature function $T(x, y)$ in the first quadrant $x > 0$, $y > 0$ that satisfies the boundary conditions in Figure 10.29. *Hint*: Use $w = \text{Arcsin } z^2$.

$$T(x, 0) = 100 \qquad \text{for } x > 1,$$
$$T(0, y) = -50 \qquad \text{for } y > 1,$$
$$\frac{\partial T}{\partial n} = T_y(x, 0) = 0 \qquad \text{for } 0 < x < 1,$$
$$\frac{\partial T}{\partial n} = T_x(0, y) = 0 \qquad \text{for } 0 < y < 1.$$

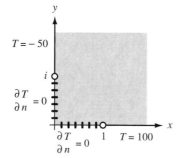

FIGURE 10.29   Accompanies Exercise 9.

10. Find the temperature function $T(x, y)$ in the infinite strip $0 < y < \pi$ that satisfies the boundary conditions in Figure 10.30. *Hint*: Use $w = e^z$.

$$T(x, 0) = 50 \qquad \text{for } x > 0,$$
$$T(x, \pi) = -50 \qquad \text{for } x > 0,$$
$$\frac{\partial T}{\partial n} = T_y(x, 0) = 0 \qquad \text{for } x < 0,$$
$$\frac{\partial T}{\partial n} = T_y(x, \pi) = 0 \qquad \text{for } x < 0.$$

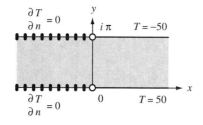

FIGURE 10.30   Accompanies Exercise 10.

11. Find the temperature function $T(x, y)$ in the upper half plane $\text{Im}(z) > 0$ that satisfies the boundary conditions in Figure 10.31. *Hint:* Use $w = 1/z$.

$$T(x, 0) = 100 \qquad \text{for } 0 < x < 1,$$
$$T(x, 0) = -100 \quad \text{for } -1 < x < 0,$$
$$\frac{\partial T}{\partial n} = T_y(x, 0) = 0 \qquad \text{for } x > 1,$$
$$\frac{\partial T}{\partial n} = T_y(x, 0) = 0 \qquad \text{for } x < -1.$$

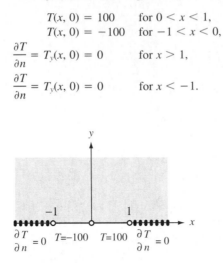

FIGURE 10.31   Accompanies Exercise 11.

12. Find the temperature function $T(x, y)$ in the first quadrant $x > 0$, $y > 0$ that satisfies the boundary conditions in Figure 10.32.

$$T(x, 0) = 50 \qquad \text{for } x > 0,$$
$$T(0, y) = -50 \quad \text{for } y > 1,$$
$$\frac{\partial T}{\partial n} = T_x(0, y) = 0 \qquad \text{for } 0 < y < 1.$$

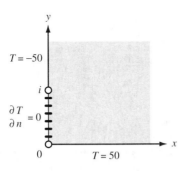

FIGURE 10.32   Accompanies Exercise 12.

13. For the temperature function

$$T(x, y) = 100 - \frac{100}{\pi} \arctan \frac{1 - x^2 - y^2}{2y}$$

in the upper half-disk $|z| < 1$, $\text{Im}(z) > 0$, show that the isothermals $T(x, y) = \alpha$ are portions of circles that pass through the points $+1$ and -1 as illustrated in Figure 10.33.

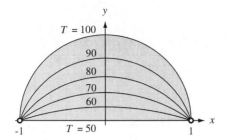

FIGURE 10.33   Accompanies Exercise 13.

14. For the temperature function

$$T(x, y) = \frac{300}{\pi} \text{Re}(\text{Arcsin } z)$$

in the upper half plane $\text{Im}(z) > 0$, show that the isothermals $T(x, y) = \alpha$ are portions of hyperbolas that have foci at the points ± 1 as illustrated in Figure 10.34.

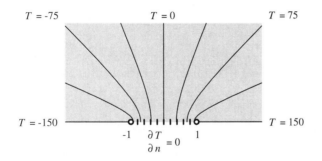

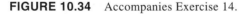

FIGURE 10.34   Accompanies Exercise 14.

15. Find the temperature function in the portion of the upper half plane $\text{Im}(z) > 0$ that lies inside the ellipse

$$\frac{x^2}{\cosh^2 2} + \frac{y^2}{\sinh^2 2} = 1$$

and satisfies the boundary conditions given in Figure 10.35. *Hint*: Use $w = \text{Arcsin } z$.

$$T(x, y) = 80 \quad \text{for } (x, y) \text{ on the ellipse,}$$
$$T(x, 0) = 40 \quad \text{for } -1 < x < 1,$$
$$\frac{\partial T}{\partial n} = T_y(x, 0) = 0 \quad \text{when } 1 < |x| < \cosh 2.$$

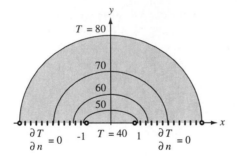

FIGURE 10.35   Accompanies Exercise 15.

10.6   Two-Dimensional Electrostatics

A two-dimensional electrostatic field is produced by a system of charged wires, plates, and cylindrical conductors that are perpendicular to the z plane. The wires, plates, and cylinders are assumed to be so long that the effects at the ends can be neglected as mentioned in Section 10.4. This sets up an electric field $\mathbf{E}(x, y)$ that can be interpreted as the force acting on a unit positive charge placed at the point (x, y). In the study of electrostatics the vector field $\mathbf{E}(x, y)$ is shown to be *conservative* and is derivable from a function $\phi(x, y)$, called the *electrostatic potential*, as expressed by the equation

(1) $\mathbf{E}(x, y) = -\operatorname{grad} \phi(x, y) = -\phi_x(x, y) - i\phi_y(x, y).$

If we make the additional assumption that there are no charges within the domain D, then Gauss' law for electrostatic fields implies that the line integral of the outward normal component of $\mathbf{E}(x, y)$ taken around any small rectangle lying inside D is identically zero. A heuristic argument similar to the one for steady state temperatures with $T(x, y)$ replaced by $\phi(x, y)$ will show that the value of the line integral is

(2) $-[\phi_{xx}(x, y) + \phi_{yy}(x, y)] \, \Delta x \, \Delta y.$

Since the quantity in expression (2) is zero, we conclude that $\phi(x, y)$ is a harmonic function. We let $\psi(x, y)$ denote the harmonic conjugate, and

(3) $F(z) = \phi(x, y) + i\psi(x, y)$

is the complex potential (not to be confused with the electrostatic potential). The curves $\phi(x, y) = K_1$ are called the *equipotential curves*, and the curves $\psi(x, y) = K_2$ are called the *lines of flux*. If a small test charge is allowed to move under the influence of the field $\mathbf{E}(x, y)$, then it will travel along a line of flux. Boundary value problems for the potential function $\phi(x, y)$ are mathematically the same as those for steady state heat flow, and they are realizations of the Dirichlet problem where the harmonic function is $\phi(x, y)$.

EXAMPLE 10.18   Consider two parallel conducting planes that pass perpendicular to the z plane through the lines $x = a$ and $x = b$, which are kept at the
potentials U_1 and U_2, respectively. Then according to the result of Example 10.1,
the electrical potential is

(4) $\phi(x, y) = U_1 + \dfrac{U_2 - U_1}{b - a}(x - a).$

EXAMPLE 10.19   Find the electrical potential $\phi(x, y)$ in the region between
two infinite coaxial cylinders $r = a$ and $r = b$, which are kept at the potentials U_1
and U_2, respectively.

Solution   The function $w = \log z = \ln |z| + i \arg z$ maps the annular
region between the circles $r = a$ and $r = b$ onto the infinite strip $\ln a < u < \ln b$
in the w plane as shown in Figure 10.36. The potential $\Phi(u, v)$ in the infinite strip
will have the boundary values

(5) $\Phi(\ln a, v) = U_1$ and $\Phi(\ln b, v) = U_2$ for all v.

If we use the result of Example 10.18, the electrical potential $\Phi(u, v)$ is

(6) $\Phi(u, v) = U_1 + \dfrac{U_2 - U_1}{\ln b - \ln a}(u - \ln a).$

Since $u = \ln |z|$, we can use equation (6) to conclude that the potential $\phi(x, y)$ is

$$\phi(x, y) = U_1 + \frac{U_2 - U_1}{\ln b - \ln a}(\ln |z| - \ln a).$$

The equipotentials $\phi(x, y) = $ constant are concentric circles centered at the origin,
and the lines of flux are portions of rays emanating from the origin. If $U_2 < U_1$,
then the situation is illustrated in Figure 10.36.

EXAMPLE 10.20   Find the electrical potential $\phi(x, y)$ produced by two
charged half planes that are perpendicular to the z plane and pass through the rays
$x < -1, y = 0$ and $x > 1, y = 0$, where the planes are kept at the fixed potentials

(7) $\phi(x, 0) = -300$ for $x < -1$ and $\phi(x, 0) = 300$ for $x > 1$.

Solution   The result of Example 9.13 shows that the function $w = \text{Arcsin } z$
is a conformal mapping of the z plane slit along the two rays $x < -1, y = 0$ and
$x > 1, y = 0$ onto the vertical strip $-\pi/2 < u < \pi/2$, where the new problem is to
find the potential $\Phi(u, v)$ that satisfies the boundary values

(8) $\Phi\left(\dfrac{-\pi}{2}, v\right) = -300$ and $\Phi\left(\dfrac{\pi}{2}, v\right) = 300$ for all v.

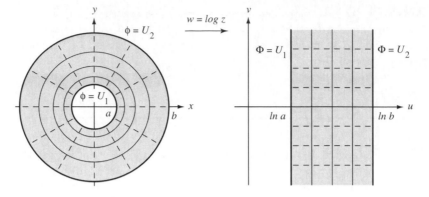

FIGURE 10.36   The electric field in a coaxial cylinder where $U_2 < U_1$.

Using the result of Example 10.1, we see that $\Phi(u, v)$ is

(9) $\Phi(u, v) = \dfrac{600}{\pi} u.$

As in the discussion of Example 10.17, the solution in the z plane is

(10) $\phi(x, y) = \dfrac{600}{\pi} \operatorname{Re}(\operatorname{Arcsin} z)$

$= \dfrac{600}{\pi} \operatorname{Arcsin}\left[\dfrac{\sqrt{(x + 1)^2 + y^2} - \sqrt{(x - 1)^2 + y^2}}{2}\right].$

Several equipotential curves are shown in Figure 10.37.

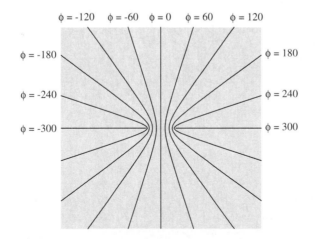

FIGURE 10.37   The electric field produced by two charged half planes that are perpendicular to the complex plane.

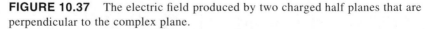

EXAMPLE 10.21   Find the electrical potential $\phi(x, y)$ in the disk D: $|z| < 1$ that satisfies the boundary values

(11) $\phi(x, y) = 80$ for z on $C_1 = \left\{ z = e^{i\theta} \colon 0 < \theta < \dfrac{\pi}{2} \right\}$ and

$\phi(x, y) = 0$ for z on $C_2 = \left\{ z = e^{i\theta} \colon \dfrac{\pi}{2} < \theta < 2\pi \right\}$.

Solution   The mapping $w = S(z) = [(1 - i)(z - i)]/(z - 1)$ is a one-to-one conformal mapping of D onto the upper half plane $\text{Im}(w) > 0$ with the property that C_1 is mapped onto the negative u axis and C_2 is mapped onto the positive u axis. The potential $\Phi(u, v)$ in the upper half plane that satisfies the new boundary values

(12) $\Phi(u, 0) = 80$ for $u < 0$ and $\Phi(u, 0) = 0$ for $u > 0$

is given by

(13) $\Phi(u, v) = \dfrac{80}{\pi} \text{Arg } w = \dfrac{80}{\pi} \text{Arctan} \dfrac{v}{u}$.

A straightforward calculation shows that

(14) $u + iv = S(z) = \dfrac{(x - 1)^2 + (y - 1)^2 - 1 + i(1 - x^2 - y^2)}{(x - 1)^2 + y^2}$.

The functions u and v in equation (14) can be substituted into equation (13) to obtain

$\phi(x, y) = \dfrac{80}{\pi} \text{Arctan} \dfrac{1 - x^2 - y^2}{(x - 1)^2 + (y - 1)^2 - 1}$.

The level curve $\Phi(u, v) = \alpha$ in the upper half plane is a ray emanating from the origin, and the preimage $\phi(x, y) = \alpha$ in the unit disk is an arc of a circle that passes through the points 1 and i. Several level curves are illustrated in Figure 10.38.

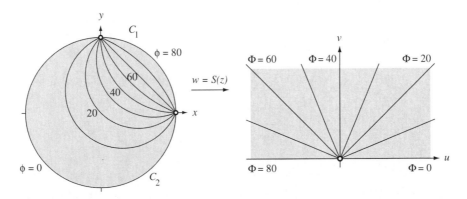

FIGURE 10.38   The potentials ϕ and Φ that are discussed in Example 10.21.

EXERCISES FOR SECTION 10.6

1. Find the electrostatic potential $\phi(x, y)$ between the two coaxial cylinders $r = 1$ and $r = 2$ that has the boundary values as shown in Figure 10.39.

    $\phi(x, y) = 100$ when $|z| = 1$,
    $\phi(x, y) = 200$ when $|z| = 2$.

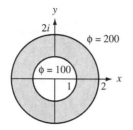

FIGURE 10.39   Accompanies Exercise 1.

2. Find the electrostatic potential $\phi(x, y)$ in the upper half plane $\text{Im}(z) > 0$ that satisfies the boundary values as shown in Figure 10.40.

    $\phi(x, 0) = 100$ for $x > 1$
    $\phi(x, 0) = 0$ for $-1 < x < 1$
    $\phi(x, 0) = -100$ for $x < -1$

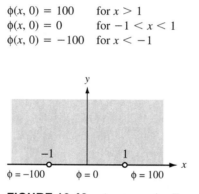

FIGURE 10.40   Accompanies Exercise 2.

3. Find the electrostatic potential $\phi(x, y)$ in the crescent-shaped region that lies inside the disk $|z - 2| < 2$ and outside the circle $|z - 1| = 1$ that satisfies the boundary values as shown in Figure 10.41. *Hint*: Use $w = 1/z$.

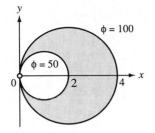

FIGURE 10.41   Accompanies Exercise 3.

$\phi(x, y) = 100$ for $|z - 2| = 2$, $z \neq 0$,
$\phi(x, y) = 50$ for $|z - 1| = 1$, $z \neq 0$.

4. Find the electrostatic potential $\phi(x, y)$ in the semi-infinite strip $-\pi/2 < x < \pi/2$, $y > 0$ that has the boundary values as shown in Figure 10.42.

$$\phi\left(\frac{\pi}{2}, y\right) = 0 \qquad \text{for } y > 0,$$

$$\phi(x, 0) = 50 \qquad \text{for } \frac{-\pi}{2} < x < \frac{\pi}{2},$$

$$\phi\left(\frac{-\pi}{2}, y\right) = 100 \quad \text{for } y > 0.$$

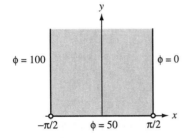

FIGURE 10.42   Accompanies Exercise 4.

5. Find the electrostatic potential $\phi(x, y)$ in the domain D in the half plane $\text{Re}(z) > 0$ that lies to the left of the hyperbola $2x^2 - 2y^2 = 1$ and satisfies the boundary values as shown in Figure 10.43. *Hint:* Use $w = \text{Arcsin } z$.

$\phi(0, y) = 50$ for all y,
$\phi(x, y) = 100$ when $2x^2 - 2y^2 = 1$.

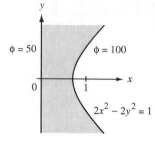

$\phi = 50$

$\phi = 100$

$2x^2 - 2y^2 = 1$

FIGURE 10.43   Accompanies Exercise 5.

6. Find the electrostatic potential $\phi(x, y)$ in the infinite strip $0 < x < \pi/2$ that satisfies the boundary values as shown in Figure 10.44. *Hint*: Use $w = \sin z$.

$$\phi(0, y) = 100 \qquad \text{for } y > 0,$$
$$\phi\left(\frac{\pi}{2}, y\right) = 0 \qquad \text{for all } y,$$
$$\phi(0, y) = -100 \quad \text{for } y < 0.$$

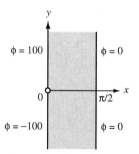

$\phi = 100$

$\phi = 0$

$\phi = -100$

$\phi = 0$

FIGURE 10.44   Accompanies Exercise 6.

7. (a) Show that the conformal mapping $w = S(z) = (2z - 6)/(z + 3)$ maps the domain D that is the portion of the right half plane $\mathrm{Re}(z) > 0$ that lies exterior to the circle $|z - 5| = 4$ onto the annulus $1 < |w| < 2$.

(b) Find the electrostatic potential $\phi(x, y)$ in the domain D that satisfies the boundary values as shown in Figure 10.45.

$$\phi(0, y) = 100 \quad \text{for all } y, \qquad \phi(x, y) = 200 \quad \text{when } |z - 5| = 4.$$

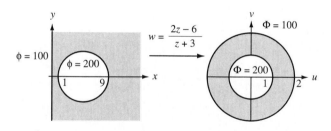

$\phi = 100$

$\phi = 200$

$w = \dfrac{2z - 6}{z + 3}$

$\Phi = 100$

$\Phi = 200$

FIGURE 10.45   Accompanies Exercise 7.

8. (a) Show that the conformal mapping $w = S(z) = (z - 10)/(2z - 5)$ maps the domain D that is the portion of the disk $|z| < 5$ that lies outside the circle $|z - 2| = 2$ onto the annulus $1 < |w| < 2$.

   (b) Find the electrostatic potential $\phi(x, y)$ in the domain D that satisfies the boundary values as shown in Figure 10.46.

$$\phi(x, y) = 100 \quad \text{when } |z| = 5, \quad \phi(x, y) = 200 \quad \text{when } |z - 2| = 2.$$

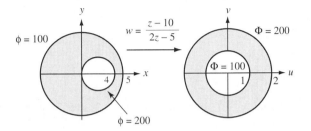

FIGURE 10.46   Accompanies Exercise 8.

10.7   Two-Dimensional Fluid Flow

Suppose that a fluid flows over the complex plane and that the velocity at the point $z = x + iy$ is given by the velocity vector

(1) $\mathbf{V}(x, y) = p(x, y) + iq(x, y)$.

We also require that the velocity does not depend on time and that the components $p(x, y)$ and $q(x, y)$ have continuous partial derivatives. The divergence of the vector field in equation (1) is given by

(2) $\operatorname{div} \mathbf{V}(x, y) = p_x(x, y) + q_y(x, y)$

and is a measure of the extent to which the velocity field diverges near the point. We will consider only fluid flows for which the divergence is zero. This is more precisely characterized by requiring that the net flow through any simply closed contour be identically zero.

    If we consider the flow out of the small rectangle in Figure 10.47, then the rate of outward flow equals the line integral of the exterior normal component of $\mathbf{V}(x, y)$ taken over the sides of the rectangle. The exterior normal component is given by $-q$ on the bottom edge, p on the right edge, q on the top edge, and $-p$ on the left edge. Integrating and setting the resulting net flow equal to zero yields

(3) $\displaystyle\int_y^{y+\Delta y} [p(x + \Delta x, t) - p(x, t)]\, dt$

$\displaystyle+ \int_x^{x+\Delta x} [q(t, y + \Delta y) - q(t, y)]\, dt = 0.$

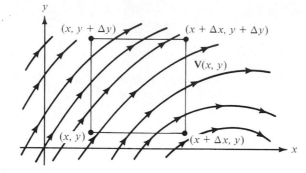

FIGURE 10.47   A two-dimensional vector field.

Since p and q are continuously differentiable, the mean value theorem can be used to show that

(4) $p(x + \Delta x, t) - p(x, t) = p_x(x_1, t) \, \Delta x$ and
$q(t, y + \Delta y) - q(t, y) = q_y(t, y_2) \, \Delta y,$

where $x < x_1 < x + \Delta x$ and $y < y_2 < y + \Delta y$. Substitution of the expressions in equation (4) into equation (3) and subsequently dividing through by $\Delta x \, \Delta y$ results in

(5) $\dfrac{1}{\Delta y} \displaystyle\int_y^{y + \Delta y} p_x(x_1, t) \, dt + \dfrac{1}{\Delta x} \displaystyle\int_x^{x + \Delta x} q_y(t, y_2) \, dt = 0.$

The mean value theorem for integrals can be used with equation (5) to show that

(6) $p_x(x_1, y_1) + q_y(x_2, y_2) = 0,$

where $y < y_1 < y + \Delta y$ and $x < x_2 < x + \Delta x$. Letting $\Delta x \to 0$ and $\Delta y \to 0$ in equation (6) results in

(7) $p_x(x, y) + q_y(x, y) = 0,$

which is called the *equation of continuity*.

The curl of the vector field in equation (1) has magnitude

(8) $\left| \text{curl } \mathbf{V}(x, y) \right| = q_x(x, y) - p_y(x, y)$

and is an indication of how the field swirls in the vicinity of a point. Imagine that a "fluid element" at the point (x, y) is suddenly frozen and then moves freely in the fluid. It can be shown that the fluid element will rotate with an angular velocity given by

(9) $\tfrac{1}{2} q_y(x, y) - \tfrac{1}{2} p_x(x, y) = \tfrac{1}{2} \left| \text{curl } \mathbf{V}(x, y) \right|.$

We will consider only fluid flows for which the curl is zero. Such fluid flows are called *irrotational*. This is more precisely characterized by requiring that the line integral of the tangential component of $\mathbf{V}(x, y)$ along any simply closed contour be identically zero. If we consider the rectangle in Figure 10.47, then the tangential component is given by p on the bottom edge, q on the right edge, $-p$ on the top edge, and $-q$ on the left edge. Integrating and setting the resulting *circulation* integral equal to zero yields the equation

$$(10) \quad \int_y^{y+\Delta y} [q(x + \Delta x, t) - q(x, t)] \, dt - \int_x^{x+\Delta x} [p(t, y + \Delta y) - p(t, y)] \, dt = 0.$$

As before, we apply the mean value theorem and divide through by $\Delta x \, \Delta y$ and obtain the equation

$$(11) \quad \frac{1}{\Delta y} \int_y^{y+\Delta y} q_x(x_1, t) \, dt - \frac{1}{\Delta x} \int_x^{x+\Delta x} p_y(t, y_2) \, dt = 0.$$

The mean value for integrals can be used with equation (11) to deduce the equation $q_x(x_1, y_1) - p_y(x_2, y_2) = 0$. Letting $\Delta x \to 0$ and $\Delta y \to 0$ yields

$$(12) \quad q_x(x, y) - p_y(x, y) = 0.$$

Equations (7) and (12) show that the function $f(z) = p(x, y) - iq(x, y)$ satisfies the Cauchy-Riemann equations and is an analytic function. Let $F(z)$ denote the antiderivative of $f(z)$. Then

$$(13) \quad F(z) = \phi(x, y) + i\psi(x, y)$$

is called the *complex potential* of the flow and has the property

$$(14) \quad \overline{F'(z)} = \phi_x(x, y) - i\psi_x(x, y) = p(x, y) + iq(x, y) = \mathbf{V}(x, y).$$

Since $\phi_x = p$ and $\phi_y = q$, we also have

$$(15) \quad \text{grad } \phi(x, y) = p(x, y) + iq(x, y) = \mathbf{V}(x, y),$$

so $\phi(x, y)$ is the *velocity potential* for the flow, and the curves

$$(16) \quad \phi(x, y) = K_1$$

are called *equipotentials*. The function $\psi(x, y)$ is called the *stream function*, and the curves

$$(17) \quad \psi(x, y) = K_2$$

are called *streamlines* and describe the paths of the fluid particles. To see this fact, we can implicitly differentiate $\psi(x, y) = K_2$ and find that the slope of a vector tangent is given by

$$(18) \quad \frac{dy}{dx} = \frac{-\psi_x(x, y)}{\psi_y(x, y)}.$$

Using the fact that $\psi_y = \phi_x$ and equation (18), we find that the tangent vector to the curve is

$$(19) \quad \mathbf{T} = \phi_x(x, y) - i\psi_x(x, y) = p(x, y) + iq(x, y) = \mathbf{V}(x, y).$$

The salient idea of the preceding discussion is the conclusion that if

$$(20) \quad F(z) = \phi(x, y) + i\psi(x, y)$$

is an analytic function, then the family of curves

(21) $\{\psi(x, y) = K_2\}$

represents the streamlines of a fluid flow.

The boundary condition for an ideal fluid flow is that $\mathbf{V}$ should be parallel to the boundary curve containing the fluid (the fluid flows parallel to the walls of a containing vessel). This means that if equation (20) is the complex potential for the flow, then the boundary curve must be given by $\psi(x, y) = K$ for some constant K; that is, the boundary curve must be a streamline.

Theorem 10.5 (Invariance of Flow)   *Let*

(22) $F_1(w) = \Phi(u, v) + i\Psi(u, v)$

denote the complex potential for a fluid flow in a domain G in the w plane where the velocity is

(23) $\mathbf{V}_1(u, v) = \overline{F_1'(w)}.$

If the function

(24) $w = S(z) = u(x, y) + iv(x, y)$

is a one-to-one conformal mapping from a domain D in the z plane onto G, then the composite function

(25) $F_2(z) = F_1(S(z)) = \Phi(u(x, y), v(x, y)) + i\Psi(u(x, y), v(x, y))$

is the complex potential for a fluid flow in D where the velocity is

(26) $\mathbf{V}_2(x, y) = \overline{F_2'(z)}.$

The situation is shown in Figure 10.48.

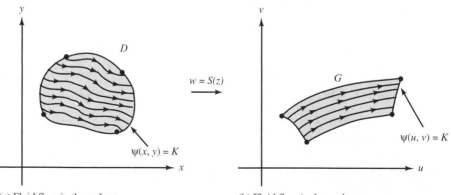

(a) Fluid flow in the z plane.

(b) Fluid flow in the w plane.

FIGURE 10.48   The image of a fluid flow under conformal mapping.

Proof   From equation (13) we see that $F_1(w)$ is an analytic function. Since the composition in equation (25) is an analytic function, $F_2(z)$ is the complex potential for an ideal fluid flow in D.

We note that the functions

(27) $\phi(x, y) = \Phi(u(x, y), v(x, y))$ and $\psi(x, y) = \Psi(u(x, y), v(x, y))$

are the new velocity potential and stream function, respectively, for the flow in D. A streamline or natural boundary curve

(28) $\psi(x, y) = K$

in the z plane is mapped onto a streamline or natural boundary curve

(29) $\Psi(u, v) = K$

in the w plane by the transformation $w = S(z)$. One method for finding a flow inside a domain D in the z plane is to conformally map D onto a domain G in the w plane in which the flow is known.

For an ideal fluid with uniform density ρ the fluid pressure $P(x, y)$ and speed $|V(x, y)|$ are related by the following special case of *Bernoulli's equation*:

(30) $\dfrac{P(x, y)}{\rho} + \dfrac{1}{2}|V(x, y)| = \text{constant}.$

It is of importance to notice that the pressure is greatest when the speed is least.

EXAMPLE 10.22   The complex potential $F(z) = (a + ib)z$ has the velocity potential and stream function given by

(31) $\phi(x, y) = ax - by$ and $\psi(x, y) = bx + ay,$

respectively, and gives rise to the fluid flow defined in the entire complex plane that has a uniform parallel velocity given by

(32) $V(x, y) = \overline{F'(z)} = a - ib.$

The streamlines are parallel lines given by the equation $bx + ay = \text{constant}$ and are inclined at an angle $\alpha = -\arctan(b/a)$ as indicated in Figure 10.49.

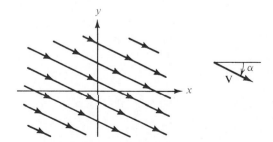

FIGURE 10.49   A uniform parallel flow.

EXAMPLE 10.23 Consider the complex potential $F(z) = (A/2)z^2$, where A is a positive real number. The velocity potential and stream function are given by

(33) $\quad \phi(x, y) = \dfrac{A}{2}(x^2 - y^2) \quad$ and $\quad \psi(x, y) = Axy,$

respectively. The streamlines $\psi(x, y) = $ constant form a family of hyperbolas with asymptotes along the coordinate axes. The velocity vector $\mathbf{V} = A\bar{z}$ indicates that in the upper half plane $\text{Im}(z) > 0$ the fluid flows down along the streamlines and spreads out along the x axis. This depicts the flow against a wall and is illustrated in Figure 10.50.

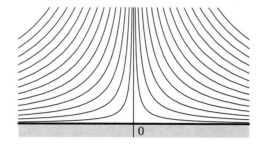

FIGURE 10.50 The fluid flow with complex potential $F(z) = (A/2)z^2$.

EXAMPLE 10.24 Find the complex potential for an ideal fluid flowing from left to right across the complex plane and around the unit circle $|z| = 1$.

Solution We will use the fact that the conformal mapping

(34) $\quad w = S(z) = z + \dfrac{1}{z}$

maps the domain $D = \{z: |z| < 1\}$ one-to-one and onto the w plane slit along the segment $-2 \le u \le 2$, $v = 0$. The complex potential for a uniform horizontal flow parallel to this slit in the w plane is

(35) $\quad F_1(w) = Aw,$

where A is a positive real number. The stream function for the flow in the w plane is $\psi(u, v) = Av$ so that the slit lies along the streamline $\Psi(u, v) = 0$.

The composite function $F_2(z) = F_1(S(z))$ will determine a fluid flow in the domain D where the complex potential is

(36) $\quad F_2(z) = A\left(z + \dfrac{1}{z}\right), \quad$ where $A > 0.$

Polar coordinates can be used to express $F_2(z)$ by the equation

(37) $F_2(z) = A\left(r + \dfrac{1}{r}\right) \cos\theta + iA\left(r - \dfrac{1}{r}\right) \sin\theta.$

The streamline $\psi(r, \theta) = A(r - 1/r) \sin\theta = 0$ consists of the rays

(38) $r > 1, \theta = 0 \quad \text{and} \quad r > 1, \theta = \pi$

along the x axis and the curve $r - 1/r = 0$, which is easily seen to be the unit circle $r = 1$. This shows that the unit circle can be considered as a boundary curve for the fluid flow.

Since the approximation $F_2(z) = A(z + 1/z) \approx Az$ is valid for large values of z, we see that the flow is approximated by a uniform horizontal flow with speed $|\mathbf{V}| = A$ at points that are distant from the origin. The streamlines $\psi(x, y) = $ constant and their images $\Psi(u, v) = $ constant under the mapping $w = S(z) = z + 1/z$ are illustrated in Figure 10.51.

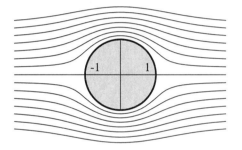

FIGURE 10.51   Fluid flow around a circle.

EXAMPLE 10.25   Find the complex potential for an ideal fluid flowing from left to right across the complex plane and around the segment from $-i$ to i.

Solution   We will use the conformal mapping

(39) $w = S(z) = (z^2 + 1)^{1/2} = (z + i)^{1/2}(z - i)^{1/2},$

where the branch of the square root of $Z = z \pm i$ in each factor is $Z^{1/2} = R^{1/2}e^{i\theta/2}$, where $R = |Z|$, and $\theta = \arg Z$, where $-\pi/2 < \arg Z \le 3\pi/2$. The function given by $w = S(z)$ is a one-to-one conformal mapping of the domain D consisting of the z plane slit along the segment $x = 0$, $-1 \le y \le 1$ onto the domain G consisting of the w plane slit along the segment $-1 \le u \le 1$, $v = 0$.

The complex potential for a uniform horizontal flow parallel to the slit in the w plane is given by $F_1(w) = Aw$, where A is a positive real number and where the slit lies along the streamline $\Psi(u, v) = Au = 0$. The composite function

$$(40) \quad F_2(z) = F_1(S(z)) = A(z^2 + 1)^{1/2}$$

is the complex potential for a fluid flow in the domain D. The streamlines given by $\psi(x, y) = cA$ for the flow in D are obtained by finding the preimage of the streamline $\Psi(u, v) = cA$ in G given by the parametric equations

$$(41) \quad v = c, \quad u = t \quad \text{for } -\infty < t < \infty.$$

The corresponding streamline in D is found by solving the equation

$$(42) \quad t + ic = (z^2 + 1)^{1/2}$$

for x and y in terms of t. Squaring both sides of equation (42) yields

$$(43) \quad t^2 - c^2 - 1 + i2ct = x^2 - y^2 + i2xy.$$

Equating the real and imaginary parts leads to the system of equations

$$(44) \quad x^2 - y^2 = t^2 - c^2 - 1 \quad \text{and} \quad xy = ct.$$

Eliminating the parameter t in equations (44) results in $c^2 = (x^2 + c^2)(y^2 - c^2)$, and we can solve for y in terms of x to obtain

$$(45) \quad y = c\sqrt{\frac{1 + c^2 + x^2}{c^2 + x^2}}$$

for streamlines in D. For large values of x this streamline approaches the asymptote $y = c$ and approximates a horizontal flow, as shown in Figure 10.52.

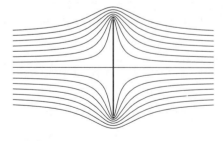

FIGURE 10.52   Flow around a segment.

EXERCISES FOR SECTION 10.7

1. Consider the ideal fluid flow where the complex potential is $F(z) = A(z + 1/z)$, where A is a positive real number.

   (a) Show that the velocity vector at the point $(1, \theta)$, $z = re^{i\theta}$ on the unit circle is given by $\mathbf{V}(1, \theta) = A(1 - \cos 2\theta - i \sin 2\theta)$.

   (b) Show that the velocity vector $\mathbf{V}(1, \theta)$ is tangent to the unit circle $|z| = 1$ at all points except -1 and $+1$. *Hint*: Show that $\mathbf{V} \cdot \mathbf{P} = 0$, where $\mathbf{P} = \cos \theta + i \sin \theta$.

   (c) Show that the speed at the point $(1, \theta)$ on the unit circle is given by $|\mathbf{V}| = 2A|\sin \theta|$ and that the speed attains the maximum of $2A$ at the points $\pm i$ and is zero at the points ± 1. Where is the pressure the greatest?

2. Show that the complex potential $F(z) = ze^{-i\alpha} + e^{i\alpha}/z$ determines the ideal fluid flow around the unit circle $|z| = 1$ where the velocity at points distant from the origin is given approximately by $\mathbf{V} \approx e^{i\alpha}$; that is, the direction of the flow for large values of z is inclined at an angle α with the x axis, as shown in Figure 10.53.

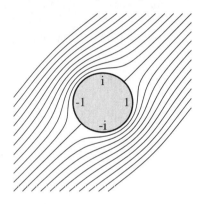

FIGURE 10.53   Accompanies Exercise 2.

3. Consider the ideal fluid flow in the channel bounded by the hyperbolas $xy = 1$ and $xy = 4$ in the first quadrant, where the complex potential is given by $F(z) = (A/2)z^2$ and A is a positive real number.

   (a) Find the speed at each point, and find the point on the boundary where the speed attains a minimum value.

   (b) Where is the pressure greatest?

4. Show that the stream function is given by $\psi(r, \theta) = Ar^3 \sin 3\theta$ for an ideal fluid flow around the angular region $0 < \theta < \pi/3$ indicated in Figure 10.54. Sketch several streamlines of the flow. *Hint*: Use the conformal mapping $w = z^3$.

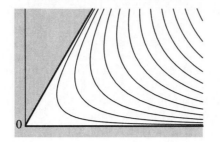

FIGURE 10.54 Accompanies Exercise 4.

5. Consider the ideal fluid flow, where the complex potential is

$$F(z) = Az^{3/2} = Ar^{3/2}\left(\cos\frac{3\theta}{2} + i\sin\frac{3\theta}{2}\right), \text{ where } 0 \le \theta \le 2\pi.$$

(a) Find the stream function $\psi(r, \theta)$.
(b) Sketch several streamlines of the flow in the angular region $0 < \theta < 4\pi/3$ as indicated in Figure 10.55.

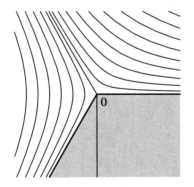

FIGURE 10.55 Accompanies Exercise 5.

6. (a) Let $A > 0$. Show that the potential $F(z) = A(z^2 + 1/z^2)$ determines an ideal fluid flow around the domain $r > 1$, $0 < \theta < \pi/2$ indicated in Figure 10.56, which shows the flow around a circle in the first quadrant. *Hint:* Use the conformal mapping $w = z^2$.
(b) Show that the speed at the point $(1, \theta)$, $z = re^{i\theta}$ on the quarter circle $r = 1$, $0 < \theta < \pi/2$ is given by $\mathbf{V} = 4A|\sin 2\theta|$.
(c) Determine the stream function for the flow and sketch several streamlines.

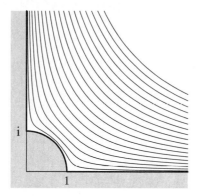

FIGURE 10.56   Accompanies Exercise 6.

7. Show that $F(z) = \sin z$ is the complex potential for the ideal fluid flow inside the semi-infinite strip $-\pi/2 < x < \pi/2$, $y > 0$, as indicated in Figure 10.57. Find the stream function.

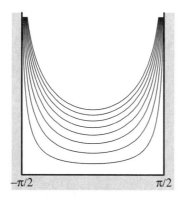

FIGURE 10.57   Accompanies Exercise 7.

8. Let $w = S(z) = \frac{1}{2}[z + (z^2 - 4)^{1/2}]$ denote the branch of the inverse of $z = w + 1/w$ that is a one-to-one mapping of the z plane slit along the segment $-2 \le x \le 2$, $y = 0$ onto the domain $|w| > 1$. Use the complex potential $F_2(w) = we^{-i\alpha} + (e^{i\alpha}/w)$ in the w plane to show that the complex potential $F_1(z) = z\cos\alpha - i(z^2 - 4)^{1/2}\sin\alpha$ determines the ideal fluid flow around the segment $-2 \le x \le 2$, $y = 0$, where the velocity at points distant from the origin is given approximately by $\mathbf{V} \approx e^{i\alpha}$, as shown in Figure 10.58.

9. (a) Show that the complex potential $F(z) = -i\operatorname{Arcsin} z$ determines the ideal fluid flow through the aperture from -1 to $+1$, as indicated in Figure 10.59.
   (b) Show that the streamline $\psi(x, y) = c$ for the flow is a portion of the hyperbola $(x^2/\sin^2 c) - (y^2/\cos^2 c) = 1$.

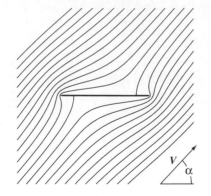

FIGURE 10.58   Accompanies Exercise 8.

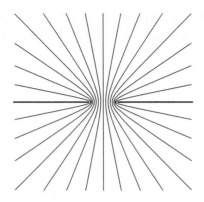

FIGURE 10.59   Accompanies Exercise 9.

10. Write a report on fluid flow and how it is related to harmonic and analytic functions. Include some ideas not mentioned in the text. Resources include bibliographical items 37, 46, 91, 98, 124, 141, 145, 158, and 166.

10.8   The Joukowski Airfoil

The function $J(z) = z + \dfrac{1}{z}$ was studied by the Russian scientist N. E. Joukowski. It will be shown that the image of a circle passing through $z_1 = 1$ and containing the point $z_2 = -1$ is mapped onto a curve that is shaped like the cross section of an airplane wing. We call this curve the *Joukowski airfoil*. If the streamlines for a flow around the circle are known, then their images under the mapping $w = J(z)$ will be streamlines for a flow around the Joukowski airfoil, as shown in Figure 10.60.

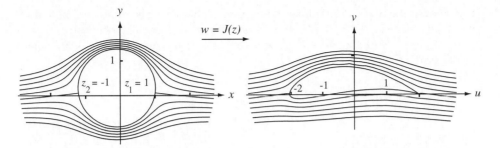

FIGURE 10.60   Image of a fluid flow under $w = J(z) = z + 1/z$.

The mapping $w = J(z)$ is two-to-one, because $J(z) = J\left(\dfrac{1}{z}\right)$, for $z \neq 0$. The region $|z| > 1$ is mapped one-to-one onto the w plane slit along the portion of the real axis $-2 \leq u \leq 2$. In order to visualize this mapping, we investigate the implicit form, which is obtained by using the substitutions

$$w - 2 = z - 2 + \frac{1}{z} = \frac{z^2 - 2z + 1}{z} = \frac{(z-1)^2}{z} \quad \text{and}$$

$$w + 2 = z + 2 + \frac{1}{z} = \frac{z^2 + 2z + 1}{z} = \frac{(z+1)^2}{z}.$$

Forming the quotient of these two quantities results in the relationship

(1) $\dfrac{w-2}{w+2} = \left(\dfrac{z-1}{z+1}\right)^2.$

The inverse of $T(w) = \dfrac{w-2}{w+2}$ is $S_3(z) = \dfrac{2 + 2z}{1 - z}$. If we use the notation $S_1(z) = \dfrac{z-1}{z+1}$ and $S_2(z) = z^2$, then $J(z)$ can be expressed as the composition of S_1, S_2, and S_3:

(2) $w = J(z) = S_3(S_2(S_1(z))).$

It is an easy calculation to show that $w = J(z) = z + \dfrac{1}{z}$ maps the four points $z_1 = -i$, $z_2 = 1$, $z_3 = i$, and $z_4 = -1$ onto $w_1 = 0$, $w_2 = 2$, $w_3 = 0$, and $w_4 = -2$, respectively. However, the composition functions in equation (2) must be considered in order to visualize the geometry involved. First, the bilinear transformation $Z = S_1(z)$ maps the region $|z| > 1$ onto the right half plane $\operatorname{Re}(Z) > 0$, and the points $z_1 = -i$, $z_2 = 1$, $z_3 = i$, and $z_4 = -1$ are mapped onto $Z_1 = -i$, $Z_2 = 0$, $Z_3 = i$, and $Z_4 = i\infty$, respectively. Second, the function $W = S_2(Z)$ maps the right half plane onto the W plane slit along its negative real axis, and the points $Z_1 = -i$, $Z_2 = 0$, $Z_3 = i$, and $Z_4 = i\infty$ are mapped onto $W_1 = -1$, $W_2 = 0$, $W_3 = -1$, and $W_4 = -\infty$, respectively. Then the bilinear transformation $w = S_3(W)$ maps the latter region onto the w plane slit along the portion of the real axis $-2 \leq u \leq 2$, and the points $W_1 = -1$, $W_2 = 0$, $W_3 = -1$, and $W_4 = -\infty$ are mapped onto $w_1 = 0$, $w_2 = 2$, $w_3 = 0$, and $w_4 = -2$, respectively. These three compositions are shown in Figure 10.61.

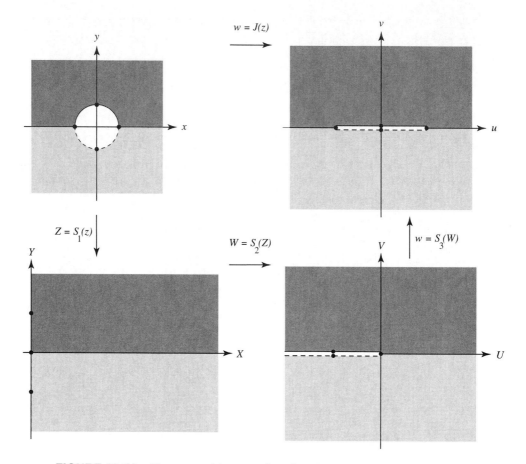

FIGURE 10.61 The composition mappings for $J(z) = S_3(S_2(S_1(z)))$.

The circle C_0 with center $c_0 = ia$ on the imaginary axis passes through the points $z_2 = 1$ and $z_4 = -1$ and has radius $r_0 = \sqrt{1 + a^2}$. If we restrict $0 < a < 1$, then this circle intersects the x axis at the point z_2 with angle $\alpha_0 = \dfrac{\pi}{2} - \arctan a$, with $\dfrac{\pi}{4} < \alpha_0 < \dfrac{\pi}{2}$. We want to track the image of C_0 in the Z, W, and w planes. First, the image of this circle C_0 under $Z = S_1(z)$ is the line L_0 that passes through the origin and is inclined at the angle α_0. Second, the function $W = S_2(Z)$ maps the line L_0 onto the ray R_0 inclined at the angle $2\alpha_0$. Finally the transformation given by $w = S_3(W)$ maps the ray R_0 onto the arc of the circle A_0 that passes through the points $w_2 = 2$ and $w_4 = -2$ and intersects the x axis at w_2 with angle $2\alpha_0$, where $\dfrac{\pi}{2} < 2\alpha_0 < \pi$. The restriction on the angle α_0, and hence $2\alpha_0$, is necessary in order for the arc A_0 to have a low profile. The arc A_0 lies in the center of the Joukowski airfoil and is shown in Figure 10.62.

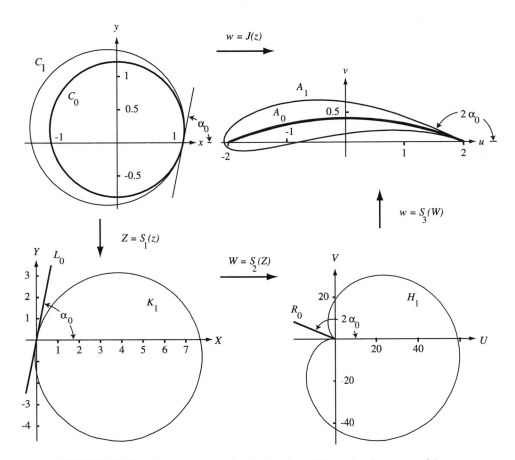

FIGURE 10.62 The images of the circles C_0 and C_1 under the composition mappings for $J(z) = S_3(S_2(S_1(z)))$.

Let b be fixed, $0 < b < 1$, then the larger circle C_1 with center given by $c_1 = -h + i(1 + h)b$ on the imaginary axis will pass through the points $z_2 = 1$ and $z_4 = -1 - 2h$ and have radius $r_1 = (1 + h)\sqrt{1 + b^2}$. The circle C_1 also intersects the x axis at the point z_2 at the angle α_0. The image of this circle C_1 under $Z = S_1(z)$

is the circle K_1 that is tangent to L_0 at the origin. The function $W = S_2(Z)$ maps the circle K_1 onto the cardioid H_1. Finally, $w = S_3(W)$ maps the cardioid H_1 onto the Joukowski airfoil A_1 that passes through the point $w_2 = 2$ and surrounds the point $w_4 = -2$, as shown in Figure 10.62. We remark that as an observer traverses C_1 in the counterclockwise direction, the image curves K_1 and H_1 will be traversed in a clockwise direction, but A_1 is traversed in the counterclockwise direction. This keeps the points z_4, Z_4, W_4, and w_4 always to the observer's left.

Now we are ready to visualize the flow around the Joukowski airfoil. We start with the fluid flow around a circle that is shown in Figure 10.51. This flow is adjusted with a linear transformation $z^* = az + b$ so that it flows horizontally around the circle C_1, as shown in Figure 10.63. Then the mapping $w = J(z^*)$ creates a flow around the Joukowski airfoil, as Figure 10.64 illustrates.

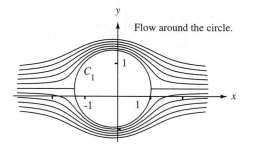

FIGURE 10.63   The horizontal flow around the circle C_1.

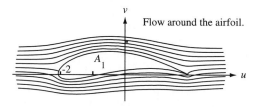

FIGURE 10.64   The horizontal flow around the Joukowski airfoil A_1.

Flow with Circulation

The function $F(z) = sz + \dfrac{s}{z} + \dfrac{k}{2\pi i} \log z$, where $s > 0$ and k is real, is the complex potential for a uniform horizontal flow past the unit circle $|z| = 1$, with circulation strength k and velocity at infinity $V_\infty = s$. For illustration purposes, we let $s = 1$ and use the substitution $a = \dfrac{-k}{2\pi}$. Now the complex potential has the form

(3) $F(z) = z + \dfrac{1}{z} + ai \log z,$

and the corresponding velocity function is

(4) $V(x, y) = \overline{F'(z)} = 1 - (\bar{z})^{-2} - ai(\bar{z})^{-1}.$

The complex potential can be expressed in $F = \phi + i\psi$ form:

(5) $F(z) = re^{i\theta} + \dfrac{1}{r}e^{i\theta} + ia(\ln r + i\theta)$

$\qquad = \left(r + \dfrac{1}{r}\right)\cos\theta - a\theta + i\left[\left(r - \dfrac{1}{r}\right)\sin\theta + a\ln r\right].$

The streamlines for the flow are given by $\psi = c$, where c is a constant:

(6) $\psi(r\cos\theta, r\sin\theta) = \left(r - \dfrac{1}{r}\right)\sin\theta + a\ln r = c$ (streamlines).

Setting $r = 1$ in equation (6) we get $\psi(\cos\theta, \sin\theta) \equiv 0$, so that the unit circle is a natural boundary curve for the flow.

Points where the flow has zero velocity are called *stagnation points*. They are found by solving $F'(z) = 0$, for the function in equation (3) this is

$\qquad 1 - \dfrac{1}{z^2} + \dfrac{a}{z} = 0.$

Multiplying through by z^2 and rearranging terms, this becomes

$\qquad z^2 + aiz - 1 = 0.$

Now the quadratic equation is invoked to obtain

$\qquad z = \dfrac{-ai \pm \sqrt{4 - a^2}}{2}$ stagnation point(s).

If $0 \le |a| < 2$, there are two stagnation points on the unit circle $|z| = 1$. If $a = 2$, there is one stagnation point on the unit circle. If $|a| > 2$, the stagnation point lies outside the unit circle. We are mostly interested in the case with two stagnation points. When $a = 0$, the two stagnation points are $z = \pm 1$, and this is the flow that was discussed in Example 10.25. The cases $a = 1$, $a = \sqrt{3}$, $a = 2$, and $a = 2.2$ are shown in Figure 10.65.

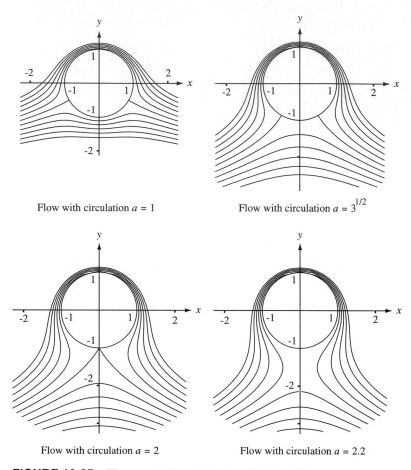

Flow with circulation $a = 1$

Flow with circulation $a = 3^{1/2}$

Flow with circulation $a = 2$

Flow with circulation $a = 2.2$

FIGURE 10.65   Flows past the unit circle with circulation a.

We are now ready to combine the preceding ideas. For illustration purposes, consider a C_1 circle with center $c_0 = -0.15 + 0.23i$ that passes through the points $z_2 = 1$ and $z_4 = -1.3$ and has radius $r_0 = 0.23 \sqrt{13/2}$. The flow with circulation $k = -0.52p$ (or $a = 0.26$) around $|z| = 1$ is mapped by the linear transformation $Z = S(z) = -0.15 + 0.23i + r_0 z$ onto the flow around the circle C_1 shown in Figure 10.66.

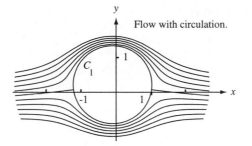

FIGURE 10.66   Flow with circulation around C_1.

Then the mapping $w = J(Z) = Z + \dfrac{1}{Z}$ is used to map this flow around the Joukowski airfoil shown in Figure 10.67. This is to be compared with the flows shown in Figures 10.63 and 10.64. If the second transformation in the composition given by $w = J(z) = S_3(S_2(S_1(z)))$ is modified to be $S_2(z) = z^{1.925}$, then the image of the flow in Figure 10.66 will be the flow around the modified airfoil in Figure 10.68. The advantage of this latter airfoil is that the sides of its tailing edge form an angle of 0.15π radians or $27°$, which is more realistic than the angle of $0°$ of the traditional Joukowski airfoil.

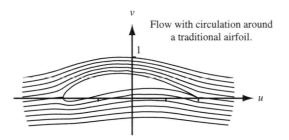

FIGURE 10.67   Flow with circulation around a traditional Joukowski airfoil.

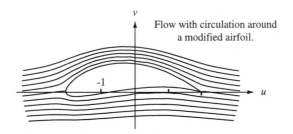

FIGURE 10.68   Flow with circulation around a modified Joukowski airfoil.

EXERCISES FOR SECTION 10.8

1. Show that the inverse of the Joukowski transformation is $z = w + (w^2 - 1)^{1/2}$.
2. Consider the Joukowski transformation is $w = z + 1/z$.

   (a) Show that the circle $C_r = \{|z| = r: r > 1\}$ is mapped onto the ellipse

   $$\frac{4 u^2}{(r + 1/r)^2} + \frac{4 v^2}{(r - 1/r)^2} = 1.$$

   (b) Show that the ray $r > 0$, $\theta = \alpha$ is mapped onto a branch of the hyperbola

   $$\frac{u^2}{\cos^2 \alpha} - \frac{v^2}{\sin^2 \alpha} = 1.$$

3. Let C_0 be a circle that passes through the points 1 and -1 and has center $c_0 = ia$.

   (a) Find the equation of the circle C_0.

   (b) Show that the image of the circle C_0 under $w = \dfrac{z - 1}{z + 1}$ is a line L_0 that passes through the origin.

   (c) Show that the line L_0 is inclined at the angle $\alpha_0 = \dfrac{\pi}{2} - \arctan a$.

4. Show that a line through the origin mapped onto a ray by the mapping $w = z^2$.
5. Let R_0 be a ray through the origin inclined at an angle β_0.

   (a) Show that the image of the ray R_0 under $w = \dfrac{2 + 2z}{1 - z}$ is an arc A_0 of a circle that passes through 2 and -2.

   (b) Show that the arc A_0 is inclined at the angle β_0.

6. Show that a circle passing through the origin is mapped onto a cardioid by $w = z^2$. Show that the cusp in the cardioid forms an angle of $0°$.

7. Let H_1 be a cardioid whose cusp is at the origin. The image of H_1 under $w = \dfrac{2 + 2z}{1 - z}$

   will be a Joukowski airfoil. Show that trailing edge forms an angle of $0°$.

8. Consider the modified Joukowski airfoil when $W = S_2(Z) = Z^{1.925}$ is used to map the Z plane onto the W plane. Use Figure 10.69 and discuss why the angle of the trailing edge of this modified Joukowski airfoil A_1 forms an angle of 0.15π radians. *Hint:* The image of the circle C_0 is the line L_0, then two rays $R_{0,1}$ and $R_{0,2}$ and then two arcs $A_{0,1}$ and $A_{0,2}$ in the respective Z, W, and w planes. The image of the circle C_1 is the circle K_1, then the "cardioid like" curve H_1, then the modified Joukowski airfoil A_1.

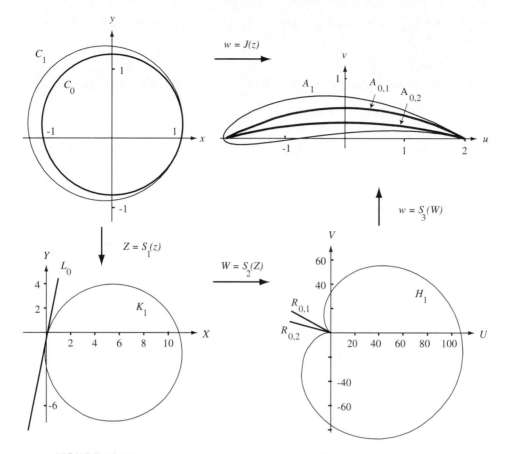

FIGURE 10.69   The images of the circles C_0 and C_1 under the modified Joukowski transformation $J(z) = S_3(S_2(S_1(z)))$.

9. Write a report on Joukowski transformation. Include ideas and examples that are not mentioned in the text. Resources include bibliographical items 37, 46, 91, 98, 124, 141, 145, 158, and 166.

10.9   The Schwarz-Christoffel Transformation

To proceed further, we must review the rotational effect of a conformal mapping $w = f(z)$ at a point z_0. If the contour C has the parameterization $z(t) = x(t) + iy(t)$, then a vector $\boldsymbol{\tau}$ tangent to C at the point z_0 is

(1) $\boldsymbol{\tau} = z'(t_0) = x'(t_0) + iy'(t_0).$

The image of C is a contour K given by $w = u(x(t), y(t)) + iv(x(t), y(t))$, and a vector $\mathbf{T}$ tangent to K at the point $w_0 = f(z_0)$ is

(2) $\mathbf{T} = w'(z_0) = f'(z_0)z'(t_0)$.

If the angle of inclination of τ is $\beta = \arg z'(t_0)$, then the angle of inclination of $\mathbf{T}$ is

(3) $\arg \mathbf{T} = \arg f'(z_0)z'(t_0) = \arg f'(z_0) + \beta$.

Hence the angle of inclination of the tangent τ to C at z_0 is rotated through the angle $\arg f'(z_0)$ to obtain the angle of inclination of the tangent $\mathbf{T}$ to K at the point w_0.

Many applications involving conformal mappings require the construction of a one-to-one conformal mapping from the upper half plane $\text{Im}(z) > 0$ onto a domain G in the w plane where the boundary consists of straight line segments. Let us consider the case where G is the interior of a polygon P with vertices $w_1, w_2, \ldots, w_n$ specified in the positive (counterclockwise) sense. We want to find a function $w = f(z)$ with the property

(4) $w_k = f(x_k)$ for $k = 1, 2, \ldots, n - 1$ and
      $w_n = f(\infty)$,   where $x_1 < x_2 < \cdots < x_{n-1} < \infty$.

Two German mathematicians Herman Amandus Schwarz (1843–1921) and Elwin Bruno Christoffel (1829–1900) independently discovered a method for finding f, and that is our next theorem.

> **Theorem 10.6 (Schwarz-Christoffel)**   *Let P be a polygon in the w plane with vertices $w_1, w_2, \ldots, w_n$ and exterior angles α_k, where $-\pi < \alpha_k < \pi$, as shown in Figure 10.70. There exists a one-to-one conformal mapping $w = f(z)$ from the upper half plane $\text{Im}(z) > 0$ onto G that satisfies the boundary conditions (4). The derivative $f'(z)$ is*
>
> (5) $f'(z) = A(z - x_1)^{-\alpha_1/\pi}(z - x_2)^{-\alpha_2/\pi} \cdots (z - x_{n-1})^{-\alpha_{n-1}/\pi}$,
>
> *and the function f can be expressed as an indefinite integral*
>
> (6) $f(z) = B + A \displaystyle\int (z - x_1)^{-\alpha_1/\pi}(z - x_2)^{-\alpha_2/\pi} \cdots (z - x_{n-1})^{-\alpha_{n-1}/\pi} \, dz$
>
> *where A and B are suitably chosen constants. Two of the points $\{x_k\}$ may be chosen arbitrarily, and the constants A and B determine the size and position of P.*

Proof   The proof relies on finding how much the tangent

(7) $\tau_j = 1 + 0i$

(which always points to the right) at the point $(x, 0)$ must be rotated by the mapping

$w = f(z)$ so that the line segment $x_{j-1} < x < x_j$ is mapped onto the edge of P that lies between the points $w_{j-1} = f(x_{j-1})$ and $w_j = f(x_j)$. Since the amount of rotation is determined by $\arg f'(x)$, formula (5) specifies $f'(z)$ in terms of the values x_j and the amount of rotation α_j that is required at the vertex $f(x_j)$.

If we let $x_0 = -\infty$ and $x_n = \infty$, then, for values of x that lie in the interval $x_{j-1} < x < x_j$, the amount of rotation is

$$(8) \quad \arg f'(x) = \arg A - \frac{1}{\pi}[\alpha_1 \arg(x - x_1) + \alpha_2 \arg(x - x_2)$$

$$+ \cdots + \alpha_{n-1} \arg(x - x_{n-1})].$$

Since $\mathrm{Arg}(x - x_k) = 0$ for $1 \le k < j$ and $\mathrm{Arg}(x - x_k) = \pi$ for $j \le k \le n - 1$, we can write equation (8) as

$$(9) \quad \arg f'(x) = \arg A - \alpha_j - \alpha_{j+1} - \cdots - \alpha_{n-1}.$$

The angle of inclination of the tangent vector $\mathbf{T}_j$ to the polygon P at the point $w = f(x)$ for $x_{j-1} < x < x_j$ is

$$(10) \quad \gamma_j = \mathrm{Arg}\, A - \alpha_j - \alpha_{j+1} - \cdots - \alpha_{n-1}.$$

The angle of inclination of the tangent vector $\mathbf{T}_{j+1}$ to the polygon P at the point $w = f(x)$ for $x_j < x < x_{j+1}$ is

$$(11) \quad \gamma_{j+1} = \mathrm{Arg}\, A - \alpha_{j+1} - \alpha_{j+2} - \cdots - \alpha_{n-1}.$$

The angle of inclination of the vector tangent to the polygon P jumps abruptly by the amount α_j as the point $w = f(x)$ moves along the side $\overset{\frown}{w_{j-1}w_j}$ through the vertex w_j to the side $\overset{\frown}{w_jw_{j+1}}$. Therefore the exterior angle to the polygon P at the vertex w_j is given by the angle α_j and satisfies the inequality $-\pi < \alpha_j < \pi$ for $j = 1, 2, \ldots,$ $n - 1$. Since the sum of the exterior angles of a polygon equals 2π, we have $\alpha_n = 2\pi - \alpha_1 - \alpha_2 - \cdots - \alpha_{n-1}$ so that only $n - 1$ angles need to be specified. This case with $n = 5$ is indicated in Figure 10.70.

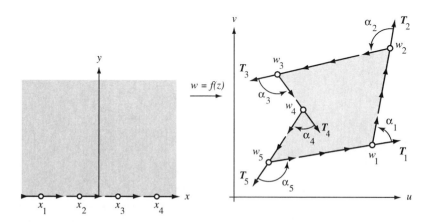

FIGURE 10.70   A Schwarz-Christoffel mapping with $n = 5$ and $\alpha_1 + \alpha_2 + \cdots + \alpha_4 > \pi$.

If the case $\alpha_1 + \alpha_2 + \cdots + \alpha_{n-1} \leq \pi$ occurs, then $\alpha_n > \pi$, and the vertices $w_1, w_2, \ldots, w_n$ cannot form a closed polygon. For this case, formulas (5) and (6) will determine a mapping from the upper half plane $\text{Im}(z) > 0$ onto an infinite region in the w plane where the vertex w_n is at infinity. The case $n = 5$ is illustrated in Figure 10.71.

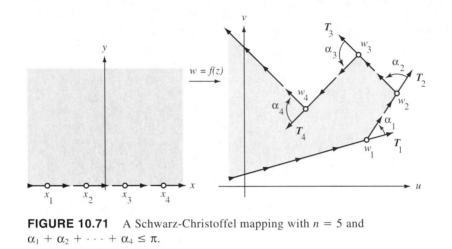

FIGURE 10.71   A Schwarz-Christoffel mapping with $n = 5$ and $\alpha_1 + \alpha_2 + \cdots + \alpha_4 \leq \pi$.

Formula (6) gives a representation for f in terms of an indefinite integral. It is important to note that these integrals do not represent elementary functions unless the image is an infinite region. Also, the integral will involve a multivalued function, and a specific branch must be selected to fit the boundary values specified in the problem. Table 10.2 is useful for our purposes.

TABLE 10.2   Indefinite Integrals

$$\int \frac{dz}{(z^2 - 1)^{1/2}} = i \arcsin z = \log(z + (z^2 - 1)^{1/2}) - \frac{i\pi}{2}$$

$$\int \frac{dz}{z^2 + 1} = \arctan z = \frac{i}{2} \log\left(\frac{i + z}{i - z}\right)$$

$$\int \frac{dz}{z(z^2 - 1)^{1/2}} = -\arcsin \frac{1}{z} = i \log\left[\frac{1}{z} + \left(\frac{1}{z^2} - 1\right)^{1/2}\right]$$

$$\int \frac{dz}{z(z + 1)^{1/2}} = -2 \arctanh[(z + 1)^{1/2}] = \log\left[\frac{1 - (z + 1)^{1/2}}{1 + (z + 1)^{1/2}}\right]$$

$$\int (1 - z^2)^{1/2} \, dz = \frac{1}{2}[z(1 - z^2)^{1/2} + \arcsin z]$$

$$= \frac{i}{2}[z(z^2 - 1)^{1/2} + \log(z + (z^2 - 1)^{1/2})]$$

EXAMPLE 10.26   Use the Schwarz-Christoffel formula to verify that the function $w = f(z) = \text{Arcsin } z$ maps the upper half plane $\text{Im}(z) > 0$ onto the semi-infinite strip $-\pi/2 < u < \pi/2$, $v > 0$ shown in Figure 10.72.

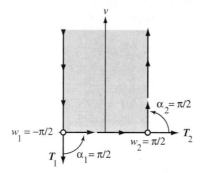

FIGURE 10.72   The region in Example 10.26.

Solution   If we choose $x_1 = -1$, $x_2 = 1$, $w_1 = -\pi/2$, and $w_2 = \pi/2$, then $\alpha_1 = \pi/2$ and $\alpha_2 = \pi/2$, and equation (5) for $f'(z)$ becomes

(12) $f'(z) = A(z + 1)^{-(\pi/2)/\pi}(z - 1)^{-(\pi/2)/\pi} = \dfrac{A}{(z^2 - 1)^{1/2}}.$

Using Table 10.2 we see that the solution to equation (12) is

(13) $f(z) = Ai \text{ Arcsin } z + B.$

Using the image values $f(-1) = -\pi/2$ and $f(1) = \pi/2$, we obtain the system

(14) $\dfrac{-\pi}{2} = A\dfrac{-i\pi}{2} + B$ and $\dfrac{\pi}{2} = A\dfrac{i\pi}{2} + B,$

which can be solved to obtain $B = 0$ and $A = -i$. Hence the required function is

(15) $f(z) = \text{Arcsin } z.$

EXAMPLE 10.27   Verify that $w = f(z) = (z^2 - 1)^{1/2}$ maps the upper half plane $\text{Im}(z) > 0$ onto the upper half plane $\text{Im}(w) > 0$ slit along the segment from 0 to i.

Solution   If we choose $x_1 = -1$, $x_2 = 0$, $x_3 = 1$, $w_1 = -d$, $w_2 = i$, and $w_3 = d$, then we see that the formula

(16) $g'(z) = A(z + 1)^{-\alpha_1/\pi}(z)^{-\alpha_2/\pi}(z - 1)^{-\alpha_3/\pi}$

will determine a mapping $w = g(z)$ from the upper half plane $\text{Im}(z) > 0$ onto the portion of the upper half plane $\text{Im}(w) > 0$ that lies outside the triangle with vertices $\pm d$, i as indicated in Figure 10.73(a). If we let $d \to 0$, then $w_1 \to 0$, $w_3 \to 0$,

$\alpha_1 \to \pi/2$, $\alpha_2 \to -\pi$, and $\alpha_3 \to \pi/2$. The limiting formula for the derivative in equation (16) becomes

(17) $\quad f'(z) = A(z+1)^{-1/2}(z)(z-1)^{-1/2},$

which will determine a mapping $w = f(z)$ from the upper half plane $\text{Im}(z) > 0$ onto the upper half plane $\text{Im}(w) > 0$ slit from 0 to i as indicated in Figure 10.73(b). An easy computation reveals that $f(z)$ is given by

(18) $\quad f(z) = A \int \dfrac{z\,dz}{(z^2-1)^{1/2}} = A(z^2-1)^{1/2} + B,$

and the boundary values $f(\pm 1) = 0$ and $f(0) = i$ lead to the solution

(19) $\quad f(z) = (z^2-1)^{1/2}.$

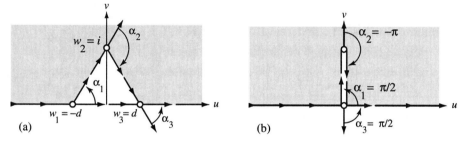

FIGURE 10.73 The regions in Example 10.27.

EXAMPLE 10.28 Show that the function

(20) $\quad w = f(z) = \dfrac{1}{\pi} \text{Arcsin}\, z + \dfrac{i}{\pi} \text{Arcsin}\, \dfrac{1}{z} + \dfrac{1+i}{2}$

maps the upper half plane $\text{Im}(z) > 0$ onto the right angle channel in the first quadrant, which is bounded by the coordinate axes and the rays $x \ge 1$, $y = 1$ and $y \ge 1$, $x = 1$ in Figure 10.74(b).

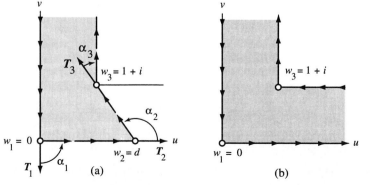

FIGURE 10.74 The regions in Example 10.28.

Solution   If we choose $x_1 = -1$, $x_2 = 0$, $x_3 = 1$, $w_1 = 0$, $w_2 = d$, and $w_3 = 1 + i$, then the formula

$$(21) \quad g'(z) = A_1(z + 1)^{-\alpha_1/\pi}(z)^{-\alpha_2/\pi}(z - 1)^{-\alpha_3/\pi}$$

will determine a mapping of the upper half plane onto the domain indicated in Figure 10.74(a).

If we let $d \to \infty$, then $\alpha_2 \to \pi$ and $\alpha_3 \to -\pi/2$. Then the limiting formula for the derivative in equation (21) becomes

$$(22) \quad f'(z) = A_1(z + 1)^{-(\pi/2)/\pi}(z)^{-(\pi)/\pi}(z - 1)^{-(-\pi/2)/\pi}$$

$$= A_1 \frac{1}{z} \frac{(z - 1)^{1/2}}{(z + 1)^{1/2}} = A_1 \frac{z - 1}{z(z^2 - 1)^{1/2}} = A \frac{z - 1}{z(1 - z^2)^{1/2}},$$

where $(A = -iA_1)$, which will determine a mapping $w = f(z)$ from the upper half plane onto the channel as indicated in Figure 10.74(b). Using Table 10.2, we obtain

$$(23) \quad f(z) = A\left[\int \frac{dz}{(1 - z^2)^{1/2}} - i \int \frac{dz}{z(z^2 - 1)^{1/2}} \right]$$

$$= A\left[\arcsin z + i \arcsin \frac{1}{z} \right] + B.$$

If the principal branch of the inverse sine function is used, then the boundary values $f(-1) = 0$ and $f(1) = 1 + i$ lead to the system

$$A\left[\frac{-\pi}{2} + i\left(\frac{-\pi}{2} \right) \right] + B = 0, \quad A\left[\frac{\pi}{2} + i\left(\frac{\pi}{2} \right) \right] + B = 1 + i,$$

which can be solved to obtain $A = 1/\pi$ and $B = (1 + i)/2$. Hence the required solution is

$$(24) \quad w = f(z) = \frac{1}{\pi} \text{Arcsin } z + \frac{i}{\pi} \text{Arcsin } \frac{1}{z} + \frac{1 + i}{2}.$$

EXERCISES FOR SECTION 10.9

1. Let a and K be real constants with $0 < K < 2$. Use the Schwarz-Christoffel formula to show that the function $w = f(z) = (z - a)^K$ maps the upper half plane $\text{Im}(z) > 0$ onto the sector $0 < \arg w < K\pi$, shown in Figure 10.75.

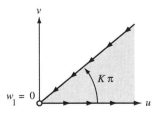

FIGURE 10.75   Accompanies Exercise 1.

2. Let a be a real constant. Use the Schwarz-Christoffel formula to show that the function $w = f(z) = \text{Log}(z - a)$ maps the upper half plane $\text{Im}(z) > 0$ onto the infinite strip $0 < v < \pi$ in Figure 10.76. *Hint:* Set $x_1 = a - 1$, $x_2 = a$, $w_1 = i\pi$, $w_2 = -d$, and let $d \to \infty$.

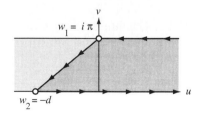

FIGURE 10.76 Accompanies Exercise 2.

3. Use the Schwarz-Christoffel formula to show that the function

$$w = f(z) = \frac{1}{\pi}\left((z^2 - 1)^{1/2} + \text{Log}[z + (z^2 - 1)^{1/2}]\right) - i$$

maps the upper half plane onto the domain indicated in Figure 10.77. *Hint:* Set $x_1 = -1$, $x_2 = 1$, $w_1 = 0$, and $w_2 = -i$.

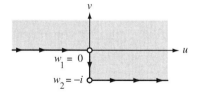

FIGURE 10.77 Accompanies Exercise 3.

4. Use the Schwarz-Christoffel formula to show that the function

$$w = f(z) = \frac{2}{\pi}\left[(z^2 - 1)^{1/2} + \text{Arcsin}\frac{1}{z}\right]$$

maps the upper half plane onto the domain indicated in Figure 10.78. *Hint:* Set $x_1 = w_1 = -1$, $x_2 = 0$, $x_3 = w_3 = 1$, and $w_2 = -id$ and let $d \to \infty$.

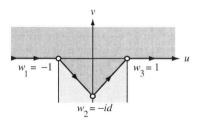

FIGURE 10.78 Accompanies Exercise 4.

5. Use the Schwarz-Christoffel formula to show that the function

$$w = f(z) = \tfrac{1}{2} \log(z^2 - 1) = \text{Log}[(z^2 - 1)^{1/2}]$$

maps the upper half plane $\text{Im}(z) > 0$ onto the infinite strip $0 < v < \pi$ slit along the ray $u \leq 0$, $v = \pi/2$, see Figure 10.79. *Hint*: Set $x_1 = -1$, $x_2 = 0$, $x_3 = 1$, $w_1 = i\pi - d$, $w_2 = i\pi/2$, and $w_3 = -d$ and let $d \to \infty$.

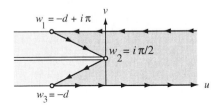

FIGURE 10.79   Accompanies Exercise 5.

6. Use the Schwarz-Christoffel formula to show that the function

$$w = f(z) = \frac{-2}{\pi}[z(1 - z^2)^{1/2} + \text{Arcsin } z]$$

maps the upper half plane onto the domain indicated in Figure 10.80. *Hint*: Set $x_1 = -1$, $x_2 = 1$, $w_1 = 1$, and $w_2 = -1$.

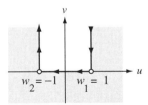

FIGURE 10.80   Accompanies Exercise 6.

7. Use the Schwarz-Christoffel formula to show that the function $w = f(z) = z + \text{Log } z$ maps the upper half plane $\text{Im}(z) > 0$ onto the upper half plane $\text{Im}(w) > 0$ slit along the ray $u \leq -1$, $v = \pi$, shown in Figure 10.81. *Hint*: Set $x_1 = -1$, $x_2 = 0$, $w_1 = -1 + i\pi$, and $w_2 = -d$ and let $d \to \infty$.

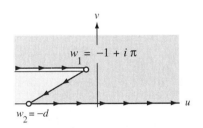

FIGURE 10.81   Accompanies Exercise 7.

8. Use the Schwarz-Christoffel formula to show that the function

$$w = f(z) = 2(z + 1)^{1/2} + \text{Log}\left[\frac{1 - (z + 1)^{1/2}}{1 + (z + 1)^{1/2}}\right] + i\pi$$

maps the upper half plane onto the domain indicated in Figure 10.82. *Hint:* Set $x_1 = -1$, $x_2 = 0$, $w_1 = i\pi$, and $w_2 = -d$ and let $d \to \infty$.

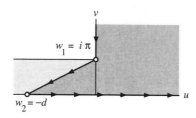

FIGURE 10.82 Accompanies Exercise 8.

9. Show that the function $w = f(z) = (z - 1)^{\alpha}[1 + \alpha z/(1 - \alpha)]^{1-\alpha}$ maps the upper half plane $\text{Im}(z) > 0$ onto the upper half plane $\text{Im}(w) > 0$ slit along the segment from 0 to $e^{i\alpha\pi}$, as shown in Figure 10.83. *Hint:* Show that $f'(z) = A[z + (1 - \alpha)/\alpha]^{-\alpha}(z)(z - 1)^{\alpha-1}$.

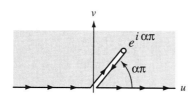

FIGURE 10.83 Accompanies Exercise 9.

10. Use the Schwarz-Christoffel formula to show that the function

$$w = f(z) = 4(z + 1)^{1/4} + \log\left[\frac{(z + 1)^{1/4} - 1}{(z + 1)^{1/4} + 1}\right] + i\log\left[\frac{i - (z + 1)^{1/4}}{i + (z + 1)^{1/4}}\right]$$

maps the upper half plane onto the domain indicated in Figure 10.84. *Hint:* Set $z_1 = -1$, $z_2 = 0$, $w_1 = i\pi$, and $w_2 = -d$ and let $d \to \infty$. Use the change of variable $z + 1 = s^4$ in the resulting integral.

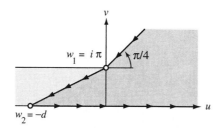

FIGURE 10.84 Accompanies Exercise 10.

11. Use the Schwarz-Christoffel formula to show that the function

$$w = f(z) = \frac{-i}{2} z^{1/2}(z - 3)$$

maps the upper half plane onto the domain indicated in Figure 10.85. *Hint*: Set $x_1 = 0$, $x_2 = 1$, $w_1 = -d$, and $w_2 = i$ and let $d \to 0$.

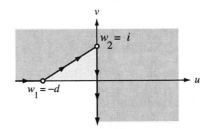

FIGURE 10.85   Accompanies Exercise 11.

12. Show that the function

$$w = f(z) = \int \frac{dz}{(1 - z^2)^{3/4}}$$

maps the upper half plane $\text{Im}(z) > 0$ onto a right triangle with angles $\pi/2$, $\pi/4$, and $\pi/4$.

13. Show that the function

$$w = f(z) = \int \frac{dz}{(1 - z^2)^{2/3}}$$

maps the upper half plane onto an equilateral triangle.

14. Show that the function

$$w = f(z) = \int \frac{dz}{(z - z^3)^{1/2}}$$

maps the upper half plane onto a square.

15. Use the Schwarz-Christoffel formula to show that the function

$$w = f(z) = 2(z + 1)^{1/2} - \text{Log}\left[\frac{1 - (z + 1)^{1/2}}{1 + (z + 1)^{1/2}}\right]$$

maps the upper half plane $\text{Im}(z) > 0$ onto the domain indicated in Figure 10.86. *Hint*: Set $x_1 = -1$, $x_2 = 0$, $x_3 = 1$, $w_1 = 0$, $w_2 = d$, and $w_3 = 2\sqrt{2} - 2\ln(\sqrt{2} - 1) + i\pi$ and let $d \to \infty$.

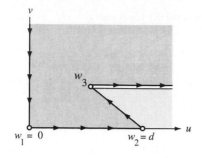

FIGURE 10.86   Accompanies Exercise 15.

16. Write a report on the Schwarz-Christoffel transformation. Include ideas and examples not mentioned in the text. Resources include bibliographical items 93, 159, and 164.

10.10   Image of a Fluid Flow

We have already examined several two-dimensional fluid flows and have discovered that the image of a flow under a conformal transformation is a flow. The conformal mapping $w = f(z) = u(x, y) + iv(x, y)$, which is obtained by using the Schwarz-Christoffel formula, will allow us to find the streamlines for flows in domains in the w plane that are bounded by straight line segments.

The first technique is finding the image of a fluid flowing horizontally from left to right across the upper half plane $\text{Im}(z) > 0$. The image of the streamline $-\infty < t < \infty$, $y = c$ will be a streamline given by the parametric equations

(1) $u = u(t, c), \quad v = v(t, c) \quad \text{for } -\infty < t < \infty$

and will be oriented in the counterclockwise (positive) sense. The streamline $u = u(t, 0)$, $v = v(t, 0)$ is considered to be a boundary wall for a containing vessel for the fluid flow.

EXAMPLE 10.29   Consider the conformal mapping

(2) $w = f(z) = \dfrac{1}{\pi}[(z^2 - 1)^{1/2} + \text{Log}(z + (z^2 - 1)^{1/2})],$

which is obtained by using the Schwarz-Christoffel formula, to map the upper half plane $\text{Im}(z) > 0$ onto the domain in the w plane that lies above the boundary curve consisting of the rays $u \leq 0$, $v = 1$ and $u \geq 0$, $v = 0$ and the segment $u = 0$, $-1 \leq v \leq 0$.

The image of horizontal streamlines in the z plane are curves in the w plane given by the parametric equation

(3) $$w = f(t + ic) = \frac{1}{\pi}(t^2 - c^2 - 1 + i2ct)^{1/2}$$

$$+ \frac{1}{\pi}\text{Log}[t + ic + (t^2 - c^2 - 1 + i2ct)^{1/2}]$$

for $-\infty < t < \infty$. The new flow is that of a step in the bed of a deep stream and is illustrated in Figure 10.87(a). The function $w = f(z)$ is also defined for values of z in the lower half plane, and the images of horizontal streamlines that lie above or below the x axis are mapped onto streamlines that flow past a long rectangular obstacle. This is illustrated in Figure 10.87(b).

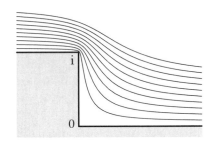

(a) Flow over a step

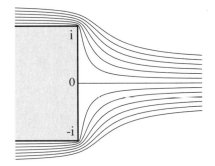

(b) Flow around a blunt object.

FIGURE 10.87   Accompanies Example 10.29.

EXERCISES FOR SECTION 10.10

For Exercises 1–4, use the Schwarz-Christoffel formula to find a conformal mapping $w = f(z)$ that will map the flow in the upper half plane $\text{Im}(z) > 0$ onto the flow indicated in each of the following figures.

1.

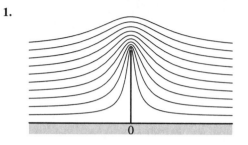

FIGURE 10.88   Accompanies Exercise 1.

2.

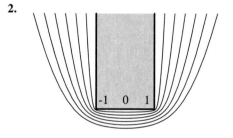

FIGURE 10.89   Accompanies Exercise 2.

3.

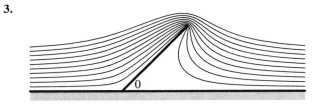

(a) Flow around an inclined segment.

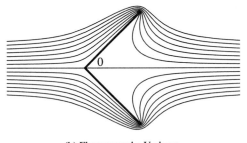

(b) Flow around a V-shape.

FIGURE 10.90   Accompanies Exercise 3.

4.

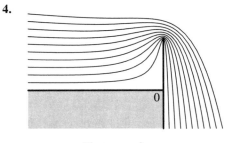

Flow over a dam.

FIGURE 10.91   Accompanies Exercise 4.

5. Use the Schwarz-Christoffel formula, and find an expression for $f'(z)$ for the transformation $w = f(z)$ that will map the upper half plane $\text{Im}(z) > 0$ onto the flow indicated in Figure 10.92(a). Extend the flow to the one indicated in Figure 10.92(b).

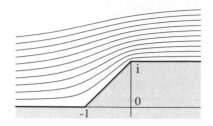

(a) Flow up an inclined step.

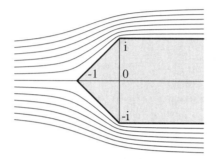

(b) Flow around a pointed object.

FIGURE 10.92   Accompanies Exercise 5.

10.11   Sources and Sinks

If the two-dimensional motion of an ideal fluid consists of an outward radial flow from a point and is symmetrical in all directions, then the point is called a *simple source*. A source at the origin can be considered as a line perpendicular to the z plane along which fluid is being created. If the rate of emission of volume of fluid per unit length is $2\pi m$, then the origin is said to be a source of strength m, the complex potential for the flow is

(1) $F(z) = m \log z$,

and the velocity $\mathbf{V}$ at the point (x, y) is given by

(2) $\mathbf{V}(x, y) = \overline{F'(z)} = \dfrac{m}{\overline{z}}$.

For fluid flows a sink is a negative source and is a point of inward radial flow at which the fluid is considered to be absorbed or annihilated. Sources and sinks for flows are illustrated in Figure 10.93.

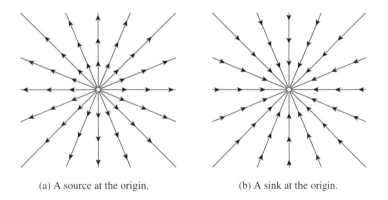

(a) A source at the origin.   (b) A sink at the origin.

FIGURE 10.93   Sources and sinks for an ideal fluid.

Source: A Charged Line

In the case of electrostatics a source will correspond to a uniformly charged line perpendicular to the z plane at the point z_0. If the line L is located at $z_0 = 0$ and carries a charge density of $\dfrac{q}{2}$ coulombs per unit length, then the magnitude electric field is $\left| \mathbf{E}(x, y) \right| = \dfrac{q}{\sqrt{x^2 + y^2}}$, hence $\mathbf{E}$ is given by

(3) $\mathbf{E}(x, y) = \dfrac{qz}{\left| z \right|^2} = \dfrac{q}{\bar{z}}$,

and the complex potential is

(4) $F(z) = -q \log z$ and $\mathbf{E}(x, y) = -\overline{F'(z)}$.

A sink for electrostatics is a negatively charged line perpendicular to the z plane. The electric field for electrostatic problems corresponds to the velocity field for fluid flow problems, except that their corresponding potentials differ by a sign change.

To establish equation (3), start with Coulomb's law, which states that two particles with charges q and Q exert a force on one another with magnitude $\dfrac{CqQ}{r^2}$, where r is the distance between particles and C is a constant that depends on the scientific units. For simplicity we assume that $C = 1$ and the test particle at the point z has charge $Q = 1$.

The contribution $\Delta\mathbf{E}_1$ induced by the element of charge $\dfrac{q\Delta h}{2}$ along the segment of length Δh situated at a height h above the plane has magnitude $\left|\Delta\mathbf{E}_1\right|$ given by

$$\left|\Delta\mathbf{E}_1\right| = \frac{(q/2)\Delta h}{r^2 + h^2}.$$

It has the same magnitude as $\Delta\mathbf{E}_2$ induced by the element $(q\Delta h)/2$ located a distance $-h$ below the plane. From the vertical symmetry involved, their sum $\Delta\mathbf{E}_2 + \Delta\mathbf{E}_2$ lies parallel to the plane along the ray from the origin, as shown in Figure 10.94.

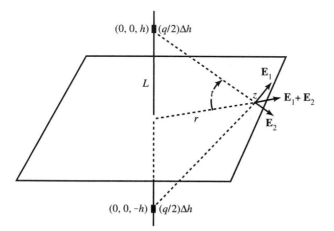

FIGURE 10.94 Contributions to $\mathbf{E}$ from the elements of charge $(q/2)\,\Delta h$ situated at $(0, 0, \pm h)$, above and below the z plane.

By the principal of superposition we add all contributions from the elements of charge along L to obtain $\mathbf{E} = \Sigma\,\Delta\mathbf{E}_k$. Using the vertical symmetry, it is evident that $\mathbf{E}$ lies parallel to the complex plane along the ray from the origin through the point z. Hence the magnitude of $\mathbf{E}$ is the sum of all components $\left|\Delta\mathbf{E}\right|\cos t$ that are parallel to the complex plane, where t is the angle between $\Delta\mathbf{E}$ and the plane. Letting $\Delta h \to 0$ in this summation process produces the definite integral

$$(5) \quad \left|\mathbf{E}(x, y)\right| = \int_{-\infty}^{\infty} \left|\Delta\mathbf{E}\right|\cos t\, dh = \int_{-\infty}^{\infty} \frac{(q/2)\cos t}{r^2 + h^2}\, dh.$$

Next use the change of variable $h = r \tan t$ and $dh = r \sec^2 t \, dt$ and the trigonometric identity $\sec^2 t = \dfrac{r^2 + h^2}{r^2}$ to obtain the equivalent integral:

$$(6) \quad |\mathbf{E}(x, y)| = \int_{-\pi/2}^{\pi/2} \frac{(q/2)\cos t}{r^2 + h^2} \frac{r^2 + h^2}{r} \, dt = \frac{q}{2r} \int_{-\pi/2}^{\pi/2} \cos t \, dt = \frac{q}{r}.$$

Multiplying this magnitude $\dfrac{q}{r}$ by the unit vector $\dfrac{z}{|z|}$ establishes formula (3). If $q > 0$ the field is directed away from $z_0 = 0$ and if $q < 0$ it is directed toward $z_0 = 0$. An electric field located at $z_0 \neq 0$ is given by

$$(7) \quad \mathbf{E}(x, y) = \frac{q(z - z_0)}{|z - z_0|^2} = \frac{q}{\overline{z} - \overline{z_0}},$$

and the corresponding complex potential is

$$(8) \quad F(z) = -q \log(z - z_0).$$

EXAMPLE 10.30 (Source and Sink of Equal Strength)

Let a source and sink of unit strength be located at the points $+1$ and -1, respectively. The complex potential for a fluid flowing from the source at $+1$ to the sink at -1 is

$$(9) \quad F(z) = \log(z - 1) - \log(z + 1) = \log\left(\frac{z - 1}{z + 1}\right).$$

The velocity potential and stream function are

$$(10) \quad \phi(x, y) = \ln\left|\frac{z - 1}{z + 1}\right| \quad \text{and} \quad \psi(x, y) = \arg\left(\frac{z - 1}{z + 1}\right),$$

respectively. Solving for the streamline $\psi(x, y) = c$, we start with

$$(11) \quad c = \arg\left(\frac{z - 1}{z + 1}\right) = \arg\left[\frac{x^2 + y^2 - 1 + i2y}{(x + 1)^2 + y^2}\right] = \arctan\left(\frac{2y}{x^2 + y^2 - 1}\right)$$

and obtain the equation $(\tan c)(x^2 + y^2 - 1) = 2y$. A straightforward calculation shows that points on the streamline must satisfy the equation

$$(12) \quad x^2 + (y - \cot c)^2 = 1 + \cot^2 c,$$

which is easily recognized as the equation of a circle with center at $(0, \cot c)$ that passes through the points $(\pm 1, 0)$. Several streamlines are indicated in Figure 10.95(a).

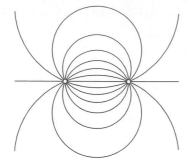

(a) Source and sink of equal strength.

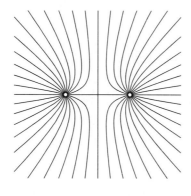

(b) Two sources of equal strength.

FIGURE 10.95   Fields depicting electrical strength.

EXAMPLE 10.31 (Two Sources of Equal Strength)   Let two
sources of unit strength be located at the points ± 1. The resulting complex potential
for a fluid flow is

(13) $F(z) = \log(z - 1) + \log(z + 1) = \log(z^2 - 1)$.

The velocity potential and stream function are

(14) $\phi(x, y) = \ln|z^2 - 1|$ and $\psi(x, y) = \arg(z^2 - 1)$,

respectively. Solving for the streamline $\psi(x, y) = c$, we start with

(15) $c = \arg(z^2 - 1) = \arg(x^2 - y^2 - 1 + i2xy) = \arctan\left(\dfrac{2xy}{x^2 - y^2 - 1}\right)$

and obtain the equation $x^2 + 2xy \cot c - y^2 = 1$. If we express this in the form $[x - y \tan(c/2)][x + y \cot(c/2)] = 1$ or

(16) $\left(x \cos \dfrac{c}{2} - y \sin \dfrac{c}{2}\right)\left(x \sin \dfrac{c}{2} + y \cos \dfrac{c}{2}\right) = \sin \dfrac{c}{2} \cos \dfrac{c}{2} = \dfrac{\sin c}{2}$

and use the rotation of axes

(17) $x^* = x \cos \dfrac{-c}{2} + y \sin \dfrac{-c}{2}$ and $y^* = -x \sin \dfrac{-c}{2} + y \cos \dfrac{-c}{2}$,

then the streamlines must satisfy the equation $x^* y^* = (\sin c)/2$ and are easily recognized to be rectangular hyperbolas with centers at the origin that pass through the points ± 1. Several streamlines are indicated in Figure 10.95(b).

Let an ideal fluid flow in a domain in the z plane be effected by a source located at the point z_0. Then the flow at points z, which lie in a small neighborhood of the point z_0, is approximated by that of a source with complex potential

(18) $\log(z - z_0) + \text{constant}.$

If $w = S(z)$ is a conformal mapping and $w_0 = S(z_0)$, then $S(z)$ has a nonzero derivative at z_0, and

(19) $w - w_0 = (z - z_0)[S'(z_0) + \eta(z)]$

where $\eta(z) \to 0$ as $z \to z_0$. Taking logarithms yields

(20) $\log(w - w_0) = \log(z - z_0) + \text{Log}[S'(z_0) + (z)].$

Since $S'(z_0) \neq 0$, the term $[\text{Log } S'(z_0) + \eta(z)]$ approaches the constant value $\text{Log}[S'(z_0)]$ as $z \to z_0$. Since $\log(z - z_0)$ is the complex potential for a source located at the point z_0, we see that the image of a source under a conformal mapping is a source.

The technique of conformal mapping can be used to determine the fluid flow in a domain D in the z plane that is produced by sources and sinks. If a conformal mapping $w = S(z)$ can be constructed so that the image of sources, sinks, and boundary curves for the flow in D are mapped onto sources, sinks, and boundary curves in a domain G where the complex potential is known to be $F_1(w)$, then the complex potential in D is given by $F_2(z) = F_1(S(z))$.

EXAMPLE 10.32 Suppose that the lines $x = \pm \pi/2$ are considered as walls of a containing vessel for a fluid flow produced by a single source of unit strength located at the origin. The conformal mapping $w = S(z) = \sin z$ maps the infinite strip bounded by the lines $x = \pm \pi/2$ onto the w plane slit along the boundary rays

$u \leq -1$, $v = 0$ and $u \geq 1$, $v = 0$, and the image of the source at $z_0 = 0$ is a source located at $w_0 = 0$. It is easy to see that the complex potential

(21) $\quad F_1(w) = \log w$

will determine a fluid flow in the w plane past the boundary curves $u \leq -1$, $v = 0$ and $u \geq 1$, $v = 0$, which lie along streamlines of the flow. Therefore the complex potential for the fluid flow in the infinite strip in the z plane is

(22) $\quad F_2(z) = \log(\sin z)$.

Several streamlines for the flow are illustrated in Figure 10.96.

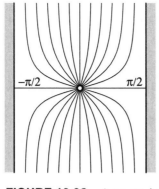

FIGURE 10.96 A source in the center of a strip.

EXAMPLE 10.33 Suppose that the lines $x = \pm\pi/2$ are considered as walls of a containing vessel for the fluid flow produced by a source of unit strength located at the point $z_1 = \pi/2$ and a sink of unit strength located at the point $z_2 = -\pi/2$. The conformal mapping $w = S(z) = \sin z$ maps the infinite strip bounded by the lines $x = \pm\pi/2$ onto the w plane slit along the boundary rays K_1: $u \leq -1$, $v = 0$ and K_2: $u \geq 1$, $v = 0$. The image of the source at z_1 is a source at $w_1 = 1$, and the image of the sink at z_2 is a sink at $w_2 = -1$. It is easy to verify that the potential

(23) $\quad F_1(w) = \log\left(\dfrac{w - 1}{w + 1}\right)$

will determine a fluid flow in the w plane past the boundary curves K_1 and K_2, which lie along streamlines of the flow. Therefore the complex potential for the fluid flow in the infinite strip in the z plane is

(24) $\quad F_2(z) = \log\left(\dfrac{\sin z - 1}{\sin z + 1}\right).$

Several streamlines for the flow are illustrated in Figure 10.97.

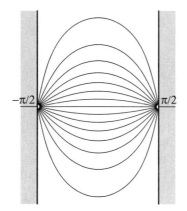

FIGURE 10.97   A source and sink on the edges of a strip.

The technique of transformation of a source can be used to determine the effluence from a channel extending from infinity. In this case, a conformal mapping $w = S(z)$ from the upper half plane $\text{Im}(z) > 0$ is constructed so that the single source located at $z_0 = 0$ is mapped to the point w_0 at infinity that lies along the channel. The streamlines emanating from $z_0 = 0$ in the upper half plane are mapped onto streamlines issuing from the channel.

EXAMPLE 10.34   Consider the conformal mapping

$$(25) \quad w = S(z) = \frac{2}{\pi}\left[(z^2 - 1)^{1/2} + \text{Arcsin}\,\frac{1}{z}\right],$$

which maps the upper half plane $\text{Im}(z) > 0$ onto the domain consisting of the upper half plane $\text{Im}(w) > 0$ joined to the channel $-1 \leq u \leq 1$, $v \leq 0$. The point $z_0 = 0$ is mapped onto the point $w_0 = -i\infty$ along the channel. Images of the rays $r > 0$, $\theta = \alpha$ are streamlines issuing from the channel as indicated in Figure 10.98.

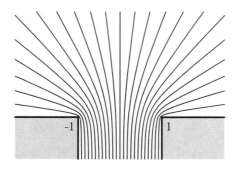

FIGURE 10.98   Effluence from a channel into a half plane.

EXERCISES FOR SECTION 10.11

1. Let the coordinate axes be walls of a containing vessel for a fluid flow in the first quadrant that is produced by a source of unit strength located at $z_1 = 1$ and a sink of unit strength located at $z_2 = i$. Show that $F(z) = \log[(z^2 - 1)/(z^2 + 1)]$ is the complex potential for the flow shown in Figure 10.99.

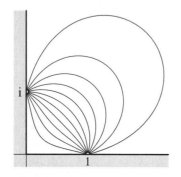

FIGURE 10.99   Accompanies Exercise 1.

2. Let the coordinate axes be walls of a containing vessel for a fluid flow in the first quadrant that is produced by two sources of equal strength located at the points $z_1 = 1$ and $z_2 = i$. Find the complex potential $F(z)$ for the flow in Figure 10.100.

3. Let the lines $x = 0$ and $x = \pi/2$ form the walls of a containing vessel for a fluid flow in the infinite strip $0 < x < \pi/2$ that is produced by a single source located at the point $z_0 = 0$. Find the complex potential for the flow in Figure 10.101.

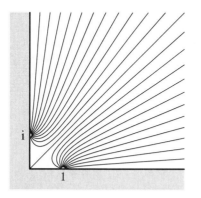

FIGURE 10.100   Accompanies Exercise 2.

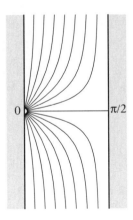

FIGURE 10.101   Accompanies Exercise 3.

4. Let the rays $x = 0$, $y > 0$ and $x = \pi$, $y > 0$ and the segment $y = 0$, $0 < x < \pi$ form the walls of a containing vessel for a fluid flow in the semi-infinite strip $0 < x < \pi$, $y > 0$ that is produced by two sources of equal strength located at the points $z_1 = 0$ and $z_2 = \pi$. Find the complex potential for the flow shown in Figure 10.102. *Hint*: Use the fact that $\sin(\pi/2 + z) = \sin(\pi/2 - z)$.

5. Let the y axis be considered a wall of a containing vessel for a fluid flow in the right half plane $\text{Re}(z) > 0$ that is produced by a single source located at the point $z_0 = 1$. Find the complex potential for the flow shown in Figure 10.103.

6. The complex potential $F(z) = 1/z$ determines an electrostatic field that is referred to as a dipole.
   (a) Show that

$$F(z) = \lim_{a \to 0} \frac{\log(z) - \log(z - a)}{a}$$

and conclude that a dipole is the limiting case of a source and sink.

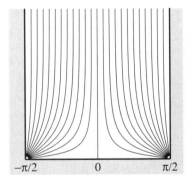

FIGURE 10.102   Accompanies Exercise 4.

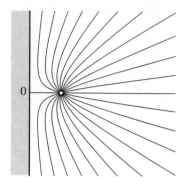

FIGURE 10.103 Accompanies Exercise 5.

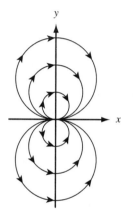

FIGURE 10.104 Accompanies Exercise 6.

(b) Show that the lines of flux of a dipole are circles that pass through the origin as shown in Figure 10.104.

7. Use a Schwarz-Christoffel transformation to find a conformal mapping $w = S(z)$ that will map the flow in the upper half plane onto the flow from a channel into a quadrant as indicated in Figure 10.105.

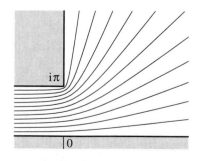

FIGURE 10.105 Accompanies Exercise 7.

8. Use a Schwarz-Christoffel transformation to find a conformal mapping $w = S(z)$ that will map the flow in the upper half plane onto the flow from a channel into a sector as indicated in Figure 10.106.

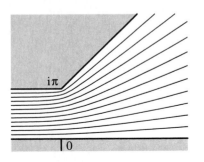

FIGURE 10.106   Accompanies Exercise 8.

9. Use a Schwarz-Christoffel transformation to find a conformal mapping $w = S(z)$ that will map the flow in the upper half plane onto the flow in a right-angled channel indicated in Figure 10.107.

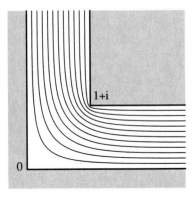

FIGURE 10.107   Accompanies Exercise 9.

10. Use a Schwarz-Christoffel transformation to find a conformal mapping $w = S(z)$ that will map the flow in the upper half plane onto the flow from a channel back into a quadrant as indicated in Figure 10.108, where $w_0 = 2\sqrt{2} - 2\ln(\sqrt{2} - 1) + i\pi$.

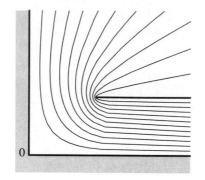

FIGURE 10.108　Accompanies Exercise 10.

11. **(a)** Show that the complex potential $F(z) = w$ given implicitly by $z = w + e^w$ determines the ideal fluid flow through an open channel bounded by the rays

$$y = \pi, \ -\infty < x < -1 \quad \text{and} \quad y = -\pi, \ -\infty < x < -1$$

into the plane.

(b) Show that the streamline $\psi(x, y) = c$ of the flow is given by the parametric equations

$$x = t + e^t \cos c, \quad y = c + e^t \sin c \quad \text{for } -\infty < t < \infty$$

as shown in Figure 10.109.

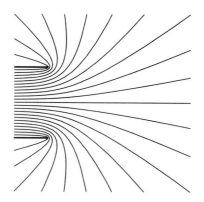

FIGURE 10.109　Accompanies Exercise 11.

11

Fourier Series and the Laplace Transform

11.1 Fourier Series

In this chapter we show how Fourier series, the Fourier transform, and the Laplace transform are related to the study of complex analysis. We develop the Fourier series representation of a real-valued function $U(t)$ of the real variable t. Complex Fourier series and Fourier transforms are then discussed. Finally, we develop the Laplace transform and the complex variable technique for finding its inverse. This chapter focuses on applying these ideas to solving problems involving real-valued functions, so many of the theorems throughout are stated without proof.

Let $U(t)$ be a real-valued function that is periodic with period 2π, that is,

(1) $U(t + 2\pi) = U(t)$ for all t.

One such function is $s = U(t) = \sin(t - \pi/2) + 0.7 \cos(2t - \pi - 1/4) + 1.7$, and its graph is obtained by repeating the portion of the graph in any interval of length 2π, as shown in Figure 11.1.

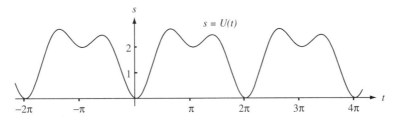

FIGURE 11.1   A function U with period 2π.

Familiar examples of real functions that have period 2π are $\sin nt$ and $\cos nt$, where n is an integer. This raises the question whether any periodic function can be represented by a sum of terms involving $a_n\cos nt$ and $b_n\sin nt$, where a_n and b_n are real constants. As we shall soon see, the answer to this question is often yes.

Definition 11.1 (Piecewise Continuous)   *The function U is piecewise continuous on the closed interval $[a, b]$, if there exists values $t_0, t_1, \ldots, t_n$ with $a = t_0 < t_1 < \cdots < t_n = b$ such that U is continuous in each of the open intervals $t_{k-1} < t < t_k$ $(k = 1, 2, \ldots, n)$ and has left- and right-hand limits at the values t_k $(k = 0, 1, \ldots, n)$.*

We use the symbols $U(a^-)$ and $U(a^+)$ for the left- and right-hand limits, respectively, of a function $U(t)$ as t approaches the point a.

The graph of a piecewise continuous function is illustrated in Figure 11.2 where the function $U(t)$ is

$$U(t) = \begin{cases} \dfrac{2}{3}\left(t - \dfrac{1}{2}\right)^2 + \dfrac{1}{4} & \text{when } 1 \leq t < 2, \\[2mm] \dfrac{5}{2} - (t - 2)^2 & \text{when } 2 < t < 3, \\[2mm] 1 + \dfrac{t - 3}{4} & \text{when } 3 < t < 4, \\[2mm] \dfrac{6}{5} - (t - 5)^3 & \text{when } 4 < t \leq 6. \end{cases}$$

The left- and right-hand limits at $t_0 = 2$, $t_1 = 3$, and $t_2 = 4$ are easy to determine:

At $t = 2$, we have $U(2^-) = \frac{7}{4}$ and $U(2^+) = \frac{5}{2}$.
At $t = 3$, we have $U(3^-) = \frac{3}{2}$ and $U(3^+) = 1$.
At $t = 4$, we have $U(4^-) = \frac{5}{4}$ and $U(4^+) = \frac{11}{5}$.

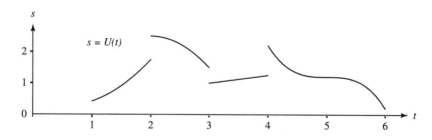

FIGURE 11.2   A piecewise continuous function U over the interval $[1, 6]$.

Definition 11.2 (Fourier Series)   *If $U(t)$ is periodic with period 2π and is piecewise continuous on $[-\pi, \pi]$, then the* Fourier series $S(t)$ *for $U(t)$ is*

(2) $$S(t) = \frac{a_0}{2} + \sum_{j=1}^{\infty} (a_j \cos jt + b_j \sin jt),$$

where the coefficients a_j and b_j are given by the so-called Euler's *formulae:*

(3) $$a_j = \frac{1}{\pi} \int_{-\pi}^{\pi} U(t) \cos jt \, dt \quad \text{for } j = 0, 1, \ldots$$

and

$$(4) \quad b_j = \frac{1}{\pi} \int_{-\pi}^{\pi} U(t) \sin jt \, dt \quad \text{for } j = 1, 2, \ldots .$$

The factor $\frac{1}{2}$ in the constant term $\frac{a_0}{2}$ on the right side of equation (2) has been introduced for convenience so that a_0 could be obtained from the general formula in equation (3) by setting $j = 0$. The reasons for this will be explained shortly. The next result discusses convergence of the Fourier series.

> **Theorem 11.1 (Fourier Expansion)** *Assume that $S(t)$ is the Fourier series for $U(t)$. If $U'(t)$ is piecewise continuous on $[-\pi, \pi]$, then $S(t)$ is convergent for all $t \in [-\pi, \pi]$. The relation $S(t) = U(t)$ holds for all $t \in [-\pi, \pi]$ where $U(t)$ is continuous. If $t = a$ is a point of discontinuity of U, then*
>
> $$S(a) = \frac{U(a^-) + U(a^+)}{2}$$
>
> *where $U(a^-)$ and $U(a^+)$ denote the left- and right-hand limits, respectively. With this understanding, we have the Fourier expansion:*

$$(5) \quad U(t) = \frac{a_0}{2} + \sum_{j=1}^{\infty} (a_j \cos jt + b_j \sin jt).$$

EXAMPLE 11.1 The function $U(t) = \dfrac{t}{2}$ for $t \in (-\pi, \pi)$, extended periodically by the equation $U(t + 2\pi) = U(t)$, has the Fourier series expansion

$$U(t) = \sum_{j=1}^{\infty} \frac{(-1)^{j+1} \sin jt}{j}.$$

Solution   Using Euler's formulae (3) and integration by parts, we obtain

$$(6) \quad a_j = \frac{1}{\pi} \int_{-\pi}^{\pi} \frac{t}{2} \cos jt \, dt = \frac{t \sin jt}{2\pi j} + \frac{\cos jt}{2\pi j^2} \Big|_{-\pi}^{\pi} = 0 \quad \text{for } j = 1, 2, \ldots$$

and

$$b_j = \frac{1}{\pi} \int_{-\pi}^{\pi} \frac{t}{2} \sin jt \, dt = \frac{-t \cos jt}{2\pi j} + \frac{\sin jt}{2\pi j^2} \Big|_{-\pi}^{\pi}$$

$$= \frac{-\cos j\pi}{j} = \frac{(-1)^{j+1}}{j} \quad \text{for } j = 1, 2, \ldots .$$

The coefficient a_0 is computed by the calculation

$$(7) \quad a_0 = \frac{1}{\pi} \int_{-\pi}^{\pi} \frac{t}{2} \, dt = \frac{t^2}{4\pi} \Big|_{-\pi}^{\pi} = 0.$$

Using the results of equations (6) and (7) in equation (2) produces the required solution. The graphs of $U(t)$ and the first three partial sums $S_1(t) = \sin t$, $S_2(t) = \sin t - \dfrac{\sin 2t}{2}$, and $S_3(t) = \sin t - \dfrac{\sin 2t}{2} + \dfrac{\sin 3t}{3}$ are shown in Figure 11.3.

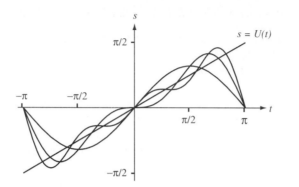

FIGURE 11.3   The function $U(t) = t/2$, and the approximations $S_1(t)$, $S_2(t)$, and $S_3(t)$.

We now state some general properties of Fourier series that are useful for calculating the coefficients. The proofs are left for the reader.

Theorem 11.2   *If $U(t)$ and $V(t)$ have Fourier series representations, then their sum $W(t) = U(t) + V(t)$ has a Fourier series representation, and the Fourier coefficients of W are obtained by adding the corresponding coefficients of U and V.*

Theorem 11.3 (Fourier Cosine Series)   *Assume that $U(x)$ is an even function. If $U(t)$ has period 2π and $U(t)$ and $U'(t)$ are piecewise continuous, then the Fourier series for $U(t)$ involves only the cosine terms (i.e., $b_j = 0$ for all j):*

(8) $$U(t) = \frac{a_0}{2} + \sum_{j=1}^{\infty} a_j \cos jt, \quad where$$

(9) $$a_j = \frac{2}{\pi} \int_0^{\pi} U(t) \cos jt\, dt \quad for\ j = 0, 1, \ldots .$$

Theorem 11.4 (Fourier Sine Series)   *Assume that $U(t)$ is an odd function. If $U(t)$ has period 2π and if $U(t)$ and $U'(x)$ are piecewise continuous, then the Fourier series for $U(t)$ involves only sine terms (i.e., $a_j = 0$ for all j):*

(10) $$U(t) = \sum_{j=1}^{\infty} b_j \sin jt, \quad where$$

(11) $$b_j = \frac{2}{\pi} \int_0^{\pi} U(t) \sin jt\, dt \quad for\ j = 1, 2, \ldots .$$

Theorem 11.5 (Termwise Integration) *If U has a Fourier series representation given in equation (5), then the integral of U has a Fourier series representation which can be obtained by termwise integration of the Fourier Series of U, that is,*

$$(12) \quad \int_0^t U(\tau)\,d\tau = \sum_{j=1}^{\infty}\left[\frac{[a_j + a_0(-1)^{j+1}]\sin jt}{j} - \frac{b_j\cos jt}{j}\right],$$

where we have used the expansion $a_0\dfrac{t}{2} = \sum_{j=1}^{\infty}\dfrac{a_0(-1)^{j+1}\sin jt}{j}$ *in Example 11.1.*

Theorem 11.6 (Termwise Differentiation) *If U′(t) has a Fourier series representation, and U(t) is given by equation (5), then*

$$(13) \quad U'(t) = \sum_{j=1}^{\infty}(jb_j\cos jt - ja_j\sin jt).$$

EXAMPLE 11.2 The function $U(t) = |t|$ for $t \in (-\pi, \pi)$, extended periodically by the equation $U(t + 2\pi) = U(t)$, has the Fourier series representation

$$U(t) = |t| = \frac{\pi}{2} - \frac{4}{\pi}\sum_{j=1}^{\infty}\frac{\cos[(2j-1)t]}{(2j-1)^2}.$$

Solution The function $U(t)$ is an even function, hence we can use Theorem 11.3 to conclude that $b_n = 0$ for all n, and

$$(14) \quad a_j = \frac{2}{\pi}\int_0^{\pi}t\cos jt\,dt = \frac{2t\sin jt}{\pi j} + \frac{2\cos jt}{\pi j^2}\Big|_0^{\pi}$$

$$= \frac{2\cos j\pi - 2}{\pi j^2} = \frac{2(-1)^j - 2}{\pi j^2} \quad \text{for } j = 1, 2, \ldots.$$

The coefficient a_0 is computed by the calculation:

$$(15) \quad a_0 = \frac{2}{\pi}\int_0^{\pi}t\,dt = \frac{t^2}{\pi}\Big|_0^{\pi} = \pi.$$

Using the results of equations (14) and (15) and Theorem 11.3 produces the required solution.

The following intuitive proof will justify the Euler formulae given in equations (3) and (4). To determine a_0 we integrate both $U(t)$ and the Fourier series representation in equation (2) from $-\pi$ to π, which results in

$$(16) \quad \int_{-\pi}^{\pi}U(t)\,dt = \int_{-\pi}^{\pi}\left[\frac{a_0}{2} + \sum_{j=1}^{\infty}(a_j\cos jt + b_j\sin jt)\right]dt.$$

We are allowed to perform integration term by term, and we obtain

(17) $\quad \int_{-\pi}^{\pi} U(t)\, dt = \dfrac{a_0}{2} \int_{-\pi}^{\pi} 1\, dt + \sum_{j=1}^{\infty} a_j \int_{-\pi}^{\pi} \cos jt\, dt + \sum_{j=1}^{\infty} b_j \int_{-\pi}^{\pi} \sin jt\, dt.$

The value of the first integral on the right side of equation (17) is 2π and all the other integrals are zero. Hence we obtain the desired result:

(18) $\quad a_0 = \dfrac{1}{\pi} \int_{-\pi}^{\pi} U(t)\, dt.$

To determine a_m, we let m ($m > 1$) denote a fixed integer and multiply both $U(t)$ and the Fourier series representation in equation (2) by the term $\cos mt$, and then we integrate and obtain

(19) $\quad \int_{-\pi}^{\pi} U(t) \cos mt\, dt = \dfrac{a_0}{2} \int_{-\pi}^{\pi} \cos mt\, dt + \sum_{j=1}^{\infty} a_j \int_{-\pi}^{\pi} \cos mt \cos jt\, dt$

$$+ \sum_{j=1}^{\infty} b_j \int_{-\pi}^{\pi} \cos mt \sin jt\, dt.$$

The value of the first term on the right side of equation (19) is easily seen to be zero:

(20) $\quad \dfrac{a_0}{2} \int_{-\pi}^{\pi} \cos mt\, dt = \dfrac{a_0 \sin mt}{2m} \bigg|_{-\pi}^{\pi} = 0.$

The value of the term involving $\cos mt \cos jt$ is found by using the trigonometric identity:

$$\cos mt \cos jt = \dfrac{1}{2}\{\cos[(m + j)t] + \cos[(m - j)t]\}.$$

Calculation reveals that if $m \neq j$ (and $m > 0$), then

(21) $\quad a_j \int_{-\pi}^{\pi} \cos mt \cos jt\, dt = \dfrac{a_m}{2}\left\{ \int_{-\pi}^{\pi} \cos[(m + j)t]\, dt \right.$

$$\left. + \int_{-\pi}^{\pi} \cos[(m - j)t]\, dt \right\} = 0.$$

When $m = j$, the value of the integral becomes:

(22) $\quad a_m \int_{-\pi}^{\pi} \cos^2 mt\, dt = \pi a_m.$

The value of the term on the right side of equation (19) involving the integrand $\cos mt \sin jt$ is found by using the trigonometric identity

$$\cos mt \sin jt = \dfrac{1}{2}\{\sin[(m + j)t] + \sin[(m - j)t]\},$$

and for all values of m and n, we obtain

$$(23) \quad b_j \int_{-\pi}^{\pi} \cos mt \sin jt \, dt = \frac{b_m}{2} \left\{ \int_{-\pi}^{\pi} \sin[(m+j)t] \, dt + \int_{-\pi}^{\pi} \sin[(m-j)t] \, dt \right\} = 0.$$

Therefore, we can use the results of equations (20)–(23) in equation (19) to obtain

$$\int_{-\pi}^{\pi} U(t) \cos mt \, dt = \pi a_m \quad \text{for } m = 0, 1, \ldots,$$

and equation (3) is established. We leave it as an exercise to establish Euler's formula for the coefficients $\{b_n\}$. A complete discussion of the details of the proof of Theorem 11.1 can be found in some advanced texts. See for instance, John W. Dettman, Chapter 8 in *Applied Complex Variables*, The Macmillan Company, New York, 1965.

EXERCISES FOR SECTION 11.1

For Exercises 1–2 and 6–11, find the Fourier series representation.

1. $U(t) = \begin{cases} 1 & \text{for } 0 < t < \pi, \\ -1 & \text{for } -\pi < t < 0. \end{cases}$ See Figure 11.4.

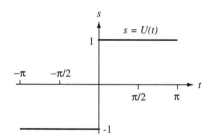

FIGURE 11.4 The graph of $U(t)$ for Exercise 1.

2. $V(t) = \begin{cases} \dfrac{\pi}{2} - t & \text{for } 0 \le t \le \pi, \\[2mm] \dfrac{\pi}{2} + t & \text{for } -\pi < t < 0. \end{cases}$ See Figure 11.5.

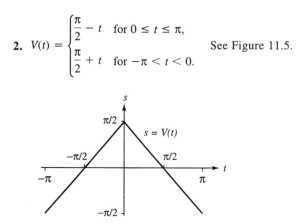

FIGURE 11.5 The graph of $V(t)$ for Exercise 2.

3. For Exercises 1 and 2, verify that $U(t) = -V'(t)$ by termwise differentiation of the Fourier series representation for $V(t)$.

4. For Exercise 1, set $t = \dfrac{\pi}{2}$ and conclude that $\dfrac{\pi}{4} = \displaystyle\sum_{j=1}^{\infty} \dfrac{(-1)^{j-1}}{2j-1}$.

5. For Exercise 2, set $t = 0$ and conclude that $\dfrac{\pi^2}{8} = \displaystyle\sum_{j=1}^{\infty} \dfrac{1}{(2j-1)^2}$.

6. $U(t) = \begin{cases} -1 & \text{for } \dfrac{\pi}{2} < t < \pi, \\[2mm] 1 & \text{for } \dfrac{-\pi}{2} < t < \dfrac{\pi}{2}, \\[2mm] -1 & \text{for } -\pi < t < \dfrac{-\pi}{2}. \end{cases}$ See Figure 11.6.

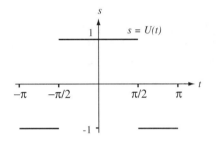

FIGURE 11.6   The graph of $U(t)$ for Exercise 6.

7. $U(t) = \begin{cases} \pi - t & \text{for } \dfrac{\pi}{2} < t < \pi, \\[2mm] t & \text{for } \dfrac{-\pi}{2} < t < \dfrac{\pi}{2}, \\[2mm] -\pi - t & \text{for } -\pi < t < \dfrac{-\pi}{2}. \end{cases}$ See Figure 11.7.

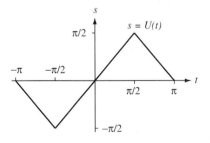

FIGURE 11.7   The graph of $U(t)$ for Exercise 7.

8. $U(t)$ is given in Figure 11.8.

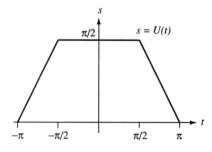

FIGURE 11.8   The graph of $U(t)$ for Exercise 8.

9. $U(t) = \begin{cases} 1 & \text{for } \dfrac{\pi}{2} < t < \pi, \\[2mm] 0 & \text{for } \dfrac{-\pi}{2} < t < \dfrac{\pi}{2}, \\[2mm] -1 & \text{for } -\pi < t < \dfrac{-\pi}{2}. \end{cases}$ See Figure 11.9.

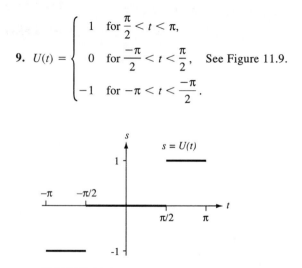

FIGURE 11.9   The graph of $U(t)$ for Exercise 9.

10. $V(t)$, given in Figure 11.10.

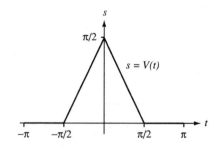

FIGURE 11.10   The graph of $V(t)$ for Exercise 10.

11. $U(t)$, given in Figure 11.11.

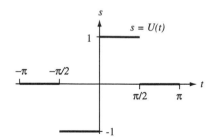

FIGURE 11.11   The graph of $U(t)$ for Exercise 11.

11.2   The Dirichlet Problem for the Unit Disk

The Dirichlet problem for the unit disk D: $|z| < 1$ is to find a real-valued function $u(x, y)$ that is harmonic in the unit disk D and that takes on the boundary values

(1) $u(\cos \theta, \sin \theta) = U(\theta)$ for $-\pi < \theta \leq \pi$,

at points $z = (\cos \theta, \sin \theta)$ on the unit circle, as shown in Figure 11.12.

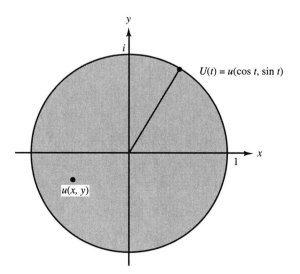

FIGURE 11.12   The Dirichlet problem for the unit disk $|z| < 1$.

Theorem 11.7   *If $U(t)$ has period 2π, and has the Fourier series representation*

(2) $U(t) = \dfrac{a_0}{2} + \displaystyle\sum_{j=1}^{\infty} (a_j \cos jt + b_j \sin jt),$

then the solution u to the Dirichlet problem in D is

(3) $u(r \cos \theta, r \sin \theta) = \dfrac{a_0}{2} + \sum\limits_{j=1}^{\infty} (a_j \, r^j \cos j\theta + b_j \, r^j \sin j\theta),$

where $x + iy = re^{i\theta}$ *denotes a complex number in the closed disk* $|z| \leq 1$.

It is easy to see that the series representation in equation (3) for u takes on the prescribed boundary values in equation (1) at points on the unit circle $|z| = 1$. Since each term $r^n \cos j\theta$ and $r^n \sin j\theta$ in series (3) is harmonic, it is reasonable to conclude that the infinite series representing u will also be harmonic. The proof follows after Theorem 11.8.

The following result gives an integral representation for a function $u(x, y)$ that is harmonic in a domain containing the closed unit disk. The result is the analog to Poisson's integral formula for the upper half plane.

Theorem 11.8 (Poisson Integral Formula for the Unit Disk)   *Let* $u(x, y)$ *be a function that is harmonic in a simply connected domain that contains the closed unit disk* $|z| \leq 1$. *If* $u(x, y)$ *takes on the boundary values*

$u(\cos \theta, \sin \theta) = U(\theta)$ *for* $-\pi < \theta \leq \pi$,

then u has the integral representation

(4) $u(r \cos \theta, r \sin \theta) = \dfrac{1}{2\pi} \displaystyle\int_{-\pi}^{\pi} \dfrac{(1 - r^2) \, U(t) \, dt}{1 + r^2 - 2r \cos(t - \theta)}$

that is valid for $|z| < 1$.

Proof   Since $u(x, y)$ is harmonic in the simply connected domain, there exists a conjugate harmonic function $v(x, y)$ such that $f(z) = u(x, y) + iv(x, y)$ is analytic. Let C denote the contour consisting of the unit circle; then Cauchy's integral formula

(5) $f(z) = \dfrac{1}{2\pi i} \displaystyle\int_C \dfrac{f(\xi) \, d\xi}{\xi - z}$

expresses the value of $f(z)$ at any point z inside C in terms of the values of $f(\xi)$ at points ξ that lie on the circle C.

If we set $z^* = (\bar{z})^{-1}$ then z^* lies outside the unit circle C and the Cauchy-Goursat theorem establishes the equation

(6) $0 = \dfrac{1}{2\pi i} \displaystyle\int_C \dfrac{f(\xi) \, d\xi}{\xi - z^*}.$

Subtracting equation (6) from equation (5) and using the parameterization $\xi = e^{it}$, $d\xi = ie^{it} \, dt$ and the substitutions $z = re^{i\theta}$, $z^* = \dfrac{1}{r} e^{i\theta}$ gives

(7) $f(z) = \dfrac{1}{2\pi} \displaystyle\int_{-\pi}^{\pi} \left(\dfrac{e^{it}}{e^{it} - re^{i\theta}} - \dfrac{e^{it}}{e^{it} - \dfrac{1}{r} e^{i\theta}} \right) f(e^{it}) \, dt.$

The expression inside the parentheses on the right side of equation (7) can be written

$$(8) \quad \frac{e^{it}}{e^{it} - re^{i\theta}} - \frac{e^{it}}{e^{it} - \dfrac{1}{r}e^{i\theta}} = \frac{1}{1 - re^{i(\theta - t)}} + \frac{re^{i(t-\theta)}}{1 - re^{i(t-\theta)}}$$

$$= \frac{1 - r^2}{1 + r^2 - 2r\cos(t - \theta)}$$

and it follows that

$$(9) \quad f(z) = \frac{1}{2\pi} \int_{-\pi}^{\pi} \frac{(1 - r^2)f(e^{it})\, dt}{1 + r^2 - 2r\cos(t - \theta)}.$$

Since $u(x, y)$ is the real part of $f(z)$ and $U(t)$ is the real part of $f(e^{it})$, we can equate the real parts in equation (9) to obtain equation (4), and proof Theorem 11.8 is complete.

We now turn our attention to the proof Theorem 11.7. The real-valued function

$$(10) \quad P(r, t - \theta) = \frac{1 - r^2}{1 + r^2 - 2r\cos(t - \theta)}$$

is known as the *Poisson kernel*. Expanding the left side of equation (8) in a geometric series gives

$$(11) \quad P(r, t - \theta) = \frac{1}{1 - re^{i(\theta-t)}} + \frac{re^{i(t-\theta)}}{1 - re^{i(t-\theta)}} = \sum_{n=0}^{\infty} r^n e^{in(\theta-t)} + \sum_{n=1}^{\infty} r^n e^{in(t-\theta)}$$

$$= 1 + \sum_{n=1}^{\infty} r^n [e^{in(\theta-t)} + e^{in(t-\theta)}] = 1 + 2 \sum_{n=1}^{\infty} r^n \cos[n(\theta - t)]$$

$$= 1 + 2 \sum_{n=1}^{\infty} r^n (\cos n\theta \cos nt + \sin n\theta \sin nt)$$

$$= 1 + 2 \sum_{n=1}^{\infty} r^n \cos n\theta \cos nt + 2 \sum_{n=1}^{\infty} r^n \sin n\theta \sin nt.$$

We now use the result of equation (11) in equation (4) to obtain

$$u(r \cos \theta, r \sin \theta) = \frac{1}{2\pi} \int_{-\pi}^{\pi} P(r, t - \theta)\, U(t)\, dt$$

$$= \frac{1}{2\pi} \int_{-\pi}^{\pi} U(t)\, dt + \frac{1}{\pi} \int_{-\pi}^{\pi} \sum_{n=1}^{\infty} r^n \cos n\theta \cos nt \, U(t)\, dt$$

$$+ \frac{1}{\pi} \int_{-\pi}^{\pi} \sum_{n=1}^{\infty} r^n \sin n\theta \sin nt \, U(t)\, dt$$

$$= \frac{1}{2\pi} \int_{-\pi}^{\pi} U(t)\, dt + \sum_{n=1}^{\infty} r^n \cos n\theta \, \frac{1}{\pi} \int_{-\pi}^{\pi} \cos nt \, U(t)\, dt$$

$$+ \sum_{n=1}^{\infty} r^n \sin n\theta \, \frac{1}{\pi} \int_{-\pi}^{\pi} \sin nt \, U(t)\, dt$$

$$= \frac{a_0}{2} + \sum_{n=1}^{\infty} a_n r^n \cos n\theta + \sum_{n=1}^{\infty} b_n r^n \sin n\theta,$$

where $\{a_n\}$ and $\{b_n\}$ are the Fourier series coefficients for $U(t)$. This establishes the representation for $u(r \cos \theta, r \sin \theta)$ stated in Theorem 11.7.

EXAMPLE 11.3   Find the function $u(x, y)$ that is harmonic in the unit disk $|z| < 1$ and takes on the boundary values

$$(12) \quad u(\cos \theta, \sin \theta) = U(\theta) = \frac{\theta}{2} \quad \text{for } -\pi < \theta < \pi.$$

Solution   Using Example 11.1, we write

$$(13) \quad U(t) = \sum_{n=1}^{\infty} \frac{(-1)^{n+1}}{n} \sin nt.$$

Using formula (3) for the solution of the Dirichlet problem, we obtain

$$(14) \quad u(r \cos \theta, r \sin \theta) = \sum_{n=1}^{\infty} \frac{(-1)^{n+1}}{n} r^n \sin n\theta.$$

We remark that the series representation (14) for u takes on the prescribed boundary values (12) at points where U is continuous. The boundary function U is discontinuous at $z = -1$, which corresponds to $\theta = \pm\pi$; and U was not prescribed at these points. Graphs of the approximations $U_7(t)$ and $u_7(x, y) = u_7(r \cos \theta, r \sin \theta)$ which involve the first seven terms in equations (13) and (14), respectively, are shown in Figure 11.13.

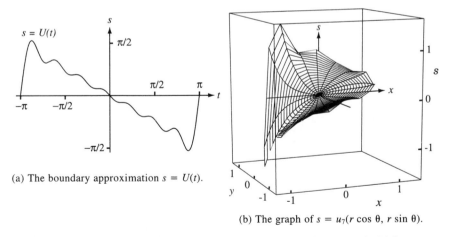

(a) The boundary approximation $s = U(t)$.

(b) The graph of $s = u_7(r \cos \theta, r \sin \theta)$.

FIGURE 11.13   Functions $U_7(t)$ and $u_7(r \cos \theta, r \sin \theta)$ for Example 11.3.

EXERCISES FOR SECTION 11.2

For problems 1–6, find the solution to the given Dirichlet problem in the unit disk D by using the Fourier series representations for the boundary functions that were derived in the examples and exercises of Section 11.1.

1. $U(\theta) = \begin{cases} 1 & \text{for } 0 < \theta < \pi, \\ -1 & \text{for } -\pi < \theta < 0. \end{cases}$

2. $U(\theta) = \begin{cases} \dfrac{\pi}{2} - \theta & \text{for } 0 \le \theta \le \pi, \\ \dfrac{\pi}{2} + \theta & \text{for } -\pi < \theta < 0. \end{cases}$ See Figure 11.14.

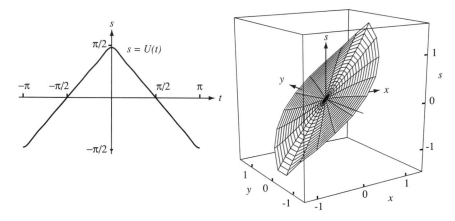

FIGURE 11.14   Approximations for $U_5(\theta)$ and $u_5(r \cos \theta, r \sin \theta)$ in Exercise 2.

3. $U(\theta) = \begin{cases} -1 & \text{for } \dfrac{\pi}{2} < \theta < \pi, \\ 1 & \text{for } \dfrac{-\pi}{2} < \theta < \dfrac{\pi}{2}, \\ -1 & \text{for } -\pi < \theta < \dfrac{-\pi}{2}. \end{cases}$

4. $U(\theta) = \begin{cases} \pi - \theta & \text{for } \dfrac{\pi}{2} \le \theta \le \pi, \\ \theta & \text{for } \dfrac{-\pi}{2} \le \theta < \dfrac{\pi}{2}, \\ -\pi - \theta & \text{for } -\pi < \theta < \dfrac{-\pi}{2}. \end{cases}$

5. $U(\theta) = \begin{cases} \pi - \theta & \text{for } \dfrac{\pi}{2} \le \theta \le \pi, \\ \dfrac{\pi}{2} & \text{for } \dfrac{-\pi}{2} \le \theta < \dfrac{\pi}{2}, \\ \pi + \theta & \text{for } -\pi < \theta < \dfrac{-\pi}{2}. \end{cases}$ See Figure 11.15.

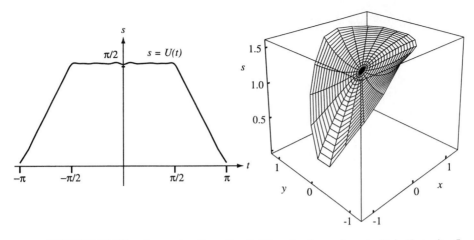

FIGURE 11.15   Approximations for $U_5(\theta)$ and $u_5(r\cos\theta, r\sin\theta)$ in Exercise 5.

6. $U(\theta) = \begin{cases} 1 & \text{for } \dfrac{\pi}{2} < \theta < \pi, \\[2mm] 0 & \text{for } \dfrac{-\pi}{2} < \theta < \dfrac{\pi}{2}, \\[2mm] -1 & \text{for } -\pi < \theta < \dfrac{-\pi}{2}. \end{cases}$ See Figure 11.16.

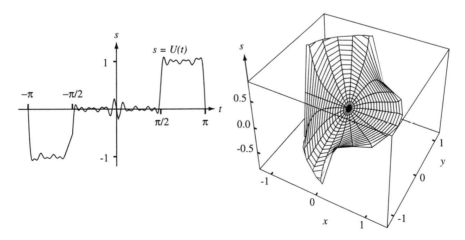

FIGURE 11.16   Approximations for $U_7(\theta)$ and $u_7(r\cos\theta, r\sin\theta)$ in Exercise 6.

7. $U(\theta) = \begin{cases} 0 & \text{for } \dfrac{\pi}{2} \leq \theta \leq \pi, \\[2mm] \dfrac{\pi - \theta}{2} & \text{for } 0 \leq \theta < \dfrac{\pi}{2}, \\[2mm] \dfrac{\pi + \theta}{2} & \text{for } \dfrac{-\pi}{2} \leq \theta < 0, \\[2mm] 0 & \text{for } -\pi < \theta < \dfrac{-\pi}{2}. \end{cases}$

8. $U(\theta) = \begin{cases} 0 & \text{for } \dfrac{\pi}{2} < \theta < \pi, \\[2mm] -1 & \text{for } 0 < \theta < \dfrac{\pi}{2}, \\[2mm] 1 & \text{for } \dfrac{-\pi}{2} < \theta < 0, \\[2mm] 0 & \text{for } -\pi < \theta < \dfrac{-\pi}{2}. \end{cases}$

9. Write a report on the Dirichlet problem and include some applications. Resources include bibliographical items 70, 71, 76, 77, 85, 98, 135, and 138.

10. Look up the article on the Poisson integral formula and discuss what you found. Use bibliographical item 115.

11.3   Vibrations in Mechanical Systems

Consider a spring that resists compression as well as extension, that is suspended vertically from a fixed support, and a body of mass m that is attached at the lower end of the spring. We make the assumption that the mass m is much larger than the mass of the spring so that we can neglect the mass of the spring. If there is no motion then the system is in static equilibrium, as illustrated in Figure 11.17(a). If the mass is pulled down further and released, then it will undergo an oscillatory motion.

Suppose there is no friction to slow down the motion of the mass, then we say that the system is *undamped*. We will determine the motion of this mechanical system by considering the forces acting on the mass during the motion. This will lead to a differential equation relating the displacement as a function of time. The most obvious force is that of *gravitational attraction* acting on the mass m and is given by

(1) $F_1 = mg,$

where g is the acceleration of gravity. The next force to be considered is the *spring force* acting on the mass and is directed upward if the spring is stretched and downward if it is compressed. It obeys Hooke's law

(2) $F_2 = ks,$

where s is the amount the spring is stretched when $s > 0$ and is the amount it is compressed when $s < 0$.

When the system is in static equilibrium and the spring is stretched by the amount s_0, the resultant of the spring force and the gravitational force is zero, which is expressed by the equation

(3) $mg - ks_0 = 0.$

Let $s = U(t)$ denote the displacement from static equilibrium with the positive s direction pointed downward as indicated in Figure 11.17(b).

The spring force can be written as

$$F_2 = -k[s_0 + U(t)] = -ks_0 - kU(t),$$

and the resultant force F_R is

(4) $F_R = F_1 + F_2 = mg - ks_0 - kU(t) = -kU(t).$

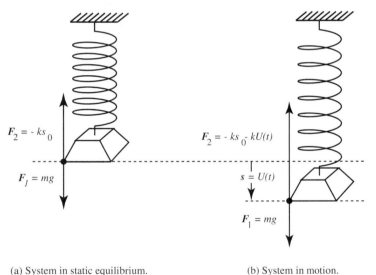

(a) System in static equilibrium.   (b) System in motion.

FIGURE 11.17   The spring mass system.

The differential equation for motion is obtained by using Newton's second law, which states that the resultant of the forces acting on the mass at any instant satisfies

(5) $F_R = ma.$

Since the distance from equilibrium at time t is measured by $U(t)$, the acceleration a is given by $a = U''(t)$, and using equations (4) and (5) we obtain

(6) $F_R = -kU(t) = mU''(t).$

Hence the undamped mechanical system is governed by the linear differential equation

(7) $mU''(t) + kU(t) = 0.$

The general solution to equation (7) is known to be

$$(8) \quad U(t) = A \cos \omega t + B \sin \omega t, \quad \text{where } \omega = \sqrt{\frac{k}{m}}.$$

Damped System

If we consider frictional forces that slow down the motion of the mass, then we say that the system is damped. This is visualized by connecting a dashpot to the mass, as indicated in Figure 11.18. For small velocities it is assumed that the frictional force F_3 is proportional to the velocity, that is,

$$(9) \quad F_3 = -cU'(t).$$

The damping constant c must be positive, for if $U'(t) > 0$, then the mass is moving downward and hence F_3 must point upward, which requires that F_3 is negative. The resultant of the three forces acting on the mass is given by

$$(10) \quad F_1 + F_2 + F_3 = -kU(t) - cU'(t) = mU''(t) = F_R.$$

Hence the damped mechanical system is governed by the differential equation

$$(11) \quad mU''(t) + cU'(t) + kU(t) = 0.$$

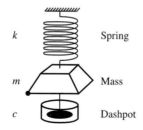

k — Spring

m — Mass

c — Dashpot

FIGURE 11.18   The spring mass dashpot system.

Forced Vibrations

The vibrations discussed earlier are called *free* vibrations because all the forces that affect the motion of the system are internal to the system. We extend our analysis to cover the case in which an external force $F_4 = F(t)$ acts on the mass, see Figure 11.19. Such a force might occur from vibrations of the support to which the top of the spring is attached, or from the effect of a magnetic field on a mass made of iron. As before, we sum the forces F_1, F_2, F_3, and F_4 and set this equal to the resultant force F_R and obtain

$$(12) \quad F_1 + F_2 + F_3 + F_4 = F_R = -KU(t) - cU'(t) + F(t) = mU''(t).$$

Therefore, the *forced motion* of the mechanical system satisfies the nonhomogeneous linear differential equation

(13) $mU''(t) + cU'(t) + kU(t) = F(t).$

The function $F(t)$ is called the *input* or *driving force* and the solution $U(t)$ is called the *output* or *response*. Of particular interest are periodic inputs $F(t)$ that can be represented by Fourier series.

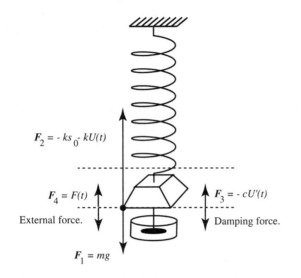

$F_2 = -ks_0 - kU(t)$

$F_4 = F(t)$ $F_3 = -cU'(t)$

External force.   Damping force.

$F_1 = mg$

FIGURE 11.19   The dashpot system with an external force.

For damped mechanical systems that are driven by a periodic input $F(t)$, the general solution involves a *transient part* that vanishes as $t \to +\infty$, and a *steady state part* that is periodic. The transient part of the solution $U_h(t)$ is found by solving the homogeneous differential equation

(14) $mU_h''(t) + cU_h'(t) + kU_h(t) = 0.$

Equation (14) leads to the characteristic equation $m\lambda^2 + c\lambda + k = 0$, which has roots $\lambda = \dfrac{-c \pm \sqrt{c^2 - 4mk}}{2m}$. The coefficients m, c, and k are all positive, and there are three cases to consider.

If $c^2 - 4mk > 0$, the roots are real and distinct, and since $\sqrt{c^2 - 4mk} < c$, it follows that the roots λ_1 and λ_2 are negative real numbers. Thus, for this case, we have

$$\lim_{t \to +\infty} U_h(t) = \lim_{t \to +\infty} (A_1 e^{\lambda_1 t} + A_2 e^{\lambda_2 t}) = 0.$$

If the roots are real and equal, then $\lambda_1 = \lambda_2 = \lambda$, where λ is a negative real number. Again, for this case we find that

$$\lim_{t \to +\infty} U_h(t) = \lim_{t \to +\infty} (A_1 e^{\lambda t} + A_2 t e^{\lambda t}) = 0.$$

If the roots are complex, then $\lambda = -\alpha \pm \beta i$, where α and β are positive real numbers, and it follows that

$$\lim_{t \to +\infty} U_h(t) = \lim_{t \to +\infty} (A_1 e^{-\alpha t}\cos \beta t + A_2 e^{-\alpha t}\sin \beta t) = 0.$$

In all three cases, we see that the homogeneous solution $U_h(t)$ decays to 0 as $t \to +\infty$.

The steady state solution $U_p(t)$ can be obtained by representing $U_p(t)$ by its Fourier series and substituting $U''_p(t)$, $U'_p(t)$, and $U_p(t)$ into the nonhomogeneous differential equation and solving the resulting system for the Fourier coefficients of $U_p(t)$. The general solution to equation (13) is then given by

$$U(t) = U_h(t) + U_p(t).$$

EXAMPLE 11.4   Find the general solution to $U''(t) + 2U'(t) + U(t) = F(t)$,

where $F(t)$ is given by the Fourier series $F(t) = \displaystyle\sum_{n=1}^{\infty} \frac{\cos[(2n-1)t]}{(2n-1)^2}$

Solution   First we solve $U''_h(t) + 2U'_h(t) + U_h(t) = 0$ for the transient solution. The characteristic equation is $\lambda^2 + 2\lambda + 1 = 0$, which has a double root $\lambda = -1$. Hence

$$U_h(t) = A_1 e^{-t} + A_2 t e^{-t}.$$

The steady state solution is obtained by assuming that $U_p(t)$ has the Fourier series representation

$$U_h(t) = \frac{a_0}{2} + \sum_{n=1}^{\infty} a_n \cos nt + \sum_{n=1}^{\infty} b_n \sin nt,$$

and that $U'_h(t)$ and $U''_h(t)$ can be obtained by termwise differentiation:

$$2U'_h(t) = 2\sum_{n=1}^{\infty} nb_n\cos nt - 2\sum_{n=1}^{\infty} na_n\sin nt, \text{ and}$$

$$U''_h(t) = -\sum_{n=1}^{\infty} n^2 a_n\cos nt - \sum_{n=1}^{\infty} n^2 b_n\sin nt.$$

Substituting these expansions into the differential equation results in

$$F(t) = \frac{a_0}{2} + \sum_{n=1}^{\infty} [(1-n^2)a_n + 2nb_n] \cos nt$$

$$+ \sum_{n=1}^{\infty} [-2na_n + (1-n^2)b_n] \sin nt.$$

Equating the coefficients with the given series for $F(t)$, we find that $\dfrac{a_0}{2} = 0$, and that

$$(1 - n^2)a_n + 2nb_n = \begin{cases} \dfrac{1}{n^2} & \text{when } n \text{ is odd,} \\ 0 & \text{when } n \text{ is even,} \end{cases}$$

$$-2na_n + (1 - n^2)b_n = 0 \quad \text{for all } n.$$

Solving this linear system for a_n and b_n, we find that

$$a_n = \begin{cases} \dfrac{1 - n^2}{n^2(1 + n^2)^2} & \text{for } n \text{ odd,} \\ 0 & \text{for } n \text{ even,} \end{cases}$$

$$b_n = \begin{cases} \dfrac{2n}{n^2(1 + n^2)^2} & \text{for } n \text{ odd,} \\ 0 & \text{for } n \text{ even.} \end{cases}$$

And the general solution is given by

$$U(t) = A_1 e^{-t} + A_2 t e^{-t} + \sum_{n=1}^{\infty} \frac{[1 - (2n - 1)^2] \cos[(2n - 1)t]}{(2n - 1)^2[1 + (2n - 1)^2]^2}$$
$$+ \sum_{n=1}^{\infty} \frac{2(2n - 1) \sin[(2n - 1)t]}{(2n - 1)^2[1 + (2n - 1)^2]^2}.$$

EXERCISES FOR SECTION 11.3

For the exercises, use the results of Section 11.1.

1. Find the general solution to $U''(t) + 3U'(t) + U(t) = F(t)$.

    (a) $F(t) = \dfrac{t}{2}$ **(b)** $F(t) = \displaystyle\sum_{n=1}^{\infty} \frac{(-1)^{n+1} \cos[(2n - 1)t]}{2n - 1}$

    (c) $F(t)$ is shown in Figure 11.20.

    $$F(t) = \begin{cases} \pi - t & \text{for } \dfrac{\pi}{2} < t < \pi, \\ t & \text{for } \dfrac{-\pi}{2} < t < \dfrac{\pi}{2}, \\ -\pi - t & \text{for } -\pi < t < \dfrac{-\pi}{2} \end{cases}$$

    Hint: $F(t) = \dfrac{4}{\pi} \displaystyle\sum_{j=1}^{\infty} \frac{(-1)^j \sin[(2j - 1)t]}{(2j - 1)^2}.$

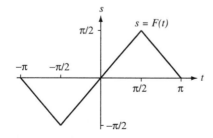

FIGURE 11.20   The graph of $F(t)$ for Exercise 1 part c.

2. Find the general solution to $2U''(t) + 2U'(t) + U(t) = F(t)$,

(a) $F(t) = \dfrac{t}{2}$

(b) $F(t) = \displaystyle\sum_{n=1}^{\infty} \dfrac{\sin[(2n-1)t]}{2n-1}$

(c) $F(t)$ is shown in Figure 11.21.

Hint: $F(t) = \dfrac{\pi}{8} + \dfrac{2}{\pi}\displaystyle\sum_{j=1}^{\infty} \dfrac{\cos[(2j-1)t]}{(2j-1)^2}$

$\qquad + \dfrac{4}{\pi}\displaystyle\sum_{j=1}^{\infty} \dfrac{\cos[2(2j-1)t]}{2^2(2j-1)^2}$,

where $a_{4n} = 0$ for all n.

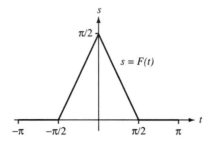

FIGURE 11.21   The graph of $F(t)$ for Exercise 2 part c.

11.4   The Fourier Transform

Let $U(t)$ be a real-valued function with period 2π which is piecewise continuous such that $U'(t)$ also exists and is piecewise continuous. Then $U(t)$ has the *complex Fourier series* representation

(1) $U(t) = \displaystyle\sum_{n=-\infty}^{\infty} c_n e^{int}$,   where

(2) $c_n = \dfrac{1}{2\pi} \displaystyle\int_{-\pi}^{\pi} U(t)e^{-int}\, dt$ for all n.

The coefficients $\{c_n\}$ are complex numbers. Previously, we expressed $U(t)$ as a real trigonometric series:

(3) $U(t) = \dfrac{a_0}{2} + \displaystyle\sum_{n=1}^{\infty} (a_n\cos nt + b_n\sin nt)$.

Hence, a relationship between the coefficients is

(4) $a_n = c_n + c_{-n}$ for $n = 0, 1, \ldots$,   and

$\qquad b_n = i(c_n - c_{-n})$ for $n = 1, 2, \ldots$.

These relations are easy to establish. We start by writing

(5) $U(t) = c_0 + \sum_{n=1}^{\infty} c_n e^{int} + \sum_{n=1}^{\infty} c_{-n} e^{-int}$

$= c_0 + \sum_{n=1}^{\infty} c_n(\cos nt + i \sin nt) + \sum_{n=1}^{\infty} c_{-n}(\cos nt - i \sin nt)$

$= c_0 + \sum_{n=1}^{\infty} [(c_n + c_{-n}) \cos nt + i(c_n - c_{-n}) \sin nt].$

Comparing equation (5) with equation (3), we see that $a_0 = 2c_0$, $a_n = c_n + c_{-n}$, and $b_n = i(c_n - c_{-n})$.

If $U(t)$ and $U'(t)$ are piecewise continuous and have period $2L$, then $U(t)$ has the *complex Fourier series* representation

(6) $U(t) = \sum_{n=-\infty}^{\infty} c_n e^{i\pi n(t/L)}$,   where

(7) $c_n = \frac{1}{2L} \int_{-L}^{L} U(t) e^{-i\pi n(t/L)} \, dt$ for all n.

We have seen how periodic functions are represented by trigonometric series. Many practical problems involve nonperiodic functions. A representation analogous to Fourier series for a nonperiodic function $U(t)$ is obtained by considering the Fourier series of $U(t)$ for $-L < t < L$ and then taking the limit as $L \to \infty$. The result is known as the Fourier transform of $U(t)$.

Start with a nonperiodic function $U(t)$ and consider the periodic function $U_L(t)$ with period $2L$, where

(8) $U_L(t) = U(t)$ for $-L < t \le L$,   and
       $U_L(t) = U_L(t + 2L)$ for all t.

Then $U_L(t)$ has the complex Fourier series representation

(9) $U_L(t) = \sum_{n=-\infty}^{\infty} c_n e^{i\pi nt/L}.$

We introduce some terminology to discuss the terms in equation (9), first

(10) $w_n = \pi n/L$ is called the frequency.

If t denotes time, then the units for w_n are radians per unit time. The set of all possible frequencies is called the *frequency spectrum*, i.e.,

$$\left\{ \cdots, \frac{-3\pi}{L}, \frac{-2\pi}{L}, \frac{-\pi}{L}, \frac{\pi}{L}, \frac{2\pi}{L}, \frac{3\pi}{L}, \cdots \right\}.$$

It is important to note that as L increases, the spectrum becomes finer and approaches a continuous spectrum of frequencies. It is reasonable to expect that the summation in the Fourier series for $U_L(t)$ will give rise to an integral over $[-\infty, \infty]$. This is stated in the following important theorem.

Theorem 11.9 (Fourier Transform)   *Let $U(t)$ and $U'(t)$ be piecewise continuous, and*

(11) $$\int_{-\infty}^{\infty} |U(t)|\, dt < M,$$

for some positive constant M. The Fourier transform $F(w)$ of $U(t)$ is defined as

(12) $$F(w) = \frac{1}{2\pi} \int_{-\infty}^{\infty} U(t)e^{-iwt}\, dt.$$

At points of continuity, $U(t)$ has the integral representation

(13) $$U(t) = \int_{-\infty}^{\infty} F(w)e^{iwt}\, dw,$$

and at a point $t = a$ of discontinuity of U, the integral in equation (13) converges to $\dfrac{U(a^-) + U(a^+)}{2}$. Remark: It is common to show that U is transformed into F by using the operator notation:

(14) $$\mathfrak{F}(U(t)) = F(w).$$

Proof   Set $\Delta w_n = w_{n+1} - w_n = \dfrac{\pi}{L}$ and $\dfrac{1}{2L} = \dfrac{1}{2\pi} \Delta w_n$. These quantities are used in conjunction with equations (7) and (9) and definition (10) to obtain

(15) $$U_L(t) = \sum_{n=-\infty}^{\infty} \left[\frac{1}{2L} \int_{-L}^{L} U(t)e^{-iw_n t}\, dt \right] e^{iw_n t}$$

$$= \sum_{n=-\infty}^{\infty} \left[\frac{1}{2\pi} \int_{-L}^{L} U(t)e^{-iw_n t}\, dt \right] e^{iw_n t} \Delta w_n.$$

If we define $F_L(w)$ by

(16) $$F_L(w) = \frac{1}{2\pi} \int_{-L}^{L} U(t)e^{-iwt}\, dt,$$

then equation (15) can be written as

(17) $$U_L(t) = \sum_{n=-\infty}^{\infty} F_L(w_n)e^{iw_n t} \Delta w_n.$$

As L gets large, $F_L(w_n)$ approaches $F(w_n)$ and Δw_n tends to zero. Thus the limit on the right-hand side of equation (17) can be viewed as an integral. This substantiates the integral representation

(18) $$U(t) = \int_{-\infty}^{\infty} F(w)e^{iwt}\, dw.$$

A more rigorous proof of this fact can be found in advanced texts. Table 11.1 gives some important properties of the Fourier transform.

TABLE 11.1   Properties of the Fourier Transform

Linearity	$\mathcal{F}(aU_1(t) + bU_2(t)) = a\,\mathcal{F}(U_1(t)) + b\,\mathcal{F}(U_2(t))$
Symmetry	If $\mathcal{F}(U(t)) = F(w)$, then $\mathcal{F}(F(t)) = \dfrac{1}{2\pi}U(-w)$
Time scaling	$\mathcal{F}(U(at)) = \dfrac{1}{|a|}F\left(\dfrac{w}{a}\right)$
Time shifting	$\mathcal{F}(U(t - t_0)) = e^{-it_0w}F(w)$
Frequency shifting	$\mathcal{F}(e^{-iw_0 t}U(t)) = F(w - w_0)$
Time differentiation	$\mathcal{F}(U'(t)) = iwF(w)$
Frequency differentiation	$\dfrac{d^n F(w)}{dw^n} = \mathcal{F}((-it)^n U(t))$
Moment Theorem	If $M_n = \displaystyle\int_{-\infty}^{\infty} t^n U(t)\,dt$, then $(-i)^n M_n = 2\pi F^{(n)}(0)$.

EXAMPLE 11.5   Show that $F(e^{-|t|}) = \dfrac{1}{\pi(1 + w^2)}$.

Solution   Using formula (12), we obtain

$$F(w) = \frac{1}{2\pi}\int_{-\infty}^{\infty} e^{-|t|}e^{-iwt}\,dt$$

$$= \frac{1}{2\pi}\int_{-\infty}^{0} e^{(1-iw)t}\,dt + \frac{1}{2\pi}\int_{0}^{\infty} e^{(-1-iw)t}\,dt$$

$$= \frac{1}{2\pi(1 - iw)}\,e^{(1-iw)t}\,\Big|_{t=-\infty}^{t=0} + \frac{1}{2\pi(-1 - iw)}\,e^{(-1-iw)t}\,\Big|_{t=0}^{t=\infty}$$

$$= \frac{1}{2\pi(1 - iw)} + \frac{1}{2\pi(1 + iw)} = \frac{1}{\pi(1 + w^2)},$$

and the result is established.

EXAMPLE 11.6   Show that $\mathcal{F}\left(\dfrac{1}{1 + t^2}\right) = \dfrac{1}{2}e^{-|w|}$.

Solution   Using the result of Example 11.5 and the symmetry property, we obtain

(19) $\mathcal{F}\left(\dfrac{1}{\pi(1 + t^2)}\right) = \dfrac{1}{2\pi}e^{-|-w|} = \dfrac{1}{2\pi}e^{-|w|}.$

The linearity property is used to multiply each term in equation (19) by π and obtain

$$\mathcal{F}\left(\frac{1}{1 + t^2}\right) = \frac{1}{2}e^{-|w|},$$

and the result is established.

EXERCISES FOR SECTION 11.4

1. Let $U(t) = \begin{cases} 1 & \text{for } |t| < 1, \\ 0 & \text{for } |t| > 1. \end{cases}$ Find $\mathfrak{F}(U(t))$.

2. Let $U(t) = \begin{cases} \sin t & \text{for } |t| \leq \pi, \\ 0 & \text{for } |t| > \pi. \end{cases}$ Show that $\mathfrak{F}(U(t)) = \dfrac{i \sin \pi w}{\pi(1 - w^2)}$.

3. Let $U(t) = \begin{cases} 1 - |t| & \text{for } |t| \leq 1, \\ 0 & \text{for } |t| > 1. \end{cases}$ Find $\mathfrak{F}(U(t))$.

4. Let $U(t) = e^{-t^2/2}$. Show that $\mathfrak{F}(U(t)) = \dfrac{1}{\sqrt{2\pi}} e^{-w^2/2}$.

   Hint: Use the integral definition and combine the terms in the exponent, then complete the square and use the fact that $\displaystyle\int_{-\infty}^{\infty} e^{-t^2/2}\, dt = \sqrt{2\pi}$.

5. Use the time scaling property and the example in the text to show that

   $$\mathfrak{F}(e^{-a|t|}) = \frac{|a|}{\pi(a^2 + w^2)}.$$

6. Use the symmetry and linearity properties and the result of Exercise 1 to show that

   $$\mathfrak{F}\left(\frac{\sin t}{t}\right) = \begin{cases} \frac{1}{2} & \text{for } |w| < 1, \\ 0 & \text{for } |w| > 1. \end{cases}$$

7. Use the symmetry and linearity properties and the result of Exercise 2 to show that

   $$\mathfrak{F}\left(\frac{i \sin \pi t}{1 - t^2}\right) = \begin{cases} \dfrac{i \sin w}{2} & \text{for } |w| \leq \pi, \\ 0 & \text{for } |w| > \pi. \end{cases}$$

8. Use the time differentiation property and the result of Exercise 4 to show that

   $$\mathfrak{F}(te^{-t^2/2}) = \frac{-iwe^{-w^2/2}}{\sqrt{2\pi}}.$$

9. Use the symmetry and linearity properties and the results of Exercise 3 to show that

   $$\mathfrak{F}\left(\frac{\sin^2 \frac{t}{2}}{t^2}\right) = \begin{cases} \dfrac{1 - |w|}{4\pi} & \text{for } |w| \leq 1, \\ 0 & \text{for } |w| > 1. \end{cases}$$

10. Write a report on the Fourier transform. Discuss some of the ideas you found in the literature that are not mentioned in the text. Resources include bibliographical items 15, 17, 100, 69, 149, and 159.

11.5   The Laplace Transform

From Fourier Transforms to Laplace Transforms

We have seen that certain real-valued functions $f(t)$ have a Fourier transform and that the integral

(1) $g(\omega) = \displaystyle\int_{-\infty}^{\infty} f(t)e^{-i\omega t}\, dt$

defines the complex function $g(\omega)$ of the real variable ω. By multiplying the integrand in formula (1) by $e^{-\sigma t}$, we extend this and define a complex function $G(\sigma + i\omega)$ of the complex variable $\sigma + i\omega$:

$$(2) \quad G(\sigma + i\omega) = \int_{-\infty}^{\infty} f(t)e^{-\sigma t}e^{-i\omega t}\, dt = \int_{-\infty}^{\infty} f(t)e^{-(\sigma + i\omega)t}\, dt.$$

The function $G(\sigma + i\omega)$ is called the *two-sided Laplace transform* of $f(t)$, and it exists when the Fourier transform of the function $f(t)e^{-\sigma t}$ exists. From the Fourier transform theory, we can state that a sufficient condition for $G(\sigma + iw)$ to exist is that

$$(3) \quad \int_{-\infty}^{\infty} |f(t)|\, e^{-\sigma t}\, dt < \infty$$

shall exist. For a given function $f(t)$, this integral is finite for values of σ that lie in some interval $a < \sigma < b$.

The two-sided Laplace transform uses the lower limit of integration $t = -\infty$, and hence requires a knowledge of the past history of the function $f(t)$, i.e., $t < 0$. For most physical applications, one is interested in the behavior of a system only for $t \geq 0$. Mathematically speaking, the initial conditions $f(0), f'(0), f''(0), \dots$, are a consequence of the past history of the system and are often all that is necessary to know. For this reason, it is useful to define the one-sided Laplace transform of $f(t)$, which is commonly referred to simply as the *Laplace transform* of $f(t)$, which is also defined as an integral:

$$(4) \quad \mathcal{L}(f(t)) = F(s) = \int_{0}^{\infty} f(t)e^{-st}\, dt, \quad \text{where } s = \sigma + i\omega.$$

If the defining integral (4) for the Laplace transform exists for $s_0 = \sigma_0 + i\omega$, then values of σ with $\sigma > \sigma_0$ imply that $e^{-\sigma t} < e^{-\sigma_0 t}$, and thus

$$(5) \quad \int_{0}^{\infty} |f(t)|\, e^{-\sigma t}\, dt < \int_{0}^{\infty} |f(t)|\, e^{-\sigma_0 t}\, dt < \infty,$$

and it follows that $F(s)$ exists for $s = \sigma + i\omega$. Therefore, the Laplace transform $\mathcal{L}(f(t))$ is defined for all points s in the right half-plane $\mathrm{Re}(s) > \sigma_0$.

Another way to view the relationship between the Fourier transform and the Laplace transform is to consider the function $U(t)$ given by

$$(6) \quad U(t) = \begin{cases} f(t) & \text{for } t \geq 0, \\ 0 & \text{for } t < 0. \end{cases}$$

Then the Fourier transform theory shows us that

$$(7) \quad U(t) = \frac{1}{2\pi} \int_{-\infty}^{\infty} \left[\int_{-\infty}^{\infty} U(t)e^{-i\omega t}\, dt \right] e^{i\omega t}\, d\omega,$$

and since the integrand $U(t)$ is zero for $t < 0$, equation (7) can be written as

$$(8) \quad f(t) = \frac{1}{2\pi} \int_{-\infty}^{\infty} \left[\int_{0}^{\infty} f(t)e^{-i\omega t}\, dt \right] e^{i\omega t}\, d\omega.$$

Use the change of variable $s = \sigma + i\omega$ and $d\omega = \dfrac{ds}{i}$, where $\sigma > \sigma_0$ is held fixed, then the new limits of integration are from $s = \sigma - i\omega$ to $s = \sigma + i\omega$. The resulting equation is

$$(9) \quad f(t) = \frac{1}{2\pi i} \int_{\sigma - i\infty}^{\sigma + i\infty} \left[\int_0^\infty f(t)e^{-st}\, dt \right] e^{st}\, ds.$$

From equation (9) it is easy to recognize that the Laplace transform is

$$(10) \quad \mathcal{L}(f(t)) = F(s) = \int_0^\infty f(t)e^{-st}\, dt,$$

and that the inverse Laplace transform is

$$(11) \quad \mathcal{L}^{-1}(F(s)) = f(t) = \frac{1}{2\pi i} \int_{\sigma - i\infty}^{\sigma + i\infty} F(s)e^{st}\, ds.$$

Properties of the Laplace Transform

Although a function $f(t)$ may be defined for all values of t, its Laplace transform is not influenced by values of $f(t)$, where $t < 0$. The Laplace transform of $f(t)$ is actually defined for the function $U(t)$ given by

$$(12) \quad U(t) = \begin{cases} f(t) & \text{for } t \geq 0, \\ 0 & \text{for } t < 0. \end{cases}$$

A sufficient condition for the existence of the Laplace transform is that $|f(t)|$ does not grow too rapidly as $t \to +\infty$. We say that the function f is of *exponential order* if there exists real constants $M > 0$ and K, such that

$$(13) \quad |f(t)| \leq Me^{Kt} \quad \text{holds for all } t \geq 0.$$

All functions in this chapter are assumed to be of exponential order. The next theorem shows that their Laplace transform $F(\sigma + i\tau)$ exists for values of s in a domain that includes the right half-plane $\mathrm{Re}(s) > K$.

> **Theorem 11.10 (Existence of the Laplace Transform)**   *If f is of exponential order, then its Laplace transform $\mathcal{L}(f(t)) = F(s)$ is given by*
>
> $$(14) \quad F(s) = \int_0^\infty f(t)e^{-st}\, dt, \text{ where } s = \sigma + i\omega.$$
>
> *The defining integral for F exists at points $s = \sigma + i\tau$ in the right half plane $\sigma > K$.*
>
> **Proof**   Using $s = \sigma + i\tau$ we see that $F(s)$ can be expressed as
>
> $$(15) \quad F(s) = \int_0^\infty f(t)e^{-\sigma t}\cos \tau t\, dt - i \int_0^\infty f(t)e^{-\sigma t}\sin \tau t\, dt.$$

Then for values of $\sigma > K$, we have

(16) $\displaystyle\int_0^\infty \left|f(t)\right| e^{-\sigma t} \left|\cos \tau t\right| \, dt \le M \int_0^\infty e^{(K-\sigma)t} \, dt \le \frac{M}{\sigma - K}$,   and

(17) $\displaystyle\int_0^\infty \left|f(t)\right| e^{-\sigma t} \left|\sin \tau t\right| \, dt \le M \int_0^\infty e^{(K-\sigma)t} \, dt \le \frac{M}{\sigma - K}$,

which imply that the integrals defining the real and imaginary parts of F exist for values of $\text{Re}(s) > K$, and the proof is complete.

Remarks   The domain of definition of the defining integral for the Laplace transform $\mathcal{L}(f(t))$ seems to be restricted to a half plane. However, the resulting formula $F(s)$ might have a domain much larger than this half plane. Later we will show that $F(s)$ is an analytic function of the complex variable s. For most applications involving Laplace transforms that we will study, the Laplace transforms are rational functions that have the form $\dfrac{P(s)}{Q(s)}$, where P and Q are polynomials, and some other important ones will have the form $\dfrac{e^{as}P(s)}{Q(s)}$.

> **Theorem 11.11 (Linearity of the Laplace Transform)**   *Let f and g have Laplace transforms F and G, respectively. If a and b are constants, then*

(18) $\mathcal{L}(af(t) + bg(t)) = aF(s) + bG(s)$.

Proof   Let K be chosen so that both F and G are defined for $\text{Re}(s) > K$, then

(19) $\begin{aligned}\mathcal{L}(af(t) + bg(t)) &= \int_0^\infty [af(t) + bg(t)]e^{-st} \, dt \\ &= a\int_0^\infty f(t)e^{-st} \, dt + b\int_0^\infty g(t)e^{-st} \, dt \\ &= aF(s) + bG(s).\end{aligned}$

> **Theorem 11.12 (Uniqueness of the Laplace Transform)**   *Let f and g have Laplace transforms, F and G, respectively. If $F(s) \equiv G(s)$, then $f(t) \equiv g(t)$.*

Proof   If σ is sufficiently large, then the integral representation, equation (10), for the inverse Laplace transform can be used to obtain

(20) $\begin{aligned}f(t) = \mathcal{L}^{-1}(F(s)) &= \frac{1}{2\pi i}\int_{\sigma - i\infty}^{\sigma + i\infty} F(s)e^{st} \, ds = \frac{1}{2\pi i}\int_{s - i\infty}^{s + i\infty} G(s)e^{st} \, ds \\ &= \mathcal{L}^{-1}(G(s)) = g(t),\end{aligned}$

and the theorem is proven.

EXAMPLE 11.7 Show that the Laplace transform of the step function given by

$$f(t) = \begin{cases} 1 & \text{for } 0 \le t < c, \\ 0 & \text{for } c < t \end{cases} \quad \text{is } \mathcal{L}(f(t)) = \frac{1 - e^{-cs}}{s}.$$

Solution Using the integral definition for $\mathcal{L}(f(t))$, we obtain

$$\mathcal{L}(f(t)) = \int_0^\infty f(t)e^{-st}\,dt = \int_0^c e^{-st}\,dt + \int_c^\infty e^{-st} \cdot 0\,dt = \frac{-e^{-st}}{s}\Big|_{t=0}^{t=c} = \frac{1 - e^{-cs}}{s}.$$

EXAMPLE 11.8 Show that $\mathcal{L}(e^{at}) = \dfrac{1}{s - a}$, where a is a real constant.

Solution We will actually show that the integral defining $\mathcal{L}(e^{at})$ is equal to the formula $F(s) = \dfrac{1}{s - a}$ for values of s with $\text{Re}(s) > a$, and the extension to other values of s is inferred by our knowledge about the domain of a rational function. Using straightforward integration techniques we find that

$$\mathcal{L}(e^{at}) = \int_0^\infty e^{at}e^{-st}\,dt = \lim_{R \to +\infty} \int_0^R e^{(a-s)t}\,dt$$

$$= \lim_{R \to +\infty} \frac{e^{(a-s)R}}{a - s} + \frac{1}{s - a}.$$

Let $s = \sigma + i\tau$ be held fixed, or where $\sigma > a$. Then since $a - \sigma$ is a negative real number we have $\lim\limits_{R \to +\infty} e^{(a-s)R} = 0$ and this is used in equation (10) to obtain the desired conclusion.

The property of linearity can be used to find new Laplace transforms from known ones.

EXAMPLE 11.9 Show that $\mathcal{L}(\sinh at) = \dfrac{a}{s^2 - a^2}$.

Solution Since $\sinh at = \frac{1}{2}e^{at} - \frac{1}{2}e^{-at}$, we obtain

$$\mathcal{L}(\sinh at) = \frac{1}{2}\mathcal{L}(e^{at}) - \frac{1}{2}\mathcal{L}(e^{-at}) = \frac{1}{2}\frac{1}{s - a} - \frac{1}{2}\frac{1}{s + a} = \frac{a}{s^2 - a^2}.$$

The technique of integration by parts is also helpful in finding new Laplace transforms.

EXAMPLE 11.10   Show that $\mathcal{L}(t) = \dfrac{1}{s^2}$.

Solution   Using integration by parts we obtain

$$\mathcal{L}(t) = \lim_{R \to +\infty} \int_0^R t e^{-st}\, dt$$

$$= \lim_{R \to +\infty} \left(\frac{-t}{s} e^{-st} - \frac{1}{s^2} e^{-st} \right) \Bigg|_{t=0}^{t=R}$$

$$= \lim_{R \to +\infty} \left(\frac{-R}{s} e^{-sR} - \frac{1}{s^2} e^{-sR} \right) + 0 + \frac{1}{s^2} .$$

For values of s in the right half plane $\text{Re}(s) > 0$, an argument similar to that in Example 11.8 shows that the limit approaches zero, and the result is established.

EXAMPLE 11.11   Show that $\mathcal{L}(\cos bt) = \dfrac{s}{s^2 + b^2}$.

Solution   A direct approach using the definition is tedious. Let us assume that the complex constants $\pm ib$ are permitted and hence following the Laplace transforms exist:

$$\mathcal{L}(e^{ibt}) = \frac{1}{s - ib} \quad \text{and} \quad \mathcal{L}(e^{-ibt}) = \frac{1}{s + ib} .$$

Using the linearity of the Laplace transform we obtain

$$\mathcal{L}(\cos bt) = \tfrac{1}{2} \mathcal{L}(e^{ibt}) + \tfrac{1}{2} \mathcal{L}(e^{-ibt})$$

$$= \frac{1}{2} \frac{1}{s - ib} + \frac{1}{2} \frac{1}{s + ib} = \frac{s}{s^2 + b^2} .$$

Inverting the Laplace transform is usually accomplished with the aid of a table of known Laplace transforms and the technique of partial fraction expansion.

EXAMPLE 11.12   Find $\mathcal{L}^{-1}\left(\dfrac{3s + 6}{s^2 + 9}\right)$.

Solution   Using linearity and lines 6 and 7 of Table 11.2, we obtain

$$\mathcal{L}^{-1}\left(\frac{3s + 6}{s^2 + 9}\right) = 3\mathcal{L}^{-1}\left(\frac{s}{s^2 + 9}\right) + 2\mathcal{L}^{-1}\left(\frac{3}{s^2 + 9}\right)$$

$$= 3 \cos 3t + 2 \sin 3t.$$

Table 11.2 gives the Laplace transforms of some well-known functions, and Table 11.3 highlights some important properties of Laplace transforms.

TABLE 11.2 Table of Laplace Transforms

Line	$f(t)$	$F(s) = \int_0^\infty f(t)e^{-st}\, dt$
1	1	$\dfrac{1}{s}$
2	t^n	$\dfrac{n!}{s^{n+1}}$
3	$U_c(t)$ unit step	$\dfrac{e^{-cs}}{s}$
4	e^{at}	$\dfrac{1}{s-a}$
5	$t^n e^{at}$	$\dfrac{n!}{(s-a)^{n+1}}$
6	$\cos bt$	$\dfrac{s}{s^2+b^2}$
7	$\sin bt$	$\dfrac{b}{s^2+b^2}$
8	$e^{at}\cos bt$	$\dfrac{s-a}{(s-a)^2+b^2}$
9	$e^{at}\sin bt$	$\dfrac{b}{(s-a)^2+b^2}$
10	$t\cos bt$	$\dfrac{s^2-b^2}{(s^2+b^2)^2}$
11	$t\sin bt$	$\dfrac{2bs}{(s^2+b^2)^2}$
12	$\cosh at$	$\dfrac{s}{s^2-a^2}$
13	$\sinh at$	$\dfrac{a}{s^2-a^2}$

TABLE 11.3   Properties of the Laplace Transform

Definition	$\mathcal{L}(f(t)) = F(s)$
Derivatives of $f(t)$	$\mathcal{L}(f'(t)) = sF(s) - f(0)$
	$\mathcal{L}(f''(t)) = s^2F(s) - sf(0) - f'(0)$
Integral of $f(t)$	$\mathcal{L}\left(\int_0^t f(\tau)\, d\tau \right) = \dfrac{F(s)}{s}$
Multiplication by t	$\mathcal{L}(tf(t)) = -F'(s)$
Division by t	$\mathcal{L}\left(\dfrac{f(t)}{t} \right) = \int_s^\infty F(\sigma)\, d\sigma$
Shifting on the s axis	$\mathcal{L}(e^{at}f(t)) = F(s - a)$
Shifting on the t axis	$\mathcal{L}(U_a(t)f(t - a)) = e^{-as}F(s)$ for $a > 0$
Convolution	$\mathcal{L}(h(t)) = F(s)G(s)$
	$h(t) = \int_0^t f(t - \tau)g(\tau)\, d\tau$

EXERCISES FOR SECTION 11.5

1. Show that $\mathcal{L}(1) = \dfrac{1}{s}$ by using the integral definition for the Laplace transform. Assume that s is restricted to values satisfying $\operatorname{Re}(s) > 0$.

2. Let $U(t) = \begin{cases} 1 & \text{for } 1 < t < 2, \\ 0 & \text{otherwise,} \end{cases}$ find $\mathcal{L}(f(t))$.

3. Let $U(t) = \begin{cases} t & \text{for } 0 \le t < c, \\ 0 & \text{otherwise,} \end{cases}$ find $\mathcal{L}(f(t))$.

4. Show that $\mathcal{L}(t^2) = \dfrac{2}{s^3}$ by using the integral definition for the Laplace transform. Assume that s is restricted to values satisfying $\operatorname{Re}(s) > 0$.

5. Let $U(t) = \begin{cases} e^{at} & \text{for } 0 \le t < 1, \\ 0 & \text{otherwise,} \end{cases}$ find $\mathcal{L}(f(t))$.

6. Let $U(t) = \begin{cases} \sin(t) & \text{for } 0 \le t \le \pi, \\ 0 & \text{otherwise,} \end{cases}$ find $\mathcal{L}(f(t))$.

For exercises 7–12 use the linearity of Laplace transform and Table 11.2.

7. Find $\mathcal{L}(3t^2 - 4t + 5)$.

8. Find $\mathcal{L}(2 \cos 4t)$.

9. Find $\mathcal{L}(e^{2t-3})$.

10. Find $\mathcal{L}(6e^{-t} + 3 \sin 5t)$.

11. Find $\mathcal{L}((t + 1)^4)$.

12. Find $\mathcal{L}(\cosh 2t)$.

For exercises 13–18 use the linearity of the inverse Laplace transform and Table 11.3.

13. $\mathcal{L}^{-1}\left(\dfrac{1}{s^2 + 25} \right)$.

14. Find $\mathcal{L}^{-1}\left(\dfrac{4}{s} - \dfrac{6}{s^2} \right)$.

15. Find $\mathcal{L}^{-1}\left(\dfrac{1 + s^2 - s^3}{s^4} \right)$.

16. Find $\mathcal{L}^{-1}\left(\dfrac{2s + 9}{s^2 + 9} \right)$.

17. Find $\mathcal{L}^{-1}\left(\dfrac{6s}{s^2 - 4} \right)$.

18. Find $\mathcal{L}^{-1}\left(\dfrac{2s + 1}{s(s + 1)} \right)$.

19. Write a report on how complex analysis is used in the study of Laplace transforms. Include ideas and examples that are not mentioned in the text. Resources include bibliographical items 17, 40, 69, 129, 149, 159, and 186.

11.6   Laplace Transforms of Derivatives and Integrals

Theorem 11.13 (Differentiation of *f(t)*)   *Let $f(t)$ and $f'(t)$ be continuous for $t \geq 0$, and of exponential order. Then,*

(1) $\mathcal{L}(f'(t)) = sF(s) - f(0),$

where $F(s) = \mathcal{L}(f(t))$.

Proof   Let K be chosen large enough so that both $f(t)$ and $f'(t)$ are of exponential order K. If $\text{Re}(s) > K$, then $\mathcal{L}(f'(t))$ is given by

(2) $\mathcal{L}(f'(t)) = \displaystyle\int_0^\infty f'(t)e^{-st}\, dt.$

Using integration by parts, we write equation (2) as

(3) $\mathcal{L}(f'(t)) = \lim_{R \to +\infty} [f(t)e^{-st}]\Big|_{t=0}^{t=R} + s\displaystyle\int_0^\infty f(t)e^{-st}\, dt.$

Since $f(t)$ is of exponential order K, and $\text{Re}(s) > K$, we have $\lim_{R \to +\infty} f(R)e^{-sR} = 0$, hence equation (3) becomes

(4) $\mathcal{L}(f'(t)) = -f(0) + s\displaystyle\int_0^\infty f(t)e^{-st}\, dt = sF(s) - f(0),$

and the theorem is proven.

Corollary 11.1   *If $f(t), f'(t)$, and $f''(t)$ are of exponential order, then*

(5) $\mathcal{L}(f''(t)) = s^2 F(s) - sf(0) - f'(0).$

EXAMPLE 11.13   Show that $\mathcal{L}(\cos^2 t) = \dfrac{s^2 + 2}{s(s^2 + 4)}.$

Solution   Let $f(t) = \cos^2 t$, then $f(0) = 1$ and $f'(t) = -2\sin t \cos t = -\sin 2t$. Using the fact that $\mathcal{L}(-\sin 2t) = \dfrac{-2}{s^2 + 4}$, Theorem 11.13 implies that

$$\frac{-2}{s^2 + 4} = \mathcal{L}(f'(t)) = s\mathcal{L}(\cos^2 t) - 1,$$

from which it follows that $\mathcal{L}(\cos^2 t) = \dfrac{-2}{s(s^2 + 4)} + \dfrac{1}{s} = \dfrac{s^2 + 2}{s(s^2 + 4)}.$

Theorem 11.14 (Integration of *f*(*t*))   *Let $f(t)$ be continuous for $t \geq 0$, and of exponential order and let $F(s)$ be its Laplace transform, then*

(6) $$\mathcal{L}\left(\int_0^t f(\tau)\, d\tau\right) = \frac{F(s)}{s}.$$

Proof   Let $g(t) = \int_0^t f(\tau)\, d\tau$, then $g'(t) = f(t)$ and $g(0) = 0$. If we can show that g is of exponential order, then Theorem 11.13 implies that

$$\mathcal{L}(f(t)) = \mathcal{L}(g'(t)) = s\mathcal{L}(g(t)) - 0 = s\mathcal{L}\left(\int_0^t f(\tau)\, d\tau\right),$$

and the proof will be complete. Since $f(t)$ is of exponential order, we can find positive values M and K, so that

$$|g(t)| \leq \int_0^t f(\tau)\, d\tau \leq M \int_0^t e^{K\tau}\, d\tau = \frac{M}{K}(e^{Kt} - 1) \leq e^{Kt},$$

so that g is of exponential order and the proof is complete.

EXAMPLE 11.14   Show that $\mathcal{L}(t^2) = \dfrac{2}{s^3}$ and $\mathcal{L}(t^3) = \dfrac{6}{s^4}$.

Solution   Using Theorem 11.14 and the fact that $\mathcal{L}(2t) = \dfrac{2}{s^2}$ we obtain

$$\mathcal{L}(t^2) = \mathcal{L}\left(\int_0^t 2\tau\, d\tau\right) = \frac{1}{s}\mathcal{L}(2t) = \frac{1}{s}\frac{2}{s^2} = \frac{2}{s^3}.$$

Now we can use the first result $\mathcal{L}(t^2) = \dfrac{2}{s^3}$ to establish the second one:

$$\mathcal{L}(t^3) = \mathcal{L}\left(\int_0^t 3\tau^2\, d\tau\right) = \frac{1}{s}\mathcal{L}(3t^2) = \frac{1}{s}\frac{6}{s^3} = \frac{6}{s^4}.$$

One of the main uses of the Laplace transform is its role in the solution of differential equations. The utility of the Laplace transform lies in the fact that the transform of the derivative $f'(t)$ corresponds to multiplication of the transform $F(s)$ by s and then the subtraction of $f(0)$. This permits us to replace the calculus operation of differentiation with simple algebraic operations on transforms.

This idea is used to develop a method for solving linear differential equations with constant coefficients. Consider the initial value problem

(7) $y''(t) + ay'(t) + by(t) = f(t),$

with initial conditions $y(0) = y_0$ and $y'(0) = d_0$. The linearity property of the Laplace transform can be used to obtain the equation

$$(8) \quad \mathcal{L}(y''(t)) + a\mathcal{L}(y'(t)) + b\mathcal{L}(y(t)) = \mathcal{L}(f(t)).$$

Let $Y(s) = \mathcal{L}(y(t))$ and $F(s) = \mathcal{L}(f(t))$ and apply Theorem 11.13 and Corollary 11.1 in the form $\mathcal{L}(y'(t)) = sY(s) - y(0)$ and $\mathcal{L}(y''(t)) = s^2Y(s) - sy(0) - y'(0)$ we can rewrite equation (8) in the form

$$(9) \quad s^2Y(s) + asY(s) + bY(s) = F(s) + sy(0) + y'(0) + ay(0).$$

The Laplace transform $Y(s)$ of the solution $y(t)$ is easily found to be

$$(10) \quad Y(s) = \frac{F(s) + sy(0) + y'(0) + ay(0)}{s^2 + as + b}.$$

For many physical problems involving mechanical systems and electric circuits, the transform $F(s)$ is known, and the inverse of $Y(s)$ can easily be computed. This process is referred to as the operational calculus and has the advantage of changing problems in differential equations into problems in algebra. Then the solution obtained will satisfy the specific initial conditions.

EXAMPLE 11.15   Solve the initial value problem

$$y''(t) + y(t) = 0 \text{ with } y(0) = 2 \text{ and } y'(0) = 3.$$

Solution   Since the right-hand side of the differential equation is $f(t) \equiv 0$ we have $F(s) \equiv 0$. The initial conditions yield $\mathcal{L}(y''(t)) = s^2Y(s) - 2s - 3$ and equation (9) becomes $s^2Y(s) + Y(s) = 2s + 3$. Solving we get $Y(s) = \dfrac{2s + 3}{s^2 + 1}$ and the solution $y(t)$ is assisted by using Table 11.2 and the computation

$$y(t) = \mathcal{L}^{-1}\left(\frac{2s + 3}{s^2 + 1}\right) = 2\mathcal{L}^{-1}\left(\frac{s}{s^2 + 1}\right) + 3\mathcal{L}^{-1}\left(\frac{1}{s^2 + 1}\right) = 2\cos t + 3\sin t.$$

EXAMPLE 11.16   Solve the initial value problem

$$y''(t) + y'(t) - 2y(t) = 0 \quad \text{with } y(0) = 1 \quad \text{and} \quad y'(0) = 4.$$

Solution   In the spirit of Example 11.15, we use the initial conditions and equation (10) becomes

$$Y(s) = \frac{s + 4 + 1}{s^2 + s - 2} = \frac{s + 5}{(s - 1)(s + 2)}.$$

Using partial fraction expansion $Y(s) = \dfrac{2}{s-1} - \dfrac{1}{s+2}$ and the solution $y(t)$ is

$$y(t) = \mathcal{L}^{-1}(Y(s)) = 2\mathcal{L}^{-1}\left(\frac{1}{s-1}\right) - \mathcal{L}^{-1}\left(\frac{1}{s+2}\right) = 2e^t - e^{-2t}.$$

EXERCISES FOR SECTION 11.6

1. Derive $\mathcal{L}(\sin t)$ from $\mathcal{L}(\cos t)$.
2. Derive $\mathcal{L}(\cosh t)$ from $\mathcal{L}(\sinh t)$.
3. Find $\mathcal{L}(\sin^2 t)$.
4. Show that $\mathcal{L}(te^t) = \dfrac{1}{(s-1)^2}$. *Hint:* Let $f(t) = te^t$ and $f'(t) = te^t + e^t$.
5. Find $\mathcal{L}^{-1}\left(\dfrac{1}{s(s-4)}\right)$.   6. Find $\mathcal{L}^{-1}\left(\dfrac{1}{s(s^2+4)}\right)$.
7. Show that $\mathcal{L}^{-1}\left(\dfrac{1}{s^2(s+1)}\right) = t - 1 + e^t$.
8. Show that $\mathcal{L}^{-1}\left(\dfrac{1}{s^2(s^2+1)}\right) = t - \sin t$.

For exercises 9–18, solve the initial value problem.

9. $y''(t) + 9y(t) = 0$, with $y(0) = 2$ and $y'(0) = 9$
10. $y''(t) + y(t) = 1$, with $y(0) = 0$ and $y'(0) = 2$
11. $y''(t) + 4y(t) = -8$, with $y(0) = 0$ and $y'(0) = 2$
12. $y'(t) + y(t) = 1$, with $y(0) = 2$
13. $y'(t) - y(t) = -2$, with $y(0) = 3$
14. $y''(t) - 4y(t) = 0$, with $y(0) = 1$ and $y'(0) = 2$
15. $y''(t) - y(t) = 1$, with $y(0) = 0$ and $y'(0) = 2$
16. $y'(t) + 2y(t) = 3e^t$, with $y(0) = 2$
17. $y''(t) + y'(t) - 2y(t) = 0$, with $y(0) = 2$ and $y'(0) = -1$
18. $y''(t) - y'(t) - 2y(t) = 0$, with $y(0) = 2$ and $y'(0) = 1$

11.7   Shifting Theorems and the Step Function

We have seen how the Laplace transform can be used to solve linear differential equations. Familiar functions that arise in solutions to differential equations are $e^{at}\cos bt$ and $e^{at}\sin bt$. The *first shifting theorem* will show how their transforms are related to those of $\cos bt$ and $\sin bt$ by shifting the variable s in $F(s)$. A companion result, called the *second shifting theorem,* will show how the transform of $f(t - a)$ can be obtained by multiplying $F(s)$ by e^{-as}. Loosely speaking, these results show that multiplication of $f(t)$ by e^{at} corresponds to shifting $F(s - a)$, and shifting $f(t - a)$ corresponds to multiplication of the transform $F(s)$ by e^{as}.

Theorem 11.15 (Shifting the Variable s)   *If $F(s)$ is the Laplace transform of $f(t)$, then*

(1) $\mathcal{L}(e^{at}f(t)) = F(s - a)$.

Proof   Using the integral definition $\mathcal{L}(f(t)) = F(s) = \int_0^\infty f(t)e^{-st}\,dt$, we see that

(2) $\mathcal{L}(e^{at}f(t)) = \int_0^\infty e^{at}f(t)e^{-st}\,dt = \int_0^\infty f(t)e^{-(s-a)t}\,dt = F(s - a)$.

Definition 11.3 (Unit Step Function)   *Let $a \geq 0$, then the unit step function $U_a(t)$ is*

(3) $U_a(t) = \begin{cases} 0 & \text{for } t < a, \\ 1 & \text{for } t > a. \end{cases}$

The graph of $U_a(t)$ is shown in Figure 11.22.

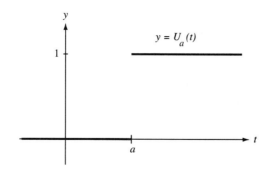

FIGURE 11.22   The graph of the unit step function $y = U_a(t)$ in Definition 11.3.

Theorem 11.16 (Shifting the Variable t)   *If $F(s)$ is the Laplace transform of $f(t)$ and $a \geq 0$, then*

(4) $\mathcal{L}(U_a(t)f(t - a)) = e^{-as}F(s)$,

where $f(t)$ and $U_a(t)(t - a)$ are illustrated in Figure 11.23.

Proof   Using the definition of Laplace transform, we write

(5) $e^{-as}F(s) = e^{-as}\int_0^\infty f(\tau)e^{-s\tau}\,d\tau = \int_0^\infty f(\tau)e^{-s(a+\tau)}\,d\tau$.

Using the change of variable $t = a + \tau$ and $dt = d\tau$, we obtain

(6) $e^{-as}F(s) = \int_a^\infty f(t - a)e^{-st}\,dt$.

Since $U_a(t)f(t - a) = 0$ for $t < a$, and $U_a(t)f(t - a) = f(t - a)$ for $t > a$, we can write equation (6) as

(7) $e^{-as}F(s) = \int_0^\infty U_a(t)f(t - a)e^{-st}\,dt = \mathcal{L}(U_a(t)f(t - a)),$

and the proof is complete.

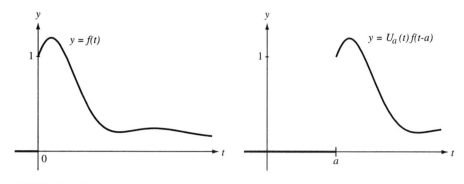

FIGURE 11.23   Comparison of the functions $f(t)$ and $U_a(t)f(t - a)$ in Theorem 11.16.

EXAMPLE 11.17   Show that $\mathcal{L}(t^n e^{at}) = \dfrac{n!}{(s - a)^{n+1}}$.

Solution   Let $f(t) = t^n$, then $F(s) = \mathcal{L}(t^n) = \dfrac{n!}{s^{n+1}}$. Applying Theorem 11.15, we obtain the desired result:

$\mathcal{L}(t^n e^{at}) = F(s - a) = \dfrac{n!}{(s - a)^{n+1}}.$

EXAMPLE 11.18   Show that $\mathcal{L}(U_c(t)) = \dfrac{e^{-cs}}{s}$.

Solution   Set $f(t) = 1$, then $F(s) = \mathcal{L}(1) = \dfrac{1}{s}$. Now apply Theorem 11.16 and get

$\mathcal{L}(U_c(t)) = \mathcal{L}(U_c(t)f(t)) = \mathcal{L}(U_c(t)\cdot 1) = e^{-cs}\,\mathcal{L}(1) = \dfrac{e^{-cs}}{s}.$

EXAMPLE 11.19   Find $\mathcal{L}(f(t))$ if $f(t)$ is given in Figure 11.24.

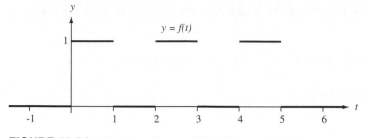

FIGURE 11.24   The function $y = f(t)$ of Example 11.19.

Solution   We can represent $f(t)$ in terms of step functions

$$f(t) = 1 - U_1(t) + U_2(t) - U_3(t) + U_4(t) - U_5(t).$$

Using the result of Example 11.18 and linearity, we obtain

$$\mathcal{L}(f(t)) = \frac{1}{s} - \frac{e^{-s}}{s} + \frac{e^{-2s}}{s} - \frac{e^{-3s}}{s} + \frac{e^{-4s}}{s} - \frac{e^{-5s}}{s}.$$

EXAMPLE 11.20   Use Laplace transforms to solve the initial value problem

$$y''(t) + y(t) = U_\pi(t) \quad \text{with } y(0) = 0 \quad \text{and} \quad y'(0) = 0.$$

Solution   As usual, let $Y(s)$ denote the Laplace transform of $y(t)$. Then we get

$$s^2 Y(s) + Y(s) = \frac{e^{-\pi s}}{s}.$$

Solving for $Y(s)$, we obtain

$$Y(s) = e^{-\pi s}\frac{1}{s(s^2 + 1)} = \frac{e^{-\pi s}}{s} - \frac{e^{-\pi s}s}{s^2 + 1}.$$

We now use Theorem 11.16 and the facts that $\dfrac{1}{s}$ and $\dfrac{s}{s^2 + 1}$ are the transforms of 1 and $\cos t$, respectively. The solution $y(t)$ is computed as follows:

$$y(t) = \mathcal{L}^{-1}\left(\frac{e^{-\pi s}}{s}\right) - \mathcal{L}^{-1}\left(\frac{e^{-\pi s}s}{s^2 + 1}\right) = U_\pi(t) - U_\pi(t)\cos(t - \pi),$$

which can be written in the more familiar form:

$$y(t) = \begin{cases} 0 & \text{for } t < \pi, \\ 1 - \cos t & \text{for } t > \pi. \end{cases}$$

EXERCISES FOR SECTION 11.7

1. Find $\mathcal{L}(e^t - te^t)$.

2. Find $\mathcal{L}(e^{-4t}\sin 3t)$.

3. Show that $\mathcal{L}(e^{at}\cos bt) = \dfrac{s - a}{(s - a)^2 + b^2}$.

4. Show that $\mathcal{L}(e^{at}\sin bt) = \dfrac{b}{(s - a)^2 + b^2}$.

For exercises 5–8, find $\mathcal{L}^{-1}(F(s))$.

5. $F(s) = \dfrac{s + 2}{s^2 + 4s + 5}$

6. $F(s) = \dfrac{8}{s^2 - 2s + 5}$

7. $F(s) = \dfrac{s + 3}{(s + 2)^2 + 1}$

8. $F(s) = \dfrac{2s + 10}{s^2 + 6s + 25}$

For exercises 9–14, find $\mathcal{L}(f(t))$.

9. $f(t) = U_2(t)(t - 2)^2$

10. $f(t) = U_1(t)e^{1-t}$

11. $f(t) = U_{3\pi}(t)\sin(t - 3\pi)$

12. $f(t) = 2U_1(t) - U_2(t) - U_3(t)$

13. Let $f(t)$ be given in Figure 11.25.

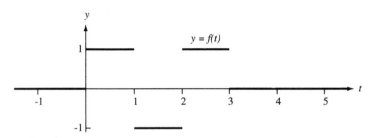

FIGURE 11.25   The graph $y = f(t)$ for Exercise 13.

14. Let $f(t)$ be given in Figure 11.26.
   Hint: The function is the integral of the one in Exercise 13.

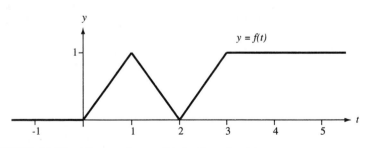

FIGURE 11.26   The graph $y = f(t)$ for Exercise 14.

15. Find $\mathcal{L}^{-1}\left(\dfrac{e^{-s} + e^{-2s}}{s}\right)$.

16. Find $\mathcal{L}^{-1}\left(\dfrac{1 - e^{-s} + e^{-2s}}{s^2}\right)$.

For exercises 17–23, solve the initial value problem.

17. $y''(t) + 2y'(t) + 2y(t) = 0$, with $y(0) = -1$ and $y'(0) = 1$
18. $y''(t) + 4y'(t) + 5y(t) = 0$, with $y(0) = 1$ and $y'(0) = -2$
19. $2y''(t) + 2y'(t) + y(t) = 0$, with $y(0) = 0$ and $y'(0) = 1$
20. $y''(t) - 2y'(t) + y(t) = 2e^t$, with $y(0) = 0$ and $y'(0) = 0$
21. $y''(t) + 2y'(t) + y(t) = 6te^{-t}$, with $y(0) = 0$ and $y'(0) = 0$
22. $y''(t) + 2y'(t) + y(t) = 2U_1(t)e^{1-t}$, with $y(0) = 0$ and $y'(0) = 0$
23. $y''(t) + y(t) = U_{\pi/2}(t)$, with $y(0) = 0$ and $y'(0) = 1$

11.8   Multiplication and Division by t

Sometimes the solutions to nonhomogeneous linear differential equations with constant coefficients involve the functions $t \cos bt$, $t \sin bt$, or $t^n e^{at}$ as part of the solution. We now show how the Laplace transforms of $tf(t)$ and $f(t)/t$ are related to the Laplace transform of $f(t)$. The transform of $tf(t)$ will be obtained via differentiation and the transform of $f(t)/t$ will be obtained via integration. To be precise, we state the following theorems.

Theorem 11.17 (Multiplication by t)   *If $F(s)$ is the Laplace transform of $f(t)$, then*

(1) $\mathcal{L}(tf(t)) = -F'(s).$

Proof   By definition we have $F(s) = \int_0^\infty f(t)e^{-st}\, dt$. Leibniz's rule for partial differentiation under the integral sign permits us to write

$$(2) \quad F'(s) = \frac{\partial}{\partial s}\int_0^\infty f(t)e^{-st}\, dt = \int_0^\infty \frac{\partial}{\partial s}[f(t)e^{-st}]\, dt$$

$$= \int_0^\infty [-tf(t)e^{-st}]\, dt = -\int_0^\infty [tf(t)e^{-st}]\, dt$$

$$= -\mathcal{L}(tf(t)),$$

and the result is established.

Theorem 11.18 (Division by t)   *Let both $f(t)$ and $f(t)/t$ have Laplace transforms and let $F(s)$ denote the transform of $f(t)$. If $\lim_{t\to 0^+} f(t)/t$ exists, then*

(3) $\mathcal{L}\left(\dfrac{f(t)}{t}\right) = \displaystyle\int_s^\infty F(\sigma)\, d\sigma.$

Proof   Since $F(\sigma) = \int_0^\infty f(t)e^{-\sigma t}\, dt$, we integrate $F(\sigma)$ from s to ∞ and obtain

$$(4) \quad \int_s^\infty F(\sigma)\, d\sigma = \int_s^\infty \left[\int_0^\infty f(t)e^{-\sigma t}\, dt\right]\, d\sigma.$$

The order of integration in equation (4) can be reversed, and we obtain

$$(5) \quad \int_s^\infty F(\sigma)\, d\sigma = \int_0^\infty \left[\int_s^\infty f(t)e^{-\sigma t}\, d\sigma \right] dt$$

$$= \int_0^\infty \left[\frac{-f(t)}{t} e^{-\sigma t} \Big|_{\sigma=s}^{\sigma=\infty} \right] dt$$

$$= \int_0^\infty \frac{f(t)}{t} e^{-st}\, dt = \mathcal{L}\!\left(\frac{f(t)}{t} \right),$$

and the proof is complete.

EXAMPLE 11.21   Show that $\mathcal{L}(t \cos bt) = \dfrac{s^2 - b^2}{(s^2 + b^2)^2}$.

Solution   Let $f(t) = \cos bt$, then $F(s) = \mathcal{L}(\cos bt) = \dfrac{s}{s^2 + b^2}$. Hence, we can differentiate $F(s)$ to obtain the desired result

$$\mathcal{L}(t \cos bt) = -F'(s) = -\frac{s^2 + b^2 - 2s^2}{(s^2 + b^2)^2} = \frac{s^2 - b^2}{(s^2 + b^2)^2}.$$

EXAMPLE 11.22   Show that $\mathcal{L}\!\left(\dfrac{\sin t}{t} \right) = \arctan \dfrac{1}{s}$.

Solution   Let $f(t) = \sin t$ and $F(s) = \dfrac{1}{s^2 + 1}$. Since $\lim\limits_{t \to 0^+} \dfrac{\sin t}{t} = 1$, we can integrate $F(s)$ to obtain the desired result:

$$\mathcal{L}\!\left(\frac{\sin t}{t} \right) = \int_s^\infty \frac{d\sigma}{\sigma^2 + 1} = -\arctan \frac{1}{\sigma} \Big|_{\sigma=s}^{\sigma=\infty} = \arctan \frac{1}{s}.$$

Some types of differential equations involve the terms $ty'(t)$ or $ty''(t)$. Laplace transforms can be used to find the solution if we use the additional substitutions

$(6) \quad \mathcal{L}(ty'(t)) = -sY'(s) - Y(s), \quad$ and

$(7) \quad \mathcal{L}(ty''(t)) = -s^2 Y'(s) - 2sY(s) + y(0).$

EXAMPLE 11.23   Use Laplace transforms to solve the initial value problem

$(8) \quad ty''(t) - ty'(t) - y(t) = 0 \quad$ with $y(0) = 0$.

Solution   Let $Y(s)$ denote the Laplace transform of $y(t)$, then using the substitutions (6) and (7) results in

$(9) \quad -s^2 Y'(s) - 2sY(s) + 0 + sY'(s) + Y(s) - Y(s) = 0.$

Equation (9) itself can be written as a first-order linear differential equation:

(10) $Y'(s) + \dfrac{2}{s-1} Y(s) = 0.$

The differential equation (10) can be solved by using an integrating factor:

(11) $\rho = \exp\left(\displaystyle\int \dfrac{2}{s-1}\, ds\right) = \exp[2\ln(s-1)] = (s-1)^2.$

Multiplying equation (10) by ρ produces

(12) $(s-1)^2 Y'(s) + 2(s-1)Y(s) = \dfrac{d}{ds}[(s-1)^2 Y(s)] = 0.$

Integrating equation (12) with respect to s results in $(s-1)^2 Y(s) = C$, where C is the constant of integration. Hence the solution to equation (9) is

(13) $Y(s) = \dfrac{C}{(s-1)^2}.$

The inverse of the transform $Y(s)$ in equation (13) is

$y(t) = Cte^t.$

EXERCISES FOR SECTION 11.8

Find the Laplace transform for Exercises 1–10.

1. Find $\mathcal{L}(te^{-2t})$.   **2.** Find $\mathcal{L}(t^2 e^{4t})$.
3. Find $\mathcal{L}(t \sin 3t)$.   **4.** Find $\mathcal{L}(t^2 \cos 2t)$.
5. Find $\mathcal{L}(t \sinh t)$.   **6.** Find $\mathcal{L}(t^2 \cosh t)$.

7. Show that $\mathcal{L}\left(\dfrac{e^t - 1}{t}\right) = \ln\left(\dfrac{s}{s-1}\right).$

8. Show that $\mathcal{L}\left(\dfrac{1 - \cos t}{t}\right) = \ln\left(\dfrac{s^2}{s^2+1}\right).$

9. Find $\mathcal{L}(t \sin bt)$.   **10.** Find $\mathcal{L}(te^{at} \cos bt)$.

11. Find $\mathcal{L}^{-1}\left(\ln\left(\dfrac{s^2+1}{(s-1)^2}\right)\right).$ **12.** Find $\mathcal{L}^{-1}\left(\ln\left(\dfrac{s}{s+1}\right)\right).$

For problems 13–21, solve the initial value problem.

13. $y''(t) + 2y'(t) + y(t) = 2e^{-t}$, with $y(0) = 0$ and $y'(0) = 1$
14. $y''(t) + y(t) = 2 \sin t$, with $y(0) = 0$ and $y'(0) = -1$
15. $ty''(t) - ty'(t) - y(t) = 0$, with $y(0) = 0$
16. $ty''(t) + (t - 1)y'(t) - 2y(t) = 0$, with $y(0) = 0$
17. $ty''(t) + ty'(t) - y(t) = 0$, with $y(0) = 0$
18. $ty''(t) + (t - 1)y'(t) + y(t) = 0$, with $y(0) = 0$
19. Solve the Laguerre equation $ty''(t) + (1 - t)y'(t) + y(t) = 0$, with $y(0) = 1$.
20. Solve the Laguerre equation $ty''(t) + (1 - t)y'(t) + 2y(t) = 0$, with $y(0) = 1$.

11.9   Inverting the Laplace Transform

So far, most of the applications involving the Laplace transform involve a transform (or part of a transform) that is expressed by

$$(1) \quad Y(s) = \frac{P(s)}{Q(s)},$$

where P and Q are polynomials that have no common factors. The inverse of $Y(s)$ is found by using its partial fraction representation and referring to Table 11.2. We now show how the theory of complex variables can be used to systematically find the partial fraction representation. The first result is an extension of Lemma 8.1 to n linear factors. The proof is left for the reader.

> **Theorem 11.19 (Nonrepeated Linear Factors)** *Let $P(s)$ be a polynomial of degree at most $n - 1$. If $Q(s)$ has degree n, and has distinct complex roots $a_1, a_2, \ldots, a_n$, then equation (1) has the representation*
>
> $$(2) \quad Y(s) = \frac{P(s)}{(s - a_1)(s - a_2) \cdots (s - a_n)} = \sum_{k=1}^{n} \frac{\text{Res}[Y, a_k]}{s - a_k}.$$

> **Theorem 11.20 (A Repeated Linear Factor)** *If $P(s)$ and $Q(s)$ are polynomials of degree μ and ν, respectively, and $\mu < \nu + n$ and $Q(a) \neq 0$, then equation (1) has the representation*
>
> $$(3) \quad Y(s) = \frac{P(s)}{(s - a)^n Q(s)} = \sum_{k=1}^{n} \frac{A_k}{(s - a)^k} + R(s),$$
>
> *where R is the sum of all partial fractions that do not involve factors of the form $(s - a)^j$. Furthermore, the coefficients A_k can be computed with the formula*
>
> $$(4) \quad A_k = \frac{1}{(n - k)!} \lim_{s \to a} \frac{d^{n-k}}{ds^{n-k}} \frac{P(s)}{Q(s)} \quad \text{for } k = 1, 2, \ldots, n.$$

Proof   We employ the method of residues. First, multiplying both sides of equation (3) by $(s - a)^n$ gives

$$(5) \quad \frac{P(s)}{Q(s)} = \sum_{j=1}^{n} A_j (s - a)^{n-j} + R(s)(s - a)^n.$$

We can differentiate both sides of equation (5) $n - k$ times to obtain

$$(6) \quad \frac{d^{n-k}}{ds^{n-k}} \frac{P(s)}{Q(s)} = \sum_{j=1}^{k} A_j \frac{(n - j)!}{(k - j)!} (s - a)^{k-j} + \frac{d^{n-k}}{ds^{n-k}} [R(s)(s - a)^n].$$

We now use the result in equation (6) and take the limit as $s \to a$. It is left as an exercise for the reader to fill in the steps to obtain

$$\lim_{s \to a} \frac{d^{n-k}}{ds^{n-k}} \frac{P(s)}{Q(s)} = (n - k)! A_k,$$

which establishes equation (4).

EXAMPLE 11.24   Let $Y(s) = \dfrac{s^3 - 4s + 1}{s(s-1)^3}$. Find $\mathcal{L}^{-1}(Y(s))$.

Solution   From equations (2) and (3) we can write

$$\frac{s^3 - 4s + 1}{s(s-1)^3} = \frac{A_3}{(s-1)^3} + \frac{A_2}{(s-1)^2} + \frac{A_1}{s-1} + \frac{B_1}{s}.$$

The coefficient B_1 is found by the calculation

$$B_1 = \operatorname{Res}[Y, 0] = \lim_{s \to 0} \frac{s^3 - 4s + 1}{(s-1)^3} = -1.$$

The coefficients A_1, A_2, and A_3 are found by using Theorem 11.20. In this case, $a = 1$ and $\dfrac{P(s)}{Q(s)} = \dfrac{s^3 - 4s + 1}{s}$, and we get

$$A_3 = \lim_{s \to 1} \frac{P(s)}{Q(s)} = \lim_{s \to 1} \frac{s^3 - 4s + 1}{s} = -2,$$

$$A_2 = \frac{1}{1!} \lim_{s \to 1} \frac{d}{ds} \frac{P(s)}{Q(s)} = \lim_{s \to 1} \left(2s - \frac{1}{s^2} \right) = 1,$$

$$A_1 = \frac{1}{2!} \lim_{s \to 1} \frac{d^2}{ds^2} \frac{P(s)}{Q(s)} = \frac{1}{2} \lim_{s \to 1} \left(2 + \frac{2}{s^3} \right) = 2.$$

Hence, the partial fraction representation is

$$Y(s) = \frac{-2}{(s-1)^3} + \frac{1}{(s-1)^2} + \frac{2}{s-1} - \frac{1}{s},$$

and the inverse is

$$y(t) = -t^2 e^t + t e^t + 2 e^t - 1.$$

Theorem 11.21 (Irreducible Quadratic Factors)   *Let P and Q be polynomials with real coefficients such that the degree of P is at most 1 larger than the degree of Q. If T does not have a factor of the form $(s - a)^2 + b^2$, then*

(7) $$Y(s) = \frac{P(s)}{Q(s)} = \frac{P(s)}{[(s-a)^2 + b^2]T(s)} = \frac{2A(s-a) - 2Bb}{[(s-a)^2 + b^2]} + R(s), \text{ where}$$

(8) $$A + iB = \frac{P(a+ib)}{Q'(a+ib)}.$$

Proof   Since P, Q, and Q' have real coefficients, it follows that

(9) $$P(a - ib) = \overline{P(a+ib)} \quad \text{and} \quad Q'(a-ib) = \overline{Q'(a+ib)}.$$

The polynomial Q has simple zeros at $s = a \pm ib$, this implies that $Q'(a \pm ib) \neq 0$. Therefore, we obtain

$$(10) \quad \text{Res}[Y, a \pm ib] = \lim_{s \to a \pm ib} \frac{s - (a \pm ib)}{Q(s) - Q(a \pm ib)} P(s) = \frac{P(a \pm ib)}{Q'(a \pm ib)},$$

from which it is easy to see that

$$(11) \quad \text{Res}[Y, a - ib] = \overline{\text{Res}[Y, a - ib]}.$$

If we set $A + iB = \text{Res}[Y, a + ib]$ and use Theorem 11.19 and equations (8), (10), and (11), then we find that

$$(12) \quad Y(s) = \frac{A + iB}{s - a - ib} + \frac{A - iB}{s - a + ib} + R(s).$$

The first two terms on the right side of equation (12) can be combined to obtain

$$\frac{(A + iB)(s - a + ib) + (A - iB)(s - a - ib)}{[(s - a)^2 + b^2]} = \frac{2A(s - a) - 2Bb}{[(s - a)^2 + b^2]},$$

and the proof of the theorem is complete.

EXAMPLE 11.25   Let $Y(s) = \dfrac{5s}{(s^2 + 4)(s^2 + 9)}$. Find $\mathcal{L}^{-1}(Y(s))$.

Solution   Here we have $P(s) = 5s$ and $Q(s) = s^4 + 13s^2 + 36$, and the roots of $Q(s)$ occur at $0 \pm 2i$ and $0 \pm 3i$. Computing the residues we find that

$$\text{Res}[Y, 2i] = \frac{P(2i)}{Q'(2i)} = \frac{5(2i)}{4(2i)^3 - 26(2i)} = \frac{1}{2}, \quad \text{and}$$

$$\text{Res}[Y, 3i] = \frac{P(3i)}{Q'(3i)} = \frac{5(3i)}{4(3i)^3 - 26(3i)} = \frac{-1}{2}.$$

We find that $A_1 + iB_1 = \dfrac{1}{2} + 0i$ and $A_2 + iB_2 = -\dfrac{1}{2} + 0i$, which correspond to $a_1 + ib_1 = 0 + 2i$ and $a_2 + ib_2 = 0 + 3i$, respectively. Thus we obtain

$$Y(s) = \frac{2(\frac{1}{2})(s - 0) - 2(0)2}{s^2 + 4} + \frac{2(-\frac{1}{2})(s - 0) - 2(0)3}{s^2 + 9} = \frac{s}{s^2 + 4} - \frac{s}{s^2 + 9},$$

and

$$\mathcal{L}^{-1}(Y(s)) = \mathcal{L}^{-1}\left(\frac{s}{s^2 + 4}\right) - \mathcal{L}^{-1}\left(\frac{s}{s^2 + 9}\right) = \cos 2t - \cos 3t.$$

EXAMPLE 11.26 Find $\mathscr{L}^{-1}(Y(s))$ if $Y(s) = \dfrac{s^3 + 3s^2 - s + 1}{s(s + 1)^2(s^2 + 1)}$.

Solution   The partial fraction expression for $Y(s)$ has the form

$$Y(s) = \frac{D}{s} + \frac{C_1}{s + 1} + \frac{C_2}{(s + 1)^2} + \frac{2A(s - 0) - 2B(1)}{(s - 0)^2 + 1^2}.$$

Since the linear factor s is nonrepeated, we have

$$D = \text{Res}[Y(s), 0] = \lim_{s \to 0} \frac{s^3 + 3s^2 - s + 1}{(s + 1)^2(s^2 + 1)} = 1.$$

Since the factor $s + 1$ is repeated, we have

$$C_1 = \text{Res}[Y(s), -1] = \lim_{s \to -1} \frac{d}{ds} \frac{s^3 + 3s^2 - s + 1}{s(s^2 + 1)} = \lim_{s \to -1} \frac{-3s^4 + 4s^3 - 1}{s^2(s + 1)^2} = -2.$$

$$C_2 = \text{Res}[(s + 1)Y(s), -1] = \lim_{s \to -1} \frac{s^3 + 3s^2 - s + 1}{s(s^2 + 1)} = -2.$$

The term $s^2 + 1$ is an irreducible quadratic, with roots $\pm i$, so that

$$A + iB = \text{Res}[Y, i] = \lim_{s \to i} \frac{s^3 + 3s^2 - s + 1}{s(s + 1)^2(s + i)} = \frac{1 - i}{2},$$

and we obtain $A = \frac{1}{2}$ and $B = -\frac{1}{2}$. Therefore,

$$Y(s) = \frac{1}{s} + \frac{-2}{s + 1} + \frac{-2}{(s + 1)^2} + \frac{2\frac{1}{2}(s - 0) - 2(-\frac{1}{2})(1)}{(s - 0)^2 + 1^2}$$

$$= \frac{1}{s} - \frac{2}{s + 1} - \frac{2}{(s + 1)^2} + \frac{s + 1}{s^2 + 1}.$$

Now we use Table 11.2 to get

$$y(t) = 1 - 2e^{-t} - 2te^{-t} + \cos t + \sin t.$$

EXAMPLE 11.27 Use Laplace transforms to solve the system

$$y'(t) = y(t) - x(t) \quad \text{with} \quad y(0) = 1,$$
$$x'(t) = 5y(t) - 3x(t) \qquad x(0) = 2.$$

Solution   Let $Y(s)$ and $X(s)$ denote the Laplace transforms of $y(t)$ and $x(t)$, respectively. If we take the transforms of the two differential equations and get

$$sY(s) - 1 = Y(s) - X(s),$$
$$sX(s) - 2 = 5Y(s) - 3X(s),$$

which can be written as

$$(s - 1)Y(s) + X(s) = 1,$$
$$5Y(s) - (s + 3)X(s) = -2.$$

Cramer's rule can be used to solve for $Y(s)$ and $X(s)$:

$$Y(s) = \frac{\begin{vmatrix} 1 & 1 \\ -2 & -s-3 \end{vmatrix}}{\begin{vmatrix} s-1 & 1 \\ 5 & -s-3 \end{vmatrix}} = \frac{-s-3+2}{(s-1)(-s-3)-5} = \frac{s+1}{(s+1)^2+1},$$

$$X(s) = \frac{\begin{vmatrix} s-1 & 1 \\ 5 & -2 \end{vmatrix}}{\begin{vmatrix} s-1 & 1 \\ 5 & -s-3 \end{vmatrix}} = \frac{-2s+2-5}{(s-1)(-s-3)-5} = \frac{2(s+1)+1}{(s+1)^2+1}.$$

The solution is obtained by computing the inverse transforms:

$$y(t) = e^{-t}\cos t,$$
$$x(t) = e^{-t}(2\cos t + \sin t).$$

According to equation (10) of Section 11.5, the inverse Laplace transform is given by the integral formula

(13) $f(t) = \mathcal{L}^{-1}(F(s)) = \dfrac{1}{2\pi i} \displaystyle\int_{\sigma_0-i\infty}^{\sigma_0+i\infty} F(s)e^{st}\,ds,$

where σ_0 is any suitably chosen large positive constant. This improper integral is a contour integral taken along the vertical line $s = \sigma_0 + i\tau$ in the complex $s = \sigma + i\tau$ plane. We shall show how the residue theory in Chapter 8 is used to evaluate it. Cases where the integrand has either infinitely many poles or has branch points is left for the reader to research in advanced texts. We state the following more elementary result.

Theorem 11.22 (Inverse Laplace Transform) *Let $F(s) = \dfrac{P(s)}{Q(s)}$, where*

$P(s)$ and $Q(s)$ are polynomials of degree m and n, respectively, and $n > m$. The inverse Laplace transformation $F(s)$ is $f(t)$ given by

(14) $f(t) = \mathcal{L}^{-1}(F(s)) = \Sigma \, \mathrm{Res}[F(s)e^{st}, s_k],$

where the sum is taken over all of the residues of the complex function $F(s)e^{st}$.

Proof   Let σ_0 be chosen so that all the poles of $F(s)e^{st}$ lie to the left of the vertical line $s = \sigma_0 + i\tau$. Let Γ_R denote the contour consisting of the vertical line segment between the points $\sigma_0 \pm iR$ and the left semicircle C_R: $s = \sigma_0 + Re^{i\theta}$, where $\dfrac{\pi}{2} \le \theta \le \dfrac{3\pi}{2}$, as shown in Figure 11.27. A slight modification of the proof of Jordan's lemma will show that

(15) $\displaystyle\lim_{R\to\infty} \int_{C_R} \frac{P(s)}{Q(s)} e^{st}\,ds = 0.$

The residue theorem and equation (15) can now be used to show that

$$\mathcal{L}^{-1}(F(s)) = \lim_{R \to \infty} \frac{1}{2\pi i} \int_{\Gamma_R} \frac{P(s)}{Q(s)} e^{st}\, ds = \Sigma\, \mathrm{Res}[F(s)e^{st}, s_k],$$

and the proof of the theorem is complete.

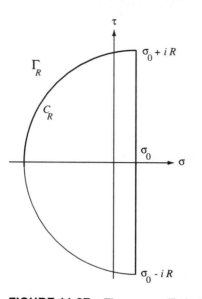

FIGURE 11.27   The contour Γ_R in the proof of Theorem 11.22.

Theorem 11.23 (Heaviside Expansion Theorem)   *Let $P(s)$ and $Q(s)$ be polynomials of degree m and n, respectively, where $n > m$. If $Q(s)$ has n distinct simple zeros at the points $s_1, s_2, \ldots, s_n$, then $\dfrac{P(s)}{Q(s)}$ is the Laplace transform of the function $f(t)$ given by*

(16) $f(t) = \mathcal{L}^{-1}\left(\dfrac{P(s)}{Q(s)}\right) = \displaystyle\sum_{k=1}^{n} \dfrac{P(s_k)}{Q'(s_k)} e^{s_k t}.$

Proof   If $P(s)$ and $Q(s)$ are polynomials and s_0 is a simple zero of $Q(s)$, then

$$\mathrm{Res}\left[\frac{P(s)}{Q(s)} e^{st}, s_0\right] = \lim_{s \to s_0} \frac{s - s_0}{Q(s) - Q(s_0)} P(s)e^{st} = \frac{P(s_0)}{Q'(s_0)} e^{s_0 t}.$$

This allows us to write the residues in equation (14) in the more convenient form given in equation (16).

EXAMPLE 11.28   Find the inverse Laplace transform of the function given by $F(s) = \dfrac{4s + 3}{s^3 + 2s^2 + s + 2}.$

Solution   Here we have $P(s) = 4s + 3$ and $Q(s) = (s + 2)(s^2 + 1)$ so that Q has simple zeros located at the points $s_1 = -2$, $s_2 = i$, and $s_3 = -i$. Using $Q'(s) = 3s^2 + 4s + 1$, calculation reveals that $\dfrac{P(-2)}{Q'(-2)} = \dfrac{-8 + 3}{12 - 8 + 1} = -1$ and $\dfrac{P(\pm i)}{Q'(\pm i)} = \dfrac{\pm 4i + 3}{-2 \pm 4i} = \dfrac{1}{2} \mp i$. Applying formula (16), we see that $f(t)$ is given by

$$f(t) = \frac{P(-2)}{Q'(-2)} e^{-2t} + \frac{P(i)}{Q'(i)} e^{it} + \frac{P(-i)}{Q'(-i)} e^{-it}$$

$$= -e^{-2t} + (\tfrac{1}{2} - i)e^{it} + (\tfrac{1}{2} + i)e^{-it}$$

$$= -e^{-2t} + \frac{e^{it} + e^{-it}}{2} + 2\,\frac{e^{it} - e^{-it}}{2i}$$

$$= -e^{-2t} + \cos t + 2 \sin t.$$

EXERCISES FOR SECTION 11.9

For exercises 1–6, use partial fractions to find the inverse Laplace transform of $Y(s)$.

1. $Y(s) = \dfrac{2s + 1}{s(s - 1)}$

2. $Y(s) = \dfrac{2s^3 - s^2 + 4s - 6}{s^4}$

3. $Y(s) = \dfrac{4s^2 - 6s - 12}{s(s + 2)(s - 2)}$

4. $Y(s) = \dfrac{s^3 - 5s^2 + 6s - 6}{(s - 2)^4}$

5. $Y(s) = \dfrac{2s^2 + s + 3}{(s + 2)(s - 1)^2}$

6. $Y(s) = \dfrac{4 - s}{s^2 + 4s + 5}$

7. Use a contour integral to find the inverse Laplace transform of $Y(s) = \dfrac{1}{s^2 + 4}$

8. Use a contour integral to find the inverse Laplace transform of $Y(s) = \dfrac{s + 3}{(s - 2)(s^2 + 1)}$

For exercises 9–12, use the heaviside expansion theorem to find the inverse Laplace transform of $Y(s)$.

9. $Y(s) = \dfrac{s^3 + s^2 - s + 3}{s^5 - s}$

10. $Y(s) = \dfrac{s^3 + 2s^2 - s + 2}{s^5 - s}$

11. $Y(s) = \dfrac{s^3 + 3s^2 - s + 1}{s^5 - s}$

12. $Y(s) = \dfrac{s^3 + s^2 + s + 3}{s^5 - s}$

13. Find the inverse of $Y(s) = \dfrac{s^3 + 2s^2 + 4s + 2}{(s^2 + 1)(s^2 + 4)}$.

For problems 14–19, solve the initial value problem.

14. $y''(t) + y(t) = 3 \sin 2t$, with $y(0) = 0$ and $y'(0) = 3$

15. $y''(t) + 2y'(t) + 5y(t) = 4e^{-t}$, with $y(0) = 1$ and $y'(0) = 1$

16. $y''(t) + 2y'(t) + 2y(t) = 2$, with $y(0) = 1$ and $y'(0) = 1$

17. $y''(t) + 4y(t) = 5e^{-t}$, with $y(0) = 2$ and $y'(0) = 1$

18. $y''(t) + 2y'(t) + y(t) = t$, with $y(0) = -1$ and $y'(0) = 0$

19. $y''(t) + 3y'(t) + 2y(t) = 2t + 5$, with $y(0) = 1$ and $y'(0) = 1$

For problems 20–25, solve the system of differential equations.

20. $x'(t) = 10y(t) - 5x(t)$, $y'(t) = y(t) - x(t)$, with $x(0) = 3$ and $y(0) = 1$
21. $x'(t) = 2y(t) - 3x(t)$, $y'(t) = 2y(t) - 2x(t)$, with $x(0) = 1$ and $y(0) = -1$
22. $x'(t) = 2x(t) + 3y(t)$, $y'(t) = 2x(t) + y(t)$, with $x(0) = 2$ and $y(0) = 3$
23. $x'(t) = 4y(t) - 3x(t)$, $y'(t) = y(t) - x(t)$, with $x(0) = -1$ and $y(0) = 0$
24. $x'(t) = 4y(t) - 3x(t) + 5$, $y'(t) = y(t) - x(t) + 1$, with $x(0) = 0$ and $y(0) = 2$
25. $x'(t) = 8y(t) - 3x(t) + 2$, $y'(t) = y(t) - x(t) - 1$, with $x(0) = 4$ and $y(0) = 2$

11.10   Convolution

Let $F(s)$ and $G(s)$ denote the transforms of $f(t)$ and $g(t)$, respectively. Then the inverse of the product $F(s)G(s)$ is given by the function $h(t) = (f * g)(t)$ and is called the *convolution* of $f(t)$ and $g(t)$ and can be regarded as a generalized product of $f(t)$ and $g(t)$. Convolution will assist us in solving integral equations.

> **Theorem 11.24 (Convolution Theorem)**   *Let $F(s)$ and $G(s)$ denote the Laplace transforms of $f(t)$ and $g(t)$, respectively. Then the product given by $H(s) = F(s)G(s)$ is the Laplace transformation of the convolution of f and g and is denoted by $h(t) = (f * g)(t)$, and has the integral representation*

(1) $h(t) = (f * g)(t) = \int_0^t f(\tau)g(t - \tau)\, d\tau$, or

(2) $h(t) = (g * f)(t) = \int_0^t g(\tau)f(t - \tau)\, d\tau$.

   Proof   The following proof is given for the special case when s is a real number. The general case is covered in advanced texts. Using the dummy variables σ and τ and the integrals defining the transforms, we can express their product as

(3) $F(s)G(s) = \left[\int_0^\infty f(\sigma)e^{-s\sigma}\, d\sigma \right]\left[\int_0^\infty g(\tau)e^{-s\tau}\, d\tau \right]$.

The product of integrals in equation (3) can be written as an iterated integral:

(4) $F(s)G(s) = \int_0^\infty \left[\int_0^\infty f(\sigma)e^{-s(\sigma+\tau)}\, d\sigma \right]g(\tau)\, d\tau$.

Hold τ fixed and use the change of variables $t = \sigma + \tau$ and $dt = d\sigma$, then the inside integral in equation (4) is rewritten to obtain

(5) $F(s)G(s) = \int_0^\infty \left[\int_\tau^\infty f(t - \tau)e^{-st}\, dt \right]g(\tau)\, d\tau$

$= \int_0^\infty \left[\int_\tau^\infty f(t - \tau)g(\tau)e^{-st}\, dt \right]d\tau$.

The iterated integral in equation (5) is taken over the wedge-shaped region in the (t, τ) plane indicated in Figure 11.28. The order of integration can be reversed to yield:

(6) $F(s)G(s) = \int_0^\infty \left[\int_0^t f(t - \tau)g(\tau)e^{-st}\, d\tau \right]dt$.

The last expression can be written as

(7) $F(s)G(s) = \int_0^\infty \left[\int_0^t f(t - \tau)g(\tau) \, d\tau \right] e^{-st} \, dt$

$= \mathcal{L}^{-1}\left(\int_0^t f(t - \tau)g(\tau) \, d\tau \right),$

which establishes the proof of equation (2). Since we can interchange the role of the functions $f(t)$ and $g(t)$, equation (1) follows immediately.

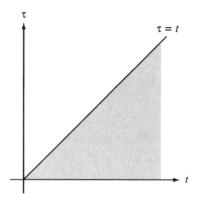

FIGURE 11.28   The region of integration in the convolution theorem.

TABLE 11.4   Properties of Convolution

Commutative	$f * g = g * f$
Distributive	$f * (g + h) = f * g + f * h$
Associative	$(f * g) * h = f * (g * h)$
Zero	$f * 0 = 0$

EXAMPLE 11.29   Show that $\mathcal{L}^{-1}\left(\dfrac{2s}{(s^2 + 1)^2} \right) = t \sin t.$

Solution   Let $F(s) = \dfrac{1}{s^2 + 1}$, $G(s) = \dfrac{2s}{s^2 + 1}$, $f(t) = \sin t$, $g(t) = 2 \cos t$, respectively. Applying the convolution theorem we get

$\mathcal{L}^{-1}\left(\dfrac{1}{s^2 + 1} \dfrac{2s}{s^2 + 1} \right) = \mathcal{L}^{-1}(F(s)G(s)) = \int_0^t 2 \sin(t - \tau) \cos \tau \, d\tau$

$= \int_0^t [2 \sin t \cos^2\tau - 2 \cos t \sin \tau \cos \tau] \, d\tau$

$= \sin t \, (\tau + \sin \tau \cos \tau) - \cos t \sin^2\tau \, \Big|_{\tau=0}^{\tau=t}$

$= t \sin t + \sin^2 t \cos t - \cos t \sin^2 t = t \sin t.$

EXAMPLE 11.30   Use the convolution theorem to solve the integral equation

$$f(t) = 2 \cos t - \int_0^t (t - \tau) f(\tau) \, d\tau.$$

Solution   Letting $F(s) = \mathcal{L}(f(t))$ and using $\mathcal{L}(t) = \dfrac{1}{s^2}$ in the convolution theorem we obtain

$$F(s) = \frac{2s}{s^2 + 1} - \frac{1}{s^2} F(s).$$

Solving for $F(s)$ we get

$$F(s) = \frac{2s^3}{(s^2 + 1)^2} = \frac{2s}{s^2 + 1} - \frac{2s}{(s^2 + 1)^2},$$

and the solution is

$$f(t) = 2 \cos t - t \sin t.$$

Engineers and physicists sometimes consider forces that produce large effects that are applied over a very short time interval. The force acting at the time an earthquake starts is an example. This leads to the idea of a *unit impulse function* $\delta(t)$. Consider the small positive constant a, then the function $\delta_a(t)$ is defined by

$$(8) \quad \delta_a(t) = \begin{cases} \dfrac{1}{a} & \text{for } 0 < t < a, \\ 0 & \text{otherwise.} \end{cases}$$

The unit impulse function is obtained by letting the interval in equation (8) go to zero, i.e.,

$$(9) \quad \delta(t) = \lim_{a \to 0} \delta_a(t).$$

Figure 11.29 shows the graph of $\delta_a(t)$ for $a = 10, 40,$ and 100. Although $\delta(t)$ is called the *Dirac delta function*, it is not an ordinary function. To be precise it is a distribution, and the theory of distributions permits manipulations of $\delta(t)$ as though it were a function. For our work, we will treat $\delta(t)$ as a function and investigate its properties.

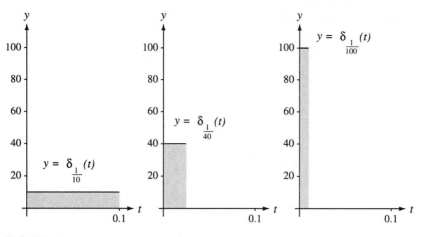

FIGURE 11.29   Graphs of $y = \delta_a(t)$ for $a = 10$, 40, and 100.

EXAMPLE 11.31   Show that $\mathscr{L}(\delta(t)) = 1$.

Solution   By definition, the Laplace transform of $\delta_a(t)$ is

(10) $\mathscr{L}(\delta_a(t)) = \int_0^\infty \delta_a(t)e^{-st}\,dt = \int_0^a \frac{1}{a}e^{-st}\,dt = \frac{1 - e^{-sa}}{sa}$.

Letting $a \to 0$ in equation (10), and using L'Hôpital's rule, we obtain

(11) $\mathscr{L}(\delta(t)) = \lim_{a \to 0} \mathscr{L}(\delta_a(t)) = \lim_{a \to 0} \frac{1 - e^{-sa}}{sa} = \lim_{a \to 0} \frac{se^{-sa}}{s} = 1$.

We now turn our attention to the unit impulse function. First, consider the function $f_a(t)$ obtained by integrating $\delta_a(t)$:

(12) $f_a(t) = \int_0^t \delta_a(\tau)\,d\tau = \begin{cases} 0 & \text{for } t < 0, \\ \dfrac{t}{a} & \text{for } 0 \le t \le a, \\ 1 & \text{for } a < t. \end{cases}$

Then it is easy to see that $U_0(t) = \lim_{a \to 0} f_a(t)$ (see Figure 11.30).

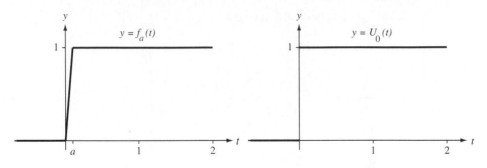

FIGURE 11.30 The integral of $\delta_a(t)$ is $f_a(t)$, which becomes $U_0(t)$ when $a \to 0$.

The response of a system to the unit impulse function is illustrated in the next example.

EXAMPLE 11.32 Solve the initial value problem

$$y''(t) + 4y'(t) + 13y(t) = 3\delta(t) \quad \text{with } y(0) = 0 \text{ and } y'(0^-) = 0.$$

Solution Taking transforms results in $(s^2 + 4s + 13)Y(s) = 3\mathcal{L}(\delta(t)) = 3$, so that

$$Y(s) = \frac{3}{s^2 + 4s + 13} = \frac{3}{(s + 2)^2 + 3^2}.$$

and the solution is

$$y(t) = e^{-2t} \sin 3t.$$

Remark The condition $y'(0^-) = 0$ is not satisfied by the "solution" $y(t)$. Recall that all solutions using the Laplace transform are to be considered zero for values of $t < 0$. Hence the graph of $y(t)$ is given in Figure 11.31. We see that $y'(t)$ has a jump discontinuity of magnitude $+3$ at the origin. This happens because either $y(t)$ or $y'(t)$ must have a jump discontinuity at the origin whenever the Dirac delta function occurs as part of the input or driving function.

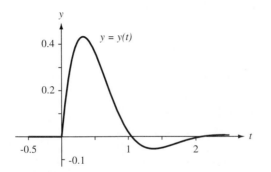

FIGURE 11.31 The solution $y = y(t)$ to Example 11.32.

The convolution method can be used to solve initial value problems. The tedious mechanical details of problem solving can be facilitated with computer software such as Maple™, Matlab™, or Mathematica™.

Theorem 11.25 (IVP Convolution Method) *The unique solution to the initial value problem*

(13) $ay''(t) + by'(t) + cy(t) = g(t)$ with $y(0) = y_0$ and $y'(0) = y_1$

is given by

(14) $y(t) = u(t) + (h * g)(t)$,

where $u(t)$ is the solution to the homogeneous equation

(15) $au''(t) + bu'(t) + cu(t) = 0$ with $u(0) = y_0$ and $u'(0) = y_1$,

and $h(t)$ has the Laplace transform given by $H(s) = \dfrac{1}{as^2 + bs + c}$.

Proof   The particular solution is found by solving the equation

(16) $av''(t) + bv'(t) + cv(t) = g(t)$ with $v(0) = 0$ and $v'(0) = 0$.

Taking the Laplace transform of both sides of equation (16) produces

(17) $as^2V(s) + bsV(s) + cV(s) = G(s)$.

Solving for $V(s)$ in equation (17) yields $V(s) = \dfrac{1}{as^2 + bs + c} G(s)$. If we set $H(s) = \dfrac{1}{as^2 + bs + c}$, then $V(s) = H(s)\,G(s)$ and the particular solution is given by the convolution

(18) $v(t) = (h * g)(t)$.

The general solution is $y(t) = u(t) + v(t) = u(t) + (h * g)(t)$. To verify that the initial conditions are met we compute

$y(0) = u(0) + v(0) = y_0 + 0 = y_0$,

and

$y'(0) = u'(0) + v'(0) = y_1 + 0 = y_1$,

and the proof of the theorem is complete.

EXAMPLE 11.33   Use the convolution method to solve the IVP

$y''(t) + y(t) = \tan t$ with $y(0) = 1$ and $y'(0) = 2$.

Solution   First solve $u''(t) + u(t) = 0$ with $u(0) = 1$ and $u'(0) = 2$. Taking the Laplace transform yields $s^2 U(s) - s - 2 + U(s) = 0$. Solving for $U(s)$ gives $U(s) = \dfrac{s + 2}{s^2 + 1}$ and it follows that

$$u(t) = \cos t + 2 \sin t.$$

Second, observe that $H(s) = \dfrac{1}{s^2 + 1}$ and $h(t) = \sin t$ so that

$$v(t) = (h * g)(t) = \int_0^t \sin(t - s) \tan(s) \, ds$$

$$= \left[\cos(t) \ln\left[\frac{\cos s}{1 + \sin s}\right] - \sin(t - s)\right]\Big|_{s=0}^{s=t}$$

$$= \cos(t) \ln\left[\frac{\cos t}{1 + \sin t}\right] + \sin(t)$$

Therefore, the solution is

$$y(t) = u(t) + v(t) = \cos t + 3 \sin t + \cos(t) \ln\left[\frac{\cos t}{1 + \sin t}\right].$$

EXERCISES FOR SECTION 11.10

For exercises 1–6, find the indicated convolution.

1. $t * t$ **2.** $t * \sin t$

3. $e^t * e^{2t}$ **4.** $\sin t * \sin 2t$

For exercises 5–8, use convolution to find $\mathcal{L}^{-1}(F(s))$.

5. $F(s) = \dfrac{2}{(s - 1)(s - 2)}$ **6.** $F(s) = \dfrac{6}{s^3}$

7. $F(s) = \dfrac{1}{s(s^2 + 1)}$ **8.** $F(s) = \dfrac{s}{(s^2 + 1)(s^2 + 4)}$

9. Prove the distributive law for convolution: $f * (g + h) = f * g + f * g$.

10. Use the convolution theorem and mathematical induction to show that

$$\mathcal{L}^{-1}\left(\frac{1}{(s - a)^n}\right) = \frac{1}{(n - 1)!} t^{n-1} e^{at}.$$

11. Find $\mathcal{L}^{-1}\left(\dfrac{s}{s - 1}\right)$.   **12.** Find $\mathcal{L}^{-1}\left(\dfrac{s^2}{s^2 + 1}\right)$.

13. Use the convolution theorem to solve the initial value problem:

$$y''(t) + y(t) = 2 \sin t \quad \text{with } y(0) = 0 \text{ and } y'(0) = 0.$$

14. Use the convolution theorem to show that the solution to the initial value problem
$y''(t) + \omega^2 y(t) = f(t)$, with $y(0) = 0$ and $y'(0) = 0$ is

$$y(t) = \frac{1}{\omega} \int_0^t f(\tau) \sin[\omega(t - \tau)] \, d\tau.$$

15. Find $\mathcal{L}\left(\int_0^t e^{-\tau} \cos(t - \tau) \, d\tau \right)$.   **16.** Find $\mathcal{L}\left(\int_0^t (t - \tau)^2 e^\tau \, d\tau \right)$.

17. Let $F(s) = \mathcal{L}(f(t))$. Use convolution to show that

$$\mathcal{L}^{-1}\left(\frac{F(s)}{s} \right) = \int_0^t f(\tau) \, d\tau.$$

For exercises 18–21, use the convolution theorem to solve the integral equation.

18. $f(t) + 4 \int_0^t (t - \tau) f(\tau) \, d\tau = 2$ **19.** $f(t) = e^t + \int_0^t e^{t-\tau} f(\tau) \, d\tau$

20. $f(t) = 2t + \int_0^t \sin(t - \tau) f(\tau) \, d\tau$ **21.** $6f(t) = 2t^3 + \int_0^t (t - \tau)^3 f(\tau) \, d\tau$

For exercises 22–25, solve the initial value problem.

22. $y''(t) - 2y'(t) + 5y(t) = 2\delta(t)$, with $y(0) = 0$ and $y'(0) = 0$
23. $y''(t) + 2y'(t) + y(t) = \delta(t)$, with $y(0) = 0$ and $y'(0) = 0$
24. $y''(t) + 4y'(t) + 3y(t) = 2\delta(t)$, with $y(0) = 0$ and $y'(0) = 0$
25. $y''(t) + 4y'(t) + 3y(t) = 2\delta(t - 1)$, with $y(0) = 0$ and $y'(0) = 0$

For exercises 26–29, use the IVP convolution method to solve the initial value problem.

26. $y''(t) - 2y'(t) + 5y(t) = 8 \exp(-t)$ with $y(0) = 1$ and $y'(0) = 2$.
27. $y''(t) + 2y'(t) + y(t) = t^4$ with $y(0) = 1$ and $y'(0) = 2$.
28. $y''(t) + 4y'(t) + 3y(t) = 24t^2 \exp(-t)$ with $y(0) = 1$ and $y'(0) = 2$.
29. $y''(t) + 4y'(t) + 3y(t) = 2t \exp(-t)$ with $y(0) = 1$ and $y'(0) = 2$.

Appendix A
Undergraduate Student
Research Projects

The following list of journal articles and books is appropriate for undergraduate students. For this reason, several advanced and graduate-level textbooks have been omitted. Journal references include those accessible to students, such as, *American Mathematical Monthly*, *Mathematics and Computer Education*, and *Two Year College Mathematics Journal*. Instructors should encourage their students regarding research in the mathematical literature. The following list of topics is a starting point for either independent or group research projects.

Analytic continuation: 4, 19, 46, 51, 52, 93, 106, 128, 129, 141, 145, and 166
Analytic function: 21, 39, 62, 72, 86, 155, and 161
Bieberbach conjecture: 49, 73, 108, 148, and 189
Bilinear transformation: 12, 23, 24, 30, 36, and 43
Cauchy integral formula: 13, 59, 107, 110, 118, 119, and 187
Cauchy-Riemann equations: 21, 39, 62, 72, 86, 155, and 161
Chaos: 11, 53, 54, 55, 57, 58, 142, and 168
Computer graphics: 33, 34, 109, and 146
Computer technology: 25, 28, 33, 34, 41, 57, 90, 92, 109, 110, 111, 120, 123, 130, 131, 133, 140, 146, 152, 160, 162, 174, and 185
Conformal mapping: 33, 34, 35, 37, 41, 47, 48, 75, 92, 93, 96, 130, 136, 146, 154, 159, 164, 176, 180, and 182
Construction of a regular pentagon: 114
Contour integral: 5, 16, 81, 82, and 157
Curvature: 12
DeMoivre's formula: 103
Dirichlet problem: 70, 71, 76, 77, 85, 98, 135, and 138
Dynamical systems: 53, 54, 55, 58, and 143
Euler's formula: 169
Fluid flow: 37, 46, 91, 98, 124, 141, 145, 158, and 166
Fourier transform: 15, 17, 69, 100, 149, and 159
Fractals: 7, 8, 9, 11, 55, 57, 58, 78, 84, 101, 125, 126, 134, 139, 143, 167, 175, and 188
Fundamental theorem of algebra: 6, 18, 29, 38, 60, 66, 150, 151, 170, and 184
Geometry: 8, 26, 35, 78, 99, 114, 121, 123, 125, and 160
Harmonic function: 2, 14, 28, 61, 69, 70, 71, 76, 77, 85, 98, 111, 113, 131, 135, 138, 158, and 165
History: 87, 105, and 179
Infinite products: 4, 19, 51, 129, 145, and 181
Joukowski transformation: 37, 46, 91, 98, 124, 141, 145, 158, and 166
Julia set: 144 and 177
Laplace transform: 17, 40, 69, 129, 149, 159, and 186
Liouville's theorem: 117
Mandelbrot set: 31, 45, 56, 74, 125, 126, and 177
Möbius transformation: 12, 23, 24, 30, 36, and 43
Morera's theorem: 163
Partial fractions: 10 and 63
Poisson integral formula: 115
Polya vector field: 25, 26, 27, and 83
Pythagorean triples: 94 and 97
Quaternions: 1, 132, 147, and 173

Bibliography

1. Altmann, Simon L. (1989), "Hamilton, Rodrigues, and the Quaternion Scandal," *Math. Mag.*, V. 62, No. 5., pp. 291–307.
2. Axler, Sheldon (1986), "Harmonic Functions from a Complex Analysis Viewpoint," *Am. Math. M.*, V. 93, No. 4, pp. 246–258.
3. Azarnia, Nozar (1987), "A Classroom Note: on Elementary Use of Complex Numbers," Math. and Comp. Ed., Vol 21, No. 2, pp. 135–136.
4. Bak, Joseph, and Donald J. Newman (1982), *Complex Analysis*, New York: Springer-Verlag.
5. Baker, I. N., and P. J. Rippon (1985), "A Note on Complex Integration," *Am. Math. M.*, V. 92, No. 7, pp. 501–504.
6. Baker, John A. (1991), "Plane Curves, Polar Coordinates and Winding Numbers," *Math. Mag.*, V. 64, No. 2, pp. 75–91.
7. Bannon, Thomas J. (1991), "Fractals and Transformations," *Math. Teach.*, V. 81, No. 3, pp. 178–185.
8. Barcellos, Anthony (1984), "The Fractal Geometry of Mandelbrot," *T. Y. C. Math. J.*, V. 15, No. 2, pp. 98–114.
9. Barnsley, Michael Fielding (1988), *Fractals Everywhere*, Boston: Academic Press.
10. Barry, P. D., and D. J. Hurley (1991), "On Series of Partial Fractions," *Am. Math. M.*, V. 98, No. 3, pp. 240–242.
11. Barton, Ray (1990), "Chaos and Fractals," *Math. Teach.*, V. 83, No. 7, pp. 524–529.
12. Beardon, Alan F. (1987), "Curvature, Circles and Conformal Maps," *Am. Math. M.*, V. 94, No. 1, pp. 48–53.
13. Berresford, Geoffrey C. (1981), "Cauchy's Theorem," *Am. Math. M.*, V. 88, No. 10, pp. 741–742.
14. Bivens, Irl C. (1992), "When Do Orthogonal Families of Curves Possess a Complex Potential?" *Math. Mag.*, V. 65, No. 4, pp. 226–235.
15. Bleick, W. E. (1956), "Fourier Analysis of Engine Unbalance by Contour Integration," *Am. Math. M.*, V. 63, No. 7, pp. 466–472.
16. Boas, L. Mary, and R. P. Boas (1985), "Simplification of Some Contour Integrals," *Am. Math. M.*, V. 92, No. 3, pp. 212–214.
17. Boas, R. P. (1962), "Inversion of Fourier and Laplace Transforms," *Am. Math. M.*, V. 69, No. 10, pp. 955–960.
18. Boas, R. P. (1964), "Yet Another Proof of the Fundamental Theorem of Algebra," *Am. Math. M.*, V. 71, No. 2, p. 180.
19. Boas, R. P. (1987), *Invitation to Complex Analysis*, New York: Random House.
20. Boas, R. P. (1987), "Selected Topics from Polya's Work in Complex Analysis," *Math. Mag.*, V. 60, No. 5, pp. 271–274.
21. Boas, R. P. (1989), "When is a C^∞ Function Analytic?" *Math. Intell.*, V. 11, No. 4, pp. 34–37.
22. Boas, R. P., and Lowell Schoenfeld (1966), "Indefinite Integration by Residues," *SIAM Review*, V. 8, No. 2, pp. 173–183.
23. Booker, T. Hoy (1989), "Bilinear Basics," *Math. Mag.*, V. 62, No. 4, pp. 262–267.
24. Boyd, James N. (1985), "A Property of Inversion in Polar Coordinates," *Math. Teach.*, V. 78, No. 1, pp. 60–61.
25. Braden, Bart (1985), "Picturing Functions of a Complex Variable," *Coll. Math. J.*, V. 16, pp. 63–72.
26. Braden, Bart (1987), "Polya's Geometric Picture of Complex Contour Integrals," *Math. Mag.*, V. 60, No. 5, pp. 321–327.
27. Braden, Bart (1991), "The Vector Field Approach in Complex Analysis," in *Visualization in Teaching and Learning Mathematics*, Providence, R.I.: Math. Assoc. of Amer., pp. 191–196.

28. Brandt, Siegmund, and Hermann Schneider (1976), ''Computer-Drawn Field Lines and Potential Surfaces for a Wide Range of Field Configurations,'' *Am. J. Phy.*, V. 44, No. 12, pp. 1160–1171.

29. Brenner, J. L., and R. C. Lyndon (1981), ''Proof of the Fundamental Theorem of Algebra,'' *Am. Math. M.*, V. 88, No. 4, pp. 253–256.

30. Brickman, Louis (1993), ''The Symmetry Principle for Möbius Transformations,'' *Am. Math. M.*, V. 100, No. 8, pp. 781–782.

31. Bridger, Mark (1988), ''Looking at the Mandelbrot Set,'' *Coll. Math. J.*, V. 19, No. 4, pp. 353–363.

32. Brown, Robert F., and Robert E. Greene (1994), ''An Interior Fixed Point Property of the Disk,'' *Am. Math. M.*, V. 101, No. 1, pp. 39–47.

33. Bruch, John C., and Roger C. Wood (1972), ''Teaching Complex Variables with an Interactive Computer System,'' *IEEE Trans. on Ed.*, E-15, No. 1, pp. 73–80.

34. Bruch, John C. (1975), ''The Use of Interactive Computer Graphics in the Conformal Mapping Area,'' in *Computers and Graphics*, V. 1, pp. 361–374.

35. Bryngdahl, Olof (1974), ''Geometrical Transformations in Optics,'' *J. of the Optical Soc. of Am.*, V. 64, No. 8, pp. 1092–1099.

36. Budden, F. J. (1969), ''Transformation Geometry in the Plane by Complex Number Methods,'' *Math. Gazette*, V. 53, No. 383, pp. 19–31.

37. Burlington, R. S. (1940), ''On the Use of Conformal Mapping in Shaping Wing Profiles,'' *Am. Math. M.*, V. 47, No. 6, pp. 362–373.

38. Campbell, Paul J. (1967), ''The Fundamental Theorem of Algebra,'' *Pi Mu Epsilon J.*, V. 4, No. 6, pp. 243–247.

39. Cater, F. S. (1984), ''Differentiable, Nowhere Analytic Functions,'' *Am. Math. M.*, V. 91, No. 10, pp. 618–624.

40. Chakrabarti, A. (1990), ''A note on the inversion of Laplace Transforms,'' *Int. J. Math. Educ. Sci. Tech.*, V. 21, No. 2, 325–342.

41. Chakravarthy, Sukumar, and Dale Anderson (1979), ''Numerical Conformal Mapping,'' *Math. of Comp.*, V. 33, No. 147, pp. 953–969.

42. Churchill, Ruel Vance, James W. Brown, and Roger F. Verhey (1990), *Complex Variables and Applications*, 5th ed., N.Y.: McGraw-Hill.

43. Cohen, Martin P. (1983), ''Inversion in a Circle: A Different Kind of Transformation,'' *Math. Teach.*, V. 86, No. 8, pp. 620–623.

44. Colwell, Peter, and Jerold C. Mathews (1973), *Introduction to Complex Variables*, Columbus, Ohio: Merrill.

45. Crowe, W. D., R. Hasson, P. J. Rippon, and P. E. D. Strain-Clark (1989), ''On the Structure of the Mandelbar Set,'' *Nonlinearity*, V. 2, pp. 541–553.

46. Cunningham, John (1965), *Complex Variable Methods in Science and Technology*, London, New York: Van Nostrand.

47. D'Angelo, John P. (1984), ''Mapping Theorems in Complex Analysis,'' *Am. Math. M.*, V. 91, No. 7, pp. 413–414.

48. D'Appen, Heinz (1987), ''Wind-Tunnel Wall Corrections on a Two-Dimensional Plate by Conformal Mapping.'' *AIAA J.*, V. 25, pp. 1527–3015.

49. DeBranges, Louis (1985), ''A Proof of the Bieberbach Conjecture,'' *Acta Mathematica*, V. 154, pp. 137–152.

50. Denton, Brian (1985), ''Roots of unity revisited!'' *Math. Gazette*, V. 69, No. 447, pp. 17–20.

51. DePree, John D., and Charles C. Oehring (1969), *Elements of Complex Analysis*, Reading, Mass.: Addison-Wesley Pub. Co.

52. Derrick, William R. (1984), *Complex Analysis and Applications*, 2nd ed., Belmont, Calif.: Wadsworth Pub. Co.

53. Devaney, Robert L. (1986), *An Introduction To Chaotic Dynamical Systems*, Menlo Park, Calif.: Benjamin/Cummings.

54. Devaney, Robert L. (1987), ''Chaotic Bursts in Nonlinear Dynamical Systems,'' *Science*, V. 235, pp. 342–345.

55. Devaney, Robert L. (1990), *Chaos, Fractals and Dynamics*: *Computer Experiments in Mathematics*, New York: Addison-Wesley Pub. Co.

56. Devaney, Robert L. (1991), ''The Orbit Diagram and the Mandelbrot Set,'' *Coll. Math. J.*, V. 22, No. 1, pp. 23–38.

57. Devaney, Robert L., and Linda Keen (Editors) (1989), *Chaos and Fractals*: *The Mathematics Behind the Computer Graphics*, Providence R.I.: Amer. Math. Soc.

58. Devaney, Robert L., and Marilyn B. Durkin (1991), ''The Exploding Exponential and Other Chaotic Bursts in Complex Dynamics,'' *Am. Math. M.*, V. 98, No. 3, pp. 217–233.

59. Dixon, John D. (1971), ''A Brief Proof of Cauchy's Integral Theorem,'' *Proc. Am. Math. Soc.*, V. 29, No. 3, pp. 625–626.

60. Eaton, J. E. (1960), ''The Fundamental Theorem of Algebra,'' *Am. Math. M.*, V. 67, No. 6, pp. 578–579.

61. Edgar, Gerald A., and Lee A. Rubel (1990), ''The Harmonic Conjugate of an Algebraic Function,'' *Am. Math. M.*, V. 97, pp. 165–166.

62. Eidswick, Jack A. (1975), ''Alternatives to Taylor's Theorem in Proving Analyticity,'' *Am. Math. M.*, V. 82, No. 9, pp. 929–931.

63. Eustice, Dan, and M. S. Klamkin (1979), ''On the Coefficients of a Partial Fraction Decomposition,'' *Am. Math. M.*, V. 86, No. 6, pp. 478–480.

64. Evard, J.-Cl., and F. Jafari (1992), ''A Complex Rolle's Theorem,'' *Am. Math. M.*, V. 99, No. 2, 858–861.

65. Farnsworth, David (1983), ''Measuring Complex Roots,'' *Math. and Comp. Ed.*, V. 17, No. 3, pp. 191–194.

66. Fefferman, Charles (1967), ''An Easy Proof of the Fundamental Theorem of Algebra,'' *Am. Math. M.*, V. 74, No. 7, pp. 854–855.

67. Fettis, Henry E. (1976), ''Complex Roots of $\sin(z) = az$, $\cos(z) = az$ and $\cosh(z) = az$,'' *Math. of Comp.*, V. 30, No. 135, pp. 541–545.

68. Filaseta, Michael (1990), ''Rouche's Theorem for Polynomials,'' *Am. Math. M.*, V. 97, No. 9, pp. 834–835.

69. Fisher, Stephen D. (1990), *Complex Variables*, 2nd ed., Monterey, Calif.: Wadsworth & Brooks/Cole.

70. Flanigan, Francis J. (1972), *Complex Variables*: *Harmonic and Analytic Functions*, Boston: Allyn and Bacon.

71. Flanigan, Francis J. (1973), ''Some Half-Plane Dirichlet Problems: A Bare Hands Approach,'' *Am. Math. M.*, V. 83, No. 1, pp. 59–61.

72. Folland, G. B. (1986), ''On Characterizations of Analyticity,'' *Am. Math. M.*, V. 93, No. 8, pp. 640–641.

73. Fomenko, O. M., and G. V. Kuz'mina, (1986), ''The Last 100 Days of the Bieberbach Conjecture,'' *Math. Intell.*, V. 8, No. 1, pp. 40–47.

74. Frantz, Marny, and Sylvia Lazarnick (1991), ''The Mandelbrot Set in the Classroom,'' *Math. Teach.*, V. 84, No. 3, pp. 173–177.

75. Frederick, Carl, and Eric L. Schwartz (1990), ''Conformal Image Warping,'' *IEEE Computer Graphics and Applications*, V. 10, pp. 54–61.

76. Fuchs, Watson (1993), *Complex Variables and Introduction*, New York: Marcel Dekker, Inc.

77. Garding, Lars (1979), ''The Dirichlet Problem,'' *Math. Intell.*, V. 2, No. 1, pp. 43–53.

78. Gilbert, William J., (1982), ''Fractal Geometry Derived from Complex Bases,'' *Math. Intell.*, V. 4, pp. 78–86.

79. Gilbert, William J. (1984), ''Arithmetic in Complex Bases,'' *Math. Mag.*, V. 57, No. 2, pp. 77–81.

80. Gilder, John (1984), ''On $\tan(n\theta)$: Exercises in Induction,'' *Math. Gazette*, V. 68, No. 445, pp. 208–210.

81. Glaister, P. (1991), ''A Method for Determining Some Integrals,'' *Math. and Comp. Ed.*, V. 25, No. 1, pp. 31–32.

82. Gluchoff, Alan D. (1991), ''A Simple Interpretation of the Complex Contour Integral,'' *Am. Math. M.*, V. 98, No. 10, pp. 641–644.

83. Gluchoff, Alan D. (1993), ''Complex Power Series—a Vector Field Visualization,'' *Math. Mag.*, V. 66, No. 3, pp. 189–191.

84. Goffinet, Daniel (1991), ''Number Systems With a Complex Base: A Fractal Tool for Teaching Topology,'' *Am. Math. M.*, V. 98, No. 3, pp. 249–255.

85. Goulet, John (1983), ''The Dirichlet Problem: A Mathematical Development,'' *Pi Mu Epsilon J.*, V. 7, No. 8, pp. 502–511.

86. Gray, J. D., and S. A. Morris (1978), ''When is a Function that Satisfies the Cauchy-Riemann Equations Analytic?'' *Am. Math. M.*, V. 85, No. 4, pp. 246–256.

87. Green, D. R. (1976), ''The Historical Development of Complex Numbers,'' *Math. Gazette*, V. 60, No. 412, pp. 99–107.

88. Greenleaf, Frederick P. (1972), *Introduction to Complex Variables*, Philadelphia: Saunders.

89. Greenstein, David S. (1963), ''On the Modulus of an Integral,'' *Am. Math. M.*, V. 70, No. 8, p. 869.

90. Harding, R. D. (1974), ''Computer Aided Teaching in Applied Mathematics,'' *Int. J. of Math. Ed. in Sci. and Tech.*, V. 5, No. 4, pp. 447–455.

91. Harlow, F. H. (1965), ''Numerical Fluid Dynamics,'' *Am. Math. M.*, V. 72, No. 2, pp. 84–91.

92. Hayes, John K., David K. Kahaner, and Richard G. Kellner (1972), ''An Improved Method for Numerical Conformal Mapping,'' *Math. of Comp.*, V. 26, No. 118, pp. 327–334.

93. Henrici, Peter (1974), *Applied and Computational Complex Analysis*, Vol. I, New York: Wiley.

94. Holden, Herbert L. (1981), ''Applying Complex Arithmetic,'' *T. Y. C. Math. J.*, V. 12, No. 3, pp. 190–194.

95. Holland, Anthony S. B. (1980), *Complex Function Theory*, New York: Elsevier North Holland.

96. Hurst, William (1985), ''Conformal mapping solves intricate problems,'' *Oil & Gas J.*, V. 83, p. 138.

97. Jackson, Margaret (1987), ''Complex Numbers and Pythagorean Triples,'' *Math. Gazette*, V. 71, No. 456, p. 127.

98. Jeffrey, Alan (1992), *Complex Analysis and Applications*, Boca Raton, Fla.: CRC Press.

99. Jones, Gareth A., and David Singerman (1987), *Complex Functions, an Algebraic and Geometric Viewpoint*, New York: Cambridge University Press.

100. Kerley, Lyndell M., and Jeff R. Knisley (1993), ''Complex Vectors and Image Identification,'' *Coll. Math. J.*, V. 24, No. 2, pp. 166–174.

101. Kern, Jane F., and Cherry C. Mauk (1990), ''Exploring Fractals—a Problem-solving Adventure Using Mathematics and Logo,'' *Math. Teach.*, V. 83, No. 3, pp. 179–185.

102. Kim, David, and Robert Travers (1982), ''Those elusive imaginary zeros,'' *Math. Teach.*, V. 75, pp. 62–64.

103. Kimberling, Clark (1987), ''Power of Complex Numbers,'' *Math. Teach.*, V. 80, No. 1, pp. 63–67.

104. Klamkin, M. S., and V. N. Murty (1989), ''Generalizations of a Complex Number Identity,'' *Coll. Math. J.*, V. 20, No. 5, pp. 415–416.

105. Kleiner, Israel (1988), ''Thinking the Unthinkable: The Story of Complex Numbers (with a Moral),'' *Math. Teach.*, V. 81, No. 7, pp. 583–592.

106. Kodaira, Kunihiko (1984), *Introduction to Complex Analysis*, New York: Cambridge University Press.

107. Konnully, A. O. (1967), ''A Proof of the Cauchy Integral Theorem,'' *Am. Math. M.*, V. 74, No. 1, Part I, pp. 54–56.

108. Korevaar, J. (1986), ''Ludwig Bieberbach's Conjecture and its Proof by Louis DeBranges,'' *Am. Math. M.*, V. 93, No. 7, pp. 505–514.

109. Kowalski, Robert, and Helen Skala (1990), ''Determining Roots of Complex Functions with Computer Graphics,'' *Coll. Micro.*, V. VIII, No. 1, pp. 51–54.

110. Kraines, David P., Vivian Y. Kraines, and David Smith (1990), ''The Cauchy Integral Formula,'' *Coll. Math. J.*, V. 21, No. 4, pp. 327–329.

111. Kranc, S. C. (1986), ''Plotting Streamlines and Pathlines on a Microcomputer,'' *Comp. in Ed. Div. of ASEE*, V. VI, No. 3, pp. 20–21.

112. Krantz, Steven G. (1987), ''What Is Several Complex Variables?'' *Am. Math. M.*, Vol 94, No. 3, pp. 236–256.

113. Laitone, E. V. (1977), ''Relation of the Conjugate Harmonic Functions to $f(z)$,'' *Am. Math. M.*, V. 84, No. 4, pp. 281–283.

114. Lambert, Gary E. (1979), ''A 'Complex' Proof for a Geometric Construction of a Regular Pentagon,'' *Math. Teach.*, V. 72, No. 1, pp. 65–66.

115. Lange, Ridgley, and Robert A. Walsh (1985), ''A Heuristic for the Poisson Integral for the Half Plane and some Caveats,'' *Am. Math. M.*, V. 92, No. 5, pp. 356–358.

116. Lee, Sanboh, and J. C. M. Li (1988), ''Summation of Infinite Series Related to Roots of Certain Functions,'' *Int. J. of Math. Ed. in Sci. and Tech.*, V. 19, No. 1, pp. 89–93.

117. Lenard, A. (1986), ''A Note on Liouville's Thoerem,'' *Am. Math. M.*, V. 93, No. 3, pp. 200–201.

118. Loeb, Peter A. (1991), ''A Note on Dixon's Proof of Cauchy's Integral Theorem,'' *Am. Math. M.*, V. 98, No. 3, pp. 242–244.

119. Loeb, Peter A. (1993), ''A Further Simplification of Dixon's Proof of the Cauchy Integral Theorem,'' *Am. Math. M.*, V. 100, No. 7, pp. 680–681.

120. Long, Cliff, and Thomas Hern (1989), ''Graphing the Complex Zeros of Polynomials Using Modulus Surfaces,'' *Coll. Math. J.*, V. 20, No. 2, pp. 98–105.

121. Long, Cliff (1971), ''A Note on the Geometry of Zeros of Polynomials,'' *Math. Mag.*, V. 44, pp. 157–159.

122. Long, Cliff (1972), ''The Quadratic Polynomial and its Zeros,'' *T. Y. C. Math. J.*, V. 3, No. 1, pp. 23–29.

123. Lounesto, Pertti, Risto Mikkola, and Vesa Vierros (1989), ''Geometric Algebra Software for Teaching Complex Numbers, Vectors and Spinors,'' *J. Comp. in Math. and Sci. Teach.*, V. 9, No. 2, pp. 93–105.

124. Mancill, J. D. (1946), ''On the Equation of Joukowski's Aerofoils,'' *Am. Math. M.*, V. 53, No. 3, pp. 147–149.

125. Mandelbrot, Benoit B. (1982), *The Fractal Geometry of Nature*, San Francisco: W. H. Freeman.

126. Mandelbrot, Benoit B. (1983), ''Self-Inverse Fractals Osculated by Sigma-Discs and the Limit Sets of Inversion Groups,'' *Math. Intell.*, V. 5, No. 2, pp. 9–17.

127. Marden, Morris (1985), ''The Search for a Rolle's Theorem in the Complex Plane,'' *Am. Math. M.*, V. 92, No. 9, pp. 643–650.

128. Markushevich, Aleksei Ivanovich (1967), *Theory of Functions of a Complex Variable*, Vol. III, Englewood Cliffs, N.J.: Prentice-Hall.

129. Marsden, Jerrold E., and Michael J. Hoffman (1987), *Basic Complex Analysis*, 2nd ed., San Francisco: W. H. Freeman.

130. Mastin, C. Wayne (1987), ''Numerical Conformal Mapping,'' *Comp. Meth. in App. Mech. and Eng.*, V. 63, pp. 209–211.

131. Mathews, John H. (1988), ''Using a Symbol Manipulation Program to Construct a Harmonic Conjugate Function,'' *Computers in Education Division of ASEE*, Vol. VIII, No. 3, pp. 67–71.

132. McAllister, Byron L. (1989), ''A Quick Introduction to Quaternions,'' *Pi Mu Epsilon J.*, V. 9, No. 1, pp. 23–25.

133. Mikkola, Risto, and Pertti Lounesto (1983), ''Computer-Aided Vector Algebra,'' *Int. J. Math. Educ. Sci. Tech.*, V. 14, No. 5, pp. 573–578.

134. Miller, Valerie A., and G. Scott Owen (1991), ''Using Fractal Images in the Visualization of Iterative Techniques from Numerical Analysis,'' in *Visualization in Teaching and Learning Mathematics*, Providence, R.I.: Math. Assoc. of Amer., pp. 197–206.

135. Minda, Carl David (1990), ''The Dirichlet Problem for a Disk (Presenting a Simpler Proof of Schwarz' Theorem by Writing Integral in Simpler Form),'' *Am. Math. M.*, V. 97, No. 3, pp. 220–223.

136. Moorer, James A. (1983), ''The Manifold Joys of Conformal Mapping: Applications to Digital Filtering in the Studio,'' *J. of the Audio Engr. Soc.*, V. 31, pp. 826–841.

137. Musser, Gary L. (1978), ''Line Reflections in the Complex Plane-A Billiards Player's Delight,'' *Math. Teach.*, V. 71, No. 1, pp. 60–64.

138. Netuka, Ivan (1980), ''The Dirichlet Problem for Harmonic Functions,'' *Am. Math. M.*, V. 87, pp. 621–622.

139. Nievergelt, Yves (1991), ''Fractals Illustrate the Mathematical Way of Thinking,'' *Coll. Math. J.*, V. 22, No. 1, pp. 60–64.

140. Norton, Alec, and Benjamin Lotto (1984), ''Complex Roots Made Visible,'' *Coll. Math. J.*, V. 15, No. 3, pp. 248–249.

141. Paliouras, John D. (1990), *Complex Variables for Scientists and Engineers*, 2nd ed., New York: Macmillan.

142. Parris, Richard (1991), ''The Root-Finding Route to Chaos,'' *Coll. Math. J.*, V. 22, No. 1, pp. 48–55.

143. Peitgen, H. O., and P. H. Richter (1986), *The Beauty of Fractals, Images of Complex Dynamical Systems*, Berlin, New York: Springer-Verlag.

144. Peitgen, H. O., D. Saupe, and F. V. Haeseler (1984), ''Cayley's Problem and Julia Sets,'' *Math. Intell.*, V. 6, No. 2, pp. 11–20.

145. Pennisi, Louis Legendre (1976), *Elements of Complex Variables*, 2nd ed., New York: Holt, Rinehart and Winston.

146. Piele, Donald T., Morris W. Firebaugh, and Robert Manulik (1977), ''Applications of Conformal Mapping to Potential Theory Through Computer Graphics,'' *Am. Math. M.*, V. 84, No. 9, pp. 677–692.

147. Pineda, Anton, and David C. Arney (1991), ''Extending the Complex Number System,'' *Math. and Comp. Ed.*, V. 25, No. 1, pp. 10–16.

148. Pommerenke, Ch. (1985), ''The Bieberbach Conjecture,'' *Math. Intell.*, V. 7, No. 2, pp. 23–32.

149. Priestley, Hilary A. (1985), *Introduction to Complex Analysis*, New York: Oxford University Press.

150. Redheffer, Raymond M. (1957), ''The Fundamental Theorem of Algebra,'' *Am. Math. M.*, V. 64, No. 8, pp. 582–585.

151. Redheffer, Raymond M. (1964), ''What! Another Note Just on the Fundamental Theorem of Algebra,'' *Am. Math. M.*, V. 71, No. 2, pp. 180–185.

152. Renka, Robert J., and Floyd Vest (1988), ''Graphical Representation of Complex Function,'' *Math. and Comp. Ed.*, V. 22, No. 1, pp. 33–45.

153. Ricardo, Henry J. (1971), ''Summation of Series by the Residue Theorem,'' *Math. Mag.*, V. 44, No. 1, pp. 24–25.

154. Richardson, S. (1989), ''An Identity Arising in a Problem of Conformal Mapping,'' *SIAM Review*, V. 31, pp. 484–485.

155. Roy, S. (1979), ''Cauchy-Riemann Differential Equations in Classical Hydrodynamics,'' *Int. J. of Math. Ed. in Sci. and Tech.*, V. 10, No. 2, p. 291.

156. Rubenfeld, Lester A. (1985), *A First Course in Applied Complex Variables*, New York: Wiley.

157. Sachdeva, Baldev K., and Bertram Ross (1982), ''Evaluation of Certain Real Integrals by Contour Integration,'' *Am. Math. M.*, V. 89, No. 4, pp. 246–249.

158. Sacksteder, Richard C. (1978), ''On Oscillatory Flows,'' *Math. Intell.*, V. 1, No. 1, pp. 45–51.

159. Saff, E. B., and A. D. Snider (1993), *Complex Analysis for Mathematics, Science and Engineering*, Englewood Cliffs, N.J.: Prentice Hall.
160. Shilgalis, Thomas W. (1982), "Geometric Transformations on a Microcomputer," *Math. Teach.*, V. 75, No. 1, pp. 16–19.
161. Snider, Arthur David (1978), "On the Definition of Analyticity," *Int. J. of Math. Ed. in Sci. and Tech.*, V. 9, No. 3, pp. 373–374.
162. Sorkin, Sylvia (1984), "Using an Interactive Computer Program in the Calculus Classroom to Find Complex Roots of Polynomial Equations," *Math. and Comp. Ed.*, V. 18, No. 2, pp. 93–99.
163. Springer, George (1957), "On Morera's Theorem," *Am. Math. M.*, V. 64, No. 5, pp. 323–331.
164. Squire, William (1975), "Computer Implementation of the Schwarz-Christoffel Transformation," *J. of the Franklin Inst.*, V. 299, No. 5, pp. 315–322.
165. Srinivasan, V. K. (1977), "A Note on Harmonic Functions and Harmonic Conjugates," *Int. J. of Math. Ed. in Sci. and Tech.*, V. 8, No. 3, pp. 323–328.
166. Stewart, Ian, and David Tall (1983), *Complex Analysis (The Hitchhiker's Guide to the Plane)*, New York: Cambridge University Press.
167. Straffin, Philip D. (1991), "Newton's Method and Fractal Patterns," *UMAP J.*, V. 12, No. 2, pp. 147–164.
168. Strang, Gilbert (1991), "The Chaotic Search for i," *Coll. Math. J.*, V. 22, No. 1, pp. 3–12.
169. Subramaniam, K. B. (1979), "On Euler's Formula $\exp(i\theta) = \cos(\theta) + i \sin(\theta)$," *Int. J. of Math. Ed. in Sci. and Tech.*, V. 10, No. 2, p. 279.
170. Terkelsen, Frode (1976), "The Fundamental Theorem of Algebra," *Am. Math. M.*, V. 83, No. 8, p. 647.
171. Travers, Robert, and David Kim (1982), "Those Elusive Imaginary Zeros," *Math. Teach.*, V. 75, No. 1, pp. 62–64.
172. Tsarpalias, A. (1989), "A Version of Rouche's Theorem for Continuous Functions," *Am. Math. M.*, V. 96, No. 10, pp. 911–913.
173. Van der Waerden, B. L. (1976), "Hamilton's Discovery of Quaternions," *Math. Mag.*, V. 49, No. 5, pp. 227–234.
174. Vest, Floyd (1985), "Graphing the Complex Roots of a Quadratic Equation," *Coll. Math. J.*, V. 16, No. 4, pp. 257–261.
175. Viczek, Tamas (1988), *Fractal Growth Phenomena*, Teaneck, N.J.: World Scientific.
176. Vincenty, T. (1987), "Conformal Transformations between Dissimilar Plane Coordinate Systems," *Surveying and Mapping*, V. 47, pp. 271–274.
177. Vrscay, Edward R. (1986), "Julia Sets and Mandelbrot-Like Sets Associated with High Order Schroder Rational Iteration Functions: A Computer Assisted Study," *Math. of Comp.*, V. 46, No. 173 pp. 151–169.
178. Wallen, Lawrence J. (1992), "The p-th Root of an Analytic Function," *Math. Mag.*, V. 65, No. 4, pp. 260–261.
179. Walsh, J. L. (1973), "History of the Riemann Mapping Theorem," *Am. Math. M.*, V. 80, No. 3, pp. 270–276.
180. Weaver, Warren (1932), "Conformal Representation, with Applications to Problems in Applied Mathematics," *Am. Math. M.*, V. XXXIX, pp. 448–473.
181. Wermuth, Edgar M. E. (1992), "Some Elementary Properties of Infinite Products," *Am. Math. M.*, V. 99, No. 6, pp. 530–537.
182. Williams, Richard K. (1973), "A Note on Conformality," *Am. Math. M.*, V. 80, No. 3, pp. 299–302.
183. Willson, William Wynne (1970), "An Approach to Complex Numbers," *Math. Gazette*, V. 54, No. 390, pp. 342–346.
184. Wolfenstein, S. (1967), "Proof of the Fundamental Theorem of Algebra," *Am. Math. M.*, V. 74, No. 7, pp. 853–854.
185. Wood, Roger C., and John C. Brunch (1972), "Teaching Complex Variables with an Interactive Computer System," *IEEE Trans. on Ed.*, V. E-15, No. 1, pp. 73–80.

186. Wunch, David A. (1994), *Complex Variables with Applications*, New York: Addison-Wesley Pub. Co.
187. Wyler, Oswald (1965), ''The Cauchy Integral Theorem,'' *Am. Math. M.*, V. 72, No. 1, pp. 50–53.
188. Zobitz, Jennifer (1987), ''Fractals: Math. Monsters,'' *Pi Mu Epsilon J.*, V. 8, No. 7, pp. 425–440.
189. Zorn, Paul (1986), ''The Bieberbach Conjecture: A Famous Unsolved Problem and the Story of de Branges' Surprising Proof,'' *Math. Mag.*, V. 59, No. 3, pp. 131–148.

Answers to Selected Problems

Section 1.1, The Origin of Complex Numbers: page 4

3. $x_1 = -\frac{1}{3}$, $x_2 = -\frac{1}{3}$, $x_3 = \frac{2}{3}$

Section 1.2, The Algebra of Complex Numbers: page 11

1. **(a)** $8 - 6i$ **(c)** $6 - 8i$ **(e)** $2 + 2i$ **(g)** $\frac{-12}{5} + \frac{4}{5}i$ **(i)** $\frac{-27}{5} + \frac{11}{5}i$ **(j)** $-4i$

2. **(a)** 1 **(c)** $\frac{11}{5}$ **(e)** 2 **(g)** $x_1^2 - y_1^2$ **(i)** $x_1^2 + y_1^2$ **(j)** $3x_1^2 y_1 - y_1^3$

8. $(2 + 3i)^{-1} = \dfrac{2}{13} - \dfrac{3}{13}i$, $(7 - 5i)^{-1} = \dfrac{7}{74} + \dfrac{5}{74}i$

Section 1.3, The Geometry of Complex Numbers: page 16

1. **(a)** $6 + 4i$ and $-2 + 2i$ **(c)** $i2\sqrt{3}$ and 2

2. **(a)** $\sqrt{10}$ **(b)** $\sqrt{5}$ **(c)** 2^{25} **(d)** $x^2 + y^2$

4. **(a)** inside **(b)** outside

Section 1.4, The Geometry of Complex Numbers, Continued: page 23

1. **(a)** $-\pi/4$ **(c)** $2\pi/3$ **(e)** $-\pi/3$ **(g)** $-\pi/6$

2. **(a)** $4(\cos \pi + i \sin \pi) = 4e^{i\pi}$ **(c)** $7\left(\cos \dfrac{-\pi}{2} + i \sin \dfrac{-\pi}{2}\right) = 7e^{-i\pi/2}$

 (e) $\dfrac{1}{2}\left(\cos \dfrac{\pi}{2} + i \sin \dfrac{\pi}{2}\right) = \dfrac{1}{2}e^{i\pi/2}$ **(h)** $5(\cos \theta + i \sin \theta) = 5e^{i\theta}$,
 where $\theta = \arctan \frac{4}{3}$

3. **(a)** i **(c)** $4 + i4\sqrt{3}$ **(e)** $\sqrt{2} - i\sqrt{2}$ **(g)** $-e^2$

6. $\operatorname{Arg}(iz) = \operatorname{Arg}(z) + (\pi/2)$, $\operatorname{Arg}(-z) = \operatorname{Arg} z - \pi$, $\operatorname{Arg}(-iz) = \operatorname{Arg}(z) - (\pi/2)$,
 when $z = \sqrt{3} + i$.

11. All z except $z = 0$ and the negative real numbers.

Section 1.5, The Algebra of Complex Numbers, Revisited: page 28

2. **(a)** $-16 - i16\sqrt{3}$ **(c)** -64

5. $\sqrt{2} \cos\left(\dfrac{\pi}{4} + \dfrac{2\pi k}{3}\right) + i\sqrt{2} \sin\left(\dfrac{\pi}{4} + \dfrac{2\pi k}{3}\right)$ for $k = 0, 1, 2$

6. $2 \pm 2i$, $-2 \pm 2i$

8. $2 \cos\left(\dfrac{\pi}{8} + \dfrac{k\pi}{2}\right) + i2 \sin\left(\dfrac{\pi}{8} + \dfrac{k\pi}{2}\right)$ for $k = 0, 1, 2, 3$

11. $1 - 2i$ and $-2 + i$ **14.** $\pm i$ and $2 \pm i$

16. $2\sqrt{3} + 2i$, $-4i$, $-2\sqrt{3} + 2i$

Section 1.6, The Topology of Complex Numbers: page 36

2. **(a)** $z(t) = t + it$ for $0 \le t \le 1$
 (b) $z(t) = t + i$ for $0 \le t \le 1$
 (d) $z(t) = 2 - t + it$ for $0 \le t \le 1$

3. **(a)** $z(t) = t + it^2$ for $0 \le t \le 2$
 (c) $z(t) = 1 - t + i(1 - t)^2$ for $0 \le t \le 1$

4. **(a)** $z(t) = \cos t + i \sin t$ for $-\pi/2 \le t \le \pi/2$
 (b) $z(t) = -\cos t + i \sin t$ for $-\pi/2 \le t \le \pi/2$

5. **(a)** $z(t) = \cos t + i \sin t$ for $0 \le t \le \pi/2$
 (b) $z(t) = \cos t - i \sin t$ for $0 \le t \le 3\pi/2$

7. The sets (a), (d), (e), (f), and (g) are open. **8.** The sets (a)–(f) are connected.

9. The sets (a), (d), (e), and (f) are domains. **10.** The sets (a)–(f) are regions.

11. The set (c) is a closed region. **12.** The sets (c), (e), and (g) are bounded.

Section 2.1, Functions of a Complex Variable: page 40

1. **(a)** $2 - 12i$ **(b)** $1 - 33i$ **2.** **(a)** $74 - 12i$ **(b)** $24 - 4i$

3. **(a)** $6 + \dfrac{i}{2}$ **(b)** $\dfrac{1}{5} - \dfrac{2i}{5}$ **5.** **(a)** $1028 - 984i$

6. $x^2 + 2x + 3y - y^2 + i(-3x - 2xy + 2y)$

9. $r^5\cos 5\theta + r^3\cos 3\theta + i(r^5\sin 5\theta - r^3\sin 3\theta)$

10. **(a)** 1 **(b)** e **(c)** $\dfrac{\sqrt{2}}{2} + i\dfrac{\sqrt{2}}{2}$ **(d)** $\dfrac{e}{\sqrt{2}} + i\dfrac{e}{\sqrt{2}}$ **(e)** $-\tfrac{1}{2} + i\tfrac{1}{2}\sqrt{3}$

   (f) $-e^2$

11. **(a)** 0 **(b)** $\tfrac{1}{2}\ln 2 + \dfrac{i\pi}{4}$ **(c)** $\tfrac{1}{2}\ln 3$ **(d)** $\ln 2 + \dfrac{i\pi}{6}$ **(e)** $\ln 2 + \dfrac{i\pi}{3}$

    (f) $\ln 5 + i\arctan \tfrac{4}{3}$

13. **(a)** 0 **(b)** $\ln \sqrt{2} + \dfrac{i\pi}{4}$ **(c)** $\ln 2 + i\pi$ **(d)** $\ln 2 + \dfrac{i5\pi}{6}$

Section 2.2, Transformations and Linear Mappings: page 47

1. **(a)** the half plane $v > 1 - u$ **2.** the line $u = -4 + 4t$, $v = 6 - 3t$
3. **(a)** the disk $|w - 1 - 5i| < 5$
4. the circle $u = -3 + 3\cos t - 4\sin t$, $v = 8 + 4\cos t + 3\sin t$
5. the triangle with vertices $-5 - 2i$ and $-6, 3 + 2i$

6. $w = f(z) = \dfrac{3 + 2i}{13}z + \dfrac{7 + 9i}{13}$

7. $w = f(z) = -5z + 3 - 2i$ **8.** $w = f(z) = \dfrac{i}{5}z + \dfrac{7 + 4i}{5}$

Section 2.3, The Mappings $w = z^n$ and $w = z^{1/n}$: page 52

3. the region in the upper half plane $\text{Im}(w) > 0$ that lies between the parabolas $u = 4 - (v^2/16)$ and $u = (v^2/4) - 1$
4. the region in the first quadrant that lies under the parabola $u = 4 - (v^2/16)$
7. **(a)** the points that lie to the extreme right or left of the branches of the hyperbola $x^2 - y^2 = 4$
   (b) the points in quadrant I above the hyperbola $xy = 3$, and the points in quadrant III below $xy = 3$
10. **(a)** $\rho > 1$, $\pi/6 < \phi < \pi/4$ **(b)** $1 < \rho < 3$, $0 < \phi < \pi/3$
    (c) $\rho < 2$, $-\pi/2 < \phi < \pi/4$
11. the region in the w plane that lies to the right of the parabola $u = 4 - (v^2/16)$
13. the horizontal strip $1 < v < 8$
15. **(a)** $1 < \rho < 8$, $-3\pi/4 < \phi < \pi$ **(b)** $\rho > 27$, $2\pi < \phi < 9\pi/4$
16. **(a)** $\rho < 8$, $3\pi/4 < \phi < \pi$ **(c)** $\rho < 64$, $3\pi/2 < \phi < 2\pi$
17. **(a)** $\rho > 0$, $-\pi/2 < \phi < \pi/3$ **(c)** $\rho > 0$, $-\pi/4 < \phi < \pi/6$

Section 2.4, Limits and Continuity: page 58

1. $-3 + 5i$ **2.** $(5 + 3i)/2$ **3.** $-4i$ **4.** $1 - 4i$ **5.** $1 - \tfrac{3}{2}i$
10. **(a)** i **(b)** $(-3 + 4i)/5$ **(c)** 1 **(d)** The limit does *not* exist.
12. Yes. The limit is zero.
14. No. Arg z is discontinuous along the negative real axis.
15. **(a)** for all z **(b)** all z except $\pm i$ **(c)** all z except -1 and -2
19. No. The limit does not exist.

Section 2.5, Branches of Functions: page 63

1. **(a)** the sector $\rho > 0$, $\pi/4 < \phi < \pi/2$ **(b)** the sector $\rho > 0$, $5\pi/4 < \phi < 3\pi/2$
   (c) the sector $\rho > 0$, $-\pi/4 < \phi < \pi/4$ **(d)** the sector $\rho > 0$, $3\pi/4 < \phi < 5\pi/4$
4. for example, $f(z) = r^{1/2}\cos(\theta/2) + ir^{1/2}\sin(\theta/2)$, where $r > 0$, $0 < \theta \leq 2\pi$
5. **(b)** the sector $\rho > 0$, $-\pi/3 < \phi \leq \pi/3$
   (c) everywhere except at the origin and at points that lie on the negative x axis

Section 2.6, The Reciprocal Transformation $w = 1/z$: page 70

1. the circle $\left| w + \frac{5}{2}i \right| = \frac{5}{2}$ **3.** the circle $\left| w + \frac{1}{6} \right| = \frac{1}{6}$
5. the circle $\left| w - 1 + i \right| = \sqrt{2}$ **7.** the circle $\left| w - \frac{6}{5} \right| = \frac{4}{5}$

Section 3.1, Differentiable Functions: page 75

1. **(a)** $f'(z) = 15z^2 - 8z + 7$ **(c)** $h'(z) = 3/(z + 2)^2$ for $z \neq -2$
3. Parts (a), (b), (e), (f) are entire, and (c) is entire provided that $g(z) \neq 0$ for all z.
7. **(a)** $-4i$ **(c)** 3 **(e)** -16

Section 3.2, The Cauchy-Riemann Equations: page 83

1. **(c)** $u_x = v_y = -2(y + 1)$ and $u_y = -v_x = -2x$. Then $f'(z) = u_x + iv_x = -2(y + 1) + i2x$.
2. $f'(z) = f''(z) = e^x\cos y + ie^x\sin y$ **3.** $a = 1$ and $b = 2$
4. $f(z) = i/z$ and $f'(z) = -i/z^2$
5. $u_x = v_y = 2e^{2xy}[y \cos(y^2 - x^2) + x \sin(y^2 - x^2)]$,
   $u_y = -v_x = 2e^{2xy}[x \cos(y^2 - x^2) - y \sin(y^2 - x^2)]$
6. **(c)** $u_x = -e^y\sin x$, $v_y = e^y\sin x$, $u_y = e^y\cos x$, $-v_x = -e^y\cos x$. The Cauchy-Riemann equations hold if and only if both $\sin x = 0$ and $\cos x = 0$, which is impossible.
8. $u_x = v_y = 2x$, $u_y = 2y$, and $v_x = 2y$. The Cauchy-Riemann equations hold if and only if $y = 0$.
10. $u_r = \dfrac{2 \ln r}{r} = \dfrac{1}{r} (2 \ln r) = \dfrac{1}{r} v_\theta$, $v_r = \dfrac{2\theta}{r} = \dfrac{-1}{r} (-2\theta) = \dfrac{-1}{r} u_\theta$.

    $f'(z) = e^{-i\theta}[u_r + iv_r] = \dfrac{2}{r} e^{-i\theta}(\ln r + i\theta)$.

Section 3.3, Analytic and Harmonic Functions: page 92

3. f is differentiable only at points on the coordinate axes. f is nowhere analytic.
4. f is differentiable only at points on the circle $|z| = 2$. f is nowhere analytic.
5. **(a)** f is differentiable inside quadrants I and III.   **(b)** f is analytic inside quadrants I and III.
8. $c = -a$ **9.** No. v is *not* harmonic.
10. **(a)** $v(x, y) = x^3 - 3xy^2$ **(c)** $u(x, y) = -e^y\cos x$
12. $U_x(x, y) = u_x(x, -y)$, $U_{xx}(x, y) = u_{xx}(x, -y)$, $U_y(x, y) = -u_y(x, -y)$, $U_{yy}(x, y) = u_{yy}(x, -y)$. Hence, $U_{xx} + U_{yy} = u_{xx} + u_{yy} = 0$.

Section 4.1, Definitions and Basic Theorems for Sequences and Series: page 108

1. **(a)** 0 **(b)** 1 **(c)** i **(d)** i **8.** No.   **12.** Yes.   **16.** Yes.

Section 4.2, Power Series Functions: page 114

4. **(a)** $R = \infty$ **(b)** $R = 0$ **(c)** $R = \frac{3}{5}$

Section 5.1, The Complex Exponential Function: page 130

5. **(b)** Horizontal lines given by the equation $y = k$ are mapped under $f(z) = \exp z$ to rays having angles $\theta = k$ with the positive real axis. In the figure, the horizontal lines are approximately given by $y = -7\pi/8$, $y = \pi/4$, and $y = 3\pi/4$.

Section 5.2, Branches of the Complex Logarithm Function: page 137

1. **(a)** $2 + i\pi/2$ **(c)** $\ln 2 + 3\pi i/4$
2. **(a)** $\ln 3 + i(1 + 2n)\pi$, where n is an integer
   (b) $\ln 4 + i(\frac{1}{2} + 2n)\pi$, where n is an integer
3. **(a)** $(e\sqrt{2}/2)(1 - i)$ **(c)** $1 + i(-1/2 + 2n)\pi$, where n is an integer
8. **(a)** $(2z - 1)/(z^2 - z + 2)$ **(b)** $1 + \log z$
12. **(a)** $\ln(x^2 + y^2) = 2\,\mathrm{Re}(\log z)$. Hence it is harmonic.   **14. (b)** No.
15. **(a)** No. The equation does not hold along the negative x axis.
16. **(a)** $f(z) = \ln |z + 2| + i \arg(z + 2)$, where $0 < \arg(z + 2) < 2\pi$
   (c) $h(z) = \ln |z + 2| + i \arg(z + 2)$, where $-\pi/2 < \arg(z + 2) < 3\pi/2$

Section 5.3, Complex Exponents: page 142

1. **(a)** $\cos(\ln 4) + i \sin(\ln 4)$ **(b)** $e^{\frac{-\pi^2}{4}}[\cos(\pi \ln \sqrt{2}) + i \sin(\pi \ln \sqrt{2})]$
   (c) $\cos 1 + i \sin 1$
2. **(a)** $e^{-(1/2+2n)\pi}$ where n is an integer
   (b) $\cos[\sqrt{2}(1 + 2n)\pi] + i \sin[\sqrt{2}(1 + 2n)\pi]$, where n is an integer
   (c) $\cos(1 + 4n) + i \sin(1 + 4n)$, where n is an integer
4. $(-1)^{3/4} = \dfrac{1 \pm i}{\sqrt{2}}, \dfrac{-1 \pm i}{\sqrt{2}}; (i)^{2/3} = -1, \dfrac{1}{2} \pm i\dfrac{1}{2}\sqrt{3}$
6. $\alpha r^{\alpha-1} \cos(\alpha - 1)\theta + i\alpha r^{\alpha-1} \sin(\alpha - 1)\theta$, where $-\pi < \theta < \pi$
13. No. $1^{a+ib} = e^{a2\pi n}\cos b2\pi n + ie^{a2\pi n}\sin b2\pi n$, where n is an integer

Section 5.4, Trigonometric and Hyperbolic Functions: page 151

9. **(a)** $\cos(1 + i) = \cos 1 \cosh 1 - i \sin 1 \sinh 1$ **(c)** $\sin 2i = i \sinh 2$
   (e) $\tan\left(\dfrac{\pi + 2i}{4}\right) = \dfrac{1 + i \sinh 1}{\cosh 1}$
10. **(a)** $[-\cos(1/z)]/z^2$ **(c)** $2z \sec z^2 \tan z^2$
14. **(a)** $z = (\frac{1}{2} + 2n)\pi \pm 4i$, where n is an integer
   (c) $z = 2\pi n + i$ and $z = (2n + 1)\pi - i$, where n is an integer
23. **(a)** $\sinh(1 + i\pi) = -\sinh 1$ **(c)** $\cosh\left(\dfrac{4 - i\pi}{4}\right) = \dfrac{\cosh 1}{\sqrt{2}} - i\dfrac{\sinh 1}{\sqrt{2}}$
26. **(a)** $z = (\pi/6 + 2\pi n)i$, and $z = (5\pi/6 + 2\pi n)i$, where n is an integer
27. **(a)** $\sinh z + z \cosh z$ **(c)** $\tanh z + z \,\mathrm{sech}^2 z$

Section 5.5, Inverse Trigonometric and Hyperbolic Functions: page 156

1. **(a)** $(\frac{1}{2} + 2n)\pi \pm i \ln 2$, where n is an integer
   (b) $2\pi n \pm i \ln 3$, where n is an integer
   (c) $(\frac{1}{2} + 2n)\pi \pm i \ln(3 + 2\sqrt{2})$, where n is an integer
   (e) $-(\frac{1}{2} + n)\pi + i \ln \sqrt{3}$, where n is an integer
2. **(a)** $i(\frac{1}{2} + 2n)\pi$, where n is an integer
   (b) $\ln 2 + i2\pi n$, and $-\ln 2 + i(2n + 1)\pi$, where n is an integer
   (c) $\ln(\sqrt{2} + 1) + i(\frac{1}{2} + 2n)\pi$ and $\ln(\sqrt{2} - 1) + i(-\frac{1}{2} + 2n)\pi$, where n is an integer
   (e) $i(\frac{1}{4} + n)\pi$, where n is an integer

Section 6.1, Complex Integrals: page 160

1. $2 - 3i$ 2. $-\frac{23}{4} - 6i$ 3. 1 4. $2 - \arctan 2 - i \ln\sqrt{5}$
5. $\sqrt{2}\pi/8 + \sqrt{2}/2 - 1 + i(\sqrt{2}/2 - \sqrt{2}\pi/8)$

Section 6.2, Contours and Contour Integrals: page 173

2. C_1: $z_1(t) = 2 \cos t + i2 \sin t$ for $0 \le t \le \pi/2$, C_2: $z_2(t) = -t + i(2 - t)$ for $0 < t < 2$

3. C_1: $z_1(t) = (-2 + t) + it$ for $0 \le t \le 2$, C_2: $z_2(t) = t + 2i$ for $0 \le t \le 2$, C_3: $z_3(t) = 2 + i(2 - t)$ for $0 \le t \le 2$

4. **(a)** The Riemann sum approximation simplifies to $-2 \cos(\pi/4) + 2 \approx -0.828427$.

　　(b) The exact value is $-\frac{2}{3}$.

6. **(a)** $3/2$ **(b)** $\pi/2$ **7.** **(a)** $-32i$ **(b)** $-8\pi i$ **8.** 0 **9.** $32\pi i$ **10.** $i - 2$

11. 1 **12.** $-1 + 2i/3$ **15.** $-4 - \pi i$ **16.** $-2\pi i$ **17.** 0 **18.** $-2e$

Section 6.3, The Cauchy-Goursat Theorem: page 188

4. **(a)** 0 **(b)** $2\pi i$ **5.** **(a)** $4\pi i$ **(b)** $2\pi i$ **6.** $4\pi i$ **7.** 0

8. **(a)** $\pi i/4$ **(b)** $-\pi i/4$ **(c)** 0 **9.** **(a)** 0 **(b)** $-2\pi i$ **11.** $-4i/3$ **12.** 0

Section 6.4, The Fundamental Theorems of Integration: page 194

1. $\frac{4}{3} + 3i$ **2.** $-1 + i[(\pi + 2)/2]$ **3.** $i - e^2$ **4.** $-7/6 + i/2$

6. $2 - i2 \sinh 1$ **7.** $(\pi/2e) - e^2 - i(e^2\pi + 2/e)$ **9.** $-1 - \sinh 1 + \cosh 1$

10. $i[1/2 - (\sinh 2)/4]$ **11.** $\ln \sqrt{2} - \pi/4 + i(\ln \sqrt{2} + \pi/4 - 1)$

13. $\ln \sqrt{10} - \ln 2 + i \arctan 3 = \ln \sqrt{5/2} + i \arctan 3$ or $\ln \sqrt{5/2} + i(\pi/4 + \arctan 1/2)$

Section 6.5, Integral Representations for Analytic Functions: page 199

1. $4\pi i$ **2.** πi **3.** $-\pi i/2$ **4.** $2\pi i/3$ **6.** $-\pi i/3$ **7.** $2\pi i$

9. $2\pi i/(n - 1)!$ **10.** **(a)** $-\pi i/8$ **(b)** $e^4(i\pi/64)$ **11.** $(\pi - i\pi)/8$

12. **(a)** π **(b)** $-\pi$ **13.** **(a)** $i\pi \sinh 1$ **(b)** $i\pi \sinh 1$ **14.** $\pi/2$

Section 6.6, The Theorems of Morera and Liouville and Some Applications: page 206

1. $(z + 1 + i)(z + 1 - i)(z - 1 + i)(z - 1 - i)$ **2.** $(z - 1 + 2i)(z + 2 - i)$

3. $(z + i)(z - i)(z - 2 + i)(z - 2 - i)$ **4.** $(z - i)(z - 1 - i)(z - 2 - i)$

7. **(a)** 18 **(b)** 5 **(c)** 8 **(d)** 4 **8.** $\sqrt{1 + \sinh^2 2}$ **9.** $\left| f^{(3)}(1) \right| < \frac{3!(10)}{3^3} = \frac{20}{9}$ **10.** $\left| f^{(3)}(0) \right| < \frac{3!(10)}{2^3} = \frac{15}{2}$

Section 7.1, Uniform Convergence: page 213

1. **(b)** Since $S_n(x) = 1 - x + x^2 - x^3 + \cdots + (1)^{n-1}x^{n-1}$, and since the graph of $S_n(x)$ is above that of $f(x)$, the last term, x^{n-1}, must have an even exponent (explain.), so the index must be odd.

Section 7.2, Taylor Series Representations: page 221

7. The series converges for all $z \in D_2(1)$. **14.** **(a)** $f^{(3)}(0) = 48$

Section 7.3, Laurent Series Representations: page 230

1. **(a)** $\displaystyle\sum_{n=0}^{\infty} z^{n-3}$ for $|z| < 1$ **(b)** $\displaystyle-\sum_{n=1}^{\infty} \frac{1}{z^{n+3}}$ for $|z| > 1$

2. $\displaystyle\sum_{n=0}^{\infty} \frac{(-1)^n 2^{2n+1} z^{2n-3}}{(2n + 1)!}$ for $|z| > 0$ **6.** $\displaystyle\sum_{n=0}^{\infty} \frac{(-1)^n}{(2n + 1)! z^{2n+1}}$ for $|z| > 0$

7. $\displaystyle\sum_{n=1}^{\infty} \frac{2z^{4n-7}}{(4n - 2)!}$ for $|z| > 0$

9. $\displaystyle\frac{1}{16z} + \sum_{n=0}^{\infty} \frac{(n + 2)z^n}{4^{n+3}}$ for $|z| < 4$, $\displaystyle\sum_{n=1}^{\infty} \frac{n(4)^{n-1}}{z^{n+2}}$ for $|z| > 4$

Section 7.4, Singularities, Zeros, and Poles: page 238

1. **(a)** zeros of order 4 at $\pm i$ **(c)** simple zeros at $-1 \pm i$
   (e) simple zeros at $\pm i$ and $\pm 3i$
2. **(a)** simple zeros at $(\sqrt{3} \pm i)/2, (-\sqrt{3} \pm i)/2$, and $\pm i$
   (c) zeros of order 2 at $(1 \pm i\sqrt{3})/2$ and -1
   (e) simple zeros at $(1 \pm i)/\sqrt{2}$ and $(-1 \pm i)/\sqrt{2}$, and a zero of order 4 at the origin
3. **(a)** poles of order 3 at $\pm i$, and a pole of order 4 at 1
   (c) simple poles at $(\sqrt{3} \pm i)/2, (-\sqrt{3} \pm i)/2$, and $\pm i$
   (e) simple poles at $\pm\sqrt{3}i$ and $\pm i/\sqrt{3}$
4. **(a)** simple poles at $z = n\pi$ for $n = \pm 1, \pm 2, \ldots$
   (c) simple poles at $z = n\pi$ for $n = \pm 1, \pm 2, \ldots$, and a pole of order 3 at the origin
   (e) simple poles at $z = 2n\pi i$ for $n = 0, \pm 1, \pm 2, \ldots$
5. **(a)** removable singularity at the origin   **(c)** essential singularity at the origin
6. **(a)** removable singularity at the origin, and a simple pole at -1
   (c) removable singularity at the origin
7. $(-1 - i)/16$ **8.** $-1/4$ **9.** 3
20. a nonisolated singularity at the origin
21. simple poles at $z = 1/n\pi$ for $n = \pm 1, \pm 2, \ldots$, and a nonisolated singularity at the origin

Section 7.5, Applications of Taylor and Laurent Series: page 243

2. No.
3. Yes.
4. No.
7. **b.** $f^{(6)}(0) = 0; f^{(7)}(0) = 272$

Section 8.2, Calculation of Residues: page 251

1. **(a)** 1   **(b)** 8   **(c)** 1   **(d)** 5   **2.** **(a)** 1   **(b)** $-\frac{1}{2}$ **(c)** 0   **(d)** 1
3. **(a)** e **(b)** 1/5!   **(c)** 0   **4.** **(a)** $\frac{1}{6}$ **(b)** 4   **(c)** $\frac{1}{3}$
5. $(\pi + i\pi)/8$ **6.** $(\pi + i\pi)/2$
7. $(1 - \cos 1)2\pi i$ **8.** i **9.** $i2\pi \sinh 1$ **10.** **(a)** 0   **(b)** $-4\pi i/25$
11. **(a)** $\pi/3$ **(b)** $(\pi/6)(3 - i\sqrt{3})$ **12.** **(a)** $-\pi/(8\sqrt{3})$ **(b)** $\pi 3\sqrt{3}/8$
13. **(a)** $\pi i/2$ **(b)** $-\pi i/6$ **14.** $\pi i/3$ **15.** $2\pi i/3$
18. **(a)** $\dfrac{1}{z + 1} - \dfrac{1}{z + 2}$ **(b)** $\dfrac{2}{z + 1} + \dfrac{1}{z - 2}$ **(c)** $\dfrac{1}{z^2} - \dfrac{2}{z} + \dfrac{3}{z + 4}$
    (d) $\dfrac{2z}{z^2 + 4} - \dfrac{2z}{z^2 + 9}$ **(e)** $\dfrac{2}{z - 1} + \dfrac{1}{(z - 1)^2} - \dfrac{2}{(z - 1)^3}$

Section 8.3, Trigonometric Integrals: page 256

2. $2\pi/3$ **4.** $\pi/3$ **6.** $2\pi/9$ **8.** $10\pi/27$ **10.** $8\pi/45$ **12.** $10\pi/27$ **14.** $4\pi/27$

Section 8.4, Improper Integrals of Rational Functions: page 260

2. $\pi/4$ **4.** $\pi/18$ **6.** $\pi/4$ **8.** $\pi/64$ **10.** $\pi/15$ **12.** $2\pi/3$

Section 8.5, Improper Integrals Involving Trigonometric Functions: page 264

2. 0 and $\dfrac{\pi}{e^3}$ **4.** $\dfrac{3\pi}{16e^2}$ **6.** $\dfrac{\pi}{3}\left(\dfrac{1}{e} - \dfrac{1}{2e^2}\right)$ **8.** $\dfrac{\pi \cos 2}{e}$ **10.** $\dfrac{\pi \cos 1}{e}$ **12.** $\dfrac{\pi \cos 2}{e^2}$

Section 8.6, Indented Contour Integrals: page 269

2. 0 **4.** $\pi/\sqrt{3}$ **6.** $\pi/\sqrt{3}$ **8.** $\pi(1 - \sin 1)$ **10.** $\frac{1}{2}$ **12.** $\pi(2 \sin 2 - \sin 1)$

Section 8.7, Integrands with Branch Points: page 273

2. π **4.** $\pi/\sqrt{2}$ **6.** $-\pi/4$ **8.** $(\ln a)/a$

Section 9.1, Basic Properties of Conformal Mapping: page 286

1. **(b)** all z except $z = \dfrac{\pi}{2} + 2n\pi$ **(d)** all z except $z = 0$ **(f)** all z except $z = 1$

2. $\alpha = \pi$, $\left| -1 \right| = 1$; $\alpha = \pi/2$, $\left| \dfrac{i}{2} \right| = \dfrac{1}{2}$; $\alpha = 0$, $\left| 1 \right| = 1$

3. $\alpha = 0$, $\left| 1 \right| = 1$; $\alpha = -\pi/4$, $\left| (1 - i)/2 \right| = \sqrt{2}/2$; $\alpha = -\pi/2$, $\left| -i \right| = 1$; $\alpha = \pi$, $\left| -1 \right| = 1$

5. $\alpha = -\pi/2$, $\left| -i \sinh 1 \right| = \sinh 1$; $\alpha = 0$, $\left| 1 \right| = 1$; $\alpha = \pi/2$, $\left| i \sinh 1 \right| = \sinh 1$

Section 9.2, Bilinear Transformations: page 294

1. $S^{-1}(w) = \dfrac{-2w + 2}{(1 + i)w - 1 + i}$ **2.** $S^{-1}(w) = \dfrac{iw - i}{w + 1}$ **3.** the disk $\left| w \right| < 1$

5. the set $\left| w \right| > 1$ **6.** $w = \dfrac{z + i}{3z - i}$ **7.** $w = \dfrac{-iz + i}{z + 1}$ **9.** $w = \dfrac{i - iz}{1 + z}$

11. the disk $\left| w \right| < 1$ **12.** $S_1(S_2(z)) = \dfrac{-z - 6}{2z + 3}$

13. the portion of the disk $\left| w \right| < 1$ that lies in the upper half plane $\mathrm{Im}(w) > 0$

Section 9.3, Mappings Involving Elementary Functions: page 302

1. the portion of the disk $\left| w \right| < 1$ that lies in the first quadrant $u > 0$, $v > 0$
2. $\{\rho e^{i\phi}: 1 < \rho < 2, 0 < \phi < \pi/2\}$ **3.** the horizontal strip $0 < \mathrm{Im}(w) < 1$
4. $\{u + iv: 0 < u < 1, -\pi < v \leq \pi\}$ **9.** the horizontal strip $0 < \mathrm{Im}(w) < \pi$
10. the horizontal strip $\pi/2 < \mathrm{Im}(w) < \pi$
12. the horizontal strip $\left| v \right| < \pi$ slit along the ray $u < 0$, $v = 0$
13. $Z = z^2 + 1$, $w = Z^{1/2}$, where the principal branch of the square root $Z^{1/2}$ is used
15. the unit disk $\left| w \right| < 1$

Section 9.4, Mapping by Trigonometric Functions: page 308

1. the portion of the disk $\left| w \right| < 1$ that lies in the second quadrant $u < 0$, $v > 0$
3. the right branch of the hyperbola $u^2 - v^2 = \frac{1}{2}$
5. the region in the first quadrant $u > 0$, $v > 0$ that lies inside the ellipse
   $[u^2/(\cosh^2 1)] + [v^2/(\sinh^2 1)] = 1$ and to the left of the hyperbola $u^2 - v^2 = \frac{1}{2}$
7. **(a)** $\pi/3$ **(b)** $-5\pi/6$
8. **(a)** $0.754249145 + i1.734324521$ **(c)** $0.307603649 - i1.864161544$
10. the right half plane $\mathrm{Re}(w) > 0$ slit along the ray $v = 0$, $u > 1$
12. the vertical strip $0 < u < \pi/2$
14. the semi-infinite strip $-\pi/2 < u < \pi/2$, $v > 0$
16. the horizontal strip $0 < v < \pi$

Section 10.2, Invariance of Laplace's Equation and the Dirichlet Problem: page 321

1. $15 - 9y$ **3.** $5 + (3/\ln 2) \ln \left| z \right|$

5. $4 - \dfrac{4}{\pi} \mathrm{Arg}(z + 3) + \dfrac{6}{\pi} \mathrm{Arg}(z + 1) - \dfrac{3}{\pi} \mathrm{Arg}(z - 2) = 4 - \dfrac{4}{\pi} \mathrm{Arctan} \dfrac{y}{x + 3}$

$\qquad + \dfrac{6}{\pi} \mathrm{Arctan} \dfrac{y}{x + 1} - \dfrac{3}{\pi} \mathrm{Arctan} \dfrac{y}{x - 2}$

6. $\dfrac{-1}{\pi} \text{Arg}(z^2 + 1) + \dfrac{1}{\pi} \text{Arg}(z^2 - 1) = \dfrac{-1}{\pi} \text{Arctan} \dfrac{2xy}{x^2 - y^2 + 1}$

$+ \dfrac{1}{\pi} \text{Arctan} \dfrac{2xy}{x^2 - y^2 - 1}$

8. $8 - \dfrac{4}{\pi} \text{Arctan} \dfrac{1 - x^2 - y^2}{2y}$

10. $1 - \dfrac{2}{\pi} \text{Arg}\left(i\dfrac{1 + 1/z}{1 - 1/z}\right) = 1 - \dfrac{2}{\pi} \text{Arctan} \dfrac{x^2 + y^2 - 1}{2y}$

12. $\dfrac{-1}{\pi} \text{Arg}\left(i\dfrac{1 - z}{1 + z} + 1\right) + \dfrac{1}{\pi} \text{Arg}\left(i\dfrac{1 - z}{1 + z} - 1\right) = \dfrac{-1}{\pi} \text{Arctan} \dfrac{1 - x^2 - y^2}{2y + (1 + x)^2 + y^2}$

$+ \dfrac{1}{\pi} \text{Arctan} \dfrac{1 - x^2 - y^2}{2y - (1 + x)^2 - y^2}$

Section 10.3, Poisson's Integral Formula for the Upper Half Plane: page 326

1. $\dfrac{y}{2\pi} \ln \dfrac{(x - 1)^2 + y^2}{(x + 1)^2 + y^2} + \dfrac{x - 1}{\pi} \text{Arctan} \dfrac{y}{x - 1} - \dfrac{x + 1}{\pi} \text{Arctan} \dfrac{y}{x + 1} + 1$

2. $\dfrac{y}{2\pi} \ln \dfrac{(x - 1)^2 + y^2}{x^2 + y^2} + \dfrac{x}{\pi} \text{Arctan} \dfrac{y}{x - 1} - \dfrac{x}{\pi} \text{Arctan} \dfrac{y}{x}$

Section 10.5, Steady State Temperatures: page 336

2. $25 + 50(x + y)$ **4.** $60 + \dfrac{40}{\pi} \text{Arg}\left(i\dfrac{1 - z}{1 + z}\right) - \dfrac{40}{\pi} \text{Arg}\left(i\dfrac{1 - z}{1 + z} - 1\right)$

6. $100 - \dfrac{200}{\pi} \text{Arctan} \dfrac{x^2 + y^2 - 1}{2y}$ **8.** $100 - \dfrac{50}{\alpha} \text{Arg } z$

10. $\dfrac{100}{\pi} \text{Re}(\text{Arcsin } e^z)$ **12.** $50 + \dfrac{200}{\pi} \text{Re}(\text{Arcsin } iz)$

Section 10.6, Two-Dimensional Electrostatics: page 346

1. $100 + \dfrac{100}{\ln 2} \ln |z|$ **2.** $100 - \dfrac{100}{\pi} \text{Arg}(z - 1) - \dfrac{100}{\pi} \text{Arg}(z + 1)$

3. $150 - \dfrac{200x}{x^2 + y^2}$ **4.** $\dfrac{50}{\pi} \text{Arg}(\sin z - 1) + \dfrac{50}{\pi} \text{Arg}(\sin z + 1)$

5. $50 + \dfrac{200}{\pi} \text{Re}(\text{Arcsin } z)$ **6.** $\dfrac{200}{\pi} \text{Arg}(\sin z)$

Section 10.7, Two-Dimensional Fluid Flow: page 357

3. (a) Speed $= A|\bar{z}|$. The minimum speed is $A|1 - i| = A\sqrt{2}$.

  (b) The maximum pressure in the channel occurs at the point $1 + i$.

5. (a) $\Psi(r, \theta) = Ar^{3/2} \sin(3\theta/2)$

Section 10.10, Image of a Fluid Flow: page 382

1. $w = (z^2 - 1)^{1/2}$ **2.** $w = \dfrac{-2}{\pi}[z(1 - z^2)^{1/2} + \text{Arcsin } z]$

3. $w = (z - 1)^{\alpha}\left(1 + \dfrac{\alpha z}{1 - \alpha}\right)^{1 - \alpha}$ **4.** $w = \dfrac{-i}{2}z^{1/2}(z - 3)$

5. $w = -1 + \displaystyle\int_{-1}^{z} \dfrac{(\xi - 1)^{1/4}}{\xi^{1/4}} d\xi$

Section 10.11, Sources and Sinks: page 392

2. $F(z) = \log(z^4 - 1)$ **3.** $F(z) = \log(\sin z)$ **4.** $F(z) = \log(\sin z)$

5. $F(z) = \log(z^2 - 1)$ **7.** $w = 2(z + 1)^{1/2} + \mathrm{Log}\left[\dfrac{1 - (z + 1)^{1/2}}{1 + (z + 1)^{1/2}}\right] + i\pi$

8. $w = 4(z + 1)^{1/4} + \mathrm{Log}\left[\dfrac{(z + 1)^{1/4} - 1}{(z + 1)^{1/4} + 1}\right] - i\,\mathrm{Log}\left[\dfrac{i + (z + 1)^{1/4}}{i - (z + 1)^{1/4}}\right]$

9. $w = \dfrac{1}{\pi}\,\mathrm{Arcsin}\,z + \dfrac{i}{\pi}\,\mathrm{Arcsin}\,\dfrac{1}{z} + \dfrac{1 + i}{2}$

10. $w = 2(z + 1)^{1/2} - \mathrm{Log}\left[\dfrac{1 - (z + 1)^{1/2}}{1 + (z + 1)^{1/2}}\right]$

Section 11.1, Fourier Series: page 403

1. $U(t) = \dfrac{4}{\pi}\displaystyle\sum_{j=1}^{\infty}\dfrac{\sin[(2j - 1)t]}{2j - 1}$

7. $U(t) = \dfrac{4}{\pi}\displaystyle\sum_{j=1}^{\infty}\dfrac{(-1)^{j-1}\sin[(2j - 1)t]}{(2j - 1)^2}$

9. $U(t) = \dfrac{2}{\pi}\displaystyle\sum_{j=1}^{\infty}\dfrac{\sin[(2j - 1)t]}{2j - 1} - \dfrac{4}{\pi}\displaystyle\sum_{j=1}^{\infty}\dfrac{\sin[2(2j - 1)t]}{2(2j - 1)}$, where $b_{4n} = 0$ for all n

Section 11.2, The Dirichlet Problem for the Unit Disk: page 409

1. $u(r\cos\theta, r\sin\theta) = \dfrac{4}{\pi}\displaystyle\sum_{j=1}^{\infty}\dfrac{r^{2j-1}\sin[(2j - 1)\theta]}{2j - 1}$

3. $u(r\cos\theta, r\sin\theta) = \dfrac{4}{\pi}\displaystyle\sum_{j=1}^{\infty}\dfrac{(-1)^{j-1}r^{2j-1}\cos[(2j - 1)\theta]}{2j - 1}$

Section 11.4, The Fourier Transform: page 422

1. $\widetilde{\mathfrak{F}}(U(t)) = \dfrac{\sin w}{\pi w}$

3. $\widetilde{\mathfrak{F}}(U(t)) = \dfrac{1 - \cos w}{\pi w^2} = \dfrac{2\sin^2\dfrac{w}{2}}{\pi w^2}$

Section 11.5, The Laplace Transform: page 429

3. $\mathscr{L}(U(t)) = \dfrac{1}{s^2} - \dfrac{ce^{-cs}}{s} - \dfrac{e^{-cs}}{s^2}$

7. $\mathscr{L}(3t^2 - 4t + 5) = \dfrac{6}{s^3} - \dfrac{4}{s^2} + \dfrac{5}{s}$

9. $\mathscr{L}(e^{2t-3}) = \dfrac{e^{-3}}{s - 2}$

11. $\mathscr{L}((t + 1)^4) = \dfrac{24}{s^5} + \dfrac{24}{s^4} + \dfrac{12}{s^3} + \dfrac{4}{s^2} + \dfrac{1}{s}$

13. $\mathscr{L}^{-1}\left(\dfrac{1}{s^2 + 25}\right) = \dfrac{1}{5}\sin 5t$

17. $\mathscr{L}^{-1}\left(\dfrac{6s}{s^2 - 4}\right) = 3e^{-2t} + 3e^{2t} = 6\cosh 2t$

Section 11.6, Laplace Transforms of Derivatives and Integrals: page 433

1. $\mathcal{L}(\sin t) = \dfrac{1}{s^2 + 1}$

3. $\mathcal{L}(\sin^2 t) = \dfrac{2}{s(s^2 + 4)}$

5. $\mathcal{L}^{-1}\left(\dfrac{1}{s(s - 4)}\right) = -\dfrac{1}{4} + \dfrac{1}{4}e^{4t}$

9. $y(t) = 2 \cos 3t + 3 \sin 3t$

11. $y(t) = -2 + 2 \cos 2t + \sin 2t$

13. $y(t) = 2 + e^t$

15. $y(t) = -1 - \dfrac{1}{2}e^{-t} + \dfrac{3}{2}e^t = -1 + \sinh t + e^t$

17. $y(t) = e^{-2t} + e^t$

Section 11.7, Shifting Theorems and the Step Function: page 437

1. $\mathcal{L}(e^t - te^t) = \dfrac{-1}{(s - 1)^2} + \dfrac{1}{s - 1}$

5. $F(s) = \mathcal{L}^{-1}\left(\dfrac{s + 2}{s^2 + 4s + 5}\right) = e^{-2t}\cos t$

7. $F(s) = \mathcal{L}^{-1}\left(\dfrac{s + 3}{(s + 2)^2 + 1}\right) = e^{-2t}\cos t + e^{-2t}\sin t$

9. $\mathcal{L}(U_2(t)(t - 2)^2) = \dfrac{2e^{-2s}}{s^3}$

11. $\mathcal{L}(U_{3\pi}(t) \sin(t - 3\pi)) = \dfrac{e^{-3\pi s}}{s^2 + 1}$

13. $\mathcal{L}(f(t)) = \dfrac{1 - 2e^{-s} + 2e^{-2s} - e^{-3s}}{s}$

15. $\mathcal{L}^{-1}\left(\dfrac{e^{-s} + e^{-2s}}{s}\right) = U_2(t) + U_1(t)$

17. $y(t) = -e^{-t}\cos t$

19. $y(t) = 2e^{-t/2}\sin(t/2)$

21. $y(t) = t^3 e^{-t}$

23. $y(t) = [1 - \delta(t - \pi/2)] \sin t + (1 - \sin t)U_{\pi/2}(t)$

Section 11.8, Multiplication and Division by t: page 440

1. $\mathcal{L}(te^{-2t}) = \dfrac{1}{(s + 2)^2}$

3. $\mathcal{L}(t \sin 3t) = \dfrac{6s}{(s^2 + 9)^2}$

9. $\mathcal{L}(t \sin bt) = \dfrac{2bs}{(s^2 + b^2)^2}$

11. $\mathcal{L}^{-1}\left(\ln\left(\dfrac{s^2 + 1}{(s - 1)^2}\right)\right) = \dfrac{2(e^t - \cos t)}{t}$

13. $y(t) = (t + t^2)e^{-t}$

15. $y(t) = Cte^t$

17. $y(t) = Ct$

19. $y(t) = 1 - t$

Section 11.9, Inverting the Laplace Transform: page 447

1. $\mathcal{L}^{-1}\left(\dfrac{2s+1}{s(s-1)}\right) = -1 + 3e^t$

3. $\mathcal{L}^{-1}\left(\dfrac{4s^2 - 6s - 12}{s(s+2)(s-2)}\right) = 3 + 2e^{-2t} - e^{2t}$

7. $\mathcal{L}^{-1}\left(\dfrac{1}{s^2+4}\right) = \dfrac{1}{2}\sin 2t$

9. $\mathcal{L}^{-1}\left(\dfrac{s^3 + s^2 - s + 3}{s^5 - s}\right) =$

   $-3 + e^t + e^{-t} + \cos t + \sin t = -3 + 2\cosh t + \cos t + \sin t$

15. $y(t) = e^{-t} + e^{-t}\sin 2t$

17. $y(t) = e^{-t} + \cos 2t + \sin 2t$

19. $y(t) = 1 + t$

21. $x(t) = 2e^{-2t} - e^t$
    $y(t) = e^{-2t} - 2e^t$

23. $x(t) = -e^{-t} + 2te^{-t}$
    $y(t) = te^{-t}$

Section 11.10, Convolution: page 454

1. $\mathcal{L}(t * t) = \dfrac{t^3}{6}$

3. $\mathcal{L}(e^t * e^{2t}) = -e^t + e^{2t}$

5. $\mathcal{L}^{-1}\left(\dfrac{2}{(s-1)(s-2)}\right) = -2e^t + 2e^{2t}$

7. $\mathcal{L}^{-1}\left(\dfrac{1}{s(s^2+1)}\right) = 1 - \cos t$

11. $\mathcal{L}^{-1}\left(\dfrac{s}{s-1}\right) = e^t + \delta(t)$

13. $y(t) = -t\cos t + \sin t$

15. $\mathcal{L}\left(\displaystyle\int_0^t e^{-\tau}\cos(t-\tau)\,d\tau\right) = \dfrac{s}{(s+1)(s^2+1)}$

19. $F(s) = \dfrac{1}{s-2}$ and $f(t) = e^{2t}$

21. $F(s) = \dfrac{2}{s^4 - 1}$ and $f(t) = -\sin t + \sinh t$

23. $y(t) = te^{-t}$

25. $y(t) = (-e^{3-3t} + e^{1-t})U_1(t)$

27. $y(t) = 120 - 96t + 36t^2 - 8t^3 + t^4 - 119e^{-t} - 21te^{-t}$

29. $y(t) = \dfrac{7}{2}e^{-t} - 2te^{-2t} - \dfrac{5}{2}e^{-3t}$

Index